The Practical Approach in Chemistry Series

SERIES EDITORS

L. M. HARWOOD
The Dyson Perrins Laboratory
University of Oxford

C. J. MOODY
Department of Chemistry
Loughborough University of Technology

The Practical Approach in Chemistry Series

Organocopper reagents
Edited by Richard J. K. Taylor

Organocopper Reagents

A Practical Approach

Edited by

RICHARD J. K. TAYLOR

Department of Chemistry,
University of York, UK

OXFORD NEW YORK TOKYO
OXFORD UNIVERSITY PRESS
1994

Oxford University Press, Walton Street, Oxford OX2 6DP
Oxford New York
Athens Auckland Bangkok Bombay
Calcutta Cape Town Dar es Salaam Delhi
Florence Hong Kong Istanbul Karachi
Kuala Lumpur Madras Madrid Melbourne
Mexico City Nairobi Paris Singapore
Taipei Tokyo Toronto
and associated companies in
Berlin Ibadan

Oxford is a trade mark of Oxford University Press

Published in the United States
by Oxford University Press Inc., New York

A catalogue record for this book is available from the British Library

Library of Congress Cataloging-in-Publication Data
(Data available)

ISBN 0 19 855757 4 (Hbk)
ISBN 0 19 855758 2 (Pbk)

Typeset by Footnote Graphics, Warminster, Wilts
Printed in Great Britain by Bookcraft Ltd., Midsomer Norton, Avon

Preface

This is the first volume in The Practical Approach in Chemistry Series. The aim of this series is to provide detailed and accessible laboratory guides suitable for researchers who are not familiar with the area in question. The authors are leading practitioners who detail key experimental procedures in a step-by-step protocol format. These protocols contain information on reagent purification, reaction equipment and conditions, work-up procedures, and other pieces of expert advice and 'tricks of the trade'. Although a comprehensive coverage of the literature on each area is not provided, each chapter contains background information and key references to aid further reading.

The choice of Organocopper Chemistry as the first volume in this new series seems appropriate. Despite the widely acknowledged synthetic utility and selectivity of organocopper reagents, and the fact that they have been extensively employed in natural product syntheses and used industrially on the multikilo scales, organocopper chemistry is still viewed with reservation by some members of the synthetic community. There appears to be a feeling that organocopper-mediated transformations are the preserve of the specialists and not of widespread applicability. The main aims of this volume are to dispel the mystery surrounding organocopper chemistry and to encourage more researchers to use these powerful synthetic reagents to maximum effect. The book contains eleven independently referenced chapters describing, in protocol form, the preparation of all the main types of organocopper reagents and their use in a range of synthetic transformations and mechanistic studies. In addition, there is a summary of the historical development and current uses of organocopper reagents, details of general procedures, starting material purification and reaction guidelines, a fully referenced compilation of organocopper reagents, and a list of suppliers. In fact, this is a complete 'DIY' guide to organocopper chemistry.

I would like to offer special acknowledgement to Professor T. Ibuka (Kyoto University) for providing all of the apparatus illustrations throughout the book and to Dr M. Furber (Fisons Pharmaceuticals) for help with early drafts of the organocopper compilation. Thanks are also owed to Professors B. H. Lipshutz (University of California, Santa Barbara) and K. Tamao (Kyoto University) for experimental advice for Chapter 12, Dr R. L. Carney (Sandoz Agro, Palo Alto) and D. A. Bleasedale (Searle Pharmaceuticals, Morpeth) for the facts and figures concerning industrial applications of organocopper chemistry referred to in Chapter 1, and Drs G. A. McNaughton-Smith and S. Gamage (University of East Anglia) for carrying out and writing up Protocol 4 of Chapter 10. I would also like to thank Professor A. McKillop and Drs B. C. Borer and P. T. de Sousa Jr (University of East Anglia) and Drs A. F.

Parsons and K. Hemming and Mr A. Grumann (University of York) for helping to check parts of the manuscript and for useful suggestions. Last, but certainly not least, I would like to express my personal gratitude to all of the invited contributors who honoured the deadlines with great efficiency and made the editorial job so straightforward.

York R.J.K.T.
July 1994

Contents

Contributors

A. ALEXAKIS
Université Pierre et Marie Curie, CNRS UA 473, Paris, France.

I. B. CAMPBELL
Glaxo Group Research, Ware, UK.

GUY CASY
Chiroscience, Cambridge, UK.

IAN FLEMING
University Chemical Laboratory, University of Cambridge, Cambridge, UK.

JOHN M. HERBERT
Sterling Winthrop Research Centre, Alnwick, UK.

TOSHIRO IBUKA
Faculty of Pharmaceutical Sciences, Kyoto University, Sakyo-ku, Kyoto 606, Japan.

W. R. KLEIN
Department of Chemistry, University of Nebraska, Lincoln, USA.

PAUL KNOCHEL
Philipps-Universität, Fachbereich Chemie, Hans-Meerwein-Strasse, D-35032 Marburg, Germany.

BRUCE H. LIPSHUTZ
Department of Chemistry, University of California, Santa Barbara, USA.

EIICHI NAKAMURA
Department of Chemistry, Tokyo Institute of Technology, Tokyo 152, Japan.

J.-F. NORMANT
Université Pierre et Marie Curie, Laboratoire de Chimie, CNRS, Paris, France.

R. NOYORI
Department of Chemistry, Nagoya University, Chikusa, Nagoya 464-01, Japan.

R. D. RIEKE
Department of Chemistry, University of Nebraska, Lincoln, USA.

MICHAEL J. ROZEMA
Philipps-Universität, Fachbereich Chemie, Hans-Meerwein-Strasse, D-35032 Marburg, Germany.

R. A. J. SMITH
Department of Chemistry, University of Otago, New Zealand.

M. SUZUKI
Department of Applied Chemistry, Faculty of Engineering, Gifu University, Gifu 501-11, Japan.

RICHARD J. K. TAYLOR
Department of Chemistry, University of York, York, UK.

CHARLES E. TUCKER
Hoechst-Celanese, Corpus Christi, Texas, USA.

YOSHINORI YAMAMOTO
Department of Chemistry, Faculty of Science, Tohoku University, Sendai 980, Japan.

Abbreviations

Ac	acetyl
acac	acetylacetonate
Bn	benzyl
BOC	*t*-butoxycarbonyl
Cp	cyclopentadienyl
de	diastereomeric excess
DMA	*N,N*-dimethylacetamide
DMAP	dimethylaminopyridine
DME	dimethoxyethane
DMF	*N,N*-dimethylformamide
DMPU	*N,N*-dimethylpropyleneurea
ee	enantiomeric excess
EXAFS	extended X-ray absorption fine structure
FG	functional group
FW	formula weight
GLC	gas liquid chromatography
HMPA	hexamethylphosphoric triamide
HPLC	high performance liquid chromatography
HO	higher order
HREIMS	high resolution electron impact mass spectrometry
Imid	1-imidazoyl
IR	infra-red
LDA	lithium diisopropylamide
LO	lower order
MOM	methoxymethyl
Ms	methanesulfonyl
NMR	nuclear magnetic resonance
Np	1-naphthyl
PCC	pyridinium chlorochromate
Pr	propyl
Pyrr	1-pyrrolyl
TBDMS	*t*-butyl-dimethylsilyl
Tf	trifluoromethanesulfonyl
2-Th	2-thienyl
THF	tetrahydrofuran
THP	tetrahydropyranyl
TIPS	triisopropylsilyl
TMEDA	*N,N,N′,N′*-tetramethylethylene diamine
TMS	trimethylsilyl
Ts	*p*-toluenesulfonyl
)))	ultrasonification

Dedicated to Ginny, Becky, Cathy, and Philip,
to my mother, and the memory of my father.

1

Organocopper chemistry: an overview

RICHARD J. K. TAYLOR

1. Introduction

Organocopper reagents are an indispensable part of the synthetic organic chemist's tool-kit. They can be employed to prepare alkanes, alkenes, alkynes, and aromatics and the transformations are usually characterized by high chemo-, regio-, and stereoselectivity. Vicinal carbon–carbon bond formation, via conjugate addition–enolate trapping or alkyne carbocupration–trapping, can be achieved and new methods are being developed for the preparation of highly functionalized organocopper reagents. Organocopper reagents have been employed to prepare innumerable novel and natural compounds. The latter category is well illustrated by a 1992 compilation[1] which lists 522 natural product syntheses which employ organocopper reagents. One of the high points in the utilization of organocopper chemistry in natural product synthesis is shown in Scheme 1.1—the Noyori, Suzuki, and Yanagisawa[2] synthesis of prostaglandin E_2, which used as its cornerstone the complex, enantiomerically pure mono-organocopper reagent **1** as part of a three-component coupling process (more details of this chemistry are given in Chapter 9). In order to avoid confusion, the term 'organocopper reagent' will be employed in a generic sense to describe *all* types of catalytic and stoichiometric copper-based organometallic systems. The specific reagent type with the stoichiometry RCu will be referred to as a mono-organocopper reagent. Throughout this book stoichiometric formulations are used to describe the organocopper reagents; this is an oversimplification as many of the reagents exist as equilibrium mixtures or as oligomeric aggregates (see Section 4).

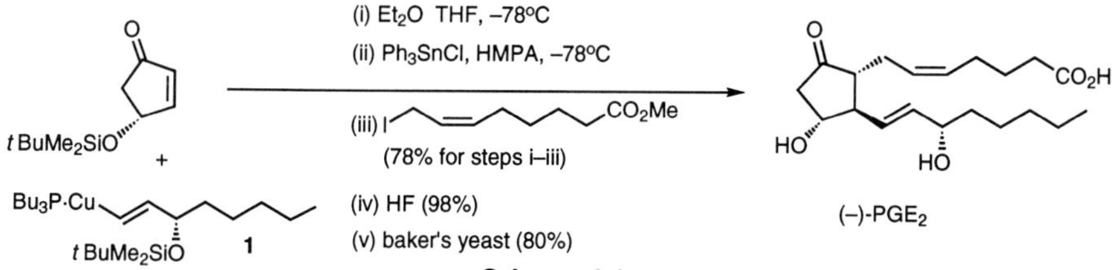

tBuMe$_2$SiO + Bu$_3$P·Cu (**1**, tBuMe$_2$SiO)

(i) Et_2O THF, –78°C
(ii) Ph_3SnCl, HMPA, –78°C
(iii) I–…–CO_2Me
(78% for steps i–iii)
(iv) HF (98%)
(v) baker's yeast (80%)

(–)-PGE_2

Scheme 1.1

Organocopper chemistry is not purely an academic preserve; there are many examples of its use in industrial laboratories. For reasons of commercial sensitivity, it is difficult to obtain in-depth information about many of these processes but Scheme 1.2 outlines two that have been described in some detail.[3,4] The first involves the preparation of the anti-ulcer compound Misoprostol **5** via conjugate addition of cuprate reagent **3**, prepared by double transmetallation of vinyltin **2**, to cyclopentenone **4** (Scheme 1.2a).[3] Careful desilylation followed by chromatography gives Misoprostol, which is marketed as a mixture of the two C-16 epimers. The process is typically carried out on a 10 molar scale with respect to enone **4** (70 molar with respect to vinylstannane **2**), producing Misoprostol in 70–75% overall yield (2.1–2.3 kg per batch). The annual production of Misoprostol is 50–200 kg with a value of *c.* £2.5–10 million at 1993 prices. More recently,[5] as discussed in Chapter 5 (Schemes 5.6 and 5.7), this route to Misoprostol **5** has been modified to incorporate vinylzirconium reagents in place of the vinyltin reagent **2** shown in Scheme 1.2a. The zirconium methodology has not yet been used for production processes, however.

The second process (Scheme 1.2b) involves the synthesis of Muscalure [(*Z*)-9-tricosene, Muscamone®, **7**], a housefly aggregation and sex pheromone used in Super Golden Malrin® fly bait for the control of flies in poultry and cattle operations.[4] Zoecon (a division of Sandoz Agro) use oleyl bromide **6** as starting material and carry out a copper(I)-catalysed coupling reaction with amylmagnesium bromide, as shown in Scheme 1.2b. It is interesting to note that when $LiCuCl_2$ (CuCl + LiCl) was employed as catalyst, plant temperatures of −10 to −20°C were required which caused a number of production problems. By changing to CuCN·LiCl as catalyst, thermal stability of the catalytically active organocopper species was improved, and temperatures of 0–5°C could be employed. This transformation is remarkably efficient: the use of one equivalent of bromide **6**, 1.15 equivalents of amylmagnesium bromide, and 0.03 equivalent of CuCN·LiCl gives near quantitative yields of Muscalure on the production scale (150 kg batches).[4]

There have been many reviews dealing with various aspects of organocopper chemistry.[1,6–16] The main aim of this book is to provide a detailed and accessible laboratory guide to enable the reader to prepare and use organocopper reagents with confidence. To this end, a large number of experimental procedures are described in a step-by-step fashion by their originators. Full details of reagent purification, apparatus, work-up procedures, and 'tricks of the trade' are given. In addition, Chapter 2 includes procedures for the preparation and purification of copper-containing starting materials, solvents, and additives as well as giving general guidelines on handling organometallic reagents and carrying out organometallic reactions. There is also an Appendix listing organocopper reagents which have been described with full experimental details in published procedures.

By way of introduction, the remainder of this chapter summarizes the

(a) Searle's Misoprostol Procedure[3]

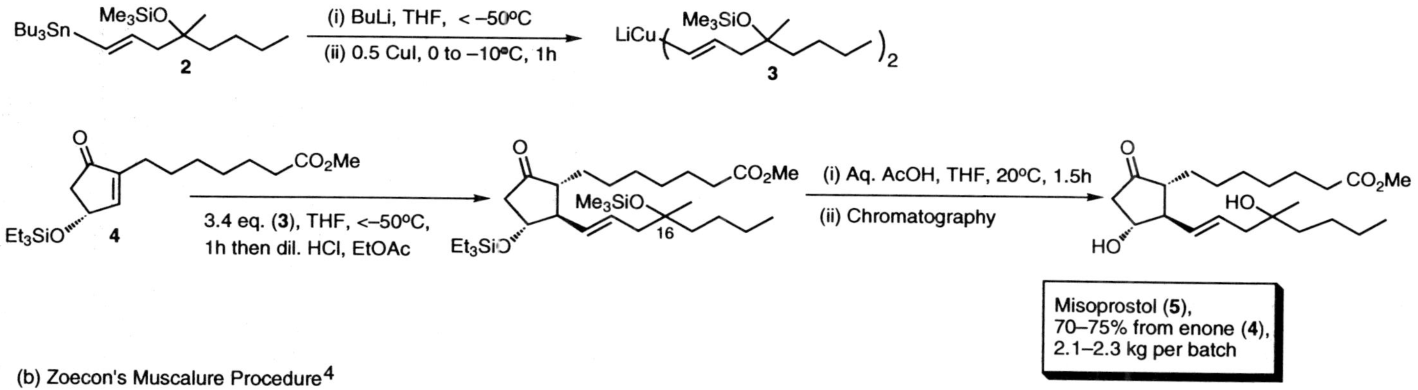

Bu_3Sn Me_3SiO **2**

(i) BuLi, THF, < −50°C

(ii) 0.5 CuI, 0 to −10°C, 1h

LiCu Me_3SiO $)_2$ **3**

CO_2Me

Et_3SiO **4**

3.4 eq. (**3**), THF, <−50°C, 1h then dil. HCl, EtOAc

CO_2Me Me_3SiO 16 Et_3SiO

(i) Aq. AcOH, THF, 20°C, 1.5h

(ii) Chromatography

CO_2Me HO HO

Misoprostol (**5**),
70–75% from enone (**4**),
2.1–2.3 kg per batch

(b) Zoecon's Muscalure Procedure[4]

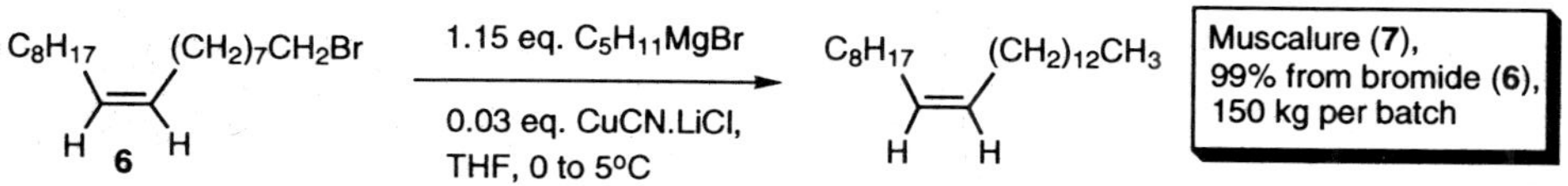

C_8H_{17} $(CH_2)_7CH_2Br$ H H **6**

1.15 eq. $C_5H_{11}MgBr$

0.03 eq. CuCN.LiCl, THF, 0 to 5°C

C_8H_{17} $(CH_2)_{12}CH_3$ H H

Muscalure (**7**),
99% from bromide (**6**),
150 kg per batch

Scheme 1.2

history of organocopper chemistry, the types of organocopper-mediated reactions and organocopper reagents, and provides a brief discussion of structure and mechanism.

2. Historical perspective

Edward Frankland and other early pioneers of organometallic chemistry investigated the reactions between diethylmercury and copper[17] and diethylzinc and copper(I) chloride,[18] but failed to isolate organocopper reagents. About the same time, in 1859, Böttger[19] prepared an explosive red precipitate (CuC≡CCu) by passing 'illuminating gas' through an ammoniacal solution of copper(I) chloride. The following year Berthelot[20] purified the hydrocarbon released when Böttger's precipitate was hydrolysed and named it acetylene. Alkynylcopper reagents (copper acetylides), being easily prepared and possessing much greater thermal and hydrolytic stability than other organocopper reagents, have a long history but there were few synthetic applications until recent times (see Scheme 1.5).[6c,21]

The copper(I)-catalysed coupling of terminal alkynes to give diynes (the Glaser reaction[22]) and the copper-promoted Ullmann biaryl[23] and aryl ether[24] syntheses were then developed. The next major advance in terms of stoichiometric organocopper reagents, however, came in 1923 when Reich[25] prepared phenylcopper from phenylmagnesium bromide and copper(I) iodide. Attempts to prepare ethylcopper by similar processes were unsuccessful owing to its lower stability. Then in 1936 Gilman and Straley[26] prepared the first mono-alkylcopper compound, ethylcopper, from ethylmagnesium iodide and copper(I) iodide. In this same pioneering publication, which also discussed the preparation and reactivity of organosilver compounds, Gilman and Straley showed that organocopper reagents do not give a Michler's ketone colour test ('Gilman test'—see Chapter 2, Section 5.4) and that ethylcopper has a much lower thermal stability than phenylcopper, undergoing significant decomposition in solution at −18°C. They also demonstrated the considerable synthetic potential of mono-organocopper reagents (Scheme 1.3), although this potential was apparently overlooked (or ignored) by the organic chemists of the time.

Gilman and Woods[27] published a synthesis of methylcopper in 1943, and in 1952 Gilman *et al.*[28] described the important observation that the insoluble, yellow methylcopper redissolved in ether on addition of a second molar equivalent of methyllithium (Scheme 1.4). This was the first literature report of what are now known as organocuprate reagents (or Gilman reagents). In the same paper the use of copper(I) thiocyanate for the preparation of organocopper reagents was described, as was the first example of an organocopper conjugate addition–enolate-trapping reaction (Scheme 1.4).

About the same time, the use of copper salts as catalysts in organometallic reactions was becoming popular. Early research had shown that copper salts

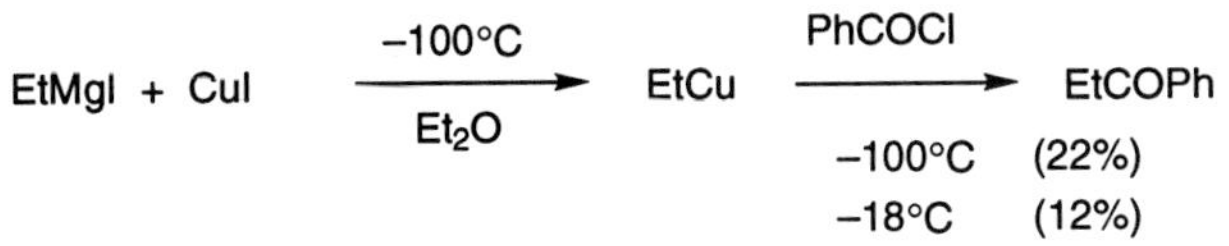

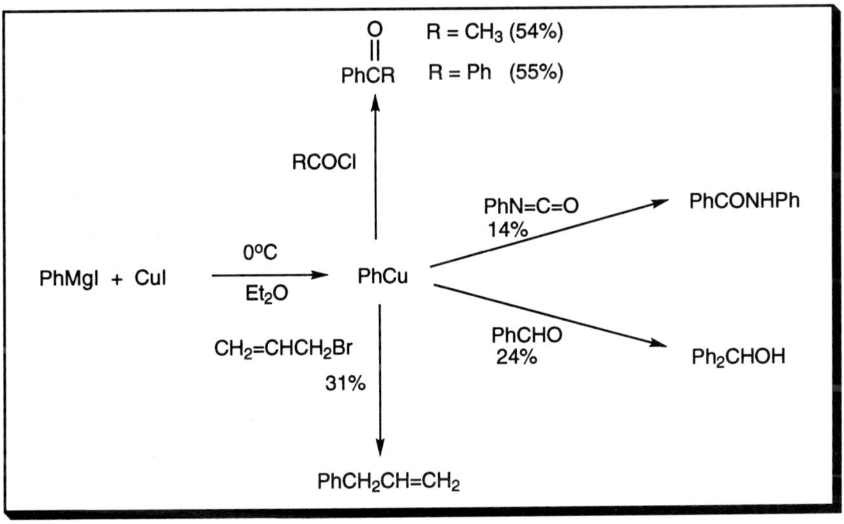

$EtMgI + CuI \xrightarrow[Et_2O]{-100°C} EtCu \xrightarrow{PhCOCl} EtCOPh$

−100°C (22%)
−18°C (12%)

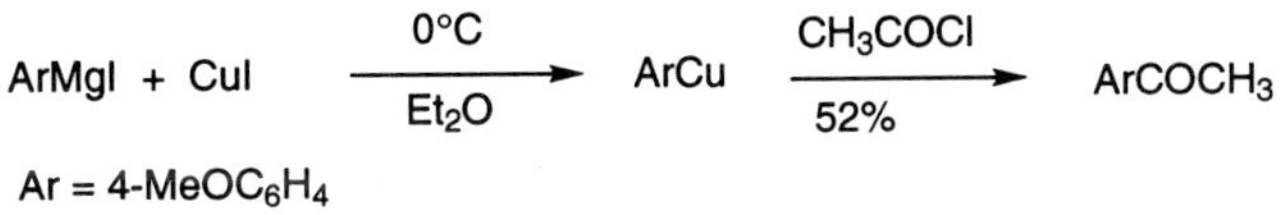

$ArMgI + CuI \xrightarrow[Et_2O]{0°C} ArCu \xrightarrow[52\%]{CH_3COCl} ArCOCH_3$

$Ar = 4\text{-}MeOC_6H_4$

$PhMgI + CuI \xrightarrow[Et_2O]{0°C} PhCu$

PhCu + RCOCl → PhC(=O)R: R = CH_3 (54%); R = Ph (55%)

PhCu + PhN=C=O (14%) → PhCONHPh

PhCu + PhCHO (24%) → Ph_2CHOH

PhCu + CH_2=$CHCH_2Br$ (31%) → $PhCH_2CH$=CH_2

Scheme 1.3

$MeLi + CuX \xrightarrow[-15°C]{ether} MeCu$ Bright yellow, ether insoluble

X = Cl, I and SCN

MeCu ⇢ (MeLi) ⇢ Me_2CuLi

$2MeLi + CuX \xrightarrow[-15°C]{ether} Me_2CuLi$ Clear, practically colourless solution

X = I and SCN

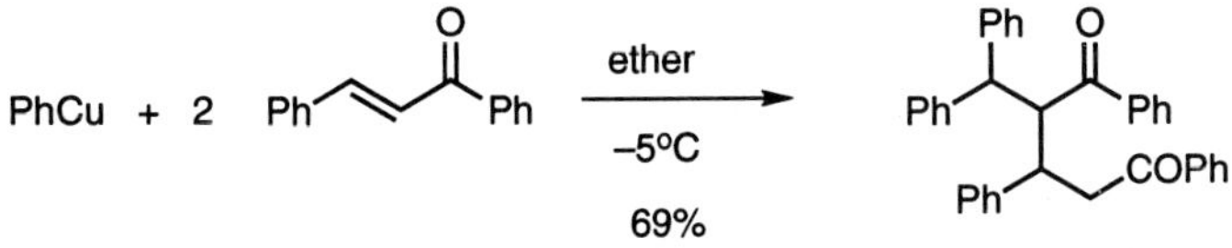

PhCu + 2 PhCH=CHCOPh $\xrightarrow[-5°C]{ether}$ (69%) → PhC(=O)CH(CHPh₂)CH(Ph)CH₂COPh

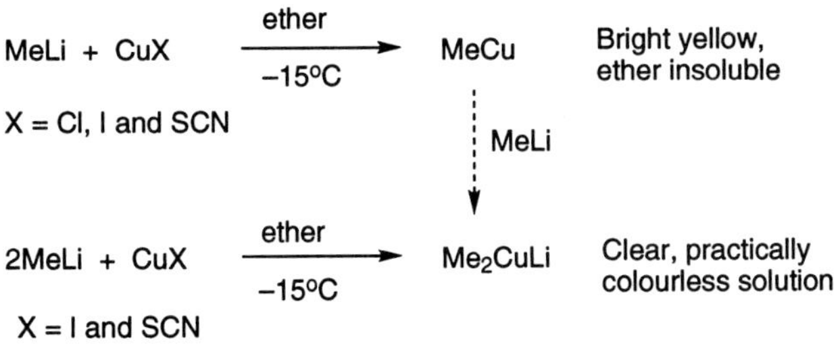

Scheme 1.4

facilitated the dimerization of aryl and vinyl Grignard reagents.[29] The most influential paper of this type, however, was Kharasch and Tawney's 1941 publication describing the dramatic effect of copper(I) catalysts on the regioselectivity of the reaction between methylmagnesium bromide and isophorone.[30] With the Grignard reagent alone only 1,2-addition was observed, but in the presence of 1% copper(I) chloride conjugate addition, or 1,4-addition, became the predominant process. Several other research groups utilized this observation,[10a] most notably Birch and Robinson[31] in their steroid studies and Munch-Petersen[32] and his colleagues working with unsaturated esters and related compounds. Other copper salts were also employed, Birch and Smith[33] preferring copper(II) acetate [which is reduced to copper(I) *in situ*] because of its greater solubility in ethereal solvents.

In 1966, House *et al.*[34] demonstrated the intermediacy of organocopper species in copper-catalysed conjugate addition reactions. They also showed that Gilman's stoichiometric mono-organocopper and organocuprate reagents could be employed to effect conjugate addition, often in higher yield and with greater reproducibility than in the copper-catalysed Grignard process. These studies brought stoichiometric organocopper reagents, particularly the more reactive organocuprates, to the attention of synthetic organic chemists, as did the work of Castro and Stephens[35] on alkynylcopper coupling reactions and Whitesides *et al.*[36] on the oxidative coupling of organocuprates (Scheme 1.5). It was soon shown that organocuprate reagents react with a range of alkyl and aryl halides in a synthetically useful manner (Scheme 1.6),[37] and in 1968 Corey *et al.*[38] described the synthesis of *Cecropia* juvenile hormone **8** using this methodology (Scheme 1.7). This synthetic route utilized organocopper reagents in three steps, most notably for the construction of two trisubstituted alkene units with total stereoselectivity, and demonstrated convincingly the value of stoichiometric organocopper reagents for the synthesis of complex natural products.

3. Scope of organocopper chemistry

Following the pioneering studies outlined in Section 2, an avalanche of publications appeared describing new types of organocopper reagents and new synthetic applications. These developments have been well covered in recent reviews,[1,6–16] notably those by Posner[10a,13a] and Lipshutz and Sengupta,[1] and so what follows is intended to be illustrative rather than comprehensive.

3.1 Types of organocopper-mediated reactions

Table 1.1 illustrates the more common applications of organocopper reagents in organic synthesis. **Conjugate addition reactions** have long been a mainstay in organic synthesis and a wide range of 'Michael acceptors' are compatible

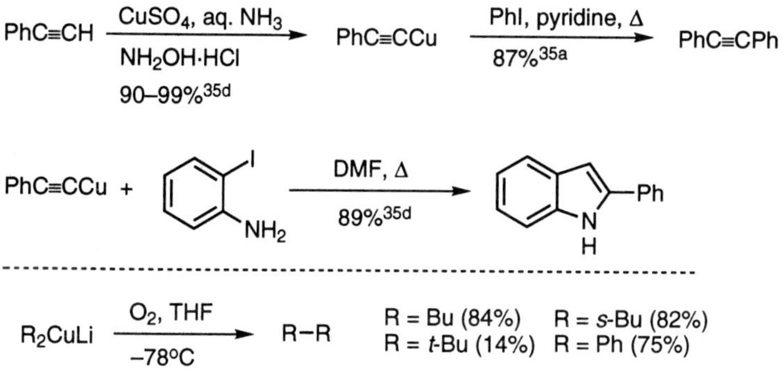

PhC≡CH → ($CuSO_4$, aq. NH_3, $NH_2OH{\cdot}HCl$; 90–99%[35d]) → PhC≡CCu → (PhI, pyridine, Δ; 87%[35a]) → PhC≡CPh

PhC≡CCu + 2-iodoaniline → (DMF, Δ; 89%[35d]) → 2-phenylindole

R_2CuLi → (O_2, THF, −78°C) → R–R R = Bu (84%) R = *s*-Bu (82%) R = *t*-Bu (14%) R = Ph (75%)

Scheme 1.5

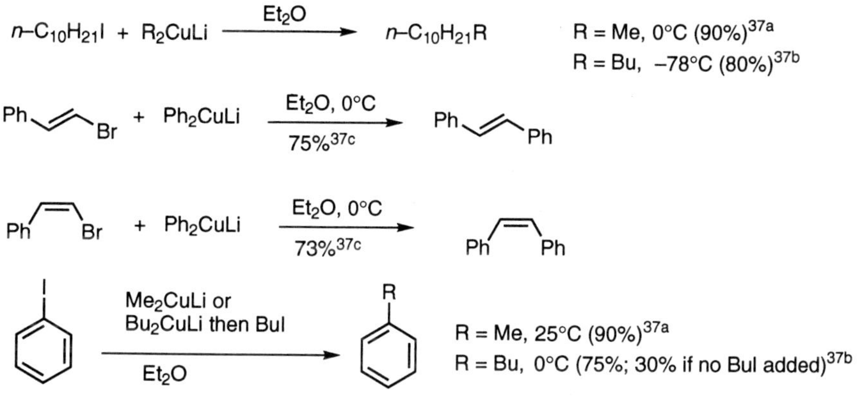

n-$C_{10}H_{21}I$ + R_2CuLi → (Et_2O) → *n*-$C_{10}H_{21}R$ R = Me, 0°C (90%)[37a] R = Bu, −78°C (80%)[37b]

(*E*)-PhCH=CHBr + Ph_2CuLi → (Et_2O, 0°C; 75%[37c]) → (*E*)-PhCH=CHPh

(*Z*)-PhCH=CHBr + Ph_2CuLi → (Et_2O, 0°C; 73%[37c]) → (*Z*)-PhCH=CHPh

PhI → (Me_2CuLi or Bu_2CuLi then BuI; Et_2O) → PhR R = Me, 25°C (90%)[37a] R = Bu, 0°C (75%; 30% if no BuI added)[37b]

Scheme 1.6

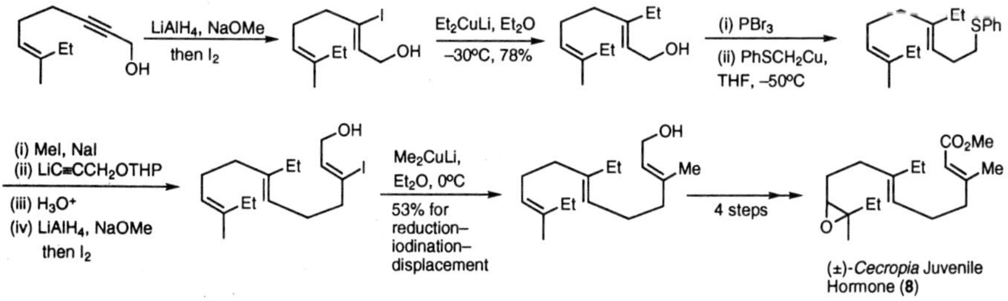

LiAlH₄, NaOMe then I_2; Et_2CuLi, Et_2O, −30°C, 78%; (i) PBr_3 (ii) $PhSCH_2Cu$, THF, −50°C; (i) MeI, NaI (ii) LiC≡CCH₂OTHP (iii) H_3O^+ (iv) $LiAlH_4$, NaOMe then I_2; Me_2CuLi, Et_2O, 0°C, 53% for reduction–iodination–displacement; 4 steps

(±)-*Cecropia* Juvenile Hormone (**8**)

Scheme 1.7

with the methodology.[1,10a] A great deal of current interest is centred on the development of asymmetric variants[11] of these processes (see Chapter 8). Difficulties encountered with conjugate addition to unsaturated aldehydes, owing to competing 1,2-addition, and to unsaturated amides, owing to their low reactivity, have been largely overcome with the introduction of

Table 1.1 Organocopper reagents in organic synthesis

Reaction	Equations	Comments
Conjugate addition	'RCu' + $CH_2{=}CH{-}Z$ $\xrightarrow{H^+}$ RCH_2CH_2Z 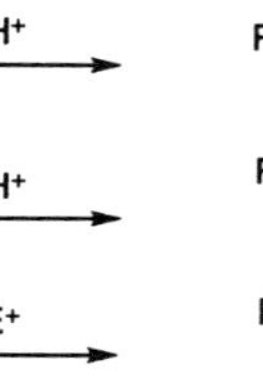	Z = ketone, aldehyde,[a] ester, acid, nitrile, amide,[a] sulfone, phosphonate, phosphine oxide, etc.
	'RCu' + $HC{\equiv}C{-}Z$ $\xrightarrow{H^+}$ $RCH{=}CH{-}Z$	Recent examples have been published[39]
Conjugate addition–enolate trapping	'RCu' + $CH_2{=}CH{-}Z$ $\xrightarrow{E^+}$ $RCH_2CH(E)Z$ 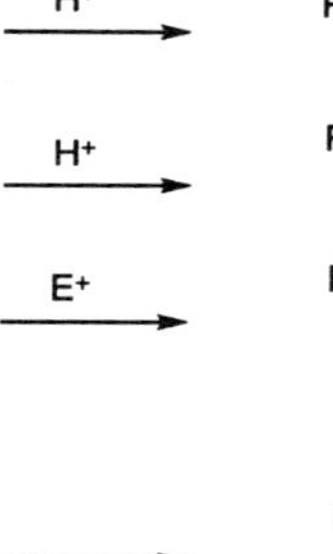	Aldehydes, alkyl halides, Michael acceptors, etc. can be used as electrophiles. Trialkylsilyl chlorides etc. give *O*-trapping
Acylation	'RCu' + R^1COX $\longrightarrow$ R^1COR  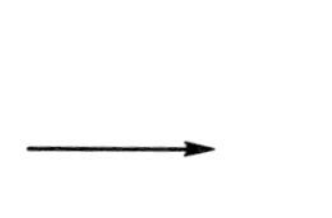 	Useful with acid halides, thioesters, CO_2, CS_2, etc. 1,2-Additions to ketones, aldehydes, imines, etc. are also useful[16,40]
Alkylation and arylation	'RCu + R^1X $\longrightarrow$ R^1R	Primary > secondary ≫ tertiary; I, Ts > Br ≫ Cl
	'RCu' + epoxide(R^1) $\xrightarrow{H^+}$ $RCH_2CH(OH)R^1$ 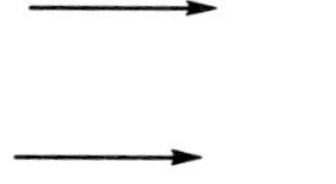	β-Lactones undergo substitution to give $RCH_2CH_2CO_2H$ (see Ch. 11, Protocol 10)
	$RC{\equiv}CH$ + ArX $\xrightarrow{\text{cat. Cu(I), Pd(0)}}$ $RC{\equiv}C{-}Ar$	Sonogashira couping of alkynes to aryl and vinyl halides is highly efficient
	'RCu' + $HC{\equiv}CCH_2X$ $\xrightarrow{H^+}$ $RCH{=}C{=}CH_2$  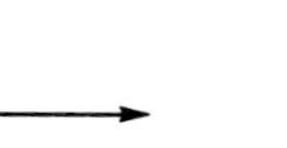 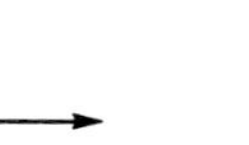	S_N2' reactions with allylic and propargylic systems are also useful
Alkyne carbocupration–trapping	'RCu' + $HC{\equiv}CH$ $\xrightarrow{E^+}$ $RCH{=}CHE$	Double acetylene carbocupration, to give (*Z,Z*)-dienes, is also possible.[41] A range of electrophiles can be employed, e.g. I_2, PhSSPh, CO_2, acid halides, Michael acceptors, etc. R_2CuLi are preferred with $HC{\equiv}CH$, $RCu{\cdot}MgBr_2{\cdot}SMe_2$ with $R^1C{\equiv}CH$
	'RCu' + $R^1C{\equiv}CH$ $\xrightarrow{E^+}$ $R(R^1)C{=}CHE$	

[a] Facilitated by Me_3SiCl addition (see text).

trimethylsilyl halide-modified organocopper reagents[8b,8c,42] (see Chapter 6). When a leaving group is present at the β- or β′-position, addition–elimination and/or double-addition processes can be effected (Scheme 1.8).[43]

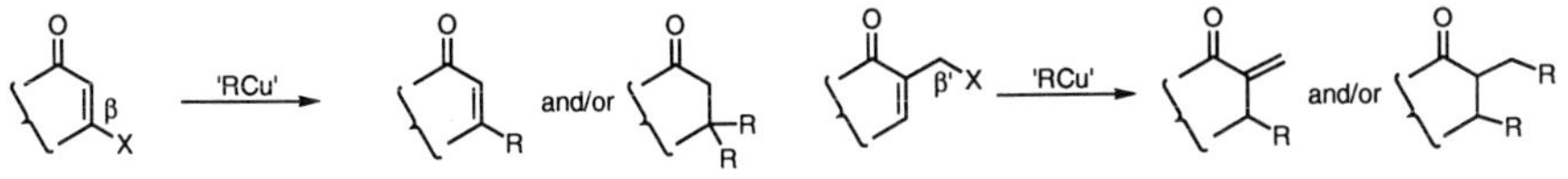

Scheme 1.8

More highly unsaturated substrates can also undergo conjugate addition reactions (1,6-addition, 1,8-addition, etc.).[1,10a,44]

Conjugate addition–enolate-trapping reactions have also attracted a great deal of attention, particularly with a view to developing tandem vicinal dialkylation processes.[12] A range of electrophiles has been employed in this sequence (see Chapter 9), but alkylation with unactivated alkyl halides has always presented problems, particularly with cyclopentanone enolates where rapid enolate equilibration, resulting in loss of regioselectivity, occurs. To overcome this problem, the enolate can be *O*-silylated, the resulting silyl enol ether isolated and purified, and then the enolate regenerated and alkylated under more favourable conditions.[12] A more convenient solution to this problem involves *in situ* conversion to the tin enolate prior to alkylation (e.g. Scheme 1.1 and Chapter 9).[2]

Acylation of organocopper reagents provides an efficient route to ketones from carboxylic acid derivatives.[1,13a] The 1,2-addition of organocopper reagents to aldehydes, ketones, and imines is of growing importance and was the topic of an excellent review in 1991.[16,40]

Alkylation/arylation reactions are of great importance to the synthetic organic chemist and allow the construction of hydrocarbon frameworks without recourse to classical functional group chemistry. Consequently, retrosynthetic disconnections do not have to be made at functional groups but instead can be made in the middle of a hydrocarbon chain, as illustrated by the synthesis of Muscalure shown in Scheme 1.2b. Again this area has been well reviewed,[13,14] although it is perhaps worth pointing out that the introduction of soluble copper(I) catalysts, especially Li_2CuCl_4 for alkylation reactions[45,46] and higher order cyanocuprates (see Section 3.2),[9] has improved the reproducibility and scope of this process enormously. Higher order cyanocuprates appear to be the reagents of choice for reactions with epoxides, primary alkyl chlorides, secondary alkyl bromides and secondary iodides (where competing elimination/reduction is often observed using lower order cuprates) (see Chapter 5). Secondary bromides and sulfonate esters react with predominant inversion of configuration, whereas iodides give mainly racemization. Problems may be encountered with the arylation of organocopper reagents owing to metal–halogen exchange, although the addition of the alkyl halide corresponding to the organocopper reagent, if available, can boost the yields (Scheme 1.6). The use of mono-organocopper reagents can minimize this exchange but it is usually better to circumvent the problem by coupling the arylcopper reagent with an alkyl halide, if possible;[13b] coupling reactions between aryl or vinyl halides and acetylides are an exception to this generalization. Highly efficient transformations can be achieved using the alkyne with catalytic quantities of copper(I)–palladium(0) catalysts and an amine base. This modification, often known as the Sonogashira coupling reaction,[14] is a marked improvement on the use of copper acetylides (Scheme 1.5) and is discussed in more detail in Chapter 10. A catalytic coupling procedure that employs only CuI and Ph_3P has recently been described.[47]

Organocopper substitution reactions with allylic and propargylic substrates are of considerable mechanistic and synthetic interest in terms of regio- and stereoselectivity. This is too complex a topic to be discussed in detail here. Instead, two recent examples are given (Scheme 1.9)[48,49] which illustrate the synthetic potential of the S_N2' sequence for the preparation of enantiomerically pure allylic and allenic alcohols. The reader is also referred to recent reviews of this topic[1,13b,50] and to further examples in Chapters 7 and 11.

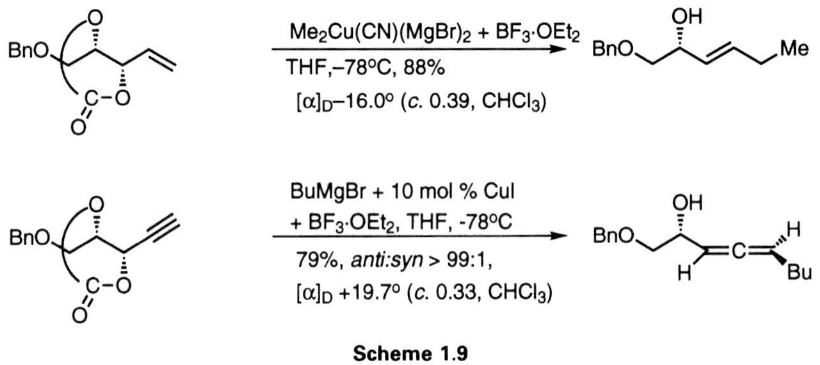

$Me_2Cu(CN)(MgBr)_2 + BF_3{\cdot}OEt_2$
THF, –78°C, 88%
$[\alpha]_D$ –16.0° (*c*. 0.39, $CHCl_3$)

BuMgBr + 10 mol % CuI
+ $BF_3{\cdot}OEt_2$, THF, -78°C
79%, *anti:syn* > 99:1,
$[\alpha]_D$ +19.7° (*c*. 0.33, $CHCl_3$)

Scheme 1.9

Alkyne carbocupration–trapping was discovered in 1971 by Normant and Bourgain[51] who realized that organocopper reagents, rather than effecting the expected deprotonation, undergo addition to acetylene and terminal alkynes with high *cis*-stereoselectivity, as shown in Table 1.1. The regioselectivity of addition to terminal alkynes is normally as shown ('Markownikov'), although chelating substituents in the alkyne side chain, or the use of silylcopper reagents, can lead to a reversal (see Chapters 11 and 12). Further studies, particularly by Normant and Alexakis, established alkyne carbocupration–trapping as a procedure to rival the Wittig reaction for the stereocontrolled preparation of di- and trisubstituted alkenes.[15] A wide range of electrophiles can be employed in the trapping reaction and the procedure can be modified to produce dienes (Chapter 2, Protocol 11 and Chapter 11).[15,41,52] Recent advances in this area include the extension of the methodology to encompass alkene[53] and allene[54] carbocupration. Not surprisingly, carbocupration chemistry has been used widely in natural product synthesis, particularly for the preparation of insect pheromones.[15b,41,50,55] Chapter 11 covers alkyne carbocupration in detail but it would be remiss to conclude this brief discussion without reference to the Marfat *et al.*[55] synthesis of the codling moth constituent (Scheme 1.10), which demonstrates the power of this methodology beautifully.

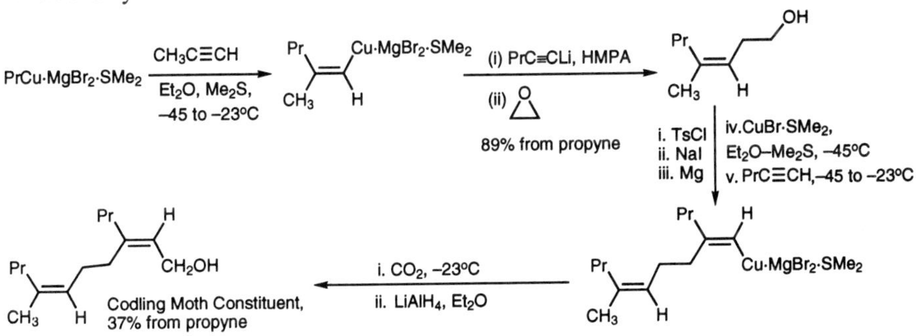

Scheme 1.10

3.2 Types of organocopper reagents: reagent choice

A summary of the organocopper reagent types available is given in Table 1.2. **Copper-catalysed Grignard reagents** are still of great utility owing to their ease of preparation and the normal requirement for only a modest excess of organometallic reagent. They would often be considered as the first choice of reagent for alkyl substitution and conjugate addition processes, stoichiometric reagents only being investigated if problems were encountered. Many of the early problems of reproducibility have been overcome by the introduction of high purity, soluble copper salts. In particular, Li_2CuCl_4,[45,46] Li_2CuCl_3,[46] $CuBr{\cdot}SMe_2$, and CuCN have proved useful. More recently, several groups have shown that copper-catalysed Grignard conjugate addition reactions (as well as reactions with the stoichiometric reagents) are accelerated, and the 1,4-regioselectivity enhanced, by the addition of Me_3SiCl and other silyl halides.[8b,8c,42] With the appropriate work-up procedure silyl enol ethers can be isolated from these reactions (Chapter 6).[56] It should be noted that when copper(II) salts are employed as catalysts, reduction to copper(I) by the Grignard reagent normally occurs *in situ*.[57] Recently, however, it has been shown[58] that in the presence of Me_3SiCl, copper(II) salts are not reduced but still act as catalysts for the conjugate addition of Grignard reagents to α,β-unsaturated esters.

Mono-organocopper reagents are usually prepared from copper(I) halides with an equivalent of RMgX or RLi. It should be noted that salt-free mono-organocopper reagents are unreactive. This implies that, in the presence of a

Table 1.2. Types of organocopper reagents[a]

Type	Preparation
Copper-catalysed Grignard reagents	RMgX (or RLi) + $\leq$25 mol% CuX
Mono-organocopper reagents (organocopper reagents)	RM + CuX $\longrightarrow$ RCu
Homocuprate reagents[b] (cuprate or Gilman reagents)	2 RM + CuX $\longrightarrow$ R_2CuM
Heterocuprate reagents	RM + CuZ $\longrightarrow$ RCu(Z)M
Higher order homocuprates, e.g. $R_{m+n}Cu_mLi_n$ ($m+n$ >2)	3 RLi + 2 CuI $\longrightarrow$ R_3Cu_2Li 5 RLi + 3 CuI $\longrightarrow$ $R_5Cu_3Li_2$
Higher order heterocuprates,[b] e.g. higher order cyanocuprates	2 RM + CuCN $\longrightarrow$ $R_2Cu(CN)M_2$

Species	Comments
R = alkyl, aryl, vinyl, etc.	Ether or THF usually used as solvent
CuX is usually CuI or $CuBr{\cdot}SMe_2$	Reactions usually carried out at −78 to 0°C under N_2 or Ar
CuZ is usually CuSPh or CuOBu-*t*	Solubilizing ligands (e.g. Me_2S, $(EtO)_3P$,Bu_3P) are often added
RM is usually RLi or RMgX	Addition of Lewis acids (e.g. BF_3, $AlCl_3$) or Me_3SiX can enhance reactivity

[a] In order to avoid confusion, the term 'organocopper reagent' will be employed in a generic sense to describe *all* types of catalytic and stoichiometric copper-based organometallic systems. The specific reagent type with the stoichiometry RCu will be referred to as a mono-organocopper reagent.
[b] Reagents with non-transferable ligands can also be prepared (see text and Table 1.3), e.g. RCu(C≡CPr)Li and $RCu(2\text{-}Th)(CN)Li_2$.

metal salt, a cuprate such as $(RCuX)^-$, is the reactive species.[13a] Mono-alkylcopper reagents tend to decompose above −15°C, whereas monovinyl- and arylcoppers are stable at much higher temperatures (PhCu to 100°C in an inert atmosphere) and copper acetylides are hydrolytically stable. Reagents of this type are often insoluble owing to their oligomeric or polymeric structures (e.g. methylcopper is a yellow polymeric precipitate in ether). Mono-organocopper reagents are usually more stable in THF than ether as solvent and the presence of amine, phosphine, or sulfide ligands further enhances solubility, thermal stability, and reactivity.[59] Indeed, it has recently been shown that dimethyl sulfide is an excellent solvent for mono-organocopper reagents and that this combination gives efficient transformations in conjugate addition and acid chloride alkylation reactions.[59] The problems of solubility/reactivity/stability referred to above have limited the synthetic applications of mono-organocopper reagents over the years, despite the fact that, in principle, they could provide an extremely efficient means of organic group transfer. However, recent developments in terms of the use of organocopper–ligand, organocopper–Lewis acid, and RCu·Me_3SiX reagents have produced an upsurge in the applications of mono-organocopper reagents. For example, Noyori and Suzuki found that tributylphosphine-complexed monovinylcopper reagents were extremely well suited to their conjugate addition–enolate alkylation approach to prostaglandin synthesis (Scheme 1.1 and Chapter 9)[2,12] and, as has already been mentioned (Table 1.1 and Scheme 1.10), dimethyl sulfide-complexed mono-organocopper reagents (RCu·$MgBr_2$·SMe_2) are the reagents of choice for terminal alkyne carbocupration. In addition, mono-organocopper reagents display excellent compatibility with a range of Lewis acids (BF_3·OEt_2, $AlCl_3$, $MgBr_2$, etc.) and these modified reagents often show unique reactivity patterns.[8] Highly diastereoselective conjugate additions to oxygenated Michael acceptors and related substrates are a particular feature of this type of reagent.[8d] It should be noted that other organocopper reagents can also be modified by the addition of Lewis acids (see Scheme 1.9); this topic is discussed in greater detail in Chapter 7. Also, the use of allylcopper reagents in combination with trimethylsilyl chloride has solved the long-standing problem of how to achieve efficient conjugate addition with allylic organocopper reagents: see Chapters 3 and 5 for more details.[60]

Homocuprate reagents (usually R_2CuLi or R_2CuMgX) are the most widely used of the organocopper reagents. In general they possess enhanced nucleophilicity and stability compared to their mono-organocopper counterparts and this, together with their greater homogeneity, often leads to more reproducible transformations. Lithium dimethylcuprate is stable at 0 °C, but usually, particularly for alkylcuprates, −78 °C under an inert atmosphere is employed. Polymer-bound cuprates have been prepared[61] and they are reported to be stable for up to three weeks when stored at room temperature under argon. Usually only one of the organic ligands is transferred from homocuprates* and if the organometallic precursors are expensive, or hard-won, this presents an obvious problem, particularly if copper-catalysed Grignards or activated mono-organocopper reagents are not suitable. To overcome this problem, in 1972 Corey and Beames[62] introduced mixed homocuprates derived from copper pentyne, having demonstrated that alkynyl ligands are transferred much more slowly than other organic groups. Since that time, a number of copper acetylides have been proposed for this purpose (see Table 1.3),[63–66] differing in terms of availability, cost, solubility, etc. In addition, (methylsulfinyl)methylcopper,[67] related sulfones,[68] thienylcopper,[69] and mesitylcopper[70] have been employed for the same purpose. The factors which influence the relative migratory aptitudes of ligands attached to copper in organocopper reagents have been reviewed.[71]

* It is possible to reactivate the resulting mono-organocopper reagent *in situ* to ensure that both of the original organic groups on copper are eventually transferred (see Chapter 11, Protocol 8, for example).

Table 1.3. Cuprates with non-transferable ('dummy') ligands (R_T = transferable ligand)

	R/Z	References
a Mixed homocuprates		
$R_TM + RCu \longrightarrow R_TCu(R)M$	C≡CPr	62
	C≡CBu-*t*	63
	$C{\equiv}CCH_2NMe_2$	64
	$C{\equiv}CCMe_2OMe$	65
	$C{\equiv}CSiMe_3$	66
	$CH_2S(O)Me$	67
	$CH_2S(O)_2Me$	68
	2-Thienyl	69
	Mesityl	70
b Heterocuprates[a]		
$R_TM + CuZ \longrightarrow R_TCu(Z)M$	CN	72
	PhS	73
	t-BuO	73
	Ph_2P	74
	$(C_6H_{11})_2N$	74
	t-Bu_2P	75
c Mixed higher order cyanocuprates		
$R_TM + RM + CuCN \longrightarrow R_TRCu(CN)M_2$	2-Thienyl	84a
	N-pyrrolyl	84b
	N-imidazolyl	84b

[a] Cyano is classed as a heteroligand in view of the similarity, in terms of preparation, stability, and reactivity, between RCu(CN)Li and the other members of this group.

Heterocuprate reagents [RCu(Z)M, including Z = CN, see Table 1.3] are also efficient in terms of organic group transfer and, although they are usually less reactive than the mixed homocuprates, they often display improved thermal stability.[72–75] The di-*t*-butylphosphidocuprates, for example, retain >96% of their activity after four hours at room temperature in THF.[75] One advantage of these reagents over the mixed homocuprates is that a number of the starting materials containing the 'dummy ligand' are commercially available (e.g. CuCN, CuSPh, $CuPPh_2$), whereas most of the mixed homocuprate precursors have to be prepared. It should also be noted that lithium phenylthiocuprates are efficient reagents for the transfer of *s*- and *t*-butyl groups, and that there are an increasing number of promising chiral heterocuprate reagents being designed to effect asymmetric conjugate addition[11] (see Chapter 8). Amidocuprate reagents which transfer the nitrogen ligand in a conjugate manner have recently been developed and employed in stereoselective synthesis.[76]

Higher order homocuprates (see Table 1.2 for examples) have been formed by varying the ratio CuX:RLi.[77] A few synthetic applications of these reagents have been developed; for example, R_3Cu_2M (M = Li or MgBr) effects carbocupration of terminal alkynes[78] and shows good diastereoselectivity on addition to chiral α,β-unsaturated esters,[79] and $Me_5Cu_3Li_2$ is useful for the conjugate methylation of α,β-unsaturated aldehydes[80] and for S_N2' addition to allylic carbamates.[81] In general, however, these reagents appear to offer little and would not be recommended as the starting point for an organocopper-mediated transformation.

Higher order heterocuprates, particularly the cyanocuprates, have proved to be extremely versatile synthetic reagents.[9,82] They are formed easily (usually from CuCN + 2RLi, although CuSCN can also be employed[83]), and combine the stability of heterocuprates with the reactivity of homocuprates. They also have the advantage that the copper precursor, CuCN, is inexpensive, non-hygroscopic, and light insensitive, and prefers the copper(I) oxidation level. The cyanocuprates have been shown to be exceptionally useful reagents for a range of transformations, particularly substitution reactions of secondary halides and epoxides, and they are probably the ideal starting-point when commencing an investigation requiring stoichiometric organocopper reagents. The exception is alkyne carbocupration, where deprotonation is observed, although the less basic silylcyanocuprates $[(R_3Si)_2Cu(CN)Li_2]$ add efficiently (Chapter 12). If the organic ligand to be transferred is valuable, mixed higher order cyanocuprates are also available (Table 1.3c),[84] and an ideal precursor, (2-Th)Cu(CN)Li, is available commercially. Chapter 5 contains a detailed discussion of the preparation and reactions of higher order cyanocuprates and their mixed counterparts.

It should be noted that, as with mono-organocopper reagents, the addition of solubilizing ligands, Lewis acids, and/or Me_3SiX can often enhance the reactivity and selectivity of the whole range of cuprate reagents.

3.3 Recent developments: functionalized organocopper reagents

Enantiomerically pure α-alkoxyorganocopper reagents have recently been generated from the corresponding stannanes and utilized in conjugate addition reactions (Scheme 1.11a).[85] It is noteworthy that complete retention of configuration was observed in both processes when the mono-organocopper reagent was employed, but with the corresponding higher order cyanocuprate a significant amount (30%) of the C-4-inverted diastereomer was obtained.[85]

The discussion so far has concentrated on the classical procedures of preparing organocopper reagents starting from organolithium or Grignard reagents. This route places an obvious limitation on the types of functionality that are compatible with the preparative procedure. In recent years some remarkable advances have been made which allow a wide range of functionalized organocopper reagents to be generated (Scheme 1.11b–d). These reagents can be prepared by the direct insertion of active (Rieke) copper into

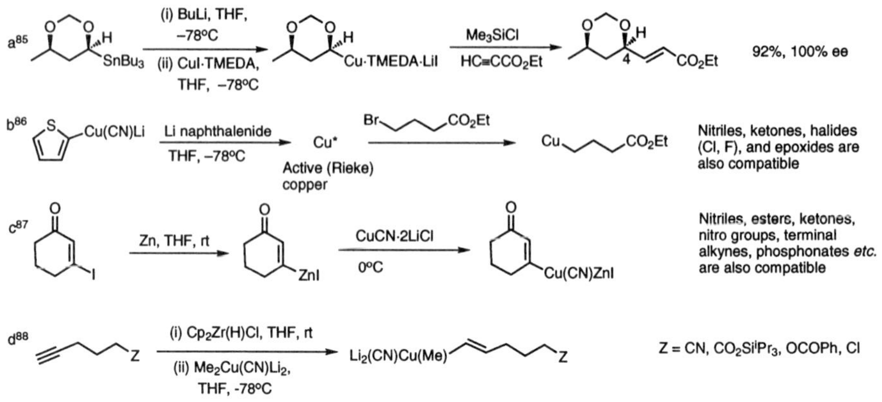
a85
(i) BuLi, THF, –78°C
(ii) CuI·TMEDA, THF, –78°C
SnBu3
Cu·TMEDA·LiI
Me3SiCl
HC≡CCO2Et
CO2Et
92%, 100% ee
b86
Cu(CN)Li
Li naphthalenide
THF, –78°C
Cu*
Active (Rieke) copper
CO2Et
Nitriles, ketones, halides (Cl, F), and epoxides are also compatible
c87
Zn, THF, rt
ZnI
CuCN·2LiCl
0°C
Cu(CN)ZnI
Nitriles, esters, ketones, nitro groups, terminal alkynes, phosphonates etc. are also compatible
d88
(i) Cp2Zr(H)Cl, THF, rt
(ii) Me2Cu(CN)Li2, THF, -78°C
Li2(CN)Cu(Me)
Z = CN, CO2SiiPr3, OCOPh, Cl

Scheme 1.11

the carbon–halogen bond (Scheme 1.11b and Chapter 3),[86] via organozinc reagents (Scheme 1.11c and Chapter 4),[9d-f,87] or by way of vinylzirconium intermediates and transmetallation using higher order cyanocuprates (Scheme 1.11d and Chapter 5).[5,88] Other vinylmetal species can also be employed in this type of direct organocopper transmetallation reaction.[89]

Finally, mention must be made of the zinc–CuI–ultrasound-mediated conjugate addition procedure devised by Luche and co-workers which proceeds under aqueous conditions.[90] This methodology, which has recently been utilized in a synthesis of vitamin D analogues (Scheme 1.12),[91] is believed to proceed by a radical pathway with an initial single-electron transfer to the carbon–iodine bond, but there is an obvious relevance to organocopper-mediated processes.

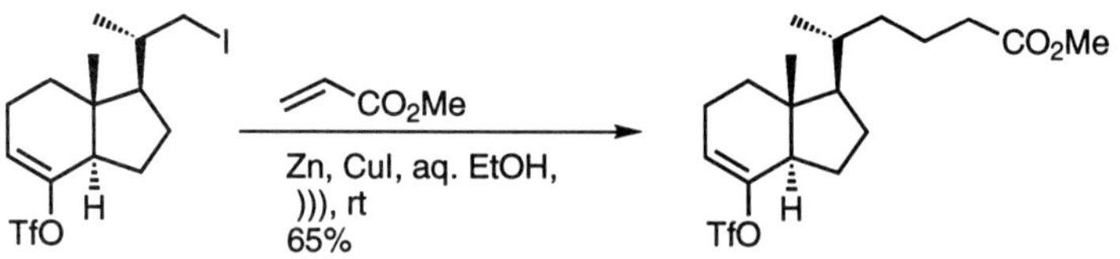
CO2Me
Zn, CuI, aq. EtOH,))), rt
65%
TfO
H
CO2Me

Scheme 1.12

4. Structure of organocopper reagents

Structural information concerning organocopper reagents has been gathered using a number of techniques including NMR spectroscopy,[92] X-ray crystallography,[93–95] and extended X-ray absorption fine structure (EXAFS) spectroscopy.[96] This subject has been well reviewed in recent years,[1,6a,6f,71,92–96] and only a brief résumé together with leading references will be presented here. A number of crystal structures have been obtained on mono-organocopper reagents, many of which exist as aggregates ranging from monomeric to octameric.[93–95] There are many examples of the importance of ligands in crystal structures: phenylcopper, for example, is tetrameric in the presence of dimethyl sulfide ligands $[Ph_4Cu_4(SMe_2)_2]$,[94] but monomeric in the presence of 1,1,1-tris[(diphenylphosphino)methyl]ethane.[95]

A variety of measurements indicate that simple homocuprates exist as dimeric species in solid and solution states, but NMR spectroscopic investigations reveal the presence of complex, solvent-dependent equilibria for higher order species.[92*] Recently, crystal structures and EXAFS data have been obtained on a range of homocuprates and higher order cuprates,[96] although the first crystal structure of a cyanocuprate is still eagerly awaited. A recent publication described a theoretical study which generated novel structures for higher order homocuprates.[97] As the amount of solid and solution data on organocopper reagents increases, hope grows that it may soon be possible to correlate structure with reactivity and mechanism.

5. Mechanisms of organocopper reactions

In efforts to elucidate the mechanisms of organocopper reactions, and thereby to explain, and further optimize, the unique reactivity/selectivity observed in these processes, a great deal of information has been gathered from spectroscopic investigations, structure–activity studies (including the use of substrates containing inbuilt radical traps), and reduction potential measurements. Despite this, it is clear that our mechanistic understanding is still at a fairly rudimentary level. There is room here only to present a précis of current ideas: for more detailed discussions the reader is referred to recent reviews.[1,10b,10c,13b]

Most investigations have concentrated on conjugate addition reactions of lithium homocuprates.[1,10b,10c] Early studies by House and his group revealed a good correlation between the reduction potential of enones and their reactivity in organocopper conjugate addition reactions.[98] House therefore proposed a mechanism involving single-electron transfer from cuprate to enone to generate a radical anion which undergoes coupling to give a copper(III) species, followed by reductive elimination to give products (Scheme 1.13a).[98] Alternative mechanisms based on the intermediacy of charge-transfer complexes,[99] or direct addition to generate the copper(III) intermediate,[100] have also been proposed. It should be noted that the intermediacy of a copper(III) species has yet to be established unambiguously, although copper(III) complexes (admittedly bearing strong σ-donor ligands) are known.[101] Berlan *et al.*[102] have proposed an alternative 1,2-carbocupration mechanism which generates an α-cupriocarbonyl intermediate and avoids the need for a

* In Et_2O, NMR studies indicate that lithium dimethylcuprate is a unique species, but with halide-free THF or THF–Et_2O as solvent, an equilibrium mixture of MeLi, Me_2CuLi, and Me_3Cu_2Li was observed.[84a]

copper(III) intermediate. The nature of the resulting enolate is also still a matter of debate.[98,103]

Studies using stopped-flow[103] and NMR spectroscopy[104] (see Chapter 13), together with theoretical calculations,[105] suggest the equilibrium formation of a copper–alkene dπ*-complex which collapses to generate a copper(III) intermediate (Scheme 1.13b). Experimental observations which are consistent with this model have been reported,[106] and solid enone–cuprate complexes, which could be dπ*-type complexes, have been observed at −78°C.[107] Finally, it has been proposed that in organocuprate-mediated 1,6- and 1,8-additions to Michael acceptors a dπ*-complex is also formed but that intermolecular attack by a second cuprate occurs to produce the observed addition products.[44b] The currently accepted mechanism for organocuprate conjugate addition is therefore the one outlined in Scheme 1.13b, but, as the above discussion indicates, there are still a number of uncertainties and ambiguities, and this model should be used with caution.

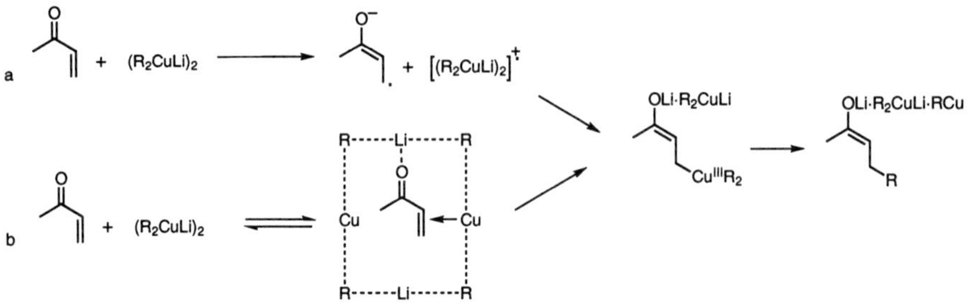

a + $(R_2CuLi)_2$ → + $[(R_2CuLi)_2]^{+}$

b + $(R_2CuLi)_2$ ⇌

OLi·R_2CuLi

$Cu^{III}R_2$

OLi·R_2CuLi·RCu

Scheme 1.13

The mechanism of organocopper substitution reactions has also been probed in some detail, although again a clear-cut picture has yet to emerge. For detailed discussions, the reader is referred to three excellent recent reviews of this topic.[1,13b,108] Mechanistic proposals have to accommodate the following observations:

(1) Reactions are usually first order in both organocuprate and substrate.
(2) Substrate reactivity is primary > secondary > tertiary.
(3) The reactivity of the group transferred from copper decreases in the order primary > secondary > tertiary.
(4) Chiral secondary bromides and tosylates react with predominant inversion of configuration.
(5) Alkenyl halides (and alkenylcopper reagents) react with predominant retention of configuration.

Coupled with the fact that radical scavengers usually have little effect on the course of this type of reaction, the above information is generally taken to support an oxidative addition–reductive elimination mechanism proceeding via a copper(III) intermediate (Scheme 1.14), rather than a direct displacement route. However, an alternative proposal has been made in which each copper atom in the dimeric cluster donates one electron to the substrate, thereby generating two copper(II) atoms rather than one copper(III).[109] This issue has still to be resolved.

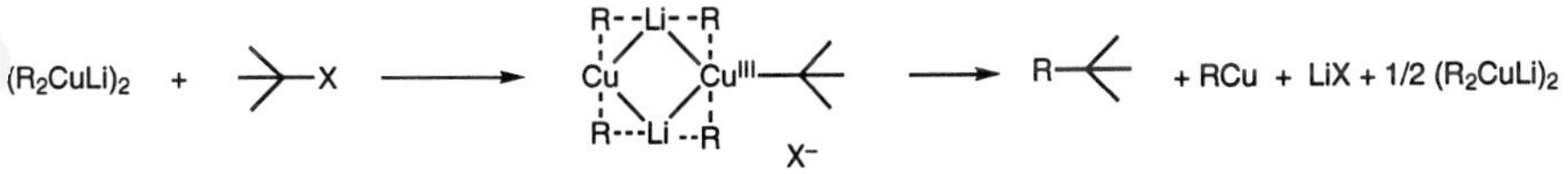

$(R_2CuLi)_2$ + ⟩—X ⟶ [R--Li--R / Cu Cu^{III}— / R---Li--R] X^- ⟶ R— + RCu + LiX + 1/2 $(R_2CuLi)_2$

Scheme 1.14

The situation is more complex than has been described above, however. Secondary iodides, unlike the corresponding bromides and tosylates, undergo almost complete racemization on treatment with organocuprates. In addition, 6-iodoheptene produces cyclopentane-based products on treatment with lithium dimethylcuprate,[110] and a number of alkyl iodides have been shown to react with $Bu_2CuLi{\cdot}LiI$ via an electron-transfer–radical-intermediate pathway.[108] It has been suggested that this alternative mechanism operates if the halide substrate has a reduction potential of ≥ -2.3 V.[108] Alkyl sulfonates are believed to react only by the inversion mechanism shown in Scheme 1.14.

Recent theoretical studies have thrown light on the carbocupration reactions of substituted alkynes[111] and of cyclopropene,[112] but much remains to be done before a clear mechanistic picture for these processes emerges.

References

1. Lipshutz, B. H.; Sengupta, S. *Org. React.* **1992**, *41*, 135–631.
2. Suzuki, M.; Yanagisawa, A.; Noyori, R. *J. Am. Chem. Soc.* **1985**, *107*, 3348–3349. Suzuki, M.; Yanagisawa, A.; Noyori, R. *J. Am. Chem. Soc.* **1988**, *110*, 4718–4726.
3. Collins, P. *Med. Res. Rev.* **1990**, *10*, 149–172. Bleasedale, D. A. Searle Pharmaceuticals, Morpeth, Northumberland, personal communication.
4. Henrick, C. A. *Tetrahedron* **1977**, *33*, 1845–1889 (pp. 1882–1885). Carney, R. L. *US Pat.* 3 948 803, **1976**.
5. Lipshutz, B. H.; Ellsworth, E. L. *J. Am. Chem. Soc.* **1990**, *112*, 7440–7441. Babiak, K. A.; Behling, J. R.; Dygos, J. H.; McLaughling, K. T.; Ng, J. S.; Kalish, V. J.; Kramer, S. W.; Shone, R. L. *J. Am. Chem. Soc.* **1990**, *112*, 7441–7442.
6. General reviews: (a) Bähr, G.; Burba, P. *Houben–Weyl Methoden der Organische Chemie*; Müller, E., ed.; Georg Thieme: Stuttgart, **1970**; Vol. 13/1, pp. 735–761. (b) Normant, J. F. *Synthesis*, **1972**, 63–80. (c) Jukes, A. E. *Adv. Organomet. Chem.* **1974**, *12*, 215–322. (d) Normant, J. F. *J. Organomet. Chem. Library* **1976**, *1*, 219–256. (e) Posner, G. H. *An Introduction to Synthesis Using Organocopper*

Reagents; Wiley: New York, **1980**. (f) Noltes, J. G.; van Koten, G. In *Comprehensive Organometallic Chemistry*; Wilkinson, G., Stone, F. G. A., Abel, E. W., eds; Pergamon Press: Oxford, **1982**; Vol. 2, Chapter 14. (g) Carruthers, W. In *Comprehensive Organometallic Chemistry*; Wilkinson, G., Stone, F. G. A., Abel, E. W., eds; Pergamon Press: Oxford, **1982**; Vol. 7, Chapter 49. (h) Alexakis, A.; Chuit, C.; Commercon-Bourgain, M.; Foulon, J. P.; Jabri, N.; Mangeney, P.; Normant, J. F. *Pure Appl. Chem.* **1984**, *56*, 91–98. (i) *Tetrahedron Symp. Print, Tetrahedron*, **1989**, *45*, 349–569. (j) Lipshutz, B. H. In *Comprehensive Organic Synthesis*; Trost, B. M., Fleming, I., eds; Pergamon Press: Oxford, **1991**; Vol. 1, Chapter 1.4; (k) Lipshutz, B. H. In *Encyclopedia of Inorganic Chemistry*; King, R. B. ed.; Wiley, New York, **1994**.

7. Reviews on copper-catalysed organometallic reactions: (a) Normant, J. F. *Pure Appl. Chem.* **1978**, *50*, 709–715; (b) Erdik, E. *Tetrahedron* **1984**, *40*, 641–657; (c) Lee, V. J. In *Comprehensive Organic Synthesis*; Trost, B. M., Fleming, I., eds; Pergamon Press: Oxford, **1991**; Vol. 4, Sections 1.2 and 1.3.
8. Reviews on Lewis acid-modified and trimethylsilyl halide-modified organocopper reactions: (a) Yamamoto, Y. *Angew. Chem., Int. Ed., Engl.* **1986**, *25*, 947–959; (b) Nakamura, E. *Synlett* **1991**, 539–547; (c) Lee, V. J. In *Comprehensive Organic Synthesis*; Trost, B. M., Fleming, I., eds; Pergamon Press: Oxford, **1991**; Vol. 4, Sections 1.2 and 1.3; (for recent results see Lipshutz, B. H.; James, B. *Tetrahedron Lett.* **1993** *34*, 6689–6692; Lipshutz, B. H.; Dimock, S. H.; James, B. *J. Am. Chem. Soc.* **1993**, *115*, 9283–9284; Bergdahl, M.; Eriksson, M.; Nilsson, M.; Olsson, T. *J. Org. Chem.* **1993**, *58*, 7238–7244; Chounan, Y.; Yamamoto, Y. *Chemtracts* **1993**, *6*, 310–314); (d) Ibuka, T.; Yamamoto, Y. *Synlett* **1992**, 769–777 (for recent results see Kant, J. *J. Org. Chem.* **1993**, *58*, 2296–2301).
9. Reviews on higher order and highly functionalized organocuprates: (a) Lipshutz, B. H.; Wilhelm, R. S.; Kozlowski, J. A. *Tetrahedron* **1984**, *40*, 5005–5038; (b) Lipshutz, B. H. *Synthesis* **1987**, 325–341; (c) Lipshutz, B. H. *Synlett* **1990**, 119–128; (d) Knochel, P.; Rozema, M. J.; Tucker, C. E.; Retherford, C.; Furlong, M.; Achyutha Rao, S. *Pure Appl. Chem.* **1992**, *64*, 361–369; (e) Knochel, P.; Singer, R. D. *Chem. Rev.* **1993**, *93*, 2117–2188; (f) Crimmins, M. T.; Nantermet, P. G. *Org. Prep. Proc. Int.* **1993**, *25*, 41–81.
10. Reviews on organocopper conjugate addition reactions: (a) Posner, G. H. *Org. React.* **1972**, *19*, 1–113; (b) Kozlowski, J. A. In *Comprehensive Organic Synthesis*; Trost, B. M., Fleming, I., eds; Pergamon Press: Oxford, **1991**; Vol. 4, Chapter 1.4; (c) Perlmutter, P. *Conjugate Addition Reactions in Organic Synthesis*; Pergamon Press: Oxford, **1992**.
11. Reviews on asymmetric organocopper conjugate addition reactions: (a) Tomioka, K.; Koga, K. In *Asymmetric Synthesis*; Morrison, J. D., ed.; Academic Press: New York, **1983**; Vol. 2, Part A, Chapter 7; (b) Tomioka, K. *Synthesis* **1990**, 541–549; (c) Alexakis, A.; Mangeney, P. *Tetrahedron Asymm.* **1990**, *1*, 477–511; (d) Schmalz, H.-G. In *Comprehensive Organic Synthesis*; Trost, B. M., Fleming, I., eds; Pergamon Press: Oxford, **1991**; Vol. 4, Chapter 1.5; (e) Rossiter, B. E.; Swingle, N. M. *Chem. Rev.* **1992**, *92*, 771–806 (for a recent publication in this area see Kanai, M.; Tomioka, K. *Tetrahedron Lett.* **1994**, *35*, 895–898).
12. Reviews on organocopper conjugate addition–enolate-trapping reactions: (a) Noyori, R.; Suzuki, M. *Angew. Chem., Int. Ed. Engl.* **1984**, *23*, 847–876; (b) Taylor, R. J. K. *Synthesis* **1985**, 364–392; (c) Noyori, R. *Chem. Br.* **1989**, 883–

888; (d) Chapdelaine, M.; Hulce, M. *Org. React.* **1990**, *38*, 225–653; (e) Noyori, R.; Suzuki, M. *Chemtracts* **1990**, 173–197; (f) Hulce, M.; Chapdelaine, M. J. In *Comprehensive Organic Synthesis*; Trost, B. M., Fleming, I., eds; Pergamon Press: Oxford, **1991**; Vol. 4, Chapter 1.6.

13. Reviews on organocopper substitution reactions: (a) Posner, G. H. *Org. React.* **1975**, *22*, 253–400 (for a recent paper on cuprate acylation see Bonini, B. F.; Capperuci, A.; Comes-Francini, M.; Degl'Innocenti, A.; Mazzanti, G.; Ricci, A.; Zani, P. *Synlett* **1993**, 937–939); (b) Klunder, J. M.; Posner, G. H. In *Comprehensive Organic Synthesis*; Trost, B. M., Fleming, I., eds; Pergamon Press: Oxford, **1991**; Vol. 3, Chapter 1.5; (c) Billington, D. C. In *Comprehensive Organic Synthesis*; Trost, B. M., Fleming, I., eds; Pergamon Press: Oxford, **1991**; Vol. 3, Chapter 2.1. Tamas, K. In *Comprehensive Organic Synthesis*; Trost, B. M., Fleming, I., eds; Pergamon Press: Oxford, **1991**; Vol. 3, Chapter 2.2. Knight, D. W. In *Comprehensive Organic Synthesis*; Trost, B. M., Fleming, I., eds; Pergamon Press: Oxford, **1991**; Vol. 3, Chapters 1.6 and 2.3. Sonogashira, K. In *Comprehensive Organic Synthesis*; Trost, B. M., Fleming, I., eds; Pergamon Press: Oxford, **1991**; Vol. 3, Chapters 2.4 and 2.5.
14. Reviews on palladium-catalysed organocopper-coupling reactions: (a) Heck, R. F. *Palladium Reagents in Organic Synthesis*; Academic Press: London, **1985**; Chapter 6; (b) Tsuji, J. *Synthesis* **1990**, 739–749; (c) Sonogashira, K. In *Comprehensive Organic Synthesis*; Trost, B. M., Fleming, I., eds; Pergamon Press: Oxford, **1991**; Vol. 3, Chapter 2.4.
15. Reviews on alkyne carbocupration: (a) Normant, J. F.; Alexakis, A. *Synthesis* **1981**, 841–870; (b) Alexakis, A. *Actual. Chim.* **1987**, 203–210; (c) Knochel, P. In *Comprehensive Organic Synthesis*; Trost, B. M., Fleming, I., eds; Pergamon Press: Oxford, **1991**; Vol. 4, Chapter 4.4; (d) Ricci, A.; Reginato, G.; Degl'Innocenti, A.; Seconi, G. *Pure Appl. Chem.* **1992**, *64*, 439–446.
16. Review on organocopper 1,2-addition reactions: Lipshutz, B. H. In *Comprehensive Organic Synthesis*; Trost, B. M., Fleming, I., eds; Pergamon Press: Oxford, **1991**; Vol. 1, Chapter 1.4.
17. Frankland, E.; Duppcr, B. F. *J. Chem. Soc.* **1864**, *17*, 31.
18. Buckton, G. *Liebigs Ann. Chem.* **1859**, *109*, 225.
19. Böttger, R. C. *Liebigs Ann. Chem.* **1859**, *109*, 351.
20. Berthelot, M. *C. R. Hebd. Seances Acad. Sci.* **1860**, *50*, 805.
21. Sladkov, A. M.; Ukhin, L.-Yu. *Russ. Chem. Rev.* **1968**, *37*, 748–763.
22. Glaser, C. *Liebigs Ann. Chem.* **1870**, *154*, 159. For reviews of this and related (Eglington and Cadiot–Chodkiewicz) reactions see: Simandi, L. I. In *The Chemistry of Functional Groups*; Patai, S., Rappoport, Z., eds; Wiley: New York, **1983**; Supplement C, Part I, pp. 529–534; Sonogashira, K. In *Comprehensive Organic Synthesis*; Trost, B. M., Fleming, I., eds; Pergamon Press: Oxford, **1991**; Vol. 3, Chapter 2.5.
23. Ullmann, F. *Liebigs Ann. Chem.* **1904**, *332*, 38. For reviews see: Fanta, P. E. *Synthesis* **1974**, 9–21; Sainsbury, M. *Tetrahedron* **1980**, *36*, 3327–3359 and references therein.
24. Ullmann, F.; Sponagel, P. *Chem. Ber.* **1905**, *38*, 2211–2212. For a review see: Bacon, R. G. R.; Hill, H. A. O. *Q. Rev. Chem. Soc.* **1965**, *19*, 95–125.
25. Reich, R. *C. R. Hebd. Seances Acad. Sci.* **1923**, *177*, 322–324.
26. Gilman, H.; Straley, J. M. *Recl. Trav. Chim. Pays-Bas*, **1936**, *55*, 821–834.

27. Gilman, H.; Woods, L. A. *J. Am. Chem. Soc.* **1943**, *65*, 435–437.
28. Gilman, H.; Jones, R. G.; Woods, L. A. *J. Org. Chem.* **1952**, *17*, 1630–1634.
29. (a) Krizewsky, J.; Turner, E. E. *J. Chem. Soc.* **1919**, *115*, 559–561. (b) Gilman, H.; Parker, H. H. *J. Am. Chem. Soc.* **1924**, *46*, 2823–2827. (c) Gilman, H.; Kirby, J. E. *Recl. Trav. Chim. Pays-Bas*, **1929**, *48*, 155–159.
30. Kharasch, M. S.; Tawney, P. O. *J. Am. Chem. Soc.* **1941**, *63*, 2308–2316. Related work showed that cobalt(II) chloride and, less efficiently, copper(I) chloride could be employed to catalyse the coupling between phenylmagnesium bromide and vinyl bromide to produce styrene: Kharasch, M. S.; Fuchs, C. F. *J. Am. Chem. Soc.* **1943**, *65*, 504–507.
31. Birch, A. J.; Robinson, R. *J. Chem. Soc.* **1943**, 501–502.
32. Munch-Petersen, J. *J. Org. Chem.* **1957**, *22*, 170–176. Munch-Petersen, J. *Bull. Soc. Chim. Fr.* **1966**, 471–480 and references therein.
33. Birch, A. J.; Smith, M. *Proc. Chem. Soc.* **1962**, 356.
34. House, H. O.; Respess, W. L.; Whitesides, G. M. *J. Org. Chem.* **1966**, *31*, 3128–3141.
35. (a) Stephens, R. D.; Castro, C. E. *J. Org. Chem.* **1963**, *28*, 3313–3320. (b) Castro, C. E.; Stephens, R. D. *J. Org. Chem.* **1963**, *28*, 2163. (c) Castro, C. E.; Gaughan, E. J.; Owsley, D. C. *J. Org. Chem.* **1966**, *31*, 4071–4078. (d) Owsley, D. C.; Castro, C. E. *Org. Synth. Coll.*. **1988**, *6*, 916–918.
36. Whitesides, G. M.; San Filippo, J.; Casey, C. P.; Panek, E. J. *J. Am. Chem. Soc.* **1967**, *89*, 5302–5303.
37. (a) Corey, E. J.; Posner, G. H. *J. Am. Chem. Soc.* **1967**, *89*, 3911–3912. (b) Corey, E. J.; Posner, G. H. *J. Am. Chem. Soc.* **1968**, *90*, 5615–5616. (c) Whitesides, G. M.; Fischer, W. F.; San Filippo, J.; Bashe, R. W.; House, H. O. *J. Am. Chem. Soc.* **1969**, *91*, 4871–4882.
38. Corey, E. J.; Katzenellenbogen, J. A.; Gilman, N. W.; Roman, S. A.; Erickson, B. W. *J. Am. Chem. Soc.* **1968**, *90*, 5618–5620. See also Corey, E. J.; Katzenellenbogen, J. A.; Posner, G. H. *J. Am. Chem. Soc.* **1967**, *90*, 4245–4247 for the use of similar chemistry for the synthesis of farnesol.
39. Crimmins, M. T.; Nantermet, P. G.; Trotter, B. W.; Vallin, I. M.; Watson, P. S.; McKerlie, L. A.; Reinhold, T. L.; Cheung, A. W.-H.; Stetson, K. A.; Dedopoulou, D.; Gray, J. L. *J. Org. Chem.* **1993**, *58*, 1038–1047.
40. For more recent references see Cainelli, G.; Giacomini, D.; Panunzio, M.; Zarantonello, P. *Tetrahedron Lett.* **1992**, *33*, 7783–7786. Metz, P.; Schoop, A. *Tetrahedron* **1993**, *49*, 10597–10608.
41. Furber, M.; Taylor, R. J. K.; Burford, S. C. *J. Chem. Soc., Perkin Trans. 1* **1986**, 1809–1815.
42. Chuit, C.; Foulon, J. P.; Normant, J. F. *Tetrahedron* **1980**, *36*, 2305–2310. For trimethylsilyl trifluoromethanesulfonate-mediated cuprate additions to epoxides see: Molander, G. A.; Bobbitt, K. L. *J. Org. Chem.* **1992**, *57*, 5031–5034.
43. Wender, P. A.; White, A. W. *J. Am. Chem. Soc.* **1988**, *110*, 2218–2223. Wender, P. A.; White, A. W.; McDonald, F. E. *Org. Synth.* **1991**, *70*, 204–214 (for a recent example in the cephalosporin area see Roth, G. S.; Peterson, S. A.; Kant, J. *Tetrahedron Lett.* **1993**, *34*, 7229–7230).
44. (a) Barbot, F.; Kadib-Elban, A.; Miginiac, P. *J. Organomet. Chem.* **1988**, *57*, 239–243. (b) Krause, N. *J. Org. Chem.* **1992**, *57*, 3509–3512 and Haubrich, A.; van Klaveren, M.; van Koten, G.; Handke, G.; Krause, N. *J. Org. Chem.* **1993**, *58*, 5849–5852 and references therein.

45. (a) Tamura, M.; Kochi, J. *Synthesis* **1971**, 303–305. (b) Fouquet, G.; Schlosser, M. *Angew. Chem., Int. Ed. Engl.* **1974**, *13*, 82–83. Schlosser, M. *Angew. Chem., Int. Ed. Engl.* **1974**, *13*, 701–706. (c) Tanis, S. P. *Tetrahedron Lett.* **1982**, *23*, 3115–3118.
46. Schlosser, M.; Bossert, H. *Tetrahedron* **1991**, *47*, 6287–6292.
47. Okuro, K.; Furuune, M.; Miura, M.; Nomura, M. *Tetrahedron Lett.* **1992**, *33*, 5363–5364.
48. Kang, S.-K.; Lee, D.-H.; Sim, H.-S.; Lim, J.-S. *Tetrahedron Lett.* **1993**, *34*, 91–94. See also Lautens, M. *Synlett* **1993**, 177–185.
49. Kang, S.-K.; Kim, S.-G.; Cho, D.-G. *Tetrahedron Asymm.* **1992**, *3*, 1509–1510.
50. Alexakis, A. *Pure Appl. Chem.* **1992**, *64*, 387–392.
51. Normant, J. F.; Bourgain, M. *Tetrahedron Lett.* **1971**, 2583–2586.
52. Gardette, M.; Jabri, N.; Alexakis, A.; Normant, J. F. *Tetrahedron* **1984**, *40*, 2741–2750.
53. Nakamura, E.; Kubota, K.; Isaka, M. *J. Org. Chem.* **1992**, *57*, 5809–5810.
54. Crandall, J. K.; Ayers, T. A. *Tetrahedron Lett.* **1992**, *33*, 5311–5314. Zimmer, R. *Synthesis* **1993**, 165–178 and references therein.
55. Marfat, A.; McGuirk, P. R.; Helquist, P. *J. Org. Chem.* **1979**, *44*, 1345–1347, 3888–3901.
56. Johnson, C. R.; Marren, T. J. *Tetrahedron Lett.* **1987**, *28*, 27–30.
57. Coates, G. E.; Glocking, F. In *Organometallic Chemistry*; Zeiss, H., ed.; Reinhold: New York, **1960**; p. 446.
58. Sakata, H.; Aoki, Y.; Kuwajima, I. *Tetrahedron Lett.* **1990**, *31*, 1161–1164.
59. Bertz, S. H.; Dabbagh, G. *Tetrahedron* **1989**, *45*, 425–434.
60. Lipshutz, B. H.; Ellsworth, E. L.; Dimock, S. H.; Smith, R. A. J. *J. Am. Chem. Soc.* **1990**, *112*, 4404–4410. Stack, D. E.; Dawson, B. T.; Rieke, R. D. *J. Am. Chem. Soc.* **1992**, *114*, 5110–5116.
61. Muralidharan, S.; Freiser, H. *Inorg. Chem.* **1988**, *27*, 3251–3253.
62. Corey, E. J.; Beames, D. J. *J. Am. Chem. Soc.* **1972**, *94*, 7210–7211.
63. House, H. O.; Umen, M. J. *J. Org. Chem.* **1973**, *38*, 3893–3901.
64. Corey, E. J.; Floyd, D.; Lipshutz, B. H. *J. Org. Chem.* **1978**, *43*, 3418–3420.
65. Ledlie, D. E.; Miller, G. *J. Org. Chem.* **1979**, *44*, 1006–1007.
66. Enda, J.; Kuwajima, I. *J. Am. Chem. Soc.* **1985**, *107*, 5495–5501.
67. Johnson, C. R.; Dhanoa, D. S. *J. Org. Chem.* **1987**, *52*, 1885–1888.
68. Johnson, C. R.; Dhanoa, D. S. *J. Chem. Soc., Chem. Commun.* **1982**, 358–359.
69. Lindstedt, E.-L.; Nilsson, M.; Olsson, T. *J. Organomet. Chem.* **1987**, *334*, 255–261.
70. Tsuda, T.; Yazawa, T.; Watanabe, K.; Fujii, T.; Saegusa, T. *J. Org. Chem.* **1981**, *46*, 192–194.
71. Hamon, L.; Levisalles, J. *Tetrahedron* **1989**, *45*, 489–494. See also reference 1, pp. 186–187 and references therein.
72. Gorlier, J. P.; Hamon, L.; Levisalles, J.; Wagnon, J. *J. Chem. Soc., Chem. Commun.* **1973**, 88. Acker, R.-D. *Tetrahedron Lett.* **1977**, 3407–3410. Hamon, L.; Levisalles, J. *J. Organomet. Chem.* **1983**, *251*, 133–138.
73. Posner, G. H.; Whitten, C. E.; Sterling, J. J. *J. Am. Chem. Soc.* **1973**, *95*, 7788–7800. Posner, G. H.; Whitten, C. E. *Org. Synth.* **1976**, *55*, 122–127. For recent work on the structure and reactivity of organocopper reagents derived from

substituted copper thiophenoxides see: Knotter, D. M.; Grove, D. M.; Smeets, W. J. J.; Spek, A. L.; van Koten, G. *J. Am. Chem. Soc.* **1992**, *114*, 3400–3410.

74. Bertz, S. H.; Dabbagh, G.; Villacorta, G. M. *J. Am. Chem. Soc.* **1982**, *104*, 5824–5826. Bertz, S. H.; Dabbagh, G. *J. Org. Chem.* **1984**, *49*, 1119–1122.
75. Martin, S. F.; Fishpaugh, J. R.; Power, J. M.; Giolando, G. M.; Jones, R. A.; Nunn, C. M.; Cowley, A. H. *J. Am. Chem. Soc.* **1988**, *110*, 7226–7228.
76. Yamamoto, Y.; Asao, N.; Uyehara, T. *J. Am. Chem. Soc.* **1992**, *114*, 5427–5429. Shida, N.; Uyehara, T.; Yamamoto, Y. *J. Org. Chem.* **1992**, *57*, 5049–5051.
77. Ashby, E. C.; Lin, J. J. *J. Org. Chem.* **1977**, *42*, 2805–2808. Ashby, E. C.; Lin, J. J.; Watkins, J. J. *J. Org. Chem.* **1977**, *42*, 1099–1102. Kojima, Y.; Kato, N. *Tetrahedron Lett.* **1980**, *21*, 4365–4368. See also reference 92a.
78. Westmijze, H.; Kleijn, H.; Meijer, M.; Vermeer, P. *Recl. Trav. Chim. Pays-Bas* **1981**, *100*, 98–102. Westmijze, H.; Kleijn, H.; Vermeer, P. *J. Organomet. Chem.* **1984**, *276*, 317–323.
79. Bergdahl, M.; Nilsson, M.; Olsson, T.; Stern, K. *Tetrahedron* **1991**, *47*, 9691–9702.
80. Clive, D. L. J.; Farina, V.; Beaulieu, P. L. *J. Org. Chem.* **1982**, *47*, 2572–2582.
81. Gallina, C. *Tetrahedron Lett.* **1982**, *23*, 3093–3096.
82. Lipshutz, B. H.; Wilhelm, R. S.; Floyd, D. M. *J. Am. Chem. Soc.* **1981**, *103*, 7672–7674.
83. Lipshutz, B. H.; Kozlowski, J. A.; Wilhelm, R. S. *J. Org. Chem.* **1983**, *48*, 546–550.
84. (a) Lipshutz, B. H.; Koerner, M.; Parker, D. A. *Tetrahedron Lett.* **1987**, *28*, 945–948. (b) Lipshutz, B. H.; Fatheree, P.; Hagen, W.; Stephens, K. L. *Tetrahedron Lett.* **1992**, *33*, 1041–1044.
85. Linderman, R. J.; Griedel, B. D. *J. Org. Chem.* **1991**, *56*, 5491–5493. Linderman, R. J.; Griedel, B. D. *J. Org. Chem.* **1990**, *55*, 5428–5430 (for a more recent Sn→Li→Cu sequence see Dieter, R. K.; Alexander, C. W. *Synlett* **1993**, 407–409).
86. Rieke, R. D.; Wu, T.-C.; Stinn, D. E.; Wehmeyer, R. M. *Synth. Commun.* **1989**, *19*, 1833–1840. Klein, W. R.; Rieke, R. D. *Synth. Commun.* **1992**, *18*, 2635–2644. See also Ebert, G. W.; Pfennig, D. R.; Suchan, S. D.; Donovan, T. A. *Tetrahedron Lett.* **1993**, *34*, 2279–2282.
87. Janakiram Rao, C.; Knochel, P. *J. Org Chem.* **1991**, *56*, 4593–4596. For a recent example producing amino acid-derived organocopper reagents see: Jackson, R. F.; Wishart, N.; Wythes, M. J. *Synlett* **1993**, 219–220.
88. Lipshutz, B. H.; Keil, R. *J. Am. Chem. Soc.* **1992**, *114*, 7919–7920. See also Venanzi, L. M.; Lehmann, R.; Keil, R.; Lipshutz, B. H. *Tetrahedron Lett.* **1992**, *33*, 5857–5860.
89. Aluminum: Wipf, P.; Smitrovitch, J. H.; Moon, C.-H. *J. Org. Chem.* **1992**, *57*, 3178–3186. Iodine: Rossi, R.; Carpita, A.; Cossi, P. *Tetrahedron Lett.* **1992**, *33*, 4495–4498. Tin: Behling, J. R.; Babiak, K. A.;Ng, J. S.; Campbell, A. L.; Mosetti, R.; Koerner, M.; Lipshutz, B. H. *J. Am. Chem. Soc.* **1988**, *110*, 2641–2643; Gooding, O. W.; Beard, C. C.; Cooper, G. F.; Jackson, D. Y. *J. Org. Chem.* **1993**, *58*, 3681–3686; Tellurium: Tucci, F. C.; Chieffi, A.; Comasseto, J. V. *Tetrahedron Lett.* **1992**, *33*, 5721–5724; Marino, J. P.; Tucci, F.; Comasseto, J. V. *Synlett* **1994**, 761–763; Chieffi, A.; Comasseto, J. V. *Tetrahedron Lett.* **1994**, *35*, 4063–4066; Zirconium: Lipshutz, B. H.; Wood, M. R. *J. Am. Chem. Soc.* **1993**, *115*, 12625–12626. This area has recently been reviewed: Wipf, P. *Synthesis* **1993**, 537–557.
90. Dupuy, C.; Petrier, C.; Sarandeses, L. A.; Luche, J.-L. *Synth. Commun.* **1991**,

21, 643–651. Sarandeses, L. A.; Mouriño, A.; Luche, J.-L. *J. Chem. Soc., Chem. Commun.* **1992**, 798–799 and references therein.

91. Sestelo, J. P.; Mascareñas, J. L.; Castedo, L.; Mouriño, A. *J. Org. Chem.* **1993**, *58*, 118–123. For applications in the carbohydrate field see: Blanchard, P.; Da Silva, A. D.; Fourrey, J.-L.; Machado, A. S.; Robert-Gero, M. *Tetrahedron Lett.* **1992**, *33*, 8069–8072 and references therein.
92. Lipshutz, B. H.; Kozlowski, J. A.; Breneman, C. M. *J. Am. Chem. Soc.* **1985**, *107*, 3197–3204. Bertz, S. H. *J. Am. Chem. Soc.* **1990**, *112*, 4031–4032. Lipshutz, B. H.; Sharma, S.; Ellsworth, E. L. *J. Am. Chem. Soc.* **1990**, *112*, 4032–4034. Bertz, S. H. *J. Am. Chem. Soc.* **1991**, *113*, 5470–5471. Singer, R. D.; Oehlschlager, A. C. *J. Org. Chem.* **1991**, *56*, 3510–3514 and references therein. Lipshutz, B. H.; Siegmann, K.; Garcia, E.; Kayser, F. *J. Am. Chem. Soc.* **1993**, *115*, 9276–9282 (see also Lipshutz, B. H.; Kayser, F.; Siegmann, K. *Tetrahedron Lett.* **1993**, *34* 6613–6696). Bertz, S. H.; Dabbagh, G.; He, X.; Power, P. P. *J. Am. Chem. Soc.* **1993**, *115*, 11640–11641. Krause, N.; Wagner, R.; Gerald, A. *J. Am. Chem. Soc.* **1994**, *116*, 381–382.
93. Power, P. P. *Prog. Inorg. Chem.* **1991**, *39*, 75–112. He, X.; Olmstead, M. M.; Power, P. P. *J. Am. Chem. Soc.* **1992**, *114*, 9668–9670. Lorenzen, N. P.; Weiss, E. *Angew. Chem., Int. Ed. Engl.* **1990**, *29*, 300–302. van Koten, *G. J. Organomet. Chem.* **1990**, *400*, 283–301.
94. Olmstead, M. M.; Power, P. P. *J. Am. Chem. Soc.* **1990**, *112*, 8008–8014.
95. Gambarotta, S.; Strologo, S.; Floriani, C.; Chiesi-Villa, A.; Guastini, C. *Organometallics* **1984**, *3*, 1444–1445.
96. Stemmler, T.; Penner-Hahn, J. E.; Knochel, P. *J. Am. Chem. Soc.* **1993**, *115*, 348–350.
97. Snyder, J. P.; Tipsword, G. E.; Spangler, D. E. *J. Am. Chem. Soc.* **1992**, *114*, 1507–1510.
98. House, H. O. *Acc. Chem. Res.* **1976**, *9*, 59–67. House, H. O.; Wilkins, J. M. *J. Org. Chem.* **1978**, *56*, 2443–2454.
99. Hannah, D. J.; Smith, R. A. J.; Teoh, I.; Weavers, R. T. *Aust. J. Chem.* **1981**, *34*, 181–188 and references therein.
100. Casey, C. P.; Cesa, M. C. *J. Am. Chem. Soc.* **1979**, *101*, 4236–4244. Johnson, C. R.; Dutra, G. A. *J. Am. Chem. Soc.* **1973**, *95*, 7777–7782, 7783–7788.
101. Anson, F. C.; Collins, T. J.; Richmond, T. G.; Santarsiero, B. D.; Toth, J. E.; Treco, B. G. R. T. *J. Am. Chem. Soc.* **1987**, *109*, 2974–2979.
102. Berlan, J.; Battioni, J.; Koosha, K. *Bull. Soc. Chim. Fr.* **1979**, 183–190. See also Yamamoto, Y.; Yamada, J.-I.; Uyehara, T. *J. Am. Chem. Soc.* **1987**, *109*, 5820–5822.
103. Krauss, S. R.; Smith, S. G. *J. Am. Chem. Soc.* **1981**, *103*, 141–148.
104. Hallnemo, G.; Olsson, T.; Ullenius, C. *J. Organomet. Chem.* **1984**, *265*, C22–C24. Hallnemo, G.; Olsson, T.; Ullenius, C. *J. Organomet. Chem.* **1985**, *282*, 133–144.
105. Budzelaar, P. H. M.; Timmermans, P. J. J. A.; Mackor, A.; Baerends, E. J. *J. Organomet. Chem.* **1987**, *331*, 387–407.
106. Corey, E. J.; Hannon, F. J. *Tetrahedron Lett.* **1990**, *31*, 1393–1396 and references therein. Christenson, B.; Hallnemo, G.; Ullenius, C. *Tetrahedron* **1991**, *47*, 4739–4752 and references therein.
107. Corey, E. J.; Boaz, N. W. *Tetrahedron Lett.* **1985**, *26*, 6015–6018.

108. Bertz, S. H.; Dabbagh, G.; Mujsce, A. M. *J. Am. Chem. Soc.* **1991**, *113*, 631–636 and references therein.
109. Pearson, R. G.; Gregory, C. D. *J. Am. Chem. Soc.* **1976**, *98*, 4098–4104.
110. Ashby, E. C.; Coleman, D. *J. Org. Chem.* **1987**, *52*, 4554–4565 and references therein.
111. Nakamura, E.; Miyachi, Y.; Koga, N.; Morokuma, K. *J. Am. Chem. Soc.* **1992**, *114*, 6686–6692.
112. Nakamura, E.; Nakamura, M.; Miyachi, Y.; Koga, N.; Morokuma, K. *J. Am. Chem. Soc.* **1993**, *115*, 99–106.

2

General procedures: starting materials and reaction guidelines

RICHARD J. K. TAYLOR and GUY CASY

1. Introduction

This chapter contains information concerning the starting materials required for organocopper reactions (commercial availability, preparative procedures, purification, etc.), together with a discussion of the apparatus and techniques commonly employed when carrying out such reactions.

2. Copper-containing starting materials

A number of copper salts and their complexes have been employed for the preparation of cuprate reagents, copper(I) iodide being the most popular. It has been shown in comparison experiments, however, that CuCN and $CuBr{\cdot}SMe_2$ are equally useful, if not better, organocuprate precursors.[1–3] Copper(II) salts used in copper-catalysed Grignard reactions are reduced to the active copper(I) species by excess Grignard reagent.[4] Table 2.1 lists copper-containing starting materials that have been employed in organocopper reactions and includes brief notes on suppliers, preparation/purification procedures, and synthetic utility. More detailed protocols for the preparation/purification of $CuBr{\cdot}SMe_2$ (and CuBr), CuI, $[CuI{\cdot}PBu_3]_4$ and $CuC{\equiv}CPr$ are given in protocols 1–4.

Table 2.1. Copper-containing starting materials

Name	Formula	FW	Commercial supplier	Preparation/purification procedures	Comments
Bis[copper(I) trifluoromethanesulfonate]–benzene complex	$(CuOSO_2CF_3)_2 \cdot C_6H_6$	503.33	Alfa, Fluka	Prepared from copper(I) oxide and $(CF_3SO_2)_2O$ in benzene[5]	Used to prepare lithio- and magnesiocuprates[3]
Bis[pentafluorophenyl copper(I)]–1,4-dioxane complex	$(C_6F_5Cu)_2 \cdot C_4H_8O_2$	549.30	Alfa		Appears not to have been used as an organocopper precursor
Copper(II) acetate monohydrate	$Cu(OAc)_2 \cdot H_2O$	199.64	Widely available; ultrapure form available	Usually used as received	Often employed in copper-catalysed Grignard reactions;[6] soluble in THF
Copper(II) acetylacetonate	$Cu(acac)_2$	261.76	Widely available	Usually used as received	Used in copper-catalysed Grignard reactions[7]
Copper(I) bromide	CuBr	143.45	Widely available; ultrapure form available	Commercial CuBr is often contaminated,[2,8] and the use of ultrapure, freshly prepared, or purified CuBr (or $CuBr \cdot SMe_2$) is recommended.[2,8–10] Prepared by reduction of $CuBr_2$[10,11] or $CuSO_4$–NaBr.[12,13a] Purified by precipitation from 48% HBr[14] or via sulfide complex[15]	Widely used as a catalyst for the preparation of stoichiometric organocopper reagents (see Protocol 1)
Copper(I) bromide–bis(dibutyl sulfide) complex	$CuBr \cdot (SBu_2)_2$	321.81	—	Liquid complex prepared from CuBr and Bu_2S[15]	Used to prepare CuBr free from iron salt impurities[15]
Copper(I) bromide–bis(trimethylphosphite) complex	$CuBr \cdot [P(OMe)_3]_2$	391.60	—	Prepared from CuBr and $(MeO)_3P$; m.p. 61–62.5°C[16]	Used to prepare organocopper reagents[16]

Copper(I) bromide–cycloocta-1,5-diene complex	$CuBr \cdot C_8H_{12}$	251.63	—	Prepared by treatment of commercial $CuBr_2$ with $(PhO)_3P$ and cycloocta-1,5-diene[17]	Used to prepare magnesiocuprates[18]
Copper(I) bromide–dimethyl sulfide complex	$CuBr \cdot SMe_2$	205.58	Aldrich, Fluka, Janssen	Prepared by treatment of commercial CuBr[8,19] or freshly purified CuBr[10] with Me_2S. Purified by recrystallization from Me_2S–pentane; m.p. 128–130°C (dec.).[10] Excessive drying and/or heating should be avoided[2]	Widely used as a catalyst for the preparation of stoichiometric organocopper reagents. Soluble in ethereal solvents, chloroform, and benzene (see Protocol 1)
Copper(I) bromide–trimethylphosphite complex	$CuBr \cdot P(OMe)_3$	267.53	—	Prepared from CuBr and $(MeO)_3P$; m.p. 212–218°C (dec.)[15]	Free from iron salt impurities: used to prepare lithiocuprates[15]
Copper(I) *t*-butoxide	CuOBu-*t*	136.66	—	Prepared from CuCl and *t*-BuOLi[20] or generated *in situ*[21,22]	Used to prepare copper acetylides[20] and heterocuprates[21,22]
Copper(I) *t*-butoxide–triethylphosphine complex	$Cu(OBu\text{-}t) \cdot PEt_3$	254.82	—	Prepared from CuOBu-*t* and Et_3P followed by sublimation of the product[20]	Used to convert cyclopentadiene into η^5-$C_5H_5Cu \cdot PEt_3$[20]
Copper(I) chloride	CuCl	98.99	Widely available; ultrapure form available	Commercial CuCl is often coloured and impure. Pure material prepared by reduction of $CuSO_4$–NaCl[13a] or $CuCl_2$.[11] Indefinitely stable when dry but light sensitive when moist[11]	Widely used in early studies of copper-catalysed Grignard reactions,[23] although its virtual insolubility in ether often leads to poor or unreproducible results.
Copper(II) chloride	$CuCl_2$	134.45	Widely available; ultrapure form available	Usually used as received	Used to promote dimerization processes ($R_2CuLi \rightarrow RR$)[24] and, occasionally, as a precursor to organocopper reagents[25]

Table 2.1. *(continued)*

Name	Formula	FW	Commercial supplier	Preparation/purification procedures	Comments
Copper(I) cyanide	CuCN	89.56	Widely available	Commercial material can be dried in a desiccator over phosphorus pentoxide,[26] in an Abderhalden at 56°C (refluxing acetone) over KOH,[27] or dried in the reaction flask immediately before use by azeotropic distillation with toluene (2 × 2 mL for 0.1 g CuCN) at room temperature under vacuum, with successive purging with argon.[28] A convenient alternative method is described in Chapter 5, Protocol 1	In contrast to copper(I) halides, CuCN is not hygroscopic or light sensitive and is stable in the copper(I) oxidation state.[29] Soluble in diethyl ether and thus useful in copper-catalysed Grignard reactions.[23] Widely used to prepare mixed lower order[22,26] and higher order[27–30] organocuprates and recently developed amidocuprate reagents.[31] Available in three forms (tan powder, green and red crystals), any one of which can be employed for cuprate formation[29]
Copper(I) cyanide–bis(lithium chloride) complex	$CuCN \cdot 2LiCl$	174.34	—	Prepared and used as a 1 M solution by dissolving a mixture of CuCN and LiCl (pre-dried under vacuum for 1 h at 120–150°C; two molar equivalents) in anhydrous THF.[32] For convenience, the reagent may be prepared in the reaction flask, as described in Chapter 4, Protocol 1	Convenient solubilized form of CuCN. Used routinely to prepare functionalized organocopper reagents from organozinc halides,[32] as a precursor to lower order magnesiocuprates,[33] and in copper-catalysed alkylation or organotitanium reagents[34]
Copper(I) cyanide–bis(triethylphosphite) complex	$CuCN \cdot [P(OEt)_3]_2$	421.88	—	Prepared from CuCN and $(EtO)_3P$; m.p. 120–120.9°C[35]	Has been used to prepare soluble mixed cuprates[35]

Copper(I) cyanide–lithium chloride complex	$CuCN\cdot LiCl$	131.95	—	Prepared for immediate use as a 1 M solution in THF, as described in Chapter 3, Protocol 3	Used as a precursor to highly reactive zero-valent copper[36]
Copper(I) iodide	CuI	190.44	Widely available; ultrapure form available	Ultrapure CuI can be used directly, although heating under vacuum prior to use is recommended.[37] Purification by continuous extraction[38] or precipitation[39,40] is often carried out prior to use. Purification can also be achieved via sulfide complexes[8,15]	Widely used precursor to organocopper reagents, including recently developed amidocuprate reagents[31] (see Protocol 2)
Copper(I) iodide–bis(dibutyl sulfide) complex	$CuI\cdot(SBu_2)_2$	483.04	—	Prepared from CuI and Bu_2S as a red-orange liquid which can be stored under nitrogen[15,35,41]	Ether-soluble liquid complex used to prepare CuI free from iron salt impurities,[15] halide-free methylcopper,[35] and organocuprates[41]
Copper(I) iodide–bis(trimethylphosphite) complex	$CuI\cdot[P(OMe)_3]_2$	438.60	—	Prepared from CuI and $(MeO)_3P$; m.p. 68–70.4°C[35]	Has been used to form a homogeneous solution of MeCu[35]
Copper(I) iodide–dimethyl sulfide complex	$CuI\cdot SMe_2$	252.58	—	Prepared from CuI and Me_2S, but readily loses Me_2S;[8] normally prepared *in situ*[42]	Used to form homogeneous solutions of cuprate reagents[42]
Copper(I) iodide–tetramethylethylenediamine complex	CuI·TMEDA	306.65	—	Prepared from CuI and TMEDA in THF; m.p. 214°C (dec.)[43]	Has been used in copper-catalysed Grignard reactions[43]

Table 2.1. *(continued)*

Name	Formula	FW	Commercial supplier	Preparation/purification procedures	Comments
Copper(I) iodide–tri-*n*-butylphosphine complex	$(CuI\cdot PBu_3)_4$	1571.04	—	Prepared from CuI and Bu_3P; can be recrystallized from ethanol–isopropanol[39] or acetone–methanol;[44] m.p. 75–75.5°C[44]	Free from copper(II) impurities and extremely soluble in organic solvents. Has been employed for the preparation of organocopper reagents[44] and as a precursor to highly reactive zero-valent copper.[45] Separation of products from the relatively high-boiling phosphine (b.p. 150°C/50 mm Hg) can present problems (see Protocol 3)
Copper(I) iodide–triethylphosphine complex	$(CuI\cdot PEt_3)_4$	1234.41	—	Prepared from CuI and Et_3P[39,45]	Used as a precursor to highly reactive zero-valent copper[45]
Copper(I) iodide–triethylphosphite complex	$CuI\cdot P(OEt)_3$	356.60	Aldrich	Prepared from CuI and $(EtO)_3P$;[46] recrystallized from diethyl ether; m.p. 114–115°C[15]	Has been used to prepare arylcopper reagents for Ullmann-type coupling reactions,[46] and for the preparation of vinylcopper reagents[47]
Copper(I) iodide–trimethylphosphite complex	$CuI\cdot P(OMe)_3$	314.52	Fluka	Prepared from CuI and $(MeO)_3P$; recrystallized from chloroform–diethyl ether; m.p. 192.5–194°C[15]	Free from copper(II) impurities and ether soluble. Has been used to prepare lithium divinylcuprate.[15] Removal of $(MeO)_3P$ from reaction products is straightforward owing to volatility (b.p. 111–112°C) and water solubility

Copper(I) iodide–triphenylphosphine complex	$CuI \cdot PPh_3$	452.73	—	Prepared in the reaction flask as a slurry in THF, as described in Chapter 3, Protocol 2	Used as a precursor to highly reactive zero-valent copper[45]
Copper(I) phenylacetylide	$CuC \equiv CPh$	164.67	Alfa	Prepared as a yellow solid from commercially available phenylacetylene by treatment with $CuSO_4$–NH_2OH[48]	Used in substitution reactions of aryl halides[48]
Copper(I) selenophenoxide (phenylselenocopper)	CuSePh	219.61	—	Prepared from diphenyl diselenide, $NaBH_4$ and copper(I) oxide[49]	Used to prepare heterocuprate reagents[49]
Copper(I) thiphenoxide (copper phenylmercaptide, copper thiophenylate, phenylthiocopper)	CuSPh	172.71	Alfa	Normally prepared from thiophenol and copper(I) oxide.[50] Can be prepared from CuI and PhSLi *in situ*[21,38]	Used to prepare heterocuprate reagents[21,38]
Copper(I) thiocyanate	CuSCN	121.62	Aldrich, Alfa	Purified by precipitation from saturated aqueous NaSCN[51]	Air- and light-stable source of copper(I). Used to prepare higher order cuprates[2b] and chiral-ligand modified organocuprates employed for enantioselective conjugate addition[52]
Di-*t*-butylphosphidocopper(I)	$(t\text{-}Bu_2PCu)_4$	546.71	—	Prepared from di-*t*-butyl(trimethylsilyl)phosphine and CuCl[53]	Used to prepare heterocuprate reagents with good thermal stability and useful reactivity towards common electrophiles[54]
Dicyclohexylamidocopper(I)	$CuN(C_6H_{11})_2$	243.86	—	Prepared *in situ* from CuI and $(C_6H_{11})_2NLi$ (available from Alfa)[55]	Forms synthetically useful heterocuprate reagents with good thermal stability[55]

Table 2.1. (*continued*)

Name	Formula	FW	Commercial supplier	Preparation/purification procedures	Comments
Dicyclohexylphosphidocopper(I)	$CuP(C_6H_{11})_2$	260.82	—	Prepared *in situ* from $(C_6H_{11})_2PH$ by sequential treatment with BuLi and $CuBr{\cdot}SMe_2$[56]	Used to prepare heterocuprate reagents with good thermal stability and reactivity. Claimed to be the 'ligand of choice' for mixed cuprates[56]
Dilithium(I) tetrachlorocuprate(II)	Li_2CuCl_4	219.25	Aldrich, Alfa (0.1 M solution in THF)	Prepared from anhydrous lithium chloride and anhydrous $CuCl_2$ as a 0.1 M solution in THF.[57,58] The deep-red solution can be stored indefinitely under dry nitrogen at room temperature	Used as a catalyst in Grignard reactions[57–59] and in alkylations of organomanganese reagents[60]
Dilithium(I) trichlorocuprate(I)	Li_2CuCl_3	183.78	—	Prepared as a 1 M solution in THF from lithium chloride and CuCl[61]	Used to mediate the coupling of alkyl tosylates with Grignard reagents[62]
3-(Dimethylamino)prop-1-ynylcopper(I)	$CuC{\equiv}CCH_2NMe_2$	145.66	—	Prepared *in situ* by treatment of the commercially available alkyne with MeLi followed by CuI[63]	Has been used to prepare mixed cuprate reagents which are of limited use owing to low solubility and reactivity[63]
3,3-Dimethylbut-1-ynylcopper(I)	$CuC{\equiv}CCMe_3$	143.67	—	Normally prepared *in situ* from the commercially available alkyne, BuLi, and CuI.[15] The crystalline acetylide can be isolated but yields are low and variable[15]	Used to prepare mixed cuprate reagents which are soluble in ethereal and hydrocarbon solvents.[15] The high cost of the alkyne limits the usefulness of this reagent
Diphenylphosphidocopper(I)	$CuPPh_2$	248.72	—	Normally prepared *in situ* from CuI and Ph_2PLi (available from Alfa)[55] although $CuPPh_2$ can be isolated[64]	Forms synthetically useful heterocuprate reagents with good thermal stability[55]

Hex-1-ynylcopper(I) [Copper(I) hexyne]	CuC≡CBu	144.67	—	Normally prepared from commercially available hexyne and $CuSO_4$–NH_2OH.[65] Can be prepared *in situ* from hexynyllithium and CuI, and this reagent has been reported to give more reproducible results[66]	Used to prepare mixed cuprate reagents.[66,67] Properties and applications very similar to the more commonly used pent-1-ynylcopper(I)
Lithium(I) dibromocuprate(II)	$LiCuBr_2$	230.30	—	Prepared from lithium bromide and CuBr as a solution in THF[68,69]	Used in copper-catalysed Grignard reactions[68] and to prepare organocopper reagents[69]
Lithium(I) 1-imidazoylcyanocuprate(I)	(Imid)Cu(CN)Li	163.59	—	Prepared from *N*-lithioimidazole and CuCN as a 0.2 M solution in THF[70]	Used to prepare mixed higher order cuprate reagents which display similar reactivity to reagents prepared from lithium(I) 2-thienylcyanocuprate(I), but allow simpler work-up owing to the absence of biaryl by-products[70]
Lithium(I) methyl(2-thienyl) cuprate(I)	$Me(C_4H_3S)CuLi$	168.65	—	Prepared as a 0.7 M solution in THF–cumene containing 0.1 M LiBr[71]	Used as a thermally stable methylcuprate reagent[71]
Lithium(I) 1-pyrrolylcyanocuprate(I)	$C_4H_4NCu(CN)Li$	162.58	—	Prepared from *N*-lithiopyrrole and CuCN as a 0.2 M solution in THF[70]	Used to prepare mixed higher order cuprate reagents which display similar reactivity to reagents prepared from lithium(I) 2-thienylcyanocuprate(I), but allow simpler work-up owing to the absence of biaryl by-products[70]

Table 2.1. *(continued)*

Name	Formula	FW	Commercial supplier	Preparation/purification procedures	Comments
Lithium(I) 2-thienylcyanocuprate(I)	$C_4H_3SCu(CN)Li$	179.63	Aldrich (0.25 M solution in THF)	Prepared from 2-lithiothiophene and CuCN as a solution in THF.[72] See Chapter 5, Protocol 3	Used to prepare higher order mixed cuprate reagents[72]
Mesitylcopper(I)	$Me_3C_6H_2Cu$	182.73	Alfa	Prepared from mesitylmagnesium bromide and CuCl[73]	Thermally stable up to 100°C. Used to prepare heterocuprates and for the metallation of amines, alcohols, and thiols[73]
3-Methoxy-3-methylbut-1-ynylcopper(I)	$CuC{\equiv}CC(OMe)Me_2$	160.67	—	Prepared *in situ* from the corresponding alkyne, BuLi, and CuI.[74] The alkyne precursor is prepared by methylation of commercially available 2-methylbut-3-yn-2-ol[74]	Used to prepare mixed cuprate reagents which are soluble in ether and THF[74,75]
(Methylsulfinyl) methylcopper(I)	$CuCH_2S(O)Me$	140.67	—	Prepared *in situ* from DMSO by sequential treatment with BuLi and CuI[76]	Used to prepare mixed cuprate reagents[76]
(Methylsulfonyl) methylcopper(I)	$CuCH_2S(O_2)Me$	156.66	—	Prepared *in situ* from dimethyl sulfone by sequential treatment with BuLi and CuI[77]	Used to prepare mixed cuprate reagents[77]

Pent-1-ynylcopper(I) [Copper(I) pentyne, propylethynylcopper]	$CuC\equiv CPr$	130.65	—	Prepared as a yellow solid from commercially available pentyne by treatment with either $CuSO_4$–NH_2OH[48,65] or CuI–NH_3[78]	Widely used to prepare mixed cuprate reagents,[79] including those formed using ultrasound.[80] One disadvantage is that two equivalents of a phosphine (usually hexamethylphosphorous triamide) are required to solubilize $CuC\equiv CPr$ in ethereal solvents[79] (see Protocol 4)
Tetrabutylammonium dicyanocuprate(I)	$Bu_4NCu(CN)_2$	358.05	—	Prepared as a colourless solid by treatment of CuCN with Bu_4NCN[81]	Has been used to prepare higher order cyanocuprates such as $RCu(CN)_2(NBu_4)Li$[81]
Trimethylsilylethynylcopper(I)	$CuC\equiv CSiMe_3$	160.75	—	Prepared *in situ* from the commercially available alkyne by sequential treatment with BuLi and $CuBr\cdot SMe_2$[82]	Used to prepare mixed cuprate reagents[82]
Tris {2-[(*R*)-1-(dimethylamino) ethyl]phenylthiolatocopper(I)}	$[CuSC_6H_4CH(Me)NMe_2]_3$	731.49	—	Prepared by direct reaction of the arenethiol and copper(I) oxide[83]	Used as a catalyst for the enantioselective conjugate addition of Grignard reagents to enones[84]

Protocol 1.
Preparation of copper(I) bromide–dimethyl sulfide complex[8,10,11,a]

Caution! Carry out all procedures in a well-ventilated hood, and wear disposable vinyl or latex gloves and chemical-resistant safety goggles.

$$CuBr_2 \xrightarrow{\text{aq. } Na_2SO_3} CuBr \xrightarrow{Me_2S} CuBr{\cdot}SMe_2$$

Equipment

- Conical flasks (2 × 100 mL, 1 × 2 L)
- Büchner flasks (2 × 500 mL)
- Two-necked round-bottomed flask (50 mL), equipped with a stopper and tap adaptor for inert gas/vacuum supply
- Magnetic stirring bars
- Addition funnel (100 mL)
- Septum for round-bottomed flask
- Dry, gas-tight syringe (10 mL)
- Sintered-glass filtration funnels
- Nitrogen supply and vacuum source

Materials

- Copper(II) bromide (FW 223.4) 10 g, 45 mmol — **irritant**
- Dry, distilled dimethyl sulfide (FW 62.1) 8 mL for complex formation, *c.* 50 mL for recrystallization — **flammable, stench**
- Anhydrous sodium sulfite 8.6 g, 68 mmol — **moisture sensitive**
- Distilled water 860 mL
- Concentrated hydrochloric acid 2 mL — **corrosive**
- Sulfuric acid 0.1 M, *c.* 100 mL — **corrosive**
- Glacial acetic acid *c.* 100 mL — **corrosive**
- Absolute ethanol *c.* 90 mL — **flammable**
- Anhydrous diethyl ether *c.* 90 mL — **flammable, irritant**
- Pentane *c.* 300 mL — **flammable**

1. In a 100 mL flask prepare a solution of Na_2SO_3 (7.6 g) in distilled water (50 mL) and add this slowly via the addition funnel to a stirred solution of $CuBr_2$ (10 g) in water (10 mL) in the second 100 mL flask at room temperature. The mixture, which is initially dark brown, rapidly deposits white CuBr.
2. After stirring vigorously for 15 min, pour the suspension into water (800 mL) containing sodium sulfite (1 g) and concentrated hydrochloric acid (2 mL) in the large flask. Stir the suspension vigorously for 10 min and then allow the CuBr to settle.
3. Decant the supernatant liquid and transfer the precipitate quickly onto a sintered-glass filtration funnel, using 0.1 M sulfuric acid to aid the transfer and prevent oxidation. Ensure that liquid covers the precipitate at all times in subsequent washing steps. While applying gentle suction, wash the precipitate with glacial acetic acid (4 × 25 mL), absolute ethanol (3 × 30 mL), and anhydrous diethyl ether (6 × 15 mL).
4. Without delay, transfer the white, crystalline CuBr to the round-bottomed flask, carefully evacuate to *c.* 1 mm Hg, and maintain the vacuum until all the solvent is removed.

5. Admit nitrogen to the flask, remove the stopper and replace with the septum, and then add dimethyl sulfide (8 mL, 6.7 g, 108 mmol) carefully (the reaction is exothermic) via syringe, with stirring, to give a clear solution with the slightest trace of green colour.
6. Dilute the solution with pentane (*c.* 30 mL) and collect the resulting precipitate by filtration under suction, washing five times with pentane. Dry the residual solid under a stream of nitrogen.[b]
7. Recrystallize the solid from pentane–dimethyl sulfide (4:1; 25 mL for each gram of complex) to obtain the pure complex as colourless prisms (*c.* 8 g, 85–90%), m.p. 128–130°C (dec.),[10] which is stored under argon[b,c] but can be handled routinely in air.

[a] $CuBr{\cdot}SMe_2$ can also be prepared from commercial CuBr, which is combined directly with dimethyl sulfide.[8,19] Before the complex is precipitated by the addition of hexane, an extra filtration step is normally required in order to remove insoluble impurities. Pre-washing the CuBr with methanol is reported[19] to reduce coloured impurities in the final product.
[b] It has been reported that prolonged drying of $CuBr{\cdot}SMe_2$ *in vacuo* results in dissociation of dimethyl sulfide and subsequent loss by evaporation.[2a] This practice should therefore be avoided, particularly if the complex is to be used for the preparation of stoichiometric organocopper reagents.
[c] During storage, $CuBr{\cdot}SMe_2$ can develop an orange/yellow colour as a result of partial oxidation to copper(II) species. It has, however, been reported that aged material can give comparable results to freshly prepared samples by the simple expedient of adding copper wire to solutions of the complex in THF.[85]

Protocol 2.
Purification of copper(I) iodide (CuI)[39,40,a,b]

Caution! Carry out all procedures in a well-ventilated hood, and wear disposable vinyl or latex gloves and chemical-resistant safety goggles.

Equipment

- Conical flask (1 L) with magnetic stirring bar
- Büchner flasks (2 × 1 L)
- Two sintered-glass filtration funnels
- Round-bottomed flask (50 mL) fitted with a three-way tap
- Nitrogen supply and vacuum source

Materials

- Commercial CuI (FW 190.4) 13.15 g, 0.069 mol — **irritant**
- Potassium iodide 130 g, 0.78 mol — **irritant**
- Distilled water *c.* 1 L
- Decolorizing charcoal — **irritant dust**
- Celite® — **irritant dust**
- Acetone 320 mL — **flammable**
- Anhydrous diethyl ether 320 mL — **flammable, irritant**

1. Add CuI to a solution of potassium iodide in distilled water (100 mL) in the conical flask and stir for 5 min. If the resulting solution is coloured, add

Protocol 2. *Continued*

decolorizing charcoal (1 g), stir the mixture for 10 min and then filter it through a layer of Celite® in a sintered-glass funnel into the first Büchner flask.

2. Dilute the resulting colourless solution with distilled water (400 mL) and collect the precipitate by suction filtration through the second sintered-glass funnel. Wash the solid in succession with distilled water (4 × 100 mL), reagent grade acetone (4 × 80 mL), and anhydrous diethyl ether (4 × 80 mL).
3. Transfer the pure, off-white CuI (*c.* 10 g) into a round-bottomed flask, evacuate carefully to 0.6 mm Hg, and leave overnight at room temperature. Finally, heat the flask at 90°C for 4 h under the same vacuum. After cooling the flask to room temperature, introduce nitrogen into the vessel.[c]

[a] Commercial CuI can also be purified by continuous extraction with freshly distilled, anhydrous tetrahydrofuran in a Soxhlet extractor for approximately 12 h in order to remove coloured impurities.[38] The resultant CuI is then dried under reduced pressure at 25°C and stored under nitrogen in a desiccator. Samples of CuI purified by either the precipitation or extraction method have been shown to be equally good in organocuprate reactions.[21]

[b] Purification of CuI via formation and decomposition of sulfide complexes gives CuI free of iron salts,[8,15] but this procedure is rarely used.

[c] Purified CuI can be stored under nitrogen for several months without noticeable deterioration. Each time a sample is removed from the container, any air introduced should be removed by several evacuation–nitrogen introduction cycles.

Protocol 3.
Preparation of copper(I) iodide–tri-*n*-butylphosphine complex[39,44]

Caution! Carry out all procedures in a well-ventilated hood, and wear disposable vinyl or latex gloves and chemical-resistant safety goggles.

$$\mathrm{CuI} \xrightarrow[\mathrm{Et_2O}]{\mathrm{KI,\ Bu_3P}} (\mathrm{CuI{\cdot}PBu_3})_4$$

Equipment

- Conical flask (2 L) with magnetic stirring bar
- Conical flask (500 mL)
- Separating funnel (2 L)
- Büchner flasks (2 × 500 mL) with sintered-glass filtration funnels
- Round-bottomed flask (1 L)
- Nitrogen supply and vacuum source

Materials

- Copper(I) iodide (FW 190.4) 95 g, 0.50 mol — **irritant**
- Saturated aqueous potassium iodide solution 800 mL
- Tri-*n*-butylphosphine (FW 202.3) 95 g, 0.47 mol; freshly distilled as described elsewhere[39] — **flammable, toxic**
- Diethyl ether 400 mL — **flammable, irritant**
- Distilled water 200 mL
- Magnesium sulfate for drying — **irritant dust**
- Acetone–methanol 90:10, 250 mL for recrystallization — **flammable, toxic**

1. In the 2 L conical flask combine solutions of copper(I) iodide in saturated aqueous potassium iodide solution (600 mL)[a] and tri-*n*-butylphosphine in diethyl ether (400 mL), and stir the mixture vigorously at room temperature for 1 h.
2. Transfer the mixture to the separating funnel, then release and discard the aqueous layer. Wash the ether layer with saturated potassium iodide solution (200 mL) and water (200 mL), then run it into the second flask, dry it over $MgSO_4$, filter off the drying agent, transfer the solution to the round-bottomed flask and remove the solvent *in vacuo*.
3. Dissolve the residual waxy white solid (*c.* 170 g, 92%) in acetone–methanol (90:10; 250 mL) at room temperature. Cool the solution to −10°C and collect the precipitated $(CuI{\cdot}PBu_3)_4$ (*c.* 100 g), m.p. 75–75.5°C, by suction filtration. To obtain a second crop (*c.* 40 g), m.p. 75°C, cool the filtrate to −78°C and collect the precipitate.[b]

[a] If the solution of copper(I) iodide in saturated aqueous potassium iodide solution is coloured it can be shaken for several minutes with decolorizing charcoal (1 g) and then filtered.[39]
[b] After drying *in vacuo*, the copper(I) iodide–tri-*n*-butylphosphine complex should be stored under nitrogen in a refrigerator or freezer. Without refrigeration it decomposes to a green solid or viscous syrup after a few days.

Protocol 4.
Preparation of pent-1-ynylcopper(I) (CuC≡CPr)[48,65]

Caution! Carry out all procedures in a well-ventilated hood, and wear disposable vinyl or latex gloves and chemical-resistant safety goggles.

$$CuSO_4 \xrightarrow[\text{(ii) HC}\equiv\text{CPr, EtOH}]{\text{(i) aq. NH}_3\text{, NH}_2\text{OH·HCl}} CuC\equiv CPr$$

Equipment

- Conical flask (2L) with magnetic stirring bar
- Büchner flask (3L)
- Coarse sintered-glass filtration funnel
- Nitrogen supply and vacuum source
- Pasteur pipette and filter funnel for nitrogen introduction
- Round-bottomed flask (100 mL) fitted with a septum

Materials

- $CuSO_4{\cdot}5H_2O$ (FW 250.0) 25 g, 0.1 mol — **irritant**
- Concentrated ammonia solution 28–30% NH_3, 100 mL — **corrosive**
- Distilled water 1.4 L
- Hydroxylamine hydrochloride (FW 69.5) 13.9 g, 0.2 mol — **corrosive**
- Pent-1-yne (FW 68.0) 6.8 g, 0.1 mol — **flammable, irritant**
- 95% ethanol 500 mL — **flammable, toxic**
- Absolute ethanol 500 mL — **flammable, toxic**
- Diethyl ether 500 mL — **flammable, irritant**

Protocol 4. *Continued*

1. In the 2 L conical flask, combine $CuSO_4{\cdot}5H_2O$ and concentrated ammonia solution (28–30% NH_3; 100 mL) with stirring. Immerse the flask in an ice–water bath and pass a stream of nitrogen over the solution via a Pasteur pipette clamped above the flask. The inert atmosphere is maintained throughout the reaction.
2. After 5 min add distilled water (400 mL) to the deep-blue solution, stir for a further 5 min, and then add solid hydroxylamine hydrochloride portionwise over 10 min. If this addition is carried out too rapidly a dark solid precipitates: if this occurs stirring should be continued until the solid redissolves.
3. To the resulting pale-blue solution rapidly add a solution of pent-1-yne in 95% ethanol (500 mL) and swirl the reaction flask by hand until a dense yellow precipitate separates. Add distilled water (500 mL), swirl the mixture, and then allow it to stand for 5 min, still under an N_2 atmosphere.
4. Collect the precipitate by filtration under suction through a coarse sintered-glass funnel swept by nitrogen through an inverted filter funnel. Wash the solid with water (5 × 100 mL), absolute ethanol (5 × 100 mL), and finally diethyl ether (5 × 100 mL). Transfer the pure CuC≡CPr into a round-bottomed flask and dry at 65°C on a rotary evaporator under reduced pressure. After cooling the flask to room temperature, fit the septum and flush the flask with nitrogen. Store the bright yellow CuC≡CPr (*c.* 90% yield) under nitrogen in the dark.

---

3. Reagent precursors, ligands, and reaction additives

Table 2.2 lists commonly used precursors for organocopper reagents together with ligands and reaction additives. Purification/preparation procedures are outlined; for liquids, distillation prior to use is normally recommended and boiling points are listed. Detailed purification procedures are available elsewhere.[51,86,87] Commercial suppliers are given for materials that are not widely available.

4. Solvents

The solvents normally employed in organocopper reactions are diethyl ether, tetrahydrofuran (THF), benzene, and toluene. Dimethoxyethane (DME), dimethylformamide (DMF), dimethylacetamide (DMA), and hexamethylphosphoric triamide (HMPA) are sometimes employed as co-solvents. As with all organometallic reactions, the solvents should be dry and oxygen free. For this purpose commercially available anhydrous solvents, packaged under nitrogen in Sure/Seal™ bottles, are usually suitable. A less costly alternative is to use technical-grade solvents which require treatment with efficient chemical drying agents, often in two stages, in order to remove the last traces of moisture. The solvents are normally left to stand over the drying agent and then heated under reflux in an inert atmosphere before being distilled immediately prior to use. Technical-grade solvents usually require preliminary treatment with a high capacity drying agent (CaH_2 or molecular sieves) for one or two days. For subsequent distillation the drying agents normally used are Na, $LiAlH_4$, and CaH_2. Since CaH_2 is rather slow-acting, $LiAlH_4$ is usually the preferred hydride reagent. Care should be taken when using $LiAlH_4$ in view of its potential hazards.[90] For this reason metallic sodium is often used in preference to the hydride reagents. When a solvent has been heated under reflux over sodium under an inert atmosphere for an hour or so the addition of a small amount of benzophenone will produce a deep-blue/purple solution of sodium benzophenone ketyl if the solvent is dry. The ketyl colour acts as an indicator; when it fades additional sodium is needed. Another advantage of this procedure is that the ketyl is an extremely efficient oxygen scavenger,[51] and so the Na–Ph_2CO method is preferred when completely oxygen-free solvents are needed. Table 2.3 outlines the procedures commonly employed to obtain dry, oxygen-free solvents; more detailed descriptions of purification methods are available elsewhere.[51,91–93]

If a number of organometallic reactions are planned a solvent still, which enables the solvent to be dried and then distilled and collected under an inert atmosphere, is extremely useful. Detailed descriptions of the use of solvent stills for drying diethyl ether, THF, and DME are available.[91,92] Still-heads are marketed by Aldrich. The still illustrated in Fig. 2.1 allows solvents to be withdrawn from the still-head collection reservoir through a septum cap by syringe, thereby minimizing its exposure to the atmosphere. If larger volumes are required, a pre-dried receiving flask, preferably the reaction vessel itself, can be connected to the Quickfit adaptor on the still-head. When using a solvent still of this or of similar design, the following precautions should be observed:

(1) The still should be situated in an efficient fume hood, and all tubing for inert gas and water supplies should be securely attached using copper wire or plastic cable ties.
(2) The heating mantle should be of such a design that there is no risk of sparks igniting the solvent. This also applies to all electric cables and plugs. The mantle should also incorporate an electricity cut-out device to operate if the water supply to the condenser fails.
(3) It is imperative that the bottom flask containing the drying agent should not be allowed to boil dry. This risk can be minimized if the flask is of greater capacity than the collection reservoir and is regularly topped up with solvent.
(4) During cooling, an adequate flow of inert gas should be maintained.

Table 2.2. Reagent precursors, ligands, and additives for organocopper reactions

Name	Formula	b.p. (°C)	Comments
Boron trifluoride etherate	$BF_3{\cdot}OEt_2$	126 (46/10 mm Hg)	To purify 500 mL: add ether (10 mL) and distil from CaH_2 (2 g) in an all-glass apparatus at 10 mm Hg
Dicyclohexylphosphine	$(C_6H_{11})_2PH$	129/8 mm Hg	Pyrophoric, stench [$(C_6H_{11})_2PLi$ available from Alfa]
Dibutyl sulfide (butyl sulfide)	Bu_2S	188–189 (74–75/14 mm Hg)	Stench
1-Dimethylaminoprop-2-yne	$HC{\equiv}CCH_2NMe_2$	79–83	From Aldrich, Fluka
3,3-Dimethylbut-1-yne (*t*-butylacetylene)	$HC{\equiv}CCMe_3$	33–35 (Fluka) 36–38[15]	Commercially available or prepared from pinacolone in 58% yield[15]
Dimethyl sulfide (methyl sulfide)	Me_2S	38	Stench
Diphenylphosphine	Ph_2PH	280	Pyrophoric (Ph_2PLi available from Alfa)
Hexamethylphosphorous triamide [tris(*N,N*-dimethylamino)phosphine]	$(Me_2N)_3P$	55–58/15 mm Hg	Distil under N_2. **Caution!** Reacts violently with water
Hex-1-yne	$HC{\equiv}CBu$	69–71	Widely available
Imidazole	$C_3H_4N_2$	m.p. 89–91	Recrystallize from benzene, dichloromethane, or acetone–petroleum ether
Lithium bromide, anhydrous	LiBr	—	Available in ultrapure form from Aldrich. Hygroscopic: store in desiccator. Solutions in ether can be prepared from LiH + Br_2[35]

Lithium chloride, anhydrous	LiCl	—	Available in ultrapure form from Aldrich. Hygroscopic: store in desiccator
Lithium iodide, anhydrous	LiI	—	Hygroscopic: store in desiccator. Solutions in ether can be prepared from LiH + I_2[35] or BuLi + MeI[88]
3-Methoxy-3-methylbut-1-yne	$HC{\equiv}CC(OMe)Me_2$	77–80	Prepared by methylation of commercially available alcohol[74]
2-Methylbut-3-yn-2-ol	$HC{\equiv}CC(OH)Me_2$	104	From Aldrich
Pent-1-yne	$PrC{\equiv}CH$	39–41	Widely available
Pyrrole	C_4H_5N	131	Fractionally distil under reduced pressure from Na or CaH_2, store under N_2. Redistil immediately before use
Thiophene	C_4H_4S	84	Fractionally distil under reduced pressure from Na
Thiophenol	PhSH	169	Severe poison, stench
Tributylphosphine	Bu_3P	156/50 mm Hg	
Triethylphosphite	$(EtO)_3P$	156	
Trimethylphosphite	$(MeO)_3P$	111–112	Stench
Trimethylsilyl chloride (chlorotrimethylsilane)	Me_3SiCl	57	Flammable liquid, corrosive. Usually used as received owing to difficulties in removing HCl. Distillation from CaH_2 or poly(4-vinylpyridine) has been recommended[89]
Trimethylsilylacetylene	$HC{\equiv}CSiMe_3$	53	

Table 2.3. Purification of reaction solvents[a,b]

Solvent	**b.p. (°C)**	**Procedure**
Benzene Toluene	80 110.8	Thiophene-free[c] solvent is dried over, and then distilled from, Na or Na–Ph_2CO under N_2/Ar. **Caution!** Benzene is a cancer suspect agent
Diethyl ether Tetrahydrofuran (THF)	34.6 65	Peroxide-free[d] solvent is dried over Na, and then distilled from Na–Ph_2CO under N_2/Ar. **Caution!** Ensure that peroxides do not accumulate[d]
Dimethoxyethane (DME)	84	Commercial DME is often rather wet and should be stirred over CaH_2 for 24 h and checked for peroxides[d] before further treatment. The decanted solvent is then dried by the procedures described for diethyl ether/THF
Dimethylacetamide (DMA)	165–167 (58/11 mm Hg)	Solvent dried over and distilled from BaO under reduced pressure
Dimethylformamide (DMF)	153 (76/39 mm Hg)	Solvent dried over 4A molecular sieves or by brief treatment with CaH_2 (prolonged exposure leads to decomposition). The decanted solvent is then distilled under reduced pressure. Note that DMF decomposes if distilled at atmospheric pressure or from CaH_2
Hexamethylphosphoric triamide (Hexamethylphosphoramide, HMPA)	235 (*c.* 90/4 mm Hg, 68–70/1 mm Hg)	Solvent dried over Na or CaH_2 under N_2/Ar, and then distilled from the drying agent under reduced pressure. **Caution!** HMPA is a severe poison and a cancer suspect agent

[a] Anhydrous, reagent-grade solvents should be employed where possible. Technical-grade solvents should be pre-dried using molecular sieves or CaH_2.

[b] Detailed purification procedures are readily available.[51,86,91,92]

[c] To remove thiophene shake with cold, concentrated H_2SO_4 (*c.* 100 mL per litre), removing the dark acid layer. This should be repeated until the acid layer shows only slight darkening (twice is usually sufficient). For toluene the acid washing must be carried out at a temperature below 30°C to avoid sulfonation. The solvent is then washed with H_2O, dilute NaOH, and H_2O again, and then dried with $CaSO_4$ or $CaCl_2$ before Na is employed.

[d] Peroxides in ethereal solvents can be detected by a number of methods.[51,86,94,95] The simplest is to mix the solvent (*c.* 1 mL) with an equal volume of glacial acetic acid containing NaI or KI crystals (*c.* 100 mg). A yellow colour indicates the presence of a small quantity of peroxides in the sample and a brown colour indicates a higher concentration. A blank determination should also be made for comparison. Peroxides can be removed in a number of ways.[51,86,94,95] The simplest is to use self-indicating impregnated molecular sieves.[94] Alternatively, shake a litre of the solvent repeatedly with acidified iron(II) sulfate solution [made from $FeSO_4$ (60 g), concentrated H_2SO_4 (6 mL), and water (110 mL)] until a negative peroxide test is obtained. The solvent should then be shaken with 0.5% $KMnO_4$ solution, then 5% NaOH solution, and finally with water. The solvent should then be dried over $CaCl_2$ for 24 h before treatment with sodium.

(5) **Caution!** *The use of a semi-permanent still for ethereal solvents can lead to a build-up of peroxides. The solvent should be checked for peroxides (see Table 2.3, footnote d) at frequent intervals and if these are detected the still should be dismantled and the drying agent and peroxides carefully destroyed.*[90a] *When renewing the still, fresh batches of the solvent and drying agent should be used.*

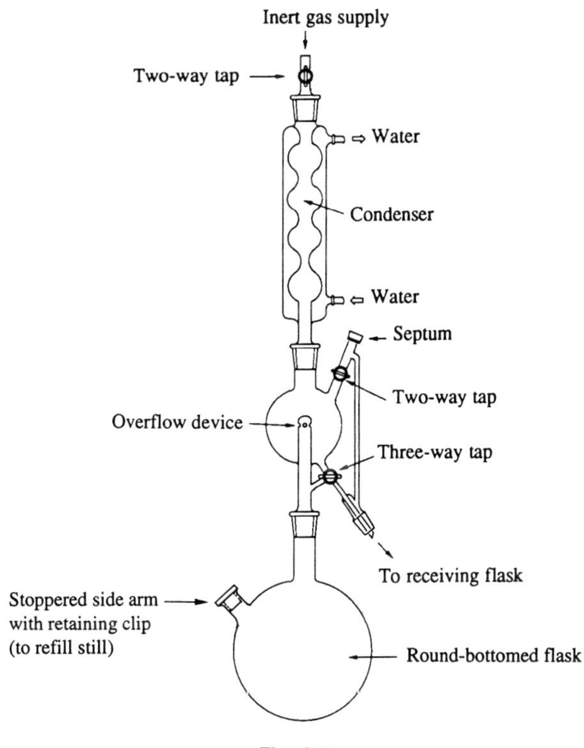

Fig. 2.1

The immediate use of dried, deoxygenated solvents is recommended, although non-ethereal solvents such as benzene, toluene, DMF, and HMPA can be stored over activated molecular sieves in thoroughly dried containers under N_2/Ar for future use.[91] This practice is less advisable in the case of ethereal solvents owing to the potential hazard of peroxide formation, since any peroxide inhibitors originally present will have been removed during distillation.

New molecular sieves can be dried by the following method:[13b,92] heat at 320 °C for 3 h, cool in an evacuated desiccator and fill the desiccator with dry N_2/Ar on releasing the vacuum. Previously used sieves should first be freed of residual solvents using a water pump and then heated in a well-ventilated oven at 320 °C.

5. Organolithium and Grignard reagents

Caution! *Organolithium reagents are pyrophoric. Solutions of organometallic reagents are flammable and corrosive. Strict safety precautions should be observed.*

5.1 Methods of preparation and commercial availability

Organocopper reagents are traditionally prepared by the reaction of copper salts with organolithium or Grignard reagents. These precursor organometallic reagents can be prepared using standard procedures,[93,96] some of which are exemplified in later sections, for example:

- organic halide + metal (see Section 6, Protocol 8)
- vinyl halide + alkyllithium (see Section 6, Protocol 9)
- vinyltin + butyllithium (see Chapter 5, Protocol 3)

Many organometallic reagents are also commercially available (Table 2.4). These are normally packed under an inert atmosphere in an airtight container fitted with a septum or a similar cap which enables withdrawal by syringe (e.g. Aldrich Sure/Seal™ system). Detailed descriptions of syringe–septa techniques

Table 2.4. Commercially available Grignard and organolithium reagents[a–c]

Carbon nucleophile	Grignard	Organolithium
C_0		Bu_3SnLi in THF[d]
C_1	MeMgBr in ether, THF, *n*-butyl ether, or THF–toluene MeMgCl in THF MeMgI in ether CD_3MgI in ether	MeLi in ether[c] or THF–cumene[e] CD_3Li in ether[c] Me_3SiCH_2Li in pentane
C_2	EtMgBr in ether, THF, or *t*-butyl methyl ether EtMgCl in ether or THF $CH_2{=}CHMgBr$ in THF $CH_2{=}CHMgCl$ in THF $HC{\equiv}CMgBr$ in THF $HC{\equiv}CMgCl$ in ether	$CH_2{=}CHLi$ in ether[f] $LiC{\equiv}CLi$ (solid)[d] $Me_3SiC{\equiv}CLi$
C_3	PrMgCl in ether *i*-PrMgCl in ether or THF $CH_2{=}CHCH_2MgBr$ in ether $CH_2{=}CHCH_2MgCl$ in THF	
C_4	BuMgCl in ether or THF *i*-BuMgBr in ether or THF *i*-BuMgCl in ether *s*-BuMgCl in ether *t*-BuMgCl in ether Me_3SiCH_2MgCl in ether	BuLi in hexanes,[g] pentane, cyclohexane, or toluene[e] *s*-BuLi in cyclohexane or heptane[e] *t*-BuLi in pentane or heptane[e]

Table 2.4. *(continued)*

Carbon nucleophile	Grignard	Organolithium
C_5	n-$C_5H_{11}MgBr$ in ether $MeCH_2CMe_2MgCl$ in ether CyclopentylMgCl in ether $BrMg(CH_2)_5MgBr$ in THF	Lithium cyclopentadienylide
C_6	n-$C_6H_{13}MgBr$ in ether CyclohexylMgCl in ether PhMgBr in ether or THF PhMgCl in THF C_6D_5MgBr in ether or THF 4-ClC_6H_4MgBr in ether 4-FC_6H_4MgBr in ether or THF 4-$MeOC_6H_4MgBr$ in THF [d]	n-$C_6H_{13}Li$ in hexane [e] PhLi in ether–cyclohexane or ether–benzene [d]
C_7	n-$C_7H_{15}MgBr$ in ether [d] $PhCH_2MgCl$ in ether or THF $C_6D_5CD_2MgCl$ in ether 2-MeC_6H_4MgBr in ether 4-MeC_6H_4MgBr in ether 4-F-3-MeC_6H_3MgBr in THF	**Organolithium**
C_8	n-$C_8H_{17}MgCl$ in THF	n-$C_8H_{17}Li$ in heptane [e] n-$BuCH(Et)CH_2Li$ in heptane [e] PhC≡CLi (solid) [d]
C_9	2,4,6-$Me_3C_6H_2MgBr$ in ether or THF	
C_{10}	n-$C_{10}H_{21}MgBr$ in ether 4 t BuC_6H_4MgBr in ether $Me_2C(Ph)CH_2MgCl$ in ether	
C_{12}	n-$C_{12}H_{25}MgBr$ in ether	
C_{14}	n-$C_{14}H_{29}MgCl$ in THF	
C_{18}	n-$C_{18}H_{37}MgCl$ in THF	

[a] Most suppliers of research chemicals stock organometallic reagents. Aldrich probably keep the widest range.
[b] Concentration is normally in the range 1–3 M.
[c] Many commercial organolithium reagents also contain dissolved lithium halide. Low halide reagents are available, however (e.g. Aldrich, 1.4 M MeLi, *c.* 0.05 M halide). Protocol 3 in Chapter 13 describes the preparation of low halide methyllithium suitable for spectroscopic studies.
[d] Available from Alfa.
[e] Available from FMC Lithium Division.
[f] Vinyllithium is often prepared from tetravinyltin and butyllithium[15,97] immediately prior to use.
[g] Also available in concentrated solution (e.g. Aldrich, 10 M in hexanes).

have been published,[90b,92,98,99] as have procedures for fitting septa to reagents that are supplied in screw-top bottles.[91] Cylinders containing several litres of organometallic reagents are also available. Procedures for transferring the reagent into smaller vessels have been described.[90b,91,92,98–100]

An important recent trend is the development of methods for preparing organocopper reagents which circumvent the usual requirement for organolithium or Grignard reagents. These are discussed in later chapters and include the reaction of activated copper with alkyl halides (Chapter 3),[101] and the use of organozinc halide (Chapter 4)[102] or organozirconium (Chapter 5)[103] precursors. These new synthetic methods are extremely valuable since they allow the preparation of organocopper reagents containing electrophilic functional groups.

5.2 Storage and disposal

Grignard reagents are normally stored at room temperature as are *n*-alkyllithium reagents in hydrocarbon solvents. *n*-Alkyllithium reagents in ether or THF and all other organolithium reagents are best stored at 0 °C (Table 2.5). Storage at −20 °C is not recommended owing to the possibility of bottle fracture.[91] For concentrated solutions (>1 M) the organometallic reagent may be partially insoluble at these lower temperatures, and in these cases warming to room temperature will be required before dispensation. If Grignard reagents crystallize out on storage the solution can be warmed to 30–40 °C in order to achieve homogeneity. To extend the shelf-life of partially used bottles of organometallic reagents, a dab of silicon grease smeared directly onto the septum or Sure/Seal™ cap before replacement of the screw top is reported to provide additional protection against the ingress of oxygen and moisture.[104] When the reagent is next required, the grease is removed with a tissue before piercing the septum with a needle.

It is advisable to determine the activity of the organometallic reagents by

Table 2.5. Loss of activity with time of some organolithium reagents[91]

Reagent	**Storage at 25 °C**	**Storage at 0 °C**
1.6 M *n*-BuLi in hexanes	5%/1 year	0%/1 year
2.5 M *n*-BuLi in hexanes	1–2%/3 months	0%/1 year
s-BuLi in cyclohexane	1%/1 week	0%/6 months
t-BuLi in *n*-pentane	2%/1 month	0%/6 months
MeLi in ether	1–2%/3 months	0%/1 year
PhLi in ether or cyclohexane	1%/1 month	0%/1 year

titration at regular intervals (see Section 5.5). If the activity of the organometallic reagent is so low that it is no longer useful, the residues should be disposed of with **extreme caution** by qualified personnel following recommended guidelines.[91] Small quantities of residues which may be pyrophoric can be deactivated by gradual exposure to the atmosphere. To do this, a small leak is introduced by piercing the septum cap of the container, which is then left undisturbed in a fume hood for several days. The septum cap is then removed and after a further one or two days it is usually safe to add carefully a large excess of isopropanol to the residue (**caution!**).

5.3 Manipulation

Transfer of reactive, moisture-sensitive solutions of organometallic reagents is usually carried out using syringes or double-tipped needles in conjunction with reagent bottle and apparatus capped with septa or their equivalent (for illustrations, see Chapter 7, Figs 7.2 and 7.3). Excellent reviews of these techniques are available.[90b,91,92,98–100] Syringes are normally employed for the transfer of up to 25 mL of reactive solutions, although it is possible to transfer up to 100 mL using this technique. A variety of syringes are available, but with pyrophoric organolithium reagents 'gas-tight' syringes are preferred. Syringes of this type should be stored in a desiccator. Prior to dispensing the solution of organometallic reagent, the syringe should be purged with inert gas (Chapter 7, Fig. 7.2) and rinsed two or three times with the anhydrous reaction solvent in order to eliminate any residual moisture. The syringe should be fitted with a needle-locking device (e.g. Luer lock) and a hypodermic needle of sufficient length (0.5–2 ft) to avoid the need to tip reagent bottles (Chapter 7, Fig. 7.3). For very sensitive compounds such as *t*-BuLi it is also advisable to use a syringe which is designed to prevent inadvertent removal of the plunger (*e.g.* a Cornwall syringe) and a syringe stopcock attachment to eliminate unwanted loss of material during transfer. Larger volumes of liquid are usually transferred under N_2/Ar pressure using double-tipped needles, although the CHEM-*FLEX*™ system recently introduced by Aldrich provides a convenient alternative.[99] This method uses flexible PTFE-lined tubing which can be cut to the desired length and fitted with transfer needles.

5.4 Qualitative tests for organolithium and Grignard reagents

Both organolithium and Grignard reagents give a positive Gilman colour test with Michler's ketone [4,4′-bis(dimethylamino)benzophenone].[105a] Since negative tests are usually* obtained with organocopper and organocuprate

* Although it has been reported by Gilman[105b] that Me_2CuLi in ether gives a positive Gilman test, in our hands this was not the case. Lipshutz *et al.*[105c] have conducted a systematic investigation of the composition of this reagent in various media, which corroborates the latter observation. This study has also shown that Me_2CuLi prepared in THF–ether, *in the absence of lithium iodide*, is in equilibrium with MeLi and Me_3Cu_2Li, and therefore gives a positive Gilman test.

reagents, this procedure can often be used to judge when the formation of the stoichiometric organocopper reagent is complete.[105b–107] Since the organocopper reagent may itself be coloured, it is advisable to carry out a 'comparison' test on the precursor organolithium or Grignard reagent in order to give a clear indication of a positive test.

5.4.1 Gilman test: procedure[105a]

The organometallic solution (0.5–1.0 mL) is withdrawn by syringe and treated with an equal volume of a 1% solution of Michler's ketone in dry benzene (**caution!**). The mixture is then hydrolysed (**caution!**) by the slow addition of water (1 mL). Gentle agitation at this stage moderates the reaction. Several drops of a 0.2% (0.01 M) solution of iodine in glacial acetic acid are then added and the mixture is shaken. The development of a characteristic greenish-blue colour confirms the presence of RMgX or RLi in the original solution.

5.5 Standardization of organolithium and Grignard reagents

A newly prepared or purchased solution of organolithium/Grignard reagent should be standardized before use: there is often a considerable difference between the expected concentration and the actual one. Organometallic reagents should also be standardized at regular intervals since they deteriorate with time (Table 2.5). A number of standardization procedures are available,[108] but the Gilman double-titration method[44,109,110] and the Watson and Eastham charge-transfer method[111] are the most popular.

The double-titration procedure using 1,2-dibromoethane can be employed to standardize alkyllithiums and phenyllithium[44,109] as well as Grignard reagents.[110] In this procedure the concentration of total base (organometallic + hydroxides + alkoxides + oxides) is determined by quenching with water and performing an acid–base titration. A second aliquot of the organometallic reagent is destroyed by reaction with an alkyl halide (usually $BrCH_2CH_2Br$) and quenching with water. An acid–base titration then gives the concentration of non-organometallic bases. Subtraction of the latter value from the former gives the concentration of organometallic reagent. There have been reports that the original procedure gives a 5–10% underestimate of the organolithium concentration. The experimental procedure described below (Protocol 5) is a modification developed by Whitesides *et al.*[44] The main difference from the original method[109] is that acidic impurities in the 1,2-dibromoethane are removed by filtration through alumina, and this leads to greater accuracy and reproducibility.

Watson and Eastham[111] devised a direct standardization procedure (Protocol 6) based on the observations that organolithium/Grignard reagents form coloured charge-transfer complexes with aromatic heterocycles such as 2,2′-biquinolyl, 1,10-phenanthroline, and 2,2′-bipyridyl.[112] These complexes can be used as indicators when the organometallic reagent is titrated with butan-2-ol (or menthol [111b]). The addition of a molar equivalent of the alcohol destroys the organometallic reagent and the colour disappears. This titration has to be carried out under an inert atmosphere.

In recent years a number of direct procedures for the standardization of organolithium reagents have been developed based on the formation of coloured dianions.[113–115] The procedure using 1-pyreneacetic acid (available from Aldrich) is one of the most convenient in view of the distinctive red end-point and the fact that the reagent can be recovered easily from titration mixtures and then reused.[113] The experimental procedure described below (Protocol 7) can be used for the standardization of organolithium and Grignard reagents. Although the end-point is less precise in the latter case, the reagent concentration can be estimated to within *c.* 0.1 M.

To ensure that accurate data are obtained during the standardization of organometallic reagents, each of the following protocols should be performed in duplicate.

Protocol 5.
Double-titration procedure[44]

Caution! Carry out all procedures in a well-ventilated hood, and wear disposable vinyl or latex gloves and chemical-resistant safety goggles.

Equipment (see Fig. 2.2a)

- Conical flask (50 mL)
- Dry, gas-tight syringe (1 mL, with 0.01 mL graduations)
- Burette (50 mL, with 0.1 mL graduations)
- Short glass column, packed with dry alumina
- Quickfit® conical flask (50 mL), pre-dried and fitted with a septum
- Inert gas supply and inlet/outlet

Materials

- Organometallic solution to be standardized 2 × 1.0 mL aliquots — **pyrophoric**
- 1,2-Dibromoethane 5 mL — **toxic, cancer suspect agent**
- Standardized dilute hydrochloric acid 0.1 M — **corrosive**
- Phenolphthalein indicator solution

1. Withdraw a 1.0 mL aliquot of the organometallic solution from its storage vessel by syringe, under positive N_2/Ar pressure, and dispense it cautiously into a conical flask containing *c.* 10 mL of water (**caution!** see Section 5.3). Add a few drops of indicator solution to give a purple solution.
2. Titrate the aqueous solution with 0.1 M hydrochloric acid until a permanent

Protocol 5. *Continued*

colourless end-point is reached in order to determine the total base concentration.

3. Pass 1,2-dibromoethane through the alumina column into the pre-dried Quickfit® conical flask. Seal the flask with the septum and flush with dry N_2. For this purpose the apparatus shown in Fig. 2.2a is suitable. Withdraw another 1.0 mL aliquot of the organometallic solution from its storage vessel and inject it through the septum. Swirl the solution for 2–10 min (the longer time is often needed for Grignard reagents) and then add water (*c.* 10 mL) and a few drops of indicator solution.
4. Titrate the mixture with 0.1 M hydrochloric acid until a permanent colourless end-point is reached in order to determine the concentration of non-organometallic bases. It is important to agitate the mixture during titration to keep the layers well mixed. Subtraction of the latter value from the total base concentration gives the true concentration of the organometallic reagent.

Protocol 6.
Charge-transfer titration[111,112,a]

Caution! Carry out all procedures in a well-ventilated hood, and wear disposable vinyl or latex gloves and chemical-resistant safety goggles.

Equipment (see Fig. 2.2b)

- A pre-dried, two-necked, round-bottomed flask (50 mL) incorporating a side-arm tap adaptor (for inert gas inlet), with magnetic stirring bar, septum, and gas outlet
- Pre-dried syringe (10 mL, with 0.1 mL graduations)
- Dry, graduated, gas-tight syringes (1/5/20 mL)
- Inert gas supply

Materials

• Organometallic solution to be standardized *c.* 2 M solution, 2.0–5.0 mL	**pyrophoric**
• 0.5–1.0 M solution of butan-2-ol (freshly distilled from CaH_2) in *p*-xylene (freshly distilled from Na), prepared in a volumetric flask and stored under N_2/Ar	**flammable**
• Indicator solution[b]	**flammable, cancer suspect agent**
• Anhydrous benzene for RLi 20 mL or anhydrous ether for RMgX 20 mL	**flammable, cancer suspect agent** **flammable, irritant**

1. Flush the round-bottomed flask and accessories with dry, oxygen-free N_2/Ar (Fig. 2.2b). Flame dry the flask and allow it to cool in a stream of inert gas. Maintain an inert atmosphere during the remaining steps.
2. Using the gas-tight syringes, inject successively into the flask the anhydrous benzene/ether, 0.2–0.5 mL of indicator solution, and an aliquot of the organometallic reagent to form a coloured solution. (The actual colour

observed is dependent on the particular combination of organometallic reagent and aromatic base. For example, MeLi reacts with 2,2′-bipyridyl to produce a purple solution.[112])

3. Charge the 10 mL syringe with the standard butan-2-ol solution and then insert the tip/needle through the septum. Stir the solution rapidly and titrate until the solution remains colourless (or when no further colour change occurs). *X* mol of butan-2-ol $\equiv$ *X* mol of organometallic reagent.

[a] See reference 111b for recently introduced improvements to this procedure.
[b] Prepared by dissolving 100 mg of the aromatic heterocycle (2,2′-bipyridyl, 2,2′-biquinolyl, or 1,10-phenanthroline) in 50 mL of anhydrous benzene (**caution!** flammable, cancer suspect agent) and stored under N_2/Ar.

Protocol 7.
1-Pyreneacetic acid procedure[113]

Caution! Carry out all procedures in a well-ventilated hood, and wear disposable vinyl or latex gloves and chemical-resistant safety goggles.

Equipment (see Fig. 2.2b)

- A pre-dried, two-necked, round-bottomed flask (50 mL) incorporating side-arm tap adaptor (for inert gas inlet), with magnetic stirring bar, septum, and gas outlet
- Dry, gas-tight syringes [1 mL, with 0.01 mL graduations (preferably fitted with a screw-driver plunger), and 10 mL]
- Inert gas supply

Materials

- Organometallic solution to be standardized **pyrophoric**
- 1-Pyreneacetic acid (FW 260.3), pre-dried to constant weight *in vacuo*, 100–200 mg)
- Anhydrous THF 10 mL **flammable, irritant**

1. Flush the round-bottomed flask and accessories with dry, oxygen-free N_2/Ar (Fig. 2.2b). Flame dry the flask and allow it to cool in a stream of inert gas. Maintain an inert atmosphere during the remaining steps.
2. Weigh out the 1-pyreneacetic acid accurately, using an analytical balance, and transfer to the round-bottomed flask. Add the THF and stir the mixture until a homogeneous solution is obtained.
3. Charge the 1 mL syringe with the organolithium reagent and then insert the needle through the septum. Add the organometallic reagent slowly and dropwise (over 3–4 min; slow addition is essential to minimize reaction between THF and RLi). The end-point is when the red colour of the dianion just persists. *X* mol of 1-pyreneacetic acid $\equiv$ *X* mol of RLi.

(a)
Septum
Nitrogen/argon inlet needle
(to bubbler or balloon reservoir)
Burette
Vent needle
to bubbler
Septum
Nitrogen or argon entry
via hypodermic needle
Stirring bar
(b)
Graduated syringe
Vent needle to bubbler
Septum
To vacuum
inert atmosphere
Two-way stopcock
Stirring bar

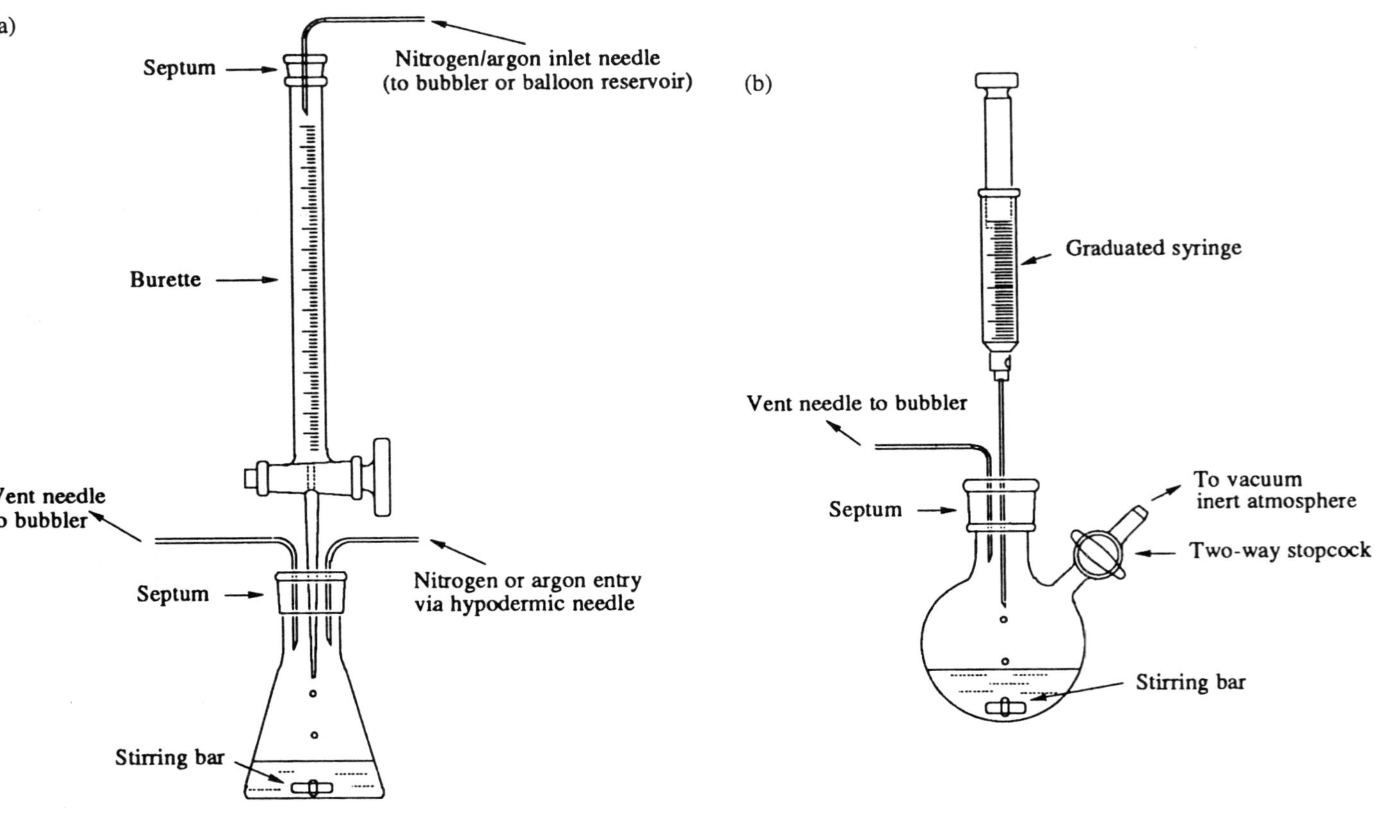

Fig. 2.2

6. Carrying out the reaction

6.1 Introduction

The majority of catalytic and stoichiometric organocopper reagents are characterized by high reactivity and low thermal stability. For these reasons they are normally prepared *in situ* at low temperatures under an inert atmosphere and reacted immediately with the chosen electrophile. In spite of this constraint, most of these reactions can be carried out using conventional Quickfit laboratory glassware and standard techniques for the manipulation of air- and moisture-sensitive reagents. Comprehensive accounts of these techniques are widely available,[91,92,98–100,116a] and guidelines for the transfer of liquids have been covered previously in this chapter (Section 5.3). The manipulation of most copper(I) salts can usually be carried out without the use of a glove box; weighing in air prior to rapid transfer to the reaction vessel is normally sufficient.[107]

The minimum requirements for equipment are outlined in Sections 6.2 and 6.3, and work-up procedures are discussed in Section 6.4. To illustrate the use of these techniques, a short selection of protocols based on research carried out in our laboratory appears at the end of the chapter in Section 6.5. Where appropriate, references to these protocols will appear in the preceding discussion. Subsequent chapters of this book provide a more comprehensive treatment of the experimental techniques associated with particular reactions and types of reagent.

6.2 Providing an inert atmosphere

Organocopper reactions are carried out under an atmosphere of anhydrous nitrogen or argon. Argon has the advantage of being heavier than air, and therefore provides a more effective barrier against the outside atmosphere, but nitrogen is more commonly used owing to its lower cost. Nitrogen sold as 'oxygen-free' grade is usually suitable for organocopper reactions, although removal of water using solid desiccants is recommended. For this purpose the gas can be passed through a glass column packed with several alternating layers of calcium chloride and self-indicating silica gel pellets. The use of an elaborate multi-stage 'drying train' is less advisable, since this can actually increase contamination levels of the gas owing to the extra glassware, joints, and tubing involved.[116b]

To enable several inert atmosphere experiments to be run simultaneously, it is convenient to provide the inert gas and a vacuum source via a purpose-built double manifold of the type shown in Fig. 2.3. The manifold should be constructed of thick-walled glass and fitted with precision-ground two-way stopcocks and a pressure-release bubbler. One port of the manifold should be provided with a Quickfit® adaptor fitted with a septum which can be used for purging syringes with inert gas. The supply of inert gas to the manifold should

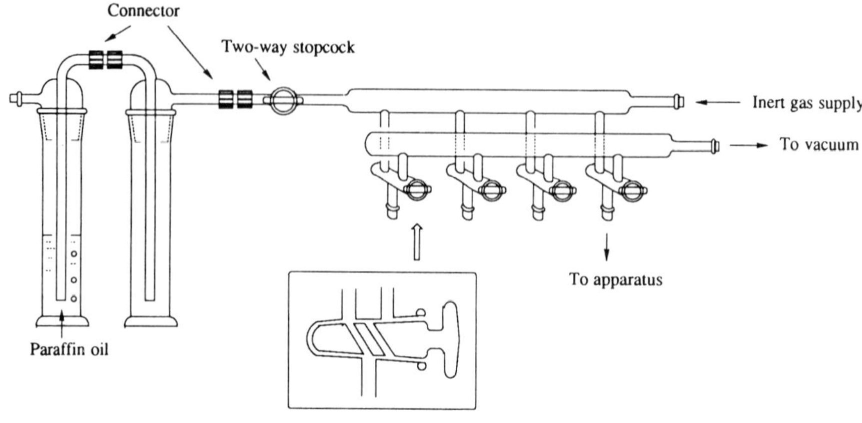

Fig. 2.3

ideally be controlled in two stages using the cylinder regulator, set to a pressure of 2–5 p.s.i., and then a needle valve.[90c] The latter can be closed in order to isolate the manifold when the cylinder is being changed.

6.3 Reaction apparatus

A variety of experimental set-ups are employed for organocopper chemistry. In many cases, an arrangement of glassware similar to that shown in Fig. 2.4a will be suitable (for similar arrangements of assembled apparatus, see Chapter 7, Figs 7.1, 7.4, and 7.5). This consists of a three-necked flask equipped with a magnetic stirring bar, a septum, a low temperature thermometer, and an inlet for inert gas and vacuum. Liquid reagents and solvents can be added via syringe through the septum. Provided that an adequate flow of inert gas is maintained, the septum can be removed to allow the addition of solids [e.g. the copper(I) salt] and the withdrawal of small samples of the reaction mixture required for TLC analysis. Before starting the reaction the apparatus should be thoroughly dry. Opinions differ as to the best way to achieve this. In many cases, the apparatus can be assembled cold, flushed with a steady stream of inert gas, and heated with a Bunsen flame (see Protocols 8–11). The heating should be started near the inert gas inlet and carried along the apparatus towards a temporary outlet, which can be provided by piercing the septum cap with a short needle. As an additional precaution after cooling, the apparatus may be evacuated before refilling with inert gas. Repeating this procedure several times should ensure that all moisture or oxygen is removed.

The reaction set-up employed depends very much on individual requirements. A reflux condenser may be needed in the early stages of the organocopper reaction, e.g. during the preparation of a Grignard reagent (see Protocol 8). For large-scale reactions, a pressure-equalizing addition funnel may be necessary to allow the addition of liquids at a controlled rate. For such

(a)

Balloon
To vacuum
Thermometer
Three-way stopcock
Fasten with an elastic band
Tubing
Nitrogen / argon entry
Needle
Septum
Graduated syringe
Dry-ice Dewar or insulated cold bath
Stirring bar

(b)

Septum
To vacuum / inert atmosphere
Two-way stopcock
Stirrer bar

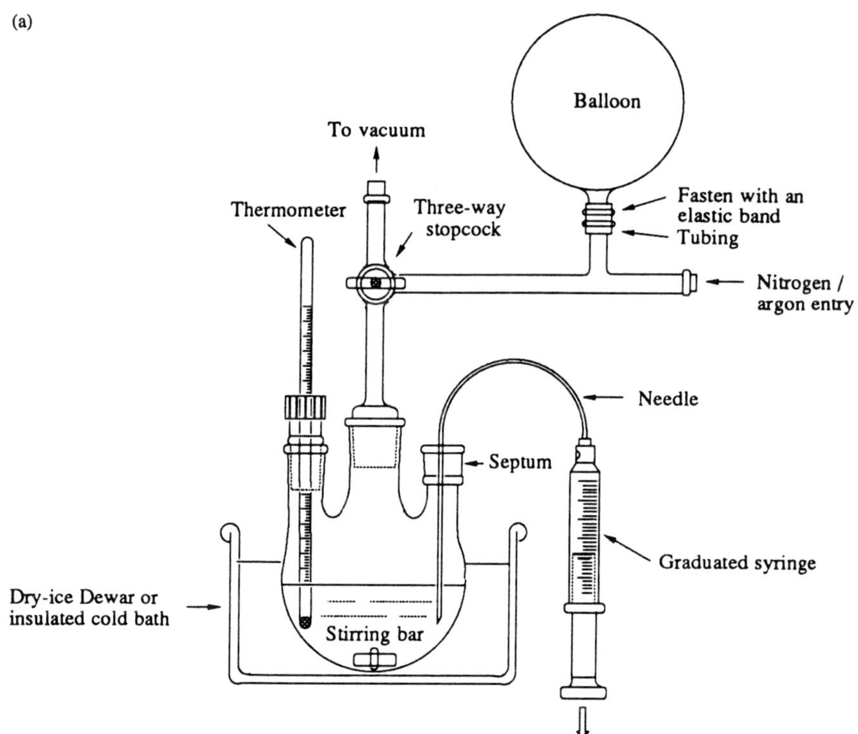

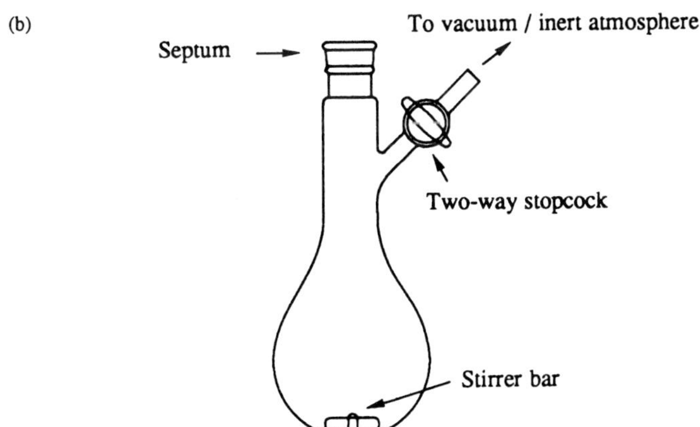

Fig. 2.4

reactions, or those in which efficient stirring is impeded by suspended solids or by the viscosity of the medium, the use of a mechanical stirrer may be required. If this is the case, the stirring rod should be positioned inside the reaction flask using a close-fitting Teflon stirrer bearing to ensure an adequate seal against the outside atmosphere. Further examples of apparatus designed for specific applications appear in Protocol 11 of this chapter, Protocol 8 in Chapter 9, and Protocol 1 in Chapter 11.

If organocopper reactions are to be carried out on a regular basis, or several reactions are to be run simultaneously, the experimentalist may find it more convenient to use Schlenk apparatus rather than conventional Quickfit glassware. A simple set-up of this type is shown in Fig. 2.4b (see also Chapter 7, Fig. 7.4). The basic requirements are an efficient 'dedicated' inert gas/vacuum manifold, a range of double-ended needles, and several purpose-built reaction vessels. A description of the associated manipulative techniques is beyond the scope of this chapter, although comprehensive accounts have been published.[90d] Protocols in which Schlenk apparatus is used appear in later chapters (for example, see Chapter 4, Protocol 9)

Most organocopper reactions are carried out in the temperature range 0 to −100 °C. Cooling systems most frequently used are ice–salt (0 to −20 °C), solvent–solid CO_2 (down to −100 °C), and liquid nitrogen 'slush' baths (down to −160 °C). Details of these systems are widely available;[86] Table 2.6 shows a representative set of frequently used cooling bath systems. The temperature of the reaction mixture should be monitored by means of an internal thermometer (or temperature probe), as shown in Fig. 2.4a, since the internal temperature may differ significantly from that of the cooling bath medium, particularly during the addition of reagents which produce an exothermic reaction. Careful temperature control is essential when working with organocopper reagents to ensure that optimum conditions for formation, stability, and reactivity are maintained. The preparation of stoichiometric organocopper reagents should be conducted at a temperature high enough to effect complete formation within a convenient time period, whilst ensuring that the

Table 2.6. Solid CO_2 cooling baths[86]

Solvent	Temperature (°C)
Ethylene glycol	−15
Carbon tetrachloride	−23
Acetonitrile	−42
2-Ethoxyethyl ether	−52
Chloroform	−61
Ethanol	−72
Acetone	−78
Diethyl ether	−100

stability limits of the particular reagent are not exceeded.[107] For example, ethereal solutions of most lithium dialkylcuprates decompose at above −20 °C. In order to suppress unwanted side-reactions, subsequent addition of an electrophile is frequently carried out at a temperature lower than that employed for cuprate formation (see Protocol 11). If the electrophilic substrate is introduced before cuprate formation is complete, an alternative mode of reaction may prevail, e.g. 1,2-addition rather than 1,4-addition to an enone. As a safeguard against such an occurrence, complete formation of a stoichiometric organocopper reagent can usually be ascertained by a negative Gilman test (see Section 5.4) or, less reliably, by complete dissolution of the copper(I) salt.[107]

6.4 Working up the reaction

In many reported cases,[117] covering a broad range of nucleophilic and electrophilic reagents, the standard work-up procedure for organocopper reactions involves pouring the reaction mixture into an excess of a saturated aqueous solution of ammonium chloride, to ensure rapid quenching of any organometallic species still present, followed by extraction of the product into an immiscible organic solvent. A common practice is to buffer the ammonium chloride solution to a pH of 8–9 by addition of a small volume of concentrated aqueous ammonia (*c.* 10% v/v). The presence of the latter also serves to sequester efficiently any copper(II) species present by forming a deep-blue, water-soluble complex. When using an organocopper reagent to effect conjugate addition to an enone, the use of dilute aqueous HCl or H_2SO_4 (*c.* 1 M) to quench the reaction may be necessary in order to ensure that the intermediate enolate anion and excess organometallic reagent are quenched simultaneously (see Protocol 9), thereby preventing 1,2-addition to the product.[118] If any insoluble inorganic material remains after quenching the reaction, it is advisable to filter the mixture through a layer of Celite® before extraction of the product into an organic solvent (see Protocol 8). If this is carried out, washing the filter cake thoroughly with the extraction solvent should minimize product loss. Once the combined organic extracts have been dried and then concentrated *in vacuo*, the product can usually be purified using standard techniques. Occasionally, product purification may be hindered by the presence of copper salts complexed to sulfides or phosphines used as additives in the original reaction. These complexes can survive aqueous work-up procedures and column chromatography, but since they tend to elute in low polarity solvents, separation from the required product is usually straightforward.

6.5 Protocols for organocopper reactions

Protocol 8.
Copper-catalysed conjugate addition of a functionalized Grignard reagent[119]

Caution! Carry out all procedures in a well-ventilated hood, and wear disposable vinyl or latex gloves and chemical-resistant safety goggles.

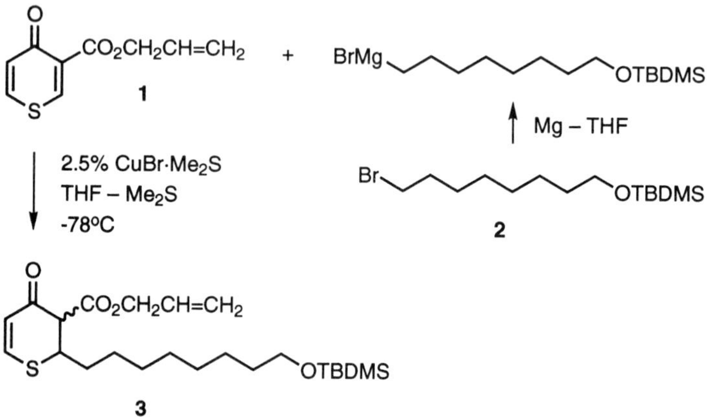

Equipment

- Three-necked, round-bottomed flask (250 mL) with magnetic stirring bar, pressure-equalizing addition funnel and stopper, reflux condenser, and septa
- Inert gas supply and inlet
- Dry, gas-tight syringes
- CO_2–acetone cooling bath

Materials

- 3-Allyloxycarbonylthiin-4-one **1** (FW 196.2) 4.00 g, 20.4 mmol — **harmful**
- 8-Bromo-1-(*t*-butyldimethylsilyloxy)octane **2** (FW 323.4) 7.92 g, 24.5 mmol — **harmful**
- Magnesium turnings (FW 24.3) 0.625 g, 25.7 mmol — **flammable solid**
- Copper(I) bromide–dimethyl sulfide complex (FW 205.6) 0.125 g, 0.61 mmol — **moisture sensitive**
- Dry, distilled THF 160 mL — **flammable, irritant**
- Dry, distilled dimethyl sulfide 4 mL — **flammable, stench**
- Saturated aqueous ammonium chloride solution 80 mL — **corrosive**
- Ether for work-up — **flammable, irritant**

1. Flame dry the reaction vessel and accessories under dry nitrogen. After cooling the apparatus to room temperature, add the magnesium turnings and repeat the flame-drying operation whilst stirring. Allow to cool.
2. Add THF (40 mL), heat the resulting suspension under reflux, and then add a solution of the bromide **2** in THF (20 mL) dropwise over 45 min via an addition funnel.

3. Heat the mixture under reflux for a further 75 min to ensure complete formation of the Grignard reagent, then cool to −78°C and add a solution of copper(I) bromide–dimethyl sulfide complex in dimethyl sulfide (4 mL).
4. Stir the mixture for 15 min, then add a solution of the enone **1** in THF (40 mL) dropwise over 45 min.
5. Continue stirring the mixture at −78°C for 1 h, then add saturated aqueous ammonium chloride (80 mL). Remove the cooling bath. Allow the mixture to warm to room temperature and then filter it through a layer of Celite®. Wash the filter cake with ether. Combine the filtrate and washings and extract three times with ether. Combine the extracts, wash with brine, dry ($MgSO_4$), and remove the solvent *in vacuo*.
6. Purify the product by column chromatography on silica gel, using gradient elution with light petroleum (b.p. 40–60°C) and 2–10% ether. 3-Allyloxycarbonyl-5,6-dihydro-2-[8-(*t*-butyldimethylsilyloxy)octyl]thiin-4-one **3** is obtained as a yellow oil (5.10 g, 57%) displaying the appropriate spectroscopic data.

Protocol 9.
Conjugate addition of an alkenyl monoorganocopper reagent stabilized by dimethyl sulfide[120]

Caution! Carry out all procedures in a well-ventilated hood, and wear disposable vinyl or latex gloves and chemical-resistant safety goggles.

I / OTBDMS **4** — i. *n*-BuLi, Et_2O, -78°C; ii. CuI – Me_2S → $Me_2S{\cdot}Cu$ / OTBDMS **5**

O / CO_2Me / S **6** — (**5**) – Me_2S → O / CO_2Me / S / OTBDMS **7**

Equipment

- Three-necked, round-bottomed flask (100 mL) with magnetic stirring bar, internal thermometer, and septum
- Inert gas supply and inlet
- Dry, gas-tight syringes
- CO_2–acetone cooling bath

Protocol 9. *Continued*

Materials

- *E*-3-(*t*-Butyldimethylsilyloxy)-1-iodo-1-octene **4** (FW 368.4) 1.58 g, 4.3 mmol — **harmful**
- 5,6-Dihydro-3-methoxycarbonylthiin-4-one **6** (FW 172.2) 0.69 g, 4.0 mmol — **harmful**
- *n*-Butyllithium 1.7 M solution in hexane, 2.53 mL, 4.3 mmol — **pyrophoric, moisture sensitive**
- Copper(I) iodide 0.80 g, 4.2 mmol — **irritant**
- Dry, distilled dimethyl sulfide 7 mL — **flammable, stench**
- Dry, distilled ether 5 mL — **flammable, irritant**
- Dilute sulfuric acid 0.2 M, 50 mL — **corrosive**
- Ether for work-up — **flammable, irritant**

1. Flame dry the reaction vessel and accessories under dry nitrogen. After cooling the apparatus to room temperature, add the iodide **4** and ether (5 mL). Cool the solution to −78°C.
2. Add the butyllithium solution dropwise by syringe. Stir the mixture for 55 min at −78°C.
3. Add a solution of copper(I) iodide in dimethyl sulfide (2 mL) over 5 min by syringe. Stir the mixture at −78°C for 15 min to give a dark-green solution of monorganocopper reagent **5**.
4. Add a solution of the enone **6** in dimethyl sulfide (5 mL) via syringe over 3 min.
5. Continue stirring the mixture at −78°C for 2 min, then add dilute sulfuric acid (0.2 M; 50 mL). Remove the cooling bath. Allow the mixture to warm to room temperature and then extract three times with ether. Combine the extracts, wash with brine, dry ($MgSO_4$), and remove the solvent *in vacuo*.
6. Purify the product by column chromatography on silica gel, using gradient elution with light petroleum (b.p. 40–60°C) and 0–10% ether. 2-[(*E*)-3-(*t*-Butyldimethylsilyloxy)oct-1-enyl]-3-methoxycarbonylthian-4-one **7** is obtained as a colourless oil (1.475 g, 89%) displaying the appropriate spectroscopic and analytical data.

Protocol 10.
Copper-catalysed coupling of a 2-halothiophene with an alkyne under Sonogashira conditions[121]

Caution! Carry out all procedures in a well-ventilated hood, and wear disposable vinyl or latex gloves and chemical-resistant safety goggles.

$CuI - (Ph_3P)_2PdCl_2$

$i\text{-}Pr_2NH$

8

Equipment

- Two-necked, round-bottomed flask (10 mL) with magnetic stirring bar and septum
- Inert gas supply and inlet

Materials

- 2-Acetyl-5-iodothiophene (FW 252.1) 149 mg, 0.59 mmol — **harmful**
- 2-Propyn-1-ol (FW 56.1) 36 mg, 0.65 mmol — **toxic, flammable**
- Copper(I) iodide (FW 190.4) 4 mg, 3 mol% — **irritant**
- Bis(triphenylphosphine)palladium(II) chloride (FW 701.9) 14 mg, 3 mol% — **hygroscopic**
- Dry, distilled diisopropylamine 3 mL — **flammable, corrosive**
- Ether for work-up — **flammable, irritant**

1. Flame dry the reaction vessel and accessories under dry nitrogen. After cooling the apparatus to room temperature, combine the reagents, adding the copper(I) iodide last.
2. Stir the mixture for 2 h at room temperature.
3. Filter the mixture to remove suspended solid and wash the filter cake with ether (3 × 5 mL). Combine the filtrate and washings and concentrate under reduced pressure.
4. Purify the product by preparative centrifugal chromatography on silica gel, eluting with light petroleum (b.p. 40–60°C)–ether. 2-Acetyl-5-(3-hydroxy-1-propynyl)thiophene **8** is obtained as a yellow solid (101 mg, 95%) displaying the appropriate spectroscopic data.
5. To obtain an analytical sample, recrystallize this material from light petroleum–ether to give yellow needles, m.p. 72°C.

Protocol 11.
Alkylation of a (*Z,Z*)-dienylcuprate generated by double acetylene carbocupration[122]

Caution! Carry out all procedures in a well-ventilated hood, and wear disposable vinyl or latex gloves and chemical-resistant safety goggles.

2 *n*-BuLi —($CuBr{\cdot}SMe_2$; Et_2O, -30°C)→ Bu_2CuLi —(excess HC≡CH; -25 to 0°C)→ (Bu–CH=CH–CH=CH)$_2$CuLi **9**

9 —(allyl bromide, HMPA; -50 to 0°C)→ Bu–CH=CH–CH=CH–CH$_2$–CH=CH$_2$ **10**

Equipment

- Three-necked, round-bottomed flask (250 mL) with magnetic stirring bar, internal thermometer, and septum (see Fig. 2.5)
- Inert gas supply and inlet
- Gas-tight syringes
- CO_2–acetone cooling bath
- Gas burette apparatus for dispensing acetylene (Fig. 2.5)

This apparatus allows acetylene gas to be purified and dispensed in accurately measured volumes. Two traps cooled in solid CO_2–acetone baths serve to remove acetone impurities. Water vapour is removed by passage through two calcium chloride drying tubes. Acetylene is introduced to the reaction mixture beneath the solvent surface by means of a wide-bore syringe needle.
Caution! *Since acetylene is a highly flammable gas the apparatus should be set up in a highly efficient fume hood.*

Materials

- *n*-Butyllithium (FW 64.1) 1.6 M solution in hexanes, 12.1 mL, 19.4 mmol — **pyrophoric, moisture sensitive**
- Copper(I) bromide–dimethyl sulfide complex (FW 205.6) 2.00 g, 9.7 mmol — **moisture sensitive**
- Acetylene (FW 26.0) 1.3 L, 60 mmol — **highly flammable**
- 3-Bromoprop-1-ene (FW 121.0) 1.17 g, 0.80 mL, 9.7 mmol — **highly toxic, flammable**
- HMPA (FW 179.2) 1.74 g, 1.70 mL, 9.7 mmol — **cancer suspect agent**
- Dry, distilled ether 45 mL — **flammable, irritant**
- Light petroleum (b.p. 40–60°C) for work-up — **flammable**
- A 2:1 mixture of saturated aqueous ammonium chloride and dilute hydrochloric acid 25 mL — **corrosive**
- Aqueous ammonium hydroxide solution 5% — **corrosive**
- Saturated aqueous ammonium chloride — **corrosive**

1. Flame dry the reaction vessel and accessories under dry nitrogen. After cooling the apparatus to room temperature, add copper(I) bromide–dimethyl sulfide complex and ether (45 mL).
2. Cool the resulting suspension to −40°C, then add the butyllithium solution dropwise. Stir at −30°C for 30 min.
3. Cool the mixture to −50°C, then pass acetylene (0.46 L, 2.2 molar equivalents) slowly into the reaction. Stir the resulting green solution at −25°C for 30 min, then warm to 0°C.
4. Pass more acetylene (0.84 L, four molar equivalents) through the reaction mixture over 10 min. Cool the resulting solution of dienylcuprate **9** to −50°C, then add 3-bromoprop-1-ene and HMPA. Allow the mixture to warm to room temperature over 1 h.
5. Add a mixture of saturated aqueous ammonium chloride and dilute hydrochloric acid (2:1; 25 mL) and then extract the mixture three times with light petroleum (b.p. 40–60°C). Combine the extracts, wash with aqueous ammonium hydroxide (5%) and saturated aqueous ammonium chloride, dry ($MgSO_4$), and remove the solvent *in vacuo*.
6. Purify the product by bulb-to-bulb distillation under reduced pressure (6 mm Hg) using a Kugelrohr apparatus. The product, (4*Z*, 6*Z*)-undeca-1,4,6-triene **10** is obtained as a colourless oil (0.905 g, 62%) displaying the appropriate spectroscopic and analytical data.

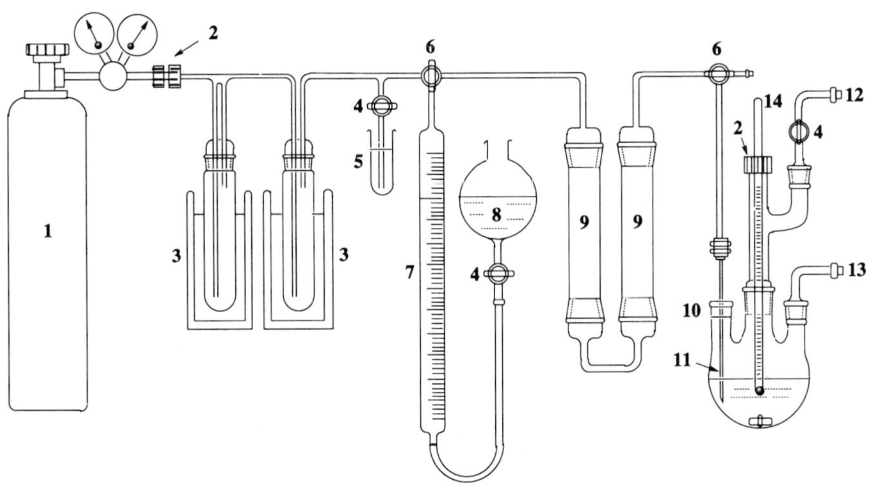

Fig. 2.5 Acetylene carbocupration apparatus

1 Acetylene tank
2 Connector
3 Acetone/dry ice trap
4 Two-way stopcock
5 Mercury bubbler
6 Three-way stopcock
7 Graduated burette
8 Water reservoir
9 Calcium chloride drying tube
10 Septum
11 Wide bore syringe needle
12 Nitrogen entry
13 Vent to bubbler
14 Low temperature thermometer

References

1. Bertz, S. H.; Gibson, C. P.; Dabbagh, G. *Tetrahedron Lett.* **1987**, *28*, 4251–4254.
2. (a) Lipshutz, B. H.; Whitney, S.; Kozlowski, J. A.; Breneman, C. M. *Tetrahedron Lett.* **1986**, *27*, 4273–4276. (b) Lipshutz, B. H.; Kozlowski, J. A.; Wilhelm, R. S. *J. Org. Chem.* **1983**, *48*, 546–550.
3. Bertz, S. H.; Dabbagh, G.; Williams, L. M. *J. Org. Chem.* **1985**, *50*, 4414–4415.
4. Coates, G. E.; Glocking, F. In *Organometallic Chemistry*; Zeiss, H., ed.; Reinhold: New York, **1960**; p. 446. See also Sakata, H.; Aoki, Y.; Kuwajima, I. *Tetrahedron Lett.* **1990**, *31*, 1161–1164.
5. Cohen, T.; Ruffner, R. J.; Shull, D. W.; Fogel, E. R.; Falck, J. R. *Org. Synth.* **1979**, *59*, 202–212.
6. Birch, A. J.; Smith, M. *Proc. Chem. Soc.* **1962**, 356. Marshall, J. A.; Fanta, W. I.; Roebke, H. *J. Org. Chem.* **1966**, *31*, 1016–1020.
7. Julia, M.; Righini-Tapie, A.; Verpeaux, J. N. *Tetrahedron* **1983**, *39*, 3283–3287. Takabe, K.; Uchiyama, Y.; Okisaka, K.; Yamada, T.; Katagiri, T.; Okazaki, T.; Oketa, Y.; Kumobayashi, H.; Akutagawa, S. *Tetrahedron Lett.* **1985**, *26*, 5153–5154.
8. House, H. O.; Chu, C.-Y.; Wilkins, J. M.; Umen, M. J. *J. Org. Chem.* **1975**, *40*, 1460–1469.
9. Marfat, A.; McGuirk, P. R.; Helquist, P. *J. Org. Chem.* **1979**, *44*, 3888–3901.
10. Theis, A. B.; Townsend, C. A. *Synth. Commun.* **1981**, *11*, 157–166.
11. Keller, R. N.; Wycoff, H. D. *Inorg. Synth.* **1946**, *2*, 1–4.
12. Hartwell, J. L. *Org. Synth. Coll.* **1955**, *3*, 185–187.
13. Furniss, B. S., Hannaford, A. J., Smith, P. W. G., Tatchell, A. R. (eds). *Vogel's Textbook of Practical Organic Chemistry*, 5th edn; Longmans: London, **1989**: (a) p. 428; (b) p. 396.
14. Dieter, R. K.; Silks, III, L. A.; Fishpaugh, J. R.; Kastner, M. E. *J. Am. Chem. Soc.* **1985**, *107*, 4679–4692.
15. House, H. O.; Umen, M. J. *J. Org. Chem.* **1973**, *38*, 3893–3901.
16. Whitesides, G. M.; Kendall, P. E. *J. Org. Chem.* **1972**, *37*, 3718–3725.
17. Cook, B. W.; Miller, R. G. J.; Todd, P. F. *J. Organomet. Chem.* **1969**, *19*, 421–430.
18. Leyendecker, F.; Jesser, F. *Tetrahedron Lett.* **1980**, *21*, 1311–1314.
19. Wuts, P. G. M. *Synth. Commun.* **1981**, *11*, 139–140.
20. Tsuda, T.; Hashimoto, T.; Saegusa, T. *J. Am. Chem. Soc.* **1972**, *94*, 658–659.
21. Posner, G. H.; Whitten, C. E.; Sterling, J. J. *J. Am. Chem. Soc.* **1973**, *95*, 7788–7800.
22. Mandeville, W. H.; Whitesides, G. M. *J. Org. Chem.* **1974**, *39*, 400–405.
23. Prout, F. S.; Abdulslam, M. E. *J. Chem. Eng. Data* **1966**, *11*, 616–617.
24. Trost, B. M.; Shimizu, M. *J. Am. Chem. Soc.* **1982**, *104*, 4299–4301.
25. Jukes, A. E. *Adv. Organomet. Chem.* **1974**, *12*, 215–322 (p. 220).
26. Gorlier, J. P.; Hamon, L.; Levisalles, J.; Wagnon, J. *J. Chem. Soc., Chem. Commun.* **1973**, 88. Acker, R.-D. *Tetrahedron Lett.* **1977**, 3407–3410.
27. Lipshutz, B. H.; Wilhelm, R. S.; Kozlowski, J. A.; Parker, D. *J. Org. Chem.* **1984**, *49*, 3928–3938.
28. Lipshutz, B. H.; Kozlowski, J.; Wilhelm, R. S. *J. Am. Chem. Soc.* **1982**, *104*, 2305–2307.

29. Lipshutz, B. H.; Wilhelm, R. S.; Floyd, D. M. *J. Am. Chem. Soc.* **1981**, *103*, 7672–7674.
30. Lipshutz, B. H. *Synlett* **1990**, 119–128. Lipshutz, B. H. *Synthesis* **1987**, 325–341. Lipshutz, B. H.; Wilhelm, R. S.; Kozlowski, J. A. *Tetrahedron* **1984**, *40*, 5005–5038.
31. Yamamoto, Y.; Asoa, N.; Uyehara, T. *J. Am. Chem. Soc.* **1992**, *114*, 5427–5429.
32. Knochel, P.; Yeh, M. C. P.; Berk, S. C.; Talbert, J. *J. Org. Chem.* **1988**, *53*, 2390–2392.
33. Yanagisawa, A.; Noritake, Y.; Nomura, N.; Yamamoto, H. *Synlett* **1991**, 251–253.
34. Arai, M.; Nakamura, E.; Lipshutz, B. H. *J. Org. Chem.* **1991**, *56*, 5489–5491.
35. House, H. O.; Fischer, Jr., W. F. *J. Org. Chem.* **1968**, *33*, 949–956.
36. Stack, D. E.; Dawson, B. T.; Rieke, R. D. *J. Am. Chem. Soc.* **1992**, *114*, 5110–5116.
37. Posner, G. H. *Org. React.* **1975**, *22*, 1, 253–400 (p. 297).
38. Posner, G. H.; Whitten, C. E. *Org. Synth.* **1976**, *55*, 122–127.
39. Kauffman, G. B.; Teter, L. A. *Inorg. Synth.* **1963**, *7*, 9–12.
40. Linstrumelle, G.; Krieger, J. K.; Whitesides, G. M. *Org. Synth.* **1976**, *55*, 103–113.
41. Binkley, E. S.; Heathcock, C. H. *J. Org. Chem.* **1975**, *40*, 2156–2160.
42. Casy, G.; Lane, S.; Taylor, R. J. K. *J. Chem. Soc., Perkin Trans. 1* **1986**, 1397–1404.
43. Johnson, C. R.; Marren, T. J. *Tetrahedron Lett.* **1987**, *28*, 27–30.
44. Whitesides, G. M.; Casey, C. P.; Krieger, J. K. *J. Am. Chem. Soc.* **1971**, *93*, 1379–1389.
45. Ebert, G. W.; Rieke, R. D. *J. Org. Chem.* **1988**, *53*, 4482–4488. Ebert, G. W.; Klein, W. R. *J. Org. Chem.* **1991**, *56*, 4744–4747 and references therein.
46. Ziegler, F. E.; Fowler, K. W.; Rodgers, W. B.; Wester, R. T. *Org. Synth.* **1987**, *65*, 108–118.
47. Ziegler, F. E.; Mikami, K. *Tetrahedron Lett.* **1984**, *25*, 131–134.
48. Owsley, D. C.; Castro, C. E. *Org. Synth.* **1972**, *52*, 128–131.
49. Back, T. G.; Collins, S.; Krishna, M. V.; Law, K.-W. *J. Org. Chem.* **1987**, *52*, 4258–4264.
50. Posner, G. H.; Brunelle, D. J.; Sinoway, L. *Synthesis* **1974**, 662–663. Adams, R.; Reifschneider, W.; Ferretti, A. *Org. Synth.* **1962**, *42*, 22–25.
51. Perrin, D. D.; Armarego, W. L. F.; Perrin, D. R. *Purification of Laboratory Chemicals*, 3rd edn; Pergamon Press: Oxford, **1986**.
52. Quinkert, G.; Müller, T.; Königer, A.; Schultheis, O.; Sickenberger, B.; Dürner, G. *Tetrahedron Lett.* **1992**, *33*, 3469–3472.
53. Cowley, A. H.; Giolando, D. M.; Jones, R. A.; Nunn, C. M.; Power, J. M. *J. Chem. Soc., Chem. Commun.* **1988**, 208–209.
54. Martin, S. F.; Fishpaugh, J. R.; Power, J. M.; Giolando, D. M.; Jones, R. A.; Nunn, C. M.; Cowley, A. H. *J. Am. Chem. Soc.* **1988**, *110*, 7226–7228.
55. Bertz, S. H.; Dabbagh, G.; Villacorta, G. M. *J. Am. Chem. Soc.* **1982**, *104*, 5824–5826. Bertz, S. H.; Dabbagh, G. *J. Chem. Soc., Chem. Commun.* **1982**, 1030.
56. Bertz, S. H.; Dabbagh, G. *J. Org. Chem.*, **1984**, *49*, 1119–1122.
57. Tamura, M.; Kochi, J. *Synthesis* **1971**, 303–305.

58. Ciaccio, J. S.; Addess, K. J.; Bell, T. W. *Tetrahedron Lett.* **1986**, *27*, 3697–3700.
59. Fouquet, G.; Schlosser, M. *Angew. Chem., Int. Ed. Engl.* **1974**, *13*, 82–83. Schlosser, M. *Angew. Chem., Int. Ed. Engl.* **1974**, *13*, 701–706. Morisaki, M.; Shibata, M.; Duque, C.; Imamura, N.; Ikekawa, N. *Chem. Pharm. Bull.* **1980**, *28*, 606–611.
60. Cahiez, G.; Marquais, S. *Synlett* **1993**, 45–47.
61. Giner, J.-L.; Margot, C.; Djerassi, C. J. *J. Org. Chem.* **1989**, *54*, 2117–2125.
62. Schlosser, M.; Bossert, H. *Tetrahedron* **1991**, *47*, 6287–6292.
63. Ledlie, D. E.; Miller, G. *J. Org. Chem.* **1979**, *44*, 1006–1007.
64. Issleib, K.; Fröhlich, H. O. *Chem. Ber.* **1962**, *95*, 375–380.
65. Castro, C. E.; Gaughan, E. J.; Owsley, D. C. *J. Org. Chem.* **1966**, *31*, 4071–4078. Stephens, R. D.; Castro, C. E. *J. Org. Chem.* **1963**, *28*, 3313–3315.
66. Marino, J. P.; Floyd, D. M. *J. Am. Chem. Soc.* **1974**, *96*, 7138–7140.
67. Bourgain, M.; Villieras, J.; Normant, J. F. *C. R. Hebd. Seances Acad. Sci., Ser. C* **1973**, *276*, 1477–1480. Normant, J. F.; Bourgain, M. *Tetrahedron Lett.* **1970**, 2659–2662.
68. Villieras, J.; Ramband, M.; Graff, M. *Synth. Commun.* **1985**, *15*, 569–580.
69. Vermeer, P.; Westmijze, P. H.; Kleijn, H.; van Dijck, L. A. *Recl. Trav. Chim. Pays-Bas* **1978**, *97*, 56–58.
70. Lipshutz, B. H.; Fatheree, P.; Hagen, W.; Stevens, K. L. *Tetrahedron Lett.* **1992**, *33*, 1041–1044.
71. FMC Lithium Division Brochure: *Organometallics in Organic Synthesis*. Wedinger, R. S.; Hatch, H. B. (FMC). Pat. Appl. Nos. USSN 344 416, **1989** and CIP USSN 404 820, **1989**.
72. Lipshutz, B. H.; Kozlowski, J. A.; Parker, D. A.; Nguyen, S. L.; McCarthy, K. E. *J. Organomet. Chem.* **1985**, *285*, 437–447. Lipshutz, B. H.; Parker, D. A.; Nguyen, S. L.; McCarthy, K. E.; Barton, J. C.; Whitney, S. E.; Kotsuki, H. *Tetrahedron* **1986**, *42*, 2873–2879.
73. Tsuda, T.; Yazawa, T.; Watanabe, K.; Fujii, T.; Saegusa, T. *J. Org. Chem.* **1981**, *46*, 192–194.
74. Corey, E. J.; Floyd, D.; Lipshutz, B. H. *J. Org. Chem.* **1978**, *43*, 3418–3420.
75. Corey, E. J.; Bock, M. G.; Kozikowski, A. P.; Rama Rao, A. V.; Floyd, D. Lipshutz, B. H. *Tetrahedron Lett.* **1978**, 1051–1054. See also reference 29 and Abelman, M. M.; Funk, R. L.; Munger, Jr., J. D. *J. Am. Chem. Soc.* **1982**, *104*, 4030–4032.
76. Johnson, C. R.; Dhanoa, D. S. *J. Org. Chem.* **1987**, *52*, 1885–1888.
77. Johnson, C. R.; Dhanoa, D. S. *J. Chem. Soc., Chem. Commun.* **1982**, 358–359.
78. Stephens, R. D.; Castro, C. E. *J. Org. Chem.* **1963**, *28*, 3313–3315.
79. Corey, E. J.; Beames, D. J. *J. Am. Chem. Soc.* **1972**, *94*, 7210–7211.
80. Luche, J. L.; Pétrier, C.; Gemal, A. L.; Zikra, N. *J. Org. Chem.* **1982**, *47*, 3805–3806.
81. Corey, E. J.; Kyler, K.; Raju, N. *Tetrahedron Lett.* **1984**, *25*, 5115–5118.
82. Enda, J.; Kuwajima, I. *J. Am. Chem. Soc.* **1985**, *107*, 5495–5501.
83. Knotter, D. M.; van Maanen, H. L.; Grove, D. M.; Spek, A. L.; van Koten, G. *Inorg. Chem.* **1991**, *30*, 3309–3317.
84. Lambert, F.; Knotter, D. M.; Janssen, M. D.; van Klaveren, M.; Boersma, J.; van Koten, G. *Tetrahedron Asymm.* **1991**, *2*, 1097–1100. Knotter, D. M.; Grove,

D. M.; Smeets, W. J. J.; Spek, A. L.; van Koten, G. *J. Am. Chem. Soc.* **1992**, *114*, 3400–3410.
85. Pan, Y.; Hutchinson, D. K.; Nantz, M. H.; Fuchs, P. L. *Tetrahedron* **1989**, *45*, 467–478.
86. Gordon, A. J.; Ford, R. A. *The Chemists Companion*; Wiley: New York, **1972**.
87. Shriver, D. S.; Drezdzon, M. A. *The Manipulation of Air-Sensitive Compounds*, 2nd edn; Wiley: New York, **1986**; Chapter 4.
88. Yamamoto, Y.; Yamamoto, S.; Yatagai, H.; Maruyama, K. *J. Am. Chem. Soc.* **1980**, *102*, 2318–2325.
89. Hutchinson, D. K. *Aldrichimica Acta* **1986**, *19*, 58.
90. See reference 87: (a) p. 239; (b) Section 1.3; (c) Section 8.3; (d) Section 1.5.
91. Gill, G. B.; Whiting, D. A. *Aldrichimica Acta* **1986**, *19*, 31–41.
92. Kramer, G. W.; Levy, A. B.; Midland, M. M. In *Organic Syntheses via Boranes*; Brown, H. C., ed.; Wiley: New York, **1975**; Chapter 9.
93. Brandsma, L.; Verkruijsse, H. D. *Preparative Polar Organometallic Chemistry*; Springer: Berlin, **1987**; Vol. 1.
94. Jackson, H. L.; McCormack, W. B.; Rondestvedt, C. S.; Smeltz, K. C.; Viele, I. E. *J. Chem. Educ.* **1970**, *47*, A175–A188.
95. Burfield, D. R. *J. Org. Chem.* **1982**, *47*, 3821–3824.
96. Schollköpf, U. In *Houben–Weyl Methoden der Organische Chemie*; Müller, E., ed.; Georg Thieme: Stuttgart, **1970**; Vol. 13/1, pp. 93–253. Wakefield, B. J. *The Chemistry of Organolithium Compounds*; Pergamon Press: New York, **1974**. Wakefield, B. *Organolithium Methods*; Academic Press: London, **1988**. Nützel, K. In *Houben–Weyl Methoden der Organische Chemie*; Müller, E., ed.; Georg Thieme: Stuttgart, **1973**; Vol. 13/2a, pp. 53–527. Kharasch, N. S.; Reinmuth, O. *Grignard Reactions of Nonmetallic Substances*; Prentice-Hall: New York, **1954**.
97. Seyferth, D.; Weiner, M. A. *J. Am. Chem. Soc.* **1961**, *83*, 3583–3586. Seyferth, D.; Johnson, C. S.; Weiner, M. A.; Waugh, J. S; Seyferth, D. *J. Am. Chem. Soc.* **1961**, *83*, 1306–1307.
98. Lane, C. F.; Kramer, G. W. *Aldrichimica Acta* **1977**, *10*, 11.
99. Aldrich Technical Information Bulletins: AL-134—Handling Air-Sensitive reagents; AL-135—Equipment for Handling Air-Sensitive Reagents; AL-136—The Aldrich Sure/Pac Cylinder Packaging System and Recommended Transfer Procedures; AL-137—The Aldrich HPLC Solvent Sure/Seal System; AL-149—Aldrich Kilo-Lab Cylinder Packaging System and Recommended Transfer Procedures; AL-150—Equipment for Transfer of Liquids from Aldrich Kilo-Lab Cylinders; AL-192—The Aldrich CHEM-FLEX System.
100. Bulletins 107 and 108, Foote Mineral Company, Pennsylvania, PA, USA.
101. Stack, D. E.; Dawson, B. T.; Rieke, R. D. *J. Am. Chem. Soc.* **1991**, *113*, 4672–4673 and references therein.
102. Yeh, M. C. P.; Chen, H. G.; Knochel, P. *Org. Synth.* **1991**, *70*, 195–203 and references therein.
103. Lipshutz, B. H.; Keil, R. *J. Am. Chem. Soc.* **1992**, *114*, 7919–7920.
104. Bochmann, M. *Aldrichimica Acta* **1986**, *19*, 2.
105. (a) Gilman, H.; Schulze, F. *J. Am. Chem. Soc.* **1925**, *47*, 2002–2005. (b) Gilman, H.; Jones, R. G.; Woods, L. A. *J. Org. Chem.* **1952**, *17*, 1630–1634. (c) Lipshutz, B. H.; Kozlowski, J. A.; Breneman, C. M. *J. Am. Chem. Soc.* **1985**, *107*, 3197–3204.

106. Corey, E. J.; Posner, G. H. *J. Am. Chem. Soc.* **1968**, *90*, 5615–5616.
107. Posner, G. H. *Org. React.* **1972**, *19*, 1–113 (pp. 54–57).
108. Crompton, T. R. *Chemical Analysis of Organometallic Compounds*; Academic Press: London, **1973**.
109. Gilman, H.; Cartledge, F. K. *J. Organomet. Chem.* **1964**, *2*, 447–454.
110. Heathcock, C. H.; Tice, C. M.; Germroth, T. C. *J. Am. Chem. Soc.* **1982**, *104*, 6081–6091.
111. (a) Watson, S. C.; Eastham, J. F. *J. Organomet. Chem.* **1967**, *9*, 165–168; (b) Lin, H.-S.; Paquette, L. A. *Synth. Commun.*, **1994**, *24*, 2503–2506.
112. Gall, M.; House, H. O. *Org. Synth.* **1972**, *52*, 39–52.
113. Kiljunen, H.; Hase, T. A. *J. Org. Chem.* **1991**, *56*, 6950–6952.
114. Kofron, W. G.; Baclawski, L. M. *J. Org. Chem.* **1976**, *41*, 1879–1880. Lipton, M. F.; Sorensen, C. M.; Sadler, A. C.; Shapiro, R. H. *J. Organomet. Chem.* **1980**, *186*, 155–158. Winkle, M. R.; Lansinger, J. M.; Ronald, R. C. *J. Chem. Soc., Chem. Commun.* **1980**, 87–88. Juaristi, E.; Martínez-Richa, A.; García-Rivera, A.; Cruz-Sánchez, J. S. *J. Org. Chem.* **1983**, *48*, 2603–2606. Suffert, J. *J. Org. Chem.* **1989**, *54*, 509–510.
115. Duhamel, L.; Plaquevent, J.-C. *J. Org. Chem.* **1979**, *44*, 3404–3405.
116. Harwood, L. M.; Moody, C. J. *Experimental Organic Chemistry*; Blackwell: Oxford, **1989**: (a) pp. 81–87; (b) pp. 79–80.
117. Lipshutz, B. H.; Sengupta, S. *Org. React.* **1992**, *41*, 135–631 (pp. 239–252 and references therein).
118. Piers, E.; Keziere, R. J. *Can. J. Chem.* **1969**, *47*, 137–144.
119. Casy, G.; Taylor, R. J. K. *Tetrahedron* **1989**, *45*, 455–466.
120. Lane, S.; Quick, S. J.; Taylor, R. J. K. *J. Chem. Soc., Perkin Trans. 1* **1985**, 893–898.
121. De Sousa, Jr., P. T.; Taylor, R. J. K. *Synlett* **1990**, 755–757.
122. Furber, M.; Taylor, R. J. K.; Burford, S. C. *J. Chem. Soc., Perkin Trans. 1* **1986**, 1809–1815.

3

Preparation of organocopper reagents using active copper

R. D. RIEKE and W. R. KLEIN

1. Introduction

Organocopper reagents have intensively been used in synthetic chemistry. The vast majority of organocopper reagents are prepared by a transmetallation reaction involving an organometallic reagent and a copper(I) salt. A variety of organometallic reagents, derived from metals more electropositive than copper, have been utilized in the preparation of organocopper reagents. The most common organometallic precursors have been organomagnesium and organolithium reagents. This approach severely limits the functionality that may be incorporated into the organocopper reagent. The use of traditional lithium or Grignard precursors can be circumvented by using a highly reactive zero-valent copper which directly undergoes oxidative addition to organic halides.[1–3]

The highly reactive zero-valent copper solution is prepared by the reduction of a soluble copper(I) salt complex by a stoichiometric amount of lithium naphthalenide. Significantly, the organocopper reagents prepared by utilizing this highly active copper may incorporate a wide variety of functionalities such as ester, nitrile, chloride, fluoride, epoxide, and ketone moieties.[1–3] These functionalized organocopper reagents undergo a variety of reactions such as cross-coupling with acid chlorides and aldehydes to give the corresponding ketones and alcohols, 1,4-conjugate addition with α,β-unsaturated carbonyl compounds, intermolecular and intramolecular epoxide openings, and allylation with allylic compounds.

2. Copper(I) salt complexes

Many copper(I) salt complexes have been reduced by lithium naphthalenide. The most common include copper(I) iodide–tri-*n*-butylphosphine, copper(I) iodide–triphenylphosphine, copper cyanide–lithium chloride, and lithium 2-thienylcyanocuprate. The choice of ligand used in preparing the active copper is a critical factor determining the reactivity of the active copper species.

Although all of the complexes used to form active copper are capable of undergoing the transformations previously mentioned, each are accompanied by their inherent advantages and disadvantages.

2.1 Phosphine-based active copper[2,3]

The use of phosphine ligands significantly enhances the reactivity of the active copper formed. In general, as the electron-donating capability of the phosphine ligand increases, the reactivity of the active copper towards oxidative addition also increases, as well as the nucleophilicity of the resulting organocopper reagent formed. However, the use of malodorous phosphine ligands often interferes with product isolation.

2.1.1 Copper(I) iodide–tri-*n*-butylphosphine

The organocopper reagents made from copper(I) iodide–tri-*n*-butylphosphine complex are very nucleophilic and are best suitable for epoxide-opening reactions, specifically intermolecular openings (Protocol 1), and 1,4-conjugate addition reactions. It should be noted that the addition of this form of active copper to alkyl bromides can produce up to 30% homocoupled products (2RBr → RR), and that higher amounts are formed from alkyl iodides.

2.1.2 General directions

Conduct all reactions under an argon atmosphere using inert atmosphere techniques (see Chapter 2, Section 6). Dry all glassware overnight in a 120°C oven and assemble while hot prior to use. Freshly distil THF over a Na–K alloy under an argon atmosphere immediately before use (see Chapter 2, Section 4). Most chemicals are commercially available and are used without further purification, with the following exceptions:

(1) Copper(I) iodide–tributylphosphine: prepare by the method described in Chapter 2, Protocol 3.

(2) Copper(I) iodide: purify according to Protocol 2 in Chapter 2.

(3) Copper(I) cyanide: purchase from Aldrich (99%) and store immediately in an argon drybox upon receipt without further drying. Otherwise, use any drying method described in Table 2.1.

(4) Lithium chloride: place in a round-bottomed flask fitted with a three-way adapter and dry under vacuum with an electric heat gun for five to 10 minutes. After cooling, introduce argon and store the flask in the drybox.

2.1.3 Copper(I) iodide–triphenylphosphine

The use of copper(I) iodide–triphenylphosphine curtails the amount of homocoupled products (less than 2%) with alkyl bromides. However, the

Protocol 1.
Intermolecular epoxide opening with a tri-*n*-butylphosphine-based active copper reagent

Caution! Carry out all procedures in a well-ventilated hood, and wear disposable vinyl or latex gloves and chemical-resistant safety goggles.

$$1.05\ Li^+ [C_{10}H_8]^{\cdot-} + CuI{\cdot}PBu_3 + 1.5\ PBu_3 \xrightarrow[THF]{0\ ^\circ C} Cu^*$$

$$Cu^* + 0.4\ Br(CH_2)_3CO_2Bu\text{-}t \xrightarrow[THF]{-78^\circ C} \xrightarrow[(2)\ -10\ ^\circ C,\ H^+]{(1)\ 0.18\ \text{1,2-epoxybutane}} Et\text{-}CH(OH)\text{-}CH_2(CH_2)_3CO_2Bu\text{-}t$$

Equipment

- Two two-necked, round-bottomed flasks (100 and 50 mL), each with a magnetic stirrer bar, two-way stopcock, and septum
- Dry glass vial with septum
- Inert argon supply double manifold and inlet
- Argon drybox
- Dry, gas-tight syringes
- Cannula

Materials

- Lithium (FW 6.9) 0.38 mm ribbon, 0.0690 g, 9.942 mmol — **flammable solid, moisture sensitive**
- Naphthalene (FW 128.1) 1.418 g, 11.06 mmol — **flammable solid, toxic**
- Copper(I) iodide–tri-*n*-butylphosphine[a] (FW 392.8) 3.592 g, 9.145 mmol — **irritant**
- Tri-*n*-butylphosphine (FW 202.3) 2.790 g, 13.79 mmol — **flammable, toxic**
- *t*-Butyl 4-bromobutyrate[4] (FW 223.0) 0.8080 g, 3.633 mmol
- 1,2-Epoxybutane (FW 72.0) 0.1170 g, 1.623 mmol — **flammable, cancer suspect agent**
- Dry, freshly distilled THF — **flammable, irritant**

1. Dry all glassware overnight in a 120°C oven. Assemble the glassware while hot and enter the argon drybox. Add the lithium, naphthalene, and magnetic stirrer bar to the 100 mL reaction vessel and seal it with the stopcock and septum. Remove the reaction vessel from the drybox and connect it to the argon/vacuum manifold.
2. Add THF (10–15 mL) to the lithium and naphthalene. Stir for two hours at room temperature until the lithium is consumed to give a green solution, then cool to 0°C.
3. Assemble the 50 mL round-bottomed flask with a stopcock, stirrer bar, and septum while hot and connect it to the argon/vacuum manifold. Under a flow of argon add the copper(I) iodide–tri-*n*-butylphosphine, then evacuate the flask and refill with argon. Add THF (10 mL) and stir until dissolved, then syringe in tri-*n*-butylphosphine.

Protocol 1. *Continued*

4. Rapidly cannulate the solution of copper(I) iodide–tri-*n*-butylphosphine and tri-*n*-butylphosphine into the preformed lithium naphthalenide solution using argon pressure. Stir for one hour at 0°C, then cool to −78°C. Then syringe in *t*-butyl 4-bromobutyrate and stir for one hour at −78°C.
5. Add 1,2-epoxybutane to a glass vial and seal with a septum. Connect the vial to the argon/vacuum manifold via a syringe needle and freeze the contents using a liquid nitrogen cold bath. Evacuate the vial and refill with argon, then gradually warm to room temperature while still connected to the argon manifold. Add THF (3 mL) to dissolve, then cannulate into the organocopper solution. Stir for one hour at −78°C, then gradually warm to −10°C.
6. Add saturated aqueous ammonium chloride (20 mL). Remove the cooling bath and allow to warm to room temperature. Extract the mixture three times with diethyl ether. Combine the extracts and wash with water and brine. Dry over $MgSO_4$ and remove the solvent *in vacuo*.
7. Purify the residue by column chromatography on silica gel, using gradient elution with hexanes and 0–10% EtOAc. *t*-Butyl 6-hydroxyoctanoate is obtained as a colourless liquid (0.3050 g, 87%) displaying the appropriate spectroscopic data.

[a] See Chapter 2, Protocol 3

---

resulting nucleophilicity of the organocopper reagent is decreased. This loss of reactivity is offset by the elimination of homocoupling. Protocol 2 shows a typical intramolecular reaction for which this formulation is ideally suited. The organocopper reagents prepared by this approach cross-couple with acid chlorides to generate functionalized ketones in excellent yields.

2.2 Copper(I) cyanide-based active copper[6]

Although the active copper derived from the low temperature (−100 to −110°C) reduction of copper(I) cyanide–lithium chloride is less reactive than that generated from the phosphine-based complexes, more importantly, this formulation allows for the direct formation of highly functionalized allylic organocopper reagents from allylic chlorides and acetates. This active copper will also undergo oxidative addition to a wide variety of functionalized alkyl, aryl, and vinyl bromides. Other advantages are also featured: (1) copper cyanide is inexpensive and is used without further purification; (2) product isolation is greatly simplified; and (3) the organocopper reagents are capable of 1,4-conjugate additions as well as cross-coupling with electrophiles. Protocol 3 shows the direct formation of a functionalized allyl organocopper reagent and subsequent cross-coupling with an acid chloride.

Protocol 2.
Intramolecular epoxide opening with a triphenylphosphine-based active copper reagent

Caution! Carry out all procedures in a well-ventilated hood, and wear disposable vinyl or latex gloves and chemical-resistant safety goggles.

1.05 Li^+ [naphthalenide] + CuI/PPh_3 $\xrightarrow[\text{THF}]{0\,^\circ\text{C}}$ Cu*

Cu* + 0.24 Br(CH$_2$)$_3$-epoxide $\xrightarrow[\text{THF}]{-40\,^\circ\text{C}}$ $\xrightarrow[\text{H}^+]{0\,^\circ\text{C}}$ cyclopentanol (OH)

Equipment

- Two two-necked, round-bottomed flasks (100 and 50 mL), each with a magnetic stirrer bar, two-way stopcock, and septum
- Dry glass vial with septum
- Inert argon supply double manifold and inlet
- Argon drybox
- Dry, gas-tight syringes
- Cannula

Materials

- Lithium (FW 6.9) 0.38 mm ribbon, 0.0705 g, 10.16 mmol — **flammable solid, moisture sensitive**
- Naphthalene (FW 128.1) 1.437 g, 11.21 mmol — **flammable solid, toxic**
- Copper(I) iodide [a] (FW 190.4) 1.769 g, 9.289 mmol — **irritant**
- Triphenylphosphine (FW 262.3) 4.87 g, 18.57 mmol
- 5-Bromo-1,2-epoxypentane[5] (FW 165.0) 0.3620 g, 2.193 mmol — **cancer suspect agent**
- Dry, freshly distilled THF — **flammable, irritant**

1. Prepare lithium naphthalenide as in Protocol 1.
2. Assemble the 50 mL round-bottomed flask with a stopcock, stirrer bar, and septum while hot and connect it to the argon/vacuum manifold. Under a flow of argon add copper(I) iodide and triphenylphosphine, then evacuate the flask and refill with argon. Add THF (10 mL) and stir for one hour at room temperature to give a white slurry, then cool to 0°C.
3. Cannulate the preformed lithium naphthalenide solution to the copper(I) iodide–triphenylphosphine solution and stir at 0°C for 30 minutes, then cool to −42°C using an acetonitrile–CO_2 cold bath.
4. Add 5-bromo-1,2-epoxypentane to a glass vial and seal with a septum. Connect the vial to the argon/vacuum manifold via a syringe needle and freeze the contents using a liquid nitrogen cold bath. Evacuate the vial and refill with argon, then gradually warm to room temperature while still connected to the argon manifold. Add THF (10 mL) then cannulate into the active

Protocol 2. *Continued*

copper solution. Stir for five minutes at −45°C, warm to −20°C and stir for three hours, then gradually warm to 0°C.

5. Add saturated aqueous ammonium chloride (15 mL). Remove the cooling bath and allow to warm to room temperature. Extract the mixture three times with diethyl ether. Combine the extracts, dry over $MgSO_4$, and remove the solvent *in vacuo*.
6. Filter the residue and wash the filter cake with hexanes–ether (1:1 v/v). Collect the organic wash and remove the solvent *in vacuo*.
7. Purify the residue by column chromatography on silica gel, using gradient elution with hexanes and 0–10% EtOAc. Cyclopentanol is obtained as a colourless liquid (0.1641 g, 87%) displaying the appropriate spectroscopic data.

[a] See Protocol 2 in Chapter 2.

Protocol 3.
Reaction of a functionalized allyl chloride with a CuCN·LiCl-based active copper and subsequent cross-coupling with an electrophile

Caution! Carry out all procedures in a well-ventilated hood, and wear disposable vinyl or latex gloves and chemical-resistant safety goggles.

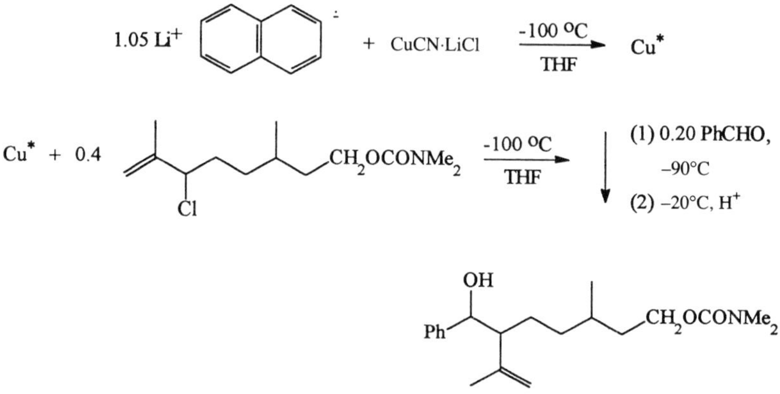

1.05 Li^+ [naphthalenide] + CuCN·LiCl $\xrightarrow[\text{THF}]{-100\,^\circ\text{C}}$ Cu*

Cu* + 0.4 [allyl chloride]–$CH_2OCONMe_2$ $\xrightarrow[\text{THF}]{-100\,^\circ\text{C}}$ (1) 0.20 PhCHO, −90°C (2) −20°C, H^+

[product: Ph–CH(OH)– ... –$CH_2OCONMe_2$]

Equipment

- Two two-necked, round-bottomed flasks (100 and 50 mL), each with a magnetic stirrer bar, two-way stopcock, and septum
- Dry glass vial with septum
- Inert argon supply double manifold and inlet
- Argon drybox
- Dry, gas-tight syringes and cannula

Materials

- Lithium (FW 6.9) 0.38 mm ribbon, 0.0377 g, 5.43 mmol — **flammable solid, moisture sensitive**
- Naphthalene (FW 128.1) 0.8267 g, 6.45 mmol — **flammable solid, toxic**
- Copper cyanide [a] (FW 89.6) 0.4504 g, 5.03 mmol — **highly toxic, irritant**
- Lithium chloride [b] (FW 42.4) 0.2615 g, 6.17 mmol — **irritant, hygroscopic**
- 6-Chloro-3,7-dimethyl-7-octenyl *N,N*-dimethylcarbamate[7] (FW 261.6) 0.5222 g, 2.01 mmol — **irritant, toxic**
- Benzaldehyde (FW 106.1) 0.1082 g, 1.02 mmol — **toxic, irritant**
- Dry, freshly distilled THF — **flammable, irritant**

1. Prepare lithium naphthalenide as in Protocol 1, then cool to −100°C using a liquid nitrogen–Et_2O cold bath.
2. Assemble the 50 mL round-bottomed flask with a stopcock, stirrer bar, and septum while hot and connect it to the argon/vacuum manifold. Under a flow of argon add copper cyanide and lithium chloride, then evacuate the flask and refill with argon. Add THF (5 mL) and stir for 30 minutes, then cool to −78°C.
3. Cannulate the copper cyanide-lithium chloride solution into the preformed lithium naphthalenide solution and stir at −100°C for five to 10 minutes.
4. Add the functionalized allyl chloride to a glass vial and seal with a septum. Connect the vial to the argon/vacuum manifold via a syringe needle and freeze the contents using a liquid nitrogen cold bath. Evacuate the vial and refill with argon, then warm to room temperature while still connected to the manifold. Add THF (4 mL), cool to −78°C, and then rapidly cannulate into the active copper solution. Stir for 10 minutes at −100°C and warm to −90°C. Then similarly rapidly cannulate a benzaldehyde solution and warm to −20°C over three hours.
5. Add saturated aqueous ammonium chloride (15 mL). Remove the cooling bath and warm to room temperature. Extract the mixture three times with diethyl ether. Combine the extracts, wash with water and brine, dry over $MgSO_4$, and remove the solvent *in vacuo*.
6. Purify the residue using column chromatography on silica gel, using gradient elution with hexanes and 0–10% EtOAc. 6-(α-Hydroxybenzyl)-3,7-dimethyl-7-octenyl *N,N*-dimethylcarbamate is obtained as a colourless liquid (0.3265 g, 96%) displaying the appropriate spectroscopic data.

[a] See Table 2.1.
[b] See Table 2.2

2.3 Thienyl-based active copper[8]

The low temperature reduction (−78 °C) of lithium 2-thienylcyanocuprate produces a zero-valent active copper which is more reactive at lower temperatures than the phosphine-based active copper. At low temperatures, the thienyl-based active copper consumes more equivalents of an alkyl chloride, accompanied with little or no homocoupled products, when compared to the phosphine-based active copper. Significantly, this form of active copper also oxidatively adds to allyl chlorides and acetates to form allyl organocopper reagents directly. Elimination of the phosphine ligands also facilitates product isolation. Although lithium 2-thienylcyanocuprate is reported to have a prolonged shelf-life, it is not as reliable a source of active copper as the copper cyanide-based active copper. The lithium 2-thienylcyanocuprate is purchased from Aldrich (0.25 M in THF) in 100 or 800 mL Sure/Seal™ bottles. The thienyl-based active copper yields are dependent upon the freshness of the cuprate or the manufacturing lot. Reactions from a freshly punctured Sure/Seal™ bottle gave the highest yields and successively decreased with each consecutive puncture within one week. Generally it is best to purchase 100 mL bottles and to use the contents entirely within a day or two. Another, although time-consuming, alternative is to prepare the cuprate freshly from literature methods (see Chapter 5, Protocol 3).[9] Protocol 4 shows an example of a typical 1,4-conjugate addition reaction.

Protocol 4.
Conjugate addition reaction with a lithium 2-thienylcyanocuprate-based active copper reagent

Caution! Carry out all procedures in a well-ventilated hood, and wear disposable vinyl or latex gloves and chemical-resistant safety goggles.

1.05 Li^+ [naphthalene]$^{\cdot-}$ + [2-thienyl]Cu(CN)Li $\xrightarrow[\text{THF}]{-78\ ^\circ C}$ Cu^*

Cu^* + 0.5 $Br(CH_2)_3CO_2Et$ $\xrightarrow[\text{THF}]{-78\ ^\circ C}$ (1) 1.0 TMSCl (2) 0.125 [cyclohexenone] (3) H^+ → 3-[$(CH_2)_3CO_2Et$]cyclohexanone

Equipment

- Two two-necked, round-bottomed flasks (100 and 50 mL), each with a magnetic stirrer bar, two-way stopcock, and septum
- Dry glass vial with septum
- Inert argon supply double manifold and inlet
- Argon drybox
- Dry, gas-tight syringes
- Cannula

Materials

- Lithium (FW 6.9) 0.38 mm ribbon, 0.0582 g, 8.40 mmol — **flammable solid, moisture sensitive**
- Naphthalene (FW 128.1) 1.1792 g, 9.20 mmol — **flammable solid, toxic**
- Lithium 2-thienylcyanocuprate[a] (FW 179.6) 0.25 M, 32 mL, 8.0 mmol — **highly toxic, flammable**
- Ethyl 4-bromobutyrate (FW 195.1) 0.7802 g, 4.00 mmol
- Trimethylsilyl chloride (FW 108.6) 0.8619 g, 8.00 mmol — **flammable, corrosive**
- Cyclohexen-2-one (FW 96.1) 0.0962 g, 1.00 mmol — **highly toxic**
- Dry, freshly distilled THF — **flammable, irritant**

1. Prepare the lithium naphthalenide as in Protocol 1, then cool to −78°C.
2. Assemble the 50 mL round-bottomed flask with a stopcock, stirrer bar, and septum while hot and connect it to the argon/vacuum manifold, evacuate, and refill with argon. Syringe the lithium 2-thienylcyanocuprate into the flask and cool to −78°C.
3. Cannulate the lithium 2-thienylcyanocuprate solution into the preformed lithium naphthalenide solution and stir at −78°C for five to 15 minutes, then cool to −100°C or below.
4. Add the ethyl 4-bromobutyrate to a glass vial and seal with a septum. Connect the vial to the argon/vacuum manifold and freeze the contents using a liquid nitrogen cold bath. Evacuate the vial and refill with argon, then warm to room temperature while still connected to the manifold. Add THF (4 mL), cool to −78°C, and then rapidly cannulate into the active copper solution. Stir for five minutes at −100°C, warm to −78°C, and syringe in the trimethylsilyl chloride.
5. Then similarly prepare and cannulate in the cyclohexen-2-one solution (THF, 15 mL) dropwise over 20–30 minutes. Then stir for 1 hour at −78°C.
6. Add saturated aqueous ammonium chloride (15 mL). Remove the cooling bath and warm to room temperature. Extract the mixture three times with diethyl ether. Combine the extracts, wash with water and brine, dry over $MgSO_4$, and remove the solvent *in vacuo*.
7. Purify the residue using column chromatography on silica gel, using gradient elution with hexanes and 0–10% EtOAc. 3-(3-Carboethoxypropyl)cyclohexanone is obtained as a colourless liquid (0.3397 g, 80%) displaying the appropriate spectroscopic data.

[a] Aldrich (or see Chapter 5, Protocol 3)

2.4 Anionic active copper[10]

The low temperature (−100 °C), two-equivalent reduction of a copper(I) salt complex with lithium naphthalenide provides a source of an anionic and highly active copper species. The resulting copper anion solution is homogeneous and readily undergoes oxidative addition. However, the copper anion is a two-electron reagent, capable of reacting with one equivalent of the halide. This provides for one of the better systems for 1,4-conjugate additions with allylic substrates. Protocol 5 shows a typical 1,4-conjugate allylation reaction.

Protocol 5. Conjugate addition reaction with an allyl organocopper reagent utilizing an active copper anion species

Caution! Carry out all procedures in a well-ventilated hood, and wear disposable vinyl or latex gloves and chemical-resistant safety goggles.

2.05 Li^+ [naphthalene]$^{\cdot -}$ + CuI · LiCl $\xrightarrow[\text{THF}]{-100°C}$ $Cu^{(-)}$

$Cu^{(-)}$ + Cl(methylallyl) $\xrightarrow[\text{THF}]{-100°C}$ $\xrightarrow[\text{THF}]{-78°C}$ (1) 1.0 TMSCl (2) 0.25 [cyclohexenone] (3) H^+ → [3-(2-methylallyl)cyclohexanone]

Equipment

- Two two-necked, round-bottomed flasks (100 and 50 mL), each with a magnetic stirrer bar, two-way stopcock, and septum
- Dry glass vial with septum
- Inert argon supply double manifold and inlet
- Argon drybox
- Dry, gas-tight syringes
- Cannula

Materials

• Lithium (FW 6.9) 0.38 mm ribbon, 0.0569 g, 8.20 mmol	**flammable solid, moisture sensitive**
• Naphthalene (FW 128.1) 1.1561 g, 9.02 mmol	**flammable solid, toxic**
• Copper (I) iodide[a] (FW 190.4) 0.718 g, 3.77 mmol	**irritant**
• Lithium chloride[b] (FW 42.4), 0.1696 g, 4.00 mmol	**irritant, hygroscopic**
• Methylallyl chloride (FW 90.6) 0.4258 g, 4.00 mmol	**flammable, cancer suspect agent**
• Trimethylsilyl chloride (FW 108.6) 0.4346 g, 4.70 mmol	**flammable, corrosive**
• Cyclohexen-2-one (FW 96.1) 0.0962 g, 1.00 mmol	**highly toxic**
• Dry, freshly distilled THF	**flammable, irritant**

1. Prepare lithium naphthalenide as in Protocol 1, then cool to −100°C using a liquid nitrogen–Et_2O cold bath.
2. Assemble the 50 mL round-bottomed flask with a stopcock, stirrer bar, and septum while hot and connect it to the argon/vacuum manifold. Under a flow of argon add the copper(I) iodide–lithium chloride, then evacuate the flask and refill with argon. Add THF (10 mL), stir for one hour, then cool to −78°C.

3. Cannulate the copper(I) iodide–lithium chloride solution into the preformed lithium naphthalenide solution and stir at −100 °C for five minutes.
4. Add the methylallyl chloride to a glass vial and seal with a septum. Connect the vial to the argon/vacuum manifold and freeze the contents using a liquid nitrogen cold bath. Evacuate the vial and refill with argon, then warm to room temperature while still connected to the manifold. Add THF (4 ml), cool to −78 °C, then rapidly cannulate into the active copper anion solution. Stir for five minutes at −100 °C, warm to −78 °C, and syringe in the trimethylsilyl chloride.
5. Then similarly prepare and cannulate in the cyclohexen-2-one solution (THF, 15 mL) dropwise over 20–30 minutes. Then stir for one hour at −78 °C.
6. Add saturated aqueous ammonium chloride (15 mL). Remove the cooling bath and warm to room temperature. Extract the mixture three times with diethyl ether. Combine the extracts, wash with water and brine, dry over $MgSO_4$, and remove the solvent *in vacuo*.
7. Purify the residue using column chromatography on silica gel, using gradient elution with hexanes and 0–10% EtOAc. 3-(2-Methylpropene)allylcyclohexanone is obtained as a colourless liquid (0.1398 g, 92%) displaying the appropriate spectroscopic data.

[a] See Protocol 2 in Chapter 2.
[b] See Table 2.2.

In summary, these organocopper reagents undergo most of the standard reactions of the traditional organocopper reagents as well as those of the lower and higher order cuprates. Moreover, these organocopper reagents tolerate a wide variety of functionality. Recent publications have further demonstrated the utility of these types of organocopper reagents.[11–13]

References

1. (a) Rieke, R. D.; Wehmeyer, R. M.; Wu, T.-C.; Ebert, G. W. *Tetrahedron* **1989**, *45*, 443–454. (b) Ebert, G. W.; Rieke, R. D. *J. Org. Chem.* **1988**, *53*, 4482–4488. (c) Wehmeyer, R. M.; Rieke, R. D. *Tetrahedron Lett.* **1988**, *29*, 4513–4516. (d) Wu, T.-C.; Rieke, R. D. *Tetrahedron Lett.* **1988**, *29*, 6753–6756. (e) Wu, T.-C.; Wehmeyer, R. M.; Rieke, R. D. *J. Org. Chem.* **1987**, *52*, 5057–5059. (f) Wehmeyer, R. M.; Rieke, R. D. *J. Org. Chem.* **1987**, *52*, 5056–5057. (g) Ebert, G. W.; Rieke, R. D. *J. Org. Chem.* **1984**, *49*, 5280–5282. (h) Rieke, R. D.; Rhyne, L. D. *J. Org. Chem.* **1979**, *44*, 3445–3446. (i) Rieke, R. D.; Kavaliunas, A. V.; Rhyne, L. D.; Frazier, D. J. *J. Am. Chem. Soc.* **1979**, *101*, 246–248.
2. (a) Rieke, R. D. *CRC Crit. Rev. Surf. Chem.* **1991**, *1*, 131–136. (b) Rieke, R. D. *Science* **1989**, 1260–1264.
3. Ebert, G. W.; Klein, W. R. *J. Org. Chem.* **1991**, *56*, 4744–4747.

4. Morin, C.; Vidal, M. *Tetrahedron* **1992**, *48*, 9277–9282.
5. Cruickshank, P. A.; Fishman, M. *J. Org. Chem.* **1969**, *34*, 4060–4065.
6. (a) Stack, D. E.; Rieke, R. D. *Tetrahedron Lett.* **1992**, *33*, 6575–6578. (b) Stack, D. E.; Dawson, B. T.; Rieke, R. D. *J. Am. Chem. Soc.* **1992**, *114*, 5110–5116. (c) Stack, D. E.; Dawson, B. T.; Rieke, R. D. *J. Am. Chem. Soc.* **1991**, *113*, 4672–4673.
7. See reference 6b and references therein.
8. (a) Klein, W. R.; Rieke, R. D. *Synth. Commun.* **1992**, *18*, 2635–2644. (b) Rieke, R. D.; Wu, T.-C.; Stinn, D. E.; Wehmeyer, R. M. *Synth. Commun.* **1989**, *19*, 1833–1840.
9. (a) Lipshutz, B. H.; Koerner, M.; Parker, D. A. *Tetrahedron Lett.* **1987**, *28*, 945–948. (b) Lipshutz, B. H.; Parker, D. A.; Ngugen, S. L.; McCarthy, K. E.; Barton, J. C.; Whitney, S. E.; Kotsuki, H. *Tetrahedron* **1986**, *42*, 2873–2879. (c) Lipshutz, B. H.; Parker, D. A.; Kozlowski, J. A.; Ngugen, S. L. *Tetrahedron Lett.* **1984**, *25*, 5959–5962.
10. Rieke, R. D.; Dawson, B. T.; Stack, D. E.; Stinn, D. E. *Synth. Commun.* **1990**, *17*, 2711–2721.
11. Rieke, R. D.; Stack, D. E.; Dawson, B. T.; Wu, T.-C. *J. Org. Chem.* **1993**, *58*, 2492–2500.
12. Rieke, R. D.; Klein, W. R.; Wu, T.-C. *J. Org. Chem.* **1993**, *58*, 2483–2491.
13. Stack, D. E.; Klein, W. R.; Rieke, R. D. *Tetrahedron Lett.* **1993**, *34*, 3063–3066.

4

Preparation of highly functionalized reagents

PAUL KNOCHEL, MICHAEL J. ROZEMA, and CHARLES E. TUCKER

1. Introduction

Many organocopper reagents are prepared from the corresponding magnesium or lithium organometallics. Unfortunately, this approach precludes the presence of most functional groups in these copper reagents, since they would not be tolerated in the starting magnesium or lithium reagent. However, functionalized copper reagents of the type FG—RCu(CN)ZnX can be prepared starting from organozincs (RZnX or R_2Zn) by performing a transmetallation using the soluble copper salt CuCN·2LiCl (eqn 1).[1,2] Since organozincs bearing a wide range of functionalities such as a cyanide (Protocol 1), an ester (Protocol 2), a ketone, an enone (Protocol 4), an amine, a terminal alkyne (Protocol 3), a phosphonate, a halide, sulfur functionalities (SCOR, SR, SOR, SO_2R), or a trialkoxysilyl[2] group $[Si(OR)_3]$ can be easily prepared, this method offers a general access to polyfunctional copper reagents.

$$\text{FG-RZnX} + \text{CuCN•2LiCl} \xrightarrow{0^\circ\text{C, 5 min}} \text{FG-RCu(CN)ZnX} \quad (1)$$

X = I, Br, Cl, R
FG = CO_2R, CN, COR, enone, NHR, NH_2, $P(O)(OR)_2$, SO_nR, $Si(OR)_3$, Cl, Br, C≡C–H

Organozinc halides are conveniently obtained by the direct insertion of zinc dust into alkyl iodides in THF (eqn 2).[1,3]

$$\text{FG-R–X} + \text{Zn} \xrightarrow{\text{THF}} \text{FG-RZnX} \quad (2)$$

35–50°C : R = primary alkyl
25–35°C : R = secondary alkyl
5–10°C : R = benzyl, allyl

In the case of benzylic or allylic substrates, the corresponding bromide or chloride can be used.[4] In addition, functionalized primary alkyl chlorides, bromides, sulfonates, and phosphates are also good substrates if the reaction is performed in the presence of a catalytic amount of LiI (0.2 equivalents) and in a polar solvent such as *N,N*-dimethylacetamide (DMA) or *N,N*-dimethylpropyleneurea (DMPU) (eqn 3).[5] This approach is especially useful for the preparation of electron-rich benzylic zinc reagents which cannot be prepared from the corresponding bromides or chlorides because of extensive Wurtz coupling reactions (eqn 4 and Protocol 5).[5] An iodine–zinc exchange reaction can also be performed using Et_2Zn in the presence of a catalytic amount of CuI (0.3 mol%). This reaction affords functionalized dialkylzincs $(FG-R)_2Zn$ in excellent yields (eqn 5 and Protocol 9).[6] These zinc compounds can also be transmetallated to the corresponding copper reagents FG–RCu(CN)ZnR–FG which, in many instances, exhibit a higher reactivity with electrophiles than the copper reagents prepared from FG–RZnX.

$$\text{FG-RX} + \text{Zn} \xrightarrow[\text{DMA or DMPU, 40–80°C, 2–12h}]{\text{LiI(0.2 equiv), LiBr (1.0 equiv)}} \text{FG-RZnX} \quad (3)$$

$X = Br, Cl, OSO_2R, OP(O)(OR)_2$

$$\text{ArCH}_2\text{OP(O)(OEt)}_2 \xrightarrow[\text{50°C,12h}]{\text{Zn, DMA, LiI (0.2 equiv)}} \text{ArCH}_2\text{Zn–OP(O)(OEt)}_2 \quad (4)$$

(Ar = 3,4-methylenedioxyphenyl)

$$\text{FG-RI} + Et_2Zn \xrightarrow[\text{2–12h, (-EtI)}]{\text{CuI (0.3 mol\%), neat}} (\text{FG-R})_2\text{Zn} \quad (5)$$

R = primary

It should be mentioned that functionalized alkenyl and aryl halides do not readily undergo a zinc insertion in THF and that the use of more polar solvents such as DMA is usually required.[7] However, if the double bond is substituted with an electron-withdrawing group, such as 3-iodo-2-cyclohexenone, then a fast zinc insertion occurs even in THF (eqn 6 and Protocol 4).[8]

$$\text{3-iodo-2-cyclohexenone} + \text{Zn} \xrightarrow[\text{25-50°C, 1-2h}]{\text{THF}} \text{3-(ZnI)-2-cyclohexenone} \quad (6)$$

Alternatively, by using low temperatures (−100°C) and the Trapp mixture (THF–ether–pentane, 4:1:1) as solvent, it is possible to prepare functionalized alkenyllithium reagents bearing functionalities such as ester, nitrile, or halide by using an iodine–lithium exchange reaction. The higher electronegativity of the carbon sp^2 bond to the lithium atom reduces the reactivity of alkenyllithiums and these functionalized lithium reagents are stable for several minutes at −100 to −90 °C. A low temperature addition of a solution of ZnI_2 in THF, followed by a THF–Me_2S (1:1) solution of CuCN·2LiCl, provides the corresponding functionalized alkenylcopper (eqn 7 and Protocol 6).[9]

$$\text{FG-R}\diagdown\!\!=\!\!\diagup\text{I} \xrightarrow[\text{THF-Et}_2\text{O-pentane (4:1:1)},\ -100\,^\circ\text{C, 3 min}]{\text{BuLi (1.05 equiv)}} \text{FG-R}\diagdown\!\!=\!\!\diagup\text{Li} \xrightarrow[\text{2) CuCN}\cdot\text{2LiCl},\ -90\,^\circ\text{C}]{\text{1) ZnI}_2,\ -100\,^\circ\text{C}} \text{FG-R}\diagdown\!\!=\!\!\diagup\text{Cu(CN)ZnI}\cdot\text{LiI} \quad (7)$$

The zinc–copper reagents FG—RCu(CN)ZnX display a considerably enhanced reactivity compared to the starting organozinc reagents and react efficiently with various classes of electrophiles such as enones (Protocol 1 and eqn 8),[1,10] nitroalkenes (Protocol 2 and eqn 9),[12] activated alkenyl iodides (Protocol 3 and eqn 10),[12] 1-halogenoalkynes (Protocol 4 and eqn 11),[8,13] aldehydes in the presence of $BF_3{\cdot}OEt_2$,[14] activated alkynes (Protocols 5 and 6 and eqns 12 and 13),[5,9,13,15] alkylidenemalonates,[9,16] and allylic halides[1,17] (Scheme 4.1).

IZn~~CN → (i) CuCN•2LiCl; (ii) 2-cyclohexen-1-one, Me_3SiCl → 3-(2-cyanoethyl)cyclohexanone, 86% (8)

EtOOC~~~Cu(CN)ZnI → Ph-CH=CH-NO_2, -78 to 0 °C, c. 5h → EtOOC~~~CH(Ph)~NO_2, 87% (9)

H-C≡C~~~Cu(CN)ZnI → 3-iodo-2-cyclohexenone, -78 to 0 °C, 3h → H-C≡C~~~(3-substituted 2-cyclohexenone), 88% (10)

(3-oxocyclohex-1-enyl)Cu(CN)ZnI → Bu−C≡C−I, -60 °C, 24h → 3-(C≡C-Bu)-2-cyclohexenone, 92% (11)

(3,4-methylenedioxyphenyl)CH_2Cu(CN)ZnI → H−C≡C−CO_2Et, DMA, -70 °C to 0 °C → (3,4-methylenedioxyphenyl)CH_2CH=CHCO_2Et, 84% (12)

t-BuC(=O)O~~~CH=CH~Cu(CN)Li•ZnI_2 → H−C≡C−CO_2Et, -70 to -20 °C, 5h → t-BuC(=O)O~~~CH=CH-CH=CH-CO_2Et, 68%; 100% (E,E) (13)

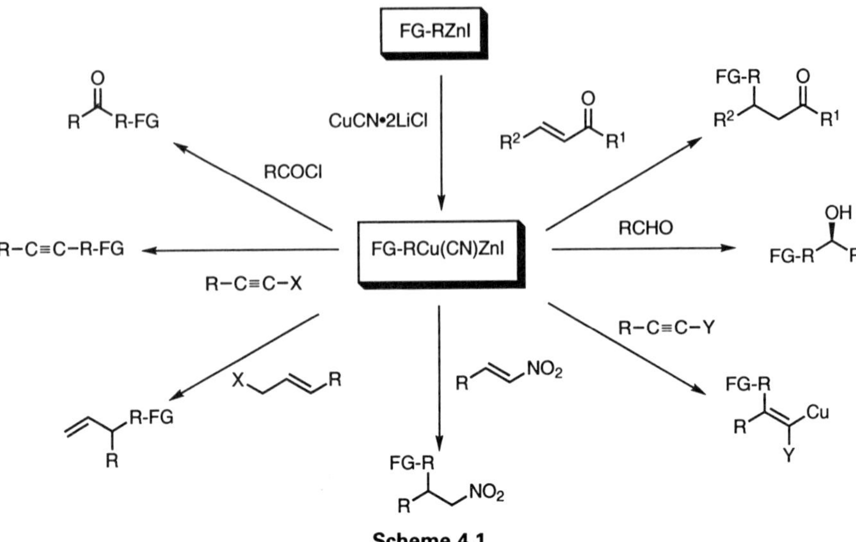

Scheme 4.1

The close proximity of the functional group and the carbon–zinc bond can lead to new and unique reaction pathways. Thus, (iodomethyl)zinc iodide can be used to homologate cleanly a variety of organocopper reagents[18–22] like alkynylcoppers (eqn 14 and Protocol 7)[21] and alkenylcoppers (eqn 15 and Protocol 8).[22]

1) BuLi
2) $CuCN{\bullet}2LiCl$
3) PhCHO
4) ICH_2ZnI

98 % (14)

$H{-}C{\equiv}C{-}CO_2Et$

1) $NC(CH_2)_3Cu(CN)ZnI$
2) PhCHO
3) ICH_2ZnI

75% (15)

The functionalized dialkylzincs $(FG{-}R)_2Zn$ prepared via an iodine–zinc exchange reaction in the presence of a catalytic amount of CuI react with an excellent enantioselectivity with aldehydes in the presence of a titanium catalyst to afford functionalized secondary alcohols (eqn 16 and Protocol 9).[6]

In conclusion, copper reagents prepared from readily available functionalized organozincs offer a unique way for preparing polyfunctional organic molecules without the need for protection–deprotection steps which are necessary when functional groups are present in the lithium-based and

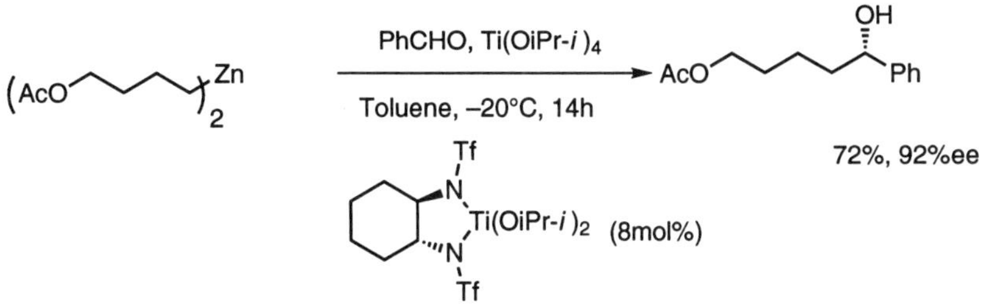

(AcO⌒⌒⌒)$_2$Zn —[PhCHO, Ti(OiPr-*i*)$_4$; Toluene, –20°C, 14h; chiral Ti(OiPr-*i*)$_2$ bis(triflamide) catalyst (8mol%)]→ AcO⌒⌒⌒CH(OH)Ph (16)

72%, 92%ee

magnesium-based copper reagents. Furthermore, most electrophiles that react with cuprates of the type RCu·LiX, R_2CuLi, and R_2CuMgX will also react with the zinc–copper reagents RCu(CN)ZnX under the appropriate reaction conditions. The synthetic utility of these reagents has been further demonstrated in recent publications.[23]

2. General

The solvents used (THF, ether, toluene, *N,N*-dimethylacetamide) were dried and distilled over CaH_2 (see Chapter 2, Section 4). Zinc dust was purchased from Aldrich (−325 mesh) or from Riedel de Haën (Zink, feines Pulver für Analyse; catalogue number 31664). This last zinc quality was found to be the most active. CuCN and CuI were used as received (Aldrich, Fluka, or Merck). All lithium and zinc halides purchased as anhydrous salts (Aldrich, Fluka, or Merck) were always dried for 1–2 h at 120–130°C under vacuum (*c.* 0.1 mm Hg) before use. Me_3SiCl, Ti(OPr-*i*)$_4$, and all organic electrophiles were distilled before use. 1,2-Dibromoethane (99% pure) was used without any purification.

3. Protocols

Protocol 1.
Preparation of 2-cyanoethylzinc iodide, its transmetallation to the corresponding copper derivative, and 1,4-addition to cyclohexenone.[10a] Preparation of 3-(3-oxocyclohexyl)propanenitrile

Caution! Carry out all procedures in a well-ventilated hood, and wear disposable vinyl or latex gloves and chemical-resistant safety goggles.

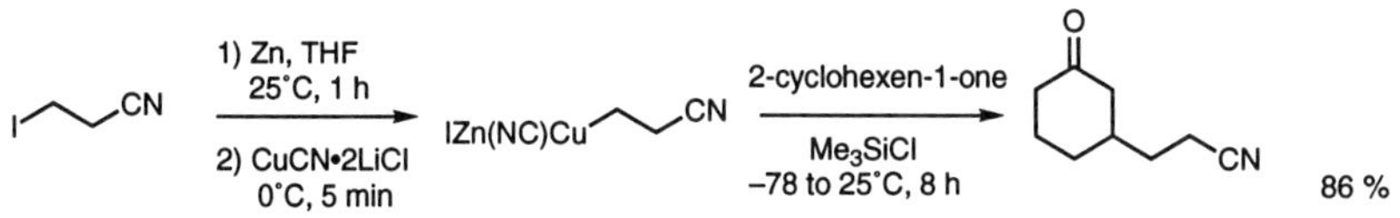

I⌒⌒CN —[1) Zn, THF, 25°C, 1 h; 2) CuCN•2LiCl, 0°C, 5 min]→ IZn(NC)Cu⌒⌒CN —[2-cyclohexen-1-one; Me_3SiCl, –78 to 25°C, 8 h]→ 3-(3-oxocyclohexyl)propanenitrile (CN) 86 %

Protocol 1. *Continued*

Equipment

- One three-necked, round-bottomed flask (100 mL) equipped with a gas inlet, a low temperature thermometer, and a septum cap
- One three-necked, round-bottomed flask (100 mL) equipped with a gas inlet, a thermometer, and an addition funnel with a septum cap
- Argon gas supply and inlet
- Dry, gas-tight syringes

Materials

- Zinc dust (FW 65.4) −325 mesh, 3.0 g, *c.* 46 mmol **flammable solid**
- LiCl (FW 42.4) *c.* 2.10 g, 50 mmol **irritant, hygroscopic**
- CuCN (FW 89.6) 2.25 g, 25 mmol **highly toxic**
- 1,2-Dibromoethane (FW 187.8) 280 mg, 1.5 mmol **toxic, cancer suspect agent**
- Me_3SiCl (FW 108.6) *c.* 0.1 mL for zinc activation and 5.5 mL **flammable, corrosive**
- *n*-Decane (internal GLC standard) 20 mg
- Dry THF **flammable, irritant**
- 3-Iodopropionitrile (FW 180.9) 4.52 g, 25 mmol
- 2-Cyclohexenone (FW 96.1) 1.766 g, 18.4 mmol **highly toxic**

1. Weigh the zinc dust into the three-necked, round-bottomed flask and flush the flask with argon. Carry out the remainder of the reaction under an argon atmosphere.
2. Add the 1,2-dibromoethane in dry THF (1 mL) and heat gently with a heat gun to boil the THF. After *c.* 1 min, the reaction mixture foams and the heating is interrupted. After *c.* 1 min repeat this heating–cooling process two more times. The foaming is indicative of a good activation of the zinc surface.
3. Add Me_3SiCl (0.1 mL) and stir the reaction mixture for *c.* 5 min at 25°C.
4. Add slowly a dry THF solution (8 mL) of 3-iodopropionitrile and *n*-decane (internal standard for GLC analysis) at such a rate that the reaction temperature remains between 25 and 28°C. Stir the reaction mixture at 30°C until completion (*c.* 1 h). Check the reaction by GLC analysis of a hydrolysed reaction sample. For this, take a reaction aliquot (*c.* 0.1 mL) with a pipette and add it to a vial containing ether (0.3 mL) and aqueous NH_4Cl (1.5 mL). The comparison between the initial alkyl iodide:decane ratio (before reaction) and the new alkyl iodide:decane ratio allows an easy determination of the conversion. A conversion of 95–98% is usually reached at the end of the reaction.
5. Add dry THF (*c.* 10 mL) and stop the stirring. Allow the zinc suspension to settle for 1.5 h.
6. During this time, dry the lithium chloride (hygroscopic!) in the second three-necked flask (equipped with two glass stoppers, a magnetic stirring bar, and a gas inlet). Connect to the vacuum (*c.* 0.1 mm Hg) and heat with an oil bath to *c.* 120–130°C for 1 h.
7. Allow the flask to cool and add the copper cyanide. Replace the two glass stoppers with an addition funnel bearing a septum cap and low temperature

thermometer. Flush with argon and add dry THF (23 mL). A slight exothermic reaction is observed and a light yellow-green solution of CuCN·2LiCl is formed. Cool to −40°C.

8. Add the solution of 2-cyanoethylzinc iodide in THF with a syringe and allow the reaction mixture to reach 0°C. After 5 min, cool back to −78°C.
9. Add within 20 min a dry ether solution (10 mL) of Me_3SiCl (5.5 mL) and 2-cyclohexenone and allow the reaction mixture to warm up to 25°C overnight (7–8 h).
10. Add a saturated aqueous NH_4Cl solution to the reaction mixture and stir for 10 min at 25°C. Pour into an Erlenmeyer containing ether (200 mL) and aqueous NH_4Cl (100 mL) and filter over Celite®.
11. Separate the organic and aqueous layers. Extract the aqueous layer with ether (2 × 50 mL). Combine the organic layers and wash the organic phase with brine (50 mL), dry over $MgSO_4$, and evaporate the solvent.
12. Purify the crude residue by flash chromatography using hexane–EtOAc (1:1). Pure 3-(3-oxocyclohexyl)propanenitrile (2.38 g, 86%) is obtained as a light-yellow oil. Characterize the product by ^{1}HNMR, ^{13}C NMR, IR spectroscopy, and high resolution mass spectrometry.

Protocol 2.
Preparation of 3-carbethoxypropylzinc iodide, its transmetallation to the corresponding copper derivative, and addition to a nitroalkene.[11c] Preparation of ethyl 5-phenyl-6-nitrohexanoate

Caution! Carry out all procedures in a well-ventilated hood, and wear disposable vinyl or latex gloves and chemical-resistant safety goggles.

EtOOC~~~I → [1) Zn, THF, 25°C, 1 h; 2) CuCN•2LiCl, 0°C, 5 min] → EtOOC~~~Cu(CN)ZnI → [Ph~NO_2; −78 to 0°C, c. 5h] → EtOOC~~~CH(Ph)~NO_2 87%

Equipment

- One three-necked round-bottomed flask (50 mL) equipped with a gas inlet, a low temperature thermometer, and a septum cap
- One three-necked round-bottomed flask (50 mL) equipped with a gas inlet, a thermometer, and an addition funnel with a septum cap
- Argon gas supply and inlet
- Dry gas-tight syringes

Materials

- Zinc dust (FW 65.4) −325 mesh, 2.0 g, *c.* 30 mmol — **flammable solid**
- LiCl (FW 42.4) *c.* 700 mg, 16 mmol — **irritant, hygroscopic**

Protocol 2. *Continued*

- CuCN (FW 89.6) 720 mg, 8 mmol — **highly toxic**
- 1,2-Dibromoethane (FW 187.8) 200 mg, 1 mmol — **toxic, cancer suspect agent**
- Me_3 SiCl (FW 108.6) *c.* 0.1 mL — **flammable, corrosive**
- *n*-Decane (internal GLC standard) 50 mg
- Dry THF — **flammable, irritant**
- Iodine (FW 253.8) *c.* 50 mg — **lachrymator, corrosive, irritant**
- Ethyl 4-iodobutyrate (FW 242.0) 2.42 g, 10 mmol
- Nitrostyrene (FW 149.2) 900 mg, 6 mmol — **irritant**

1. Weigh the zinc dust into the three-necked round-bottomed flask and flush the flask with argon.
2. Add the 1,2-dibromoethane in dry THF (3 mL) and heat gently with a heat gun to activate the zinc as described in Protocol 1.
3. Add Me_3SiCl with a syringe and stir the reaction mixture for *c.* 5 min at 25°C.
4. Add ethyl 4-iodobutyrate and *n*-decane (internal standard for GLC analysis) in dry THF (3 mL) through the addition funnel. The reaction temperature rises to 50–55°C and the reaction is usually complete after 30 min. If no exothermic reaction is observed, the reaction mixture is heated for 2 h at 55°C (oil bath temperature).
5. Check the conversion of the reaction by GLC as described in Protocol 1.
6. Confirm the organozinc iodide formation by performing an iodolysis of a reaction aliquot [addition to a vial containing iodine (50 mg) in dry THF (1 mL)] and analysing the sample by GLC. This experiment, together with the result obtained from the hydrolysed sample, allows an accurate estimate of the organozinc reagent concentration.
7. When the conversion has reached 95% (or better), add dry THF (*c.* 5 mL), stop the stirring, and allow the excess zinc dust to settle for 1–2 h.
8. Dry the lithium chloride and prepare the CuCN·2LiCl solution in dry THF as described in Protocol 1.
9. Cool the flask to *c.* −40°C and add with a syringe the clear, colourless solution of 3-carboethoxypropylzinc iodide prepared as described above.
10. Remove the cooling bath and allow the reaction mixture (a greenish milky solution) to warm to 0°C. A greenish solution is obtained.
11. Cool the reaction mixture back to −78°C and add the nitrostyrene solution in dry THF (3 mL).
12. Allow the reaction mixture to warm slowly to 0°C and stir at this temperature until completion of the reaction (*c.* 2 h, checked by GLC).
13. Dilute the reaction mixture with ether (200 mL) and pour into an Erlenmeyer containing a saturated aqueous NH_4Cl solution (200 mL).
14. Stir for 2–3 min and filter over Celite®. Work-up the reaction mixture as

described in Protocol 1 and purify the crude residue by flash chromatography (ether–hexane, 1:10). Pure ethyl 5-phenyl-6-nitrohexanoate (1.4 g, 87%) is obtained. Characterize the product by ^{1}HNMR, ^{13}C NMR, IR spectroscopy, and high resolution mass spectrometry.

Protocol 3. Preparation of a zinc–copper reagent bearing a relatively acidic hydrogen and its coupling with 3-iodo-2-cyclohexen-1-one.[12a] Preparation of 3-(4-pentynyl)-2-cyclohexen-1-one

Caution! Carry out all procedures in a well-ventilated hood, and wear disposable vinyl or latex gloves and chemical-resistant safety goggles.

1) Zn, THF 25°C, 0.5 to1 h; 2) CuCN•2LiCl -10°C, 5 min → Cu(CN)ZnI; 3-iodo-2-cyclohexenone, -78 to 0°C, 3h → 88%

Equipment

- One three-necked round-bottomed flask (50 mL) equipped with a gas inlet, a low temperature thermometer, and a septum cap
- One three-necked, round-bottomed flask (50 mL) equipped with a gas inlet, a thermometer, and an addition funnel with a septum cap
- Argon gas supply and inlet
- Dry, gas-tight syringes

Materials

• Zinc dust (FW 65.4) – 325 mesh, *c.* 1.60 g, 25 mmol	**flammable solid**
• LiCl (FW 42.4) *c.* 0.84 g, 20 mmol	**irritant, hygroscopic**
• CuCN (FW 89.6) 0.90 g, 10 mmol	**highly toxic**
• 1,2-Dibromoethane (FW 187.8) 200 mg, 1 mmol	**toxic, cancer suspect agent**
• Me_3SiCl (FW 108.6) *c.* 0.1 mL	**flammable, corrosive**
• *n*-Decane (internal GLC standard) 10 mg	
• Dry THF	**flammable, irritant**
• 5-Iodo-1-pentyne (FW 194.0) 2.23 g, 11.5 mmol	**toxic**
• 3-Iodo-2-cyclohexen-1-one (FW 222.0) 1.55 g, 7 mmol	**highly toxic**

1. Weigh the zinc dust into the three-necked, round-bottomed flask, flush with argon, and add the 1,2-dibromoethane in dry THF (2 mL).
2. Activate the zinc dust by heating as described in Protocol **1**. Add Me_3SiCl and stir the reaction mixture for *c.* 5 min at 25°C.
3. Add a THF solution (8 mL) of 5-iodo-1-pentyne and *n*-decane (internal GLC standard). Maintain the reaction temperature below 40°C during the addition.
4. Stir the reaction mixture for 0.5–1 h at 25°C. Check the completion of the

Protocol 3. *Continued*

reaction by performing a GLC analysis of a hydrolysed reaction aliquot (see Protocol 1).

5. Add dry THF (5 mL), stop the stirring, and allow the excess zinc dust to settle for 1–2h.
6. Dry the lithium chloride and prepare the CuCN·2LiCl solution in dry THF as described in Protocol 1.
7. Cool the flask to −40°C and add, via syringe, the 4-pentynylzinc iodide solution. A dark-red solution is immediately formed. Warm up to −20°C and stir at this temperature for 5 min.
8. Cool back to −78°C and add the 3-iodo-2-cyclohexen-1-one at once. Warm up to −30°C and stir the reaction mixture for 1 h at this temperature. Allow the reaction mixture to warm to 0°C and stir for 2 h at this temperature.
9. Check the completion of the reaction by performing a GLC analysis of a reaction aliquot.
10. Work-up the reaction mixture as described in Protocol 1.
11. Purify the crude residue obtained after evaporation of the solvent by flash chromatography (CH_2Cl_2–ether–hexane, 1:2:90). Pure 3-(4-pentynyl)-2-cyclohexen-1-one (1.0 g, 88%) is obtained as a colourless oil. Characterize the product by ^1H NMR, ^{13}C NMR, IR spectroscopy, and high resolution mass spectrometry.

Protocol 4. Preparation of a functionalized vinylic zinc–copper reagent and its coupling with a 1-iodoalkyne.[8] Preparation of 3-(1-hexynyl)-2-cyclohexen-1-one

Caution! Carry out all procedures in a well-ventilated hood, and wear disposable vinyl or latex gloves and chemical-resistant safety goggles.

1) Zn, THF 25°C, 1–2 h; 2) CuCN•2LiCl 0°C, 5 min → Cu(CN)ZnI; Bu–C≡C–I, −60°C, 24h → C≡C–Bu

Equipment

- One three-necked, round-bottomed flask (50 mL) equipped with a gas inlet, a low temperature thermometer, and a septum cap
- One three-necked, round-bottomed flask (50 mL) equipped with a gas inlet, a thermometer, and an addition funnel with a septum cap
- Argon gas supply and inlet
- Dry, gas-tight syringes

Materials

- Zinc dust (FW 65.4) −325 mesh, *c.* 1.95 g, 30 mmol — **flammable solid**
- LiCl (FW 42.4) 0.85 g, 20 mmol — **irritant, hygroscopic**
- CuCN (FW 89.6) 0.89 g, 10 mmol — **highly toxic**
- 1,2-Dibromoethane (FW 187.8) 200 mg, 1 mmol — **toxic, cancer suspect agent**
- Me_3SiCl (FW 108.6) 0.15 mL, 1.2 mmol — **flammable, corrosive**
- *n*-Decane (internal GLC standard) 20 mg
- Dry THF — **flammable, irritant**
- Me_2S — **flammable, stench**
- 3-Iodo-2-cyclohexen-1-one (FW 222.0) 2.22 g, 10 mmol — **highly toxic**
- 1-Iodohexyne (FW 208.0) 1.45 g, 7 mmol — **toxic**

1. Weigh the zinc dust into the three-necked round-bottomed flask, flush with argon, and add the 1,2-dibromoethane in dry THF (3 mL).
2. Activate the zinc dust by heating as described in Protocol 1. Add Me_3SiCl and stir the reaction mixture for *c.* 5 min at 25 °C.
3. Add a THF solution (3 mL) of 3-iodo-2-cyclohexen-1-one and *n*-decane (*c.* 20 mg) over 15 min. During the addition, the temperature rises to 55 °C.
4. Stir the reaction mixture for 1–2 h at 25 °C. Check for reaction completion by a GLC analysis of a hydrolysed reaction mixture aliquot (see Protocol 1). Note also the presence of *c.* 8% of a heavy by-product (3,3′-dioxo-1,1′-bicyclohexenyl).
5. Add dry THF (8 mL), stop the stirring, and allow the excess zinc dust to settle for 1–2 h.
6. Dry the lithium chloride as described in Protocol 1, add CuCN, and flush with argon. Add dry THF (8 mL) and dry Me_2S (12 mL) and cool the resulting solution to −70 °C.
7. Add the THF solution of 3-oxocyclohexenylzinc iodide with a syringe and warm the reaction mixture to −30 °C. After 5 min, cool back to −70 °C, add 1-iodohexyne, and stir for 24 h at −60 °C.
8. Work up as described in Protocol 1. Purify the crude residue obtained after evaporation of the solvent by flash chromatography (EtOAc–hexane, 1:10) to afford the desired product (1.13 g, 92%). Characterize the product by ^1H NMR, ^{13}C NMR, IR spectroscopy, and high resolution mass spectrometry.

Protocol 5.
Preparation of a functionalized benzylic zinc–copper reagent from the corresponding benzylic phosphate, and its addition to ethyl propiolate.[5] Preparation of (*E*)-ethyl 4-(3,4-methylenedioxyphenyl)-2-butenoate

Caution! Carry out all procedures in a well-ventilated hood, and wear disposable vinyl or latex gloves and chemical-resistant safety goggles.

OP(O)(OEt)$_2$ — 1) Zn, DMA, LiI (0.2 equiv), 50°C, 12h; 2) CuCN•2LiCl, −10°C, 5 min. → Cu(CN)ZnOP(O)(OEt)$_2$ — H−C≡C−CO$_2$Et, DMA, −70°C to 0°C → CO$_2$Et, 84%

Equipment

- One three-necked, round-bottomed flask (50 mL) equipped with a gas inlet, a low temperature thermometer, and a septum cap
- One three-necked round-bottomed flask (50 mL) equipped with a gas inlet, a thermometer, and an addition funnel with a septum cap
- Argon gas supply and inlet
- Dry, gas-tight syringes

Materials

- Zinc dust (FW 65.4) −325 mesh, 0.98 g, *c.* 15 mmol **flammable solid**
- LiCl (FW 42.4) 420 mg, 10 mmol **irritant, hygroscopic**
- CuCN (FW 89.6) 450 mg, 5 mmol **highly toxic**
- 1,2-Dibromoethane (FW 187.8) *c.* 200 mg, 1 mmol **toxic, cancer suspect agent**
- Me_3SiCl (FW 108.6) 0.1 mL, 0.8 mmol **flammable, corrosive**
- LiI (FW 133.8) 130 mg, 1 mmol
- *n*-Decane (internal GLC standard) 20 mg
- Dry THF **flammable, irritant**
- Dry DMA **irritant**
- Ethyl propiolate (FW 98.1) 0.39 g, 4 mmol **flammable, lachrymator**
- Diethyl 3,4-methylenedioxybenzenephosphate (FW 288.2) 1.44 g, 5 mmol

1. Weigh the zinc dust into the three-necked flask, flush with argon, and add the 1,2-dibromoethane in dry DMA (2 mL, distilled over CaH_2). Activate the zinc by heating as described in Protocol 1.

2. Add Me_3SiCl and stir for 5 min at 25°C. Add the diethyl 3,4-methylenedioxybenzenephosphate, *c.* 20 mg *n*-decane, and LiI in DMA (4 mL).
3. Stir for 12 h at 50°C and check for the completion of the reaction by GLC analysis of a hydrolysed reaction aliquot (see Protocol 1).
4. Add dry THF (8 mL), stop the stirring, and allow the excess zinc to settle for 1–2 h.
5. Dry the lithium chloride as described in Protocol 1, add CuCN, and flush with argon. Add dry THF (5 mL) and cool to −50°C.
6. Add the solution of the benzylic zinc reagent and allow to warm to −10°C. After 5 min, cool back to −70°C and add the ethyl propiolate.
7. Warm to 0°C and work-up after 5 min at this temperature (see Protocol 1).
8. Purify the crude residue obtained after evaporation of the solvent by flash chromatography (hexane–Et_2O, 19:1). The product obtained is the pure (*E*)-isomer (0.79 g, 84%) as a clear oil. The product is characterized by ^1H NMR, ^{13}C NMR, IR spectroscopy, and high resolution mass spectrometry.

Protocol 6.
Alkenylzinc and alkenylcopper organometallics by lithium–iodine exchange.[9] Preparation of (*E,E*)-ethyl 8-pivaloyloxy-2,4-octadienoate

Caution! Carry out all procedures in a well-ventilated hood, and wear disposable vinyl or latex gloves and chemical-resistant safety goggles.

1) BuLi
THF–Et_2O–pentane (4:1:1)
-100°C, 3 min
2) ZnI_2, THF,–90°C
3) CuCN•2LiCl, THF–Me_2S (1:1)
–90 to –60°C

H–C≡C–CO_2Et
–70 to –20°C, 5h

68%, 100% (*E,E*)

Equipment

- Two three-necked, round-bottomed flasks (100 mL), each equipped with a gas inlet and a magnetic stirring bar
- A dry, three-necked, round-bottomed flask (100 mL) equipped with a magnetic stirring bar, an argon inlet, an addition funnel with a small septum, and a low temperature thermometer
- Argon gas supply and inlet
- Dry, gas-tight syringes

Protocol 6. *Continued*

Materials

- ZnI_2 (FW 319.4) 0.83 g, 2.60 mmol
- LiCl (FW 42.4) 0.22 g, 5.20 mmol — **irritant, hygroscopic**
- CuCN (FW 89.6) 0.23 g, 2.60 mmol — **highly toxic**
- Dry THF — **flammable, irritant**
- Me_2S 5 mL — **flammable, stench**
- (*E*)-5-Iodo-4-pentenyl 2,2-dimethylpropanoate (FW 296.1) 0.74 g, 2.50 mmol — **harmful**
- *n*-Butyllithium (FW 64.1) 1.5 M in hexane, 1.73 mL, 2.60 mmol — **flammable, moisture sensitive**
- Ethyl propiolate (FW 98.1) 0.20 g, 2.00 mmol — **flammable, lachrymator**

1. In a three-necked flask, dry the zinc iodide under 0.1 mm Hg at 140 °C for 1 h. After this, flush with argon and dissolve the ZnI_2 in THF (5 mL).
2. In the second flask, dry the LiCl in the same way (see Protocol 1). Add the CuCN and flush again with argon. Dissolve these salts in a mixture of dimethyl sulfide and THF (5 mL each).
3. Charge the third flask (equipped with an addition funnel) with the vinylic iodide dissolved in THF–ether–pentane (8:2:2; 12 mL) and cool to −100 °C, being careful not to freeze the solution.
4. Add the butyllithium solution dropwise, taking care not to let the reaction temperature exceed −90 °C. Stir the resulting yellow solution for 3 min at −100 °C.
5. Add the ZnI_2 solution, keeping the temperature below −95 °C. Stir at −100 °C for 5 min.
6. Add the CuCN·LiCl solution dropwise and warm to −60 °C, then stir for 5 min.
7. Cool the reaction mixture to −78 °C, and add the ethyl propiolate as a solution in THF (3 mL) dropwise. Warm to −20 °C and stir until the reaction is complete (*c.* 5 h).
8. Perform the work-up in ether with saturated aqueous ammonium chloride (Protocol 1). The product (0.37 g, 68%) obtained is the pure (*E,E*)-stereoisomer after flash column chromatography (15% ether in hexane). The product is characterized by ^{1}H NMR, ^{13}C NMR, IR spectroscopy, and high resolution mass spectrometry.

Protocol 7.
Methylene homologation of a copper acetylide and the *in situ* trapping of the intermediate propargylic organometallic with an aldehyde.[21] Preparation of 4-(1-methoxycyclohexyl)-1-phenyl-3-butyn-1-ol

Caution! Carry out all procedures in a well-ventilated hood, and wear disposable vinyl or latex gloves and chemical-resistant safety goggles.

OMe, H; 1) BuLi, THF, –78 to 0°C; 2) $CuCN \cdot 2LiCl$, 0°C, 5 min; 3) PhCHO, –60°C; 4) ICH_2ZnI, –60 to 0°C, 0.5 h; OMe, OH, Ph; 98 %

Equipment

- One three-necked, round-bottomed flask (50 mL) equipped with a gas inlet, a low temperature thermometer, and a septum cap
- Two three-necked, round-bottomed flasks (50 mL), each equipped with a gas inlet, a thermometer, and an addition funnel with a septum cap
- Argon gas supply and inlet
- Dry, gas-tight syringes

Materials

• Zinc foil (FW 65.4) m3N purity, 1.67 g, *c.* 25 mmol	**flammable solid**
• 1,2-Dibromoethane (FW 187.8) 94 mg, 0.5 mmol	**toxic, cancer suspect agent**
• Me_3SiCl (FW 108.6) 0.05 mL	**flammable, corrosive**
• Dry THF	**flammable, irritant**
• *n*-BuLi (FW 64.1) 1.6 M in hexane, 3.1 mL, 5.0 mmol	**flammable, moisture sensitive**
• 1-Ethynyl-1-methoxycyclohexane (FW 138.2) 690 mg, 5.0 mmol	**harmful**
• Diiodomethane (FW 268.0) 6.7 g, 25 mmol	**corrosive, light sensitive**
• Benzaldehyde (FW 106.1) 320 mg, 3.0 mmol	**toxic, irritant**

1. Cut the zinc foil 0.62 mm thick) into 2 × 2 mm pieces and add them to a dry, three-necked flask. Flush with argon, and add the dibromoethane and dry THF (2 mL). The use of zinc foil is essential since zinc dust catalyses the decomposition of (iodomethyl)zinc iodide.
2. Activate the zinc foil by the same method as described in Protocol 1 for the zinc dust.
3. After 5 min at 25°C, add the THF solution (7.5 mL) of diiodomethane slowly (*c.* 15 min), while keeping the temperature below 28°C.
4. Keep the reaction mixture between 25 and 26°C for 4 h. The use of a cooling bath may be necessary. After this time, almost no zinc is left and the yield is estimated to be *c.* 85% (GLC analysis of an iodolysed reaction aliquot; see Protocol 2).

Protocol 7. *Continued*

5. During this time, place the 1-ethynyl-1-methoxycyclohexane and THF (5 mL) into the three-necked flask bearing a low temperature thermometer and cool to −75°C.
6. Add slowly the butyllithium solution in hexane and warm up to 0°C.
7. Cool back to −78°C and add a THF solution of CuCN·2LiCl prepared as described in Protocol 1.
8. Warm the reaction mixture to 0°C for 5 min and then cool back to −60°C.
9. Add the benzaldehyde and the THF solution of (iodomethyl)zinc iodide via syringe.
10. Allow the reaction mixture to warm to 0°C. **Caution!** An exothermic reaction occurs at about −40°C; use a dry ice–acetone bath to keep the temperature below 0°C.
11. Stir the reaction mixture for 0.5 h at 0°C and work up as described in Protocol 1.
12. Purify the crude residue obtained after evaporation of the solvent by flash chromatography (EtOAc–hexane, 1:10). The desired product is obtained as a clear oil (0.76 g, 98%), which is characterized by ^{1}H NMR, ^{13}C NMR, IR spectroscopy, and high resolution mass spectrometry.

Protocol 8.
Stereoselective one-pot preparation of α-methylene-γ-butyrolactones mediated by (iodomethyl)zinc iodide.[22] Preparation of 4-(3-cyanopropyl)-4,5-dihydro-3-methylene-5-phenyl-2-(3*H*)-furanone

Caution! Carry out all procedures in a well-ventilated hood, and wear disposable vinyl or latex gloves and chemical-resistant safety goggles.

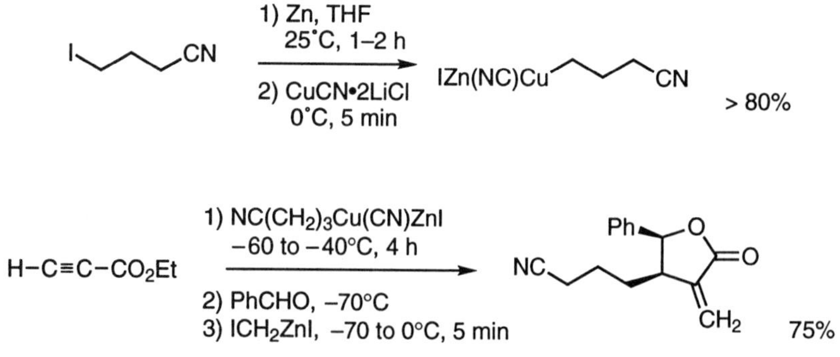

I(CH₂)₃CN → [1) Zn, THF, 25°C, 1–2 h; 2) CuCN•2LiCl, 0°C, 5 min] → IZn(NC)Cu(CH₂)₃CN > 80%

$H-C\equiv C-CO_2Et$ → [1) $NC(CH_2)_3Cu(CN)ZnI$, −60 to −40°C, 4 h; 2) PhCHO, −70°C; 3) ICH_2ZnI, −70 to 0°C, 5 min] → NC(CH₂)₃–lactone (Ph, O, =O, =CH₂) 75%

Equipment

- Two three-necked, round-bottomed flasks (50 mL), each equipped with a gas inlet, a low temperature thermometer, and a septum cap
- One three-necked, round-bottomed flask (50 mL) equipped with a gas inlet, a thermometer, and an addition funnel with a septum cap
- Argon gas supply and inlet
- Dry, gas-tight syringes

Materials

• Zinc dust (FW 65.4) −325 mesh 1.3 g, 20 mmol	**flammable solid**
• Zinc foil (FW 65.4) m3N purity, 0.62 mm thick, 0.98 g, 15 mmol	**flammable solid**
• LiCl (FW 42.4) 0.6 g, 14 mmol	**irritant, hygroscopic**
• CuCN (FW 89.6) 635 mg, 7.1 mmol	**highly toxic**
• 1,2-Dibromoethane (FW 187.8) 2 × 200 mg, 2 × 1 mmol	**toxic, cancer suspect agent**
• Me_3SiCl (FW 108.6) 2 × 0.1 mL	**flammable, corrosive**
• Dry THF	**flammable, irritant**
• *n*-Decane (internal GLC standard) 20 mg	
• Ethyl propiolate (FW 98.1) 588 mg, 6 mmol	**flammable, lachrymator**
• 4-Iodobutyronitrile (FW 195.0) 1.38 g, 7.1 mmol	**harmful**
• Benzaldehyde (FW 106.1) 0.53 g, 5.0 mmol	**toxic, irritant**
• Diiodomethane (FW 268.0) 3.75 g, 14 mmol	**corrosive, light sensitive**

1. Prepare a THF solution (8 mL) of ICH_2ZnI as described in Protocol 7.
2. Weigh zinc dust into a three-necked, round-bottomed flask and flush the flask with argon.
3. Add the 1,2-dibromoethane in dry THF (2 mL) and activate the zinc as described in Protocol 1. Add Me_3SiCl and stir the reaction mixture for *c.* 5 min at 25 C.
4. Add 4-iodobutyronitrile and *n*-decane in dry THF (4 mL). The temperature rises to 40–50°C. Stir the reaction mixture at 40°C for 1.5 h.
5. Check for the completion of the reaction by GLC analysis of a hydrolysed reaction aliquot (see Protocol 1).
6. Add dry THF (5 mL), stop the stirring, and allow the excess zinc dust to settle for 1.5–2 h.
7. During this time, dry the LiCl and prepare a THF solution (10 mL) of CuCN·2LiCl as described in Protocol 1.
8. Cool the flask containing CuCN·2LiCl to −60°C and add the THF solution of 3-cyanopropylzinc iodide. Warm up to 0°C, stir for 5 min at this temperature, and cool back to −60°C.
9. Add ethyl propiolate and stir for 4 h between −60 and −40°C. Check by GLC analysis for the completion of the carbocupration reaction.
10. Add at −70°C a THF solution (2 mL) of benzaldehyde to the vinylic copper solution.
11. Then add dropwise the THF solution of ICH_2ZnI and allow the reaction mixture to warm up to 0°C.

Protocol 8. *Continued*

12. Stir for 0.5 h at this temperature and work up according to the procedure described in Protocol 1.
13. Purify the crude oil obtained after evaporation of the solvent by flash chromatography (hexane–EtOAc, 7:3). A clear oil of the pure α-methylene-γ-butyrolactone is obtained [0.90 g, 75%; (*Z*): (*E*) ratio 9:19 as determined by GLC and ^{1}H NMR analyses], which is characterized by ^{1}H NMR, ^{13}C NMR, IR spectroscopy, and high resolution mass spectrometry.

Protocol 9. Enantioselective addition of a functionalized dialkylzinc to an aldehyde.[6] Preparation of (*S*)-(−)-5-hydroxy-5-phenypentyl acetate

Caution! Carry out all procedures in a well-ventilated hood, and wear disposable vinyl or latex gloves and chemical-resistant safety goggles.

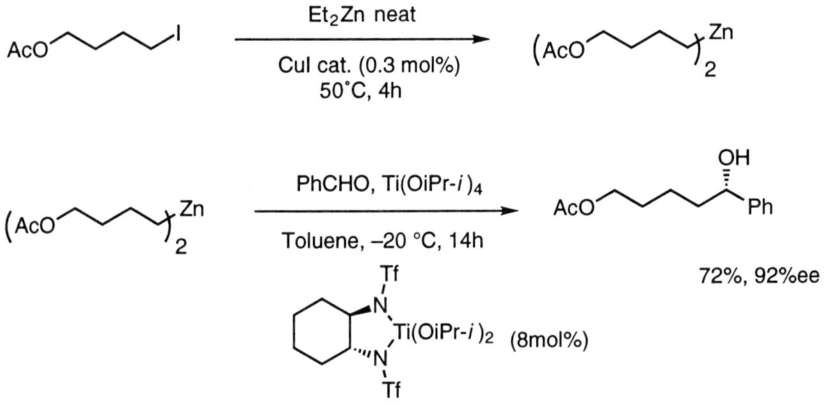

AcO(CH₂)₄I → (Et_2Zn neat, CuI cat. (0.3 mol%), 50°C, 4h) → $(AcO(CH_2)_4)_2Zn$

$(AcO(CH_2)_4)_2Zn$ → (PhCHO, Ti(OiPr-*i*)$_4$, Toluene, –20 °C, 14h; bis(Tf-amido)cyclohexane–Ti(OiPr-*i*)$_2$ (8mol%)) → AcO(CH₂)₄CH(OH)Ph

72%, 92%ee

Equipment

- A Schlenk flask (*c.* 12 cm long, 50 mL) equipped with a magnetic stirring bar and a septum cap (or a glass stopper)
- One three-necked, round-bottomed flask (150 mL) equipped with a gas inlet, a low temperature thermometer, and a septum cap
- Argon gas supply and inlet
- Dry, gas-tight syringes

Materials

- Et_2Zn (FW 123.4) neat, 1.2 mL, 12 mmol — **pyrophoric, moisture sensitive**
- CuI (FW 190.4) *c.* 5 mg, 0.024 mmol — **irritant**
- 4-Iodobutyl acetate (FW 242.1) 1.94 g, 8 mmol — **harmful**
- *n*-Decane (internal GLC standard)
- Benzaldehyde (FW 106.1) 210 mg, 2.0 mmol — **toxic, irritant**
- *Trans*-(1*R*,2*R*)-bis(trifluoromethanesulfonamido)cyclohexane[6] (FW 378.3) 63 mg, 0.16 mmol
- Ti(O-Pr-*i*)$_4$ (FW 284.3) 1.2 mL, 4.0 mmol — **flammable, moisture sensitive**
- Dry toluene — **flammable, toxic**

1. Place the copper iodide into the Schlenk flask and flush several times with argon.
2. Add 4-iodobutyl acetate, *n*-decane, and Et_2Zn (**caution**! highly pyrophoric!).
3. Stir for 4 h at 50°C. Check for the completion of the reaction by taking a reaction aliquot and comparing the current RI:decane ratio with the original RI:decane ratio.
4. Connect the gas inlet of the Schlenk flask to the vacuum pump and collect the excess Et_2Zn and EtI formed in a trap cooled with liquid N_2. Heat the Schlenk flask with an oil bath (50°C) during this operation (1 h). Avoid overheating and immediately quench the diethylzinc in the trap with acetone as soon as it is disconnected (Et_2Zn is pyrophoric, but Et_2Zn–acetone does not burn directly in air).
5. Dissolve the bis(4-acetoxybutyl)zinc in dry toluene (4 mL).
6. In a second flask, add under argon *trans*-(1*R*,2*R*)-bis(trifluoromethanesulfonamido)cyclohexane and $Ti(OPr\text{-}i)_4$ and heat the mixture to 40°C for 0.5 h. Cool to −60°C.
7. Add the toluene solution of bis(4-acetoxybutyl)zinc, followed by benzaldehyde, and stir the reaction mixture for 14 h at −20°C.
8. Check the completion of the reaction by GLC and work-up the reaction mixture as described in Protocol 1, substituting aqueous NH_4Cl with 10% aqueous HCl.
9. Purify the crude oil by flash chromatography using EtOAc–hexane (1:5). A clear oil (0.32 g, 72%) of (*S*)-(−)-5-hydroxy-5-phenylpentyl acetate is obtained, which is characterized by ^1H NMR, ^{13}C NMR, IR spectroscopy, and high resolution mass spectrometry.
10. Measure the $[\alpha]^{25}_D$ ($[\alpha]^{25}_D = -24.56°$; $c = 3.29$, benzene).
11. Make a derivative with (*S*)-(+)-*O*-acetylmandelic acid[24] and determine the enantiomeric excess using the ^1H NMR integrations of the triplets at 4.02 and 3.93 p.p.m. (96:4 ratio, 92% ee).

References

1. Knochel, P.; Yeh, M. C. P.; Berk, S. C.; Talbert, J. *J. Org. Chem.* **1988**, *53*, 2390–2392.
2. For reviews see: Knochel, P.; Rozema, M. J.; Tucker, C. E.; Retherford, C.; Furlong, M.; AchyuthaRao, S. *Pure Appl. Chem.* **1992**, *64*, 361–369. Knochel, P.; Singer, R. D. *Chem. Rev.* **1993**, *93*, 2117–2188.
3. Gaudemar, M. *Bull. Soc. Chim. Fr.* **1962**, 974–987.
4. (a) Berk, S. C.; Knochel, P.; Yeh, M. C. P. *J. Org. Chem.* **1988**, *53*, 5789–5791. (b) Berk, S. C.; Yeh, M. C. P.; Jeong, N.; Knochel, P. *Organometallics* **1990**, *9*, 3053–3063.
5. Jubert, C.; Knochel, P. *J. Org. Chem.* **1992**, *57*, 5425–5431.

6. Rozema, M. J.; AchyuthaRao, S.; Knochel, P. *J. Org. Chem.* **1992**, *57*, 1956–1958.
7. Majid, T. N.; Knochel, P. *Tetrahedron Lett.* **1990**, *31*, 4413–4416.
8. JanakiramRao, C.; Knochel, P. *J. Org. Chem.* **1991**, *56*, 4593–4596.
9. (a) Tucker, C. E.; Majid, T. N.; Knochel, P. *J. Am. Chem. Soc.* **1992**, *114*, 3983–3985. (b) Venegas, P.; Cahiez, G.; Tucker, C. E.; Majid, T. N.; Knochel, P. *J. Chem. Soc., Chem. Commun.* **1992**, 1406–1408.
10. (a) Yeh, M. C. P.; Knochel, P. *Tetrahedron Lett.* **1988**, *29*, 2395–2396. (b) Majid, T. N.; Yeh, M. C. P.; Knochel, P. *Tetrahedron Lett.* **1989**, *30*, 5069–5072.
11. (a) Retherford, C.; Chou, T.-S.; Schelkun, R. M.; Knochel, P. *Tetrahedron Lett.* **1990**, *31*, 1833–1836. (b) Retherford, C.; Knochel, P. *Tetrahedron Lett.* **1991**, *32*, 441–444. (c) Jubert, C.; Knochel, *P. J. Org. Chem.* **1992**, *57*, 5431–5438.
12. (a) Knoess, H. P.; Furlong, M. T.; Rozema, M. J.; Knochel, P. *J. Org. Chem.* **1991**, *56*, 5974–5978. (b) Yeh, M. C. P.; Knochel, P.; Butler, W. M.; Berk, S. C. *Tetrahedron Lett.* **1988**, *29*, 6693–6696.
13. Yeh, M. C. P.; Knochel, P. *Tetrahedron Lett.* **1989**, *30*, 4799–4802.
14. (a) Yeh, M. C. P.; Knochel, P.; Santa, L. E. *Tetrahedron Lett.* **1988**, *29*, 3887–3890. (b) Yeh, M. C. P.; Chen, H. G.;Knochel, P. *Org. Synth.* **1991**, *70*, 195–203.
15. AchyuthaRao, S.; Knochel, P. *J. Am. Chem. Soc.* **1991**, *113*, 5735–5741.
16. Tucker, C. E.; AchyuthaRao, S.; Knochel, P. *J. Org. Chem.* **1990**, *55*, 5446–5448.
17. Chen, H. G.; Gage, J. L.; Barrett, S. D.; Knochel, P. *Tetrahedron Lett.* **1990**, *31*, 1829–1832.
18. Knochel, P.; Jeong, N.; Rozema, M. J.; Yeh, M. C. P. *J. Am. Chem. Soc.* **1989**, *111*, 6474–6476.
19. Knochel, P.; Chou, T.-S.; Chen, H.-G.; Yeh, M. C. P.; Rozema, M. J. *J. Org. Chem.* **1989**, *54*, 5202–5204.
20. Knochel, P.; AchyuthaRao, S. *J. Am. Chem. Soc.* **1990**, *112*, 6146–6148.
21. Rozema, M. J.; Knochel, P. *Tetrahedron Lett.* **1991**, *32*, 1855–1858.
22. AchyuthaRao, S.; Knochel, P. *J. Am. Chem. Soc.* **1992**, *114*, 7579–7581.
23. (a) Jubert, C.; Nowotny, S.; Kornemann, D.; Antes, I.; Tucker, C. E.; Knochel, P. *J. Org. Chem.* **1992**, *57*, 6384–6386. (b) Waas, J. R.; Sidduri, A.; Knochel, P. *Tetrahedron Lett.* **1992**, *33*, 3717–3720. (c) Sidduri, A.; Budries, N.; Laine, R. M.; Knochel, P. *Tetrahedron Lett.* **1992**, *33*, 7515–7518. (d) Stemmler, T.; Penner-Hahn, J. E.; Knochel, P. *J. Am. Chem. Soc.* **1993**, *115*, 348–350. (e) Knochel, P.; Chou, T.-S.; Jubert, C.; Rajagopal, D. *J. Org. Chem.* **1993**, *58*, 588–599. (f) AchyuthaRao, S.; Rozema, M. J.; Knochel, P. *J. Org. Chem.* **1993**, *58*, 2694–2713. (g) Tucker, C. E.; Knochel, P. *Synthesis* **1993**, 530–536. (h) JanakiramRao, C.; Knochel, P. *Tetrahedron* **1993**, *49*, 29–48.
24. Parker, S. *J. Chem. Soc., Perkin Trans. 2* **1983**, 83–88.

5

Higher order cyanocuprates

BRUCE H. LIPSHUTZ

1. Introduction

Higher order (HO) cyanocuprates,[1] stoichiometrically represented as '$R_2Cu(CN)Li_2$', are distinctly different from the analogous lower order (LO) or Gilman cuprates R_2CuLi[2] by virtue of two distinguishing features: (1) they are believed to be copper(I) dianionic species, implying that three ligands are bound to a copper(I) centre; and (2) they contain a cyano group, which presumably imparts the required π-acidity to the lower order cyanocuprate RCu(CN)Li such that another equivalent of an organolithium can be accepted by the copper centre to afford $[R_2Cu(CN)^{2-}2Li^+]$. Although there are reports on non-cyano-containing higher order cuprates, written according to the general formula 'R_3CuLi_2'[3] relatively few synthetic uses of these have appeared to date.[4] On the other hand, higher order cyanocuprates continue to evolve,[1b,5] displaying unique properties including, as examples: (1) their ability to effect transmetallations with other organometallic species (see later); and (2) their ability to form kinetic reagents at low temperatures which lead via their oxidation to unsymmetrical biaryls with excellent selectivities.[6]

2. Reactions of higher order cyanocuprates '$R_2Cu(CN)Li_2$'

The majority of uses of HO cyanocuprates involve rather 'traditional' sorts of cuprate chemistry. That is, for cases requiring transfer of simpler, usually commercially available ligands (e.g. Me, Bu, vinyl, Ph, etc.), the standard Gilman recipe of 2RLi:CuX (with X = CN) is normally followed for reagent preparation (eqn 1 and Protocol 1).[2] Ethereal solvents (usually THF or Et_2O) are used, although cuprates **1** do not have similar appearances in each of these media. Indeed, while homogeneous in THF at most working temperatures and concentrations, even a simple example such as n-$Bu_2Cu(CN)Li_2$ is a cloudy, murky, two-phase-looking mixture in Et_2O when cooled to −78 °C. Chemistry in this medium, however, proceeds as expected. Since CuCN can be obtained in several different forms and, hence, different colours, the reagent solution resulting from each form will have a different appearance.

Which precursor salt is chosen, however, is of no consequence to the subsequent cuprate coupling. Moreover, if white CuCN is not used, these mixtures may look dark. It is good practice, however, *not* to be influenced by colours as to whether a reagent has been formed unless considerable experience with a particular species has already been gained. Once a reaction with a HO cyanocuprate has been carried out, a standard quench normally will involve adding a preformed solution consisting of 10% concentrated NH_4OH and 90% saturated NH_4Cl (v/v), which maintains the pH around 9–10. *Clearly, acidic work-ups are to be avoided given the presence of cyanide ions in solution.*

$$2RLi + CuCN \xrightarrow[Et_2O]{THF\ or} R_2Cu(CN)Li_2\ \mathbf{1} \quad (1)$$

The existence of the cyanide ligand in the cuprate cluster, whether or not formally on the copper atom,[7] can have a dramatic effect on the reactivity pattern of cuprates, the implications of which can be significant in synthetic situations. Thus, aside from large discrepancies in cost between copper(I) precursors CuCN and CuX (X = I, Br), the fact that, for example, distinct stereochemical outcomes between reactions of HO cuprates and those of the corresponding Gilman reagents have been observed provides a welcome element of flexibility in total synthesis undertakings. One case in point concerns some model studies aimed at a synthesis of aplysiatoxin **2**,[8] where spiroenone **3** was envisioned as a Michael acceptor towards Me_2CuLi. Although the 1,4-addition does occur, low selectivity was noted as a mix of diastereomers was obtained. Treatment of **3** with $Me_2Cu(CN)Li_2$, however, proceeded to give product **4** not only in quantitative yield, but with 100% diastereoselectivity (eqn 2).

The greater reactivity of HO cyanocuprates relative to that of Gilman reagents is especially noticeable in substitution reactions of alkyl halides[9,10]

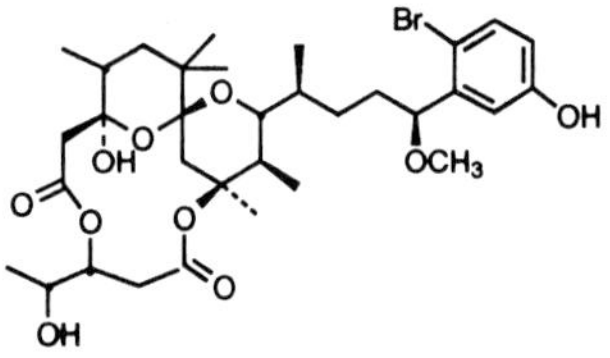

Br OH OH OCH3 O O O O O OH

2, aplysiatoxin

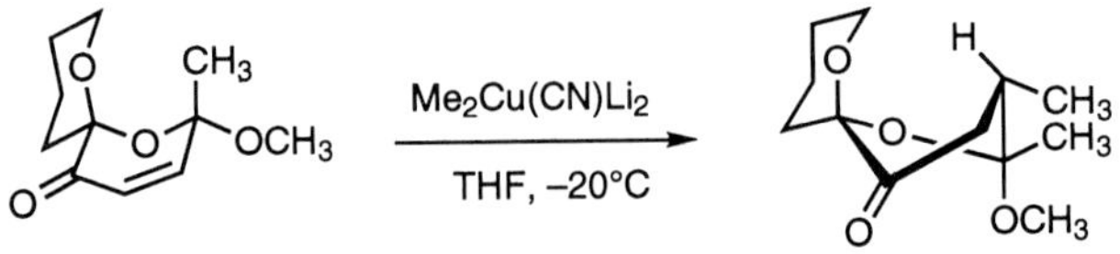

O CH3 O OCH3 O — $Me_2Cu(CN)Li_2$ / THF, –20°C → O H O CH3 CH3 O OCH3 (2)

3   **4** (100%)

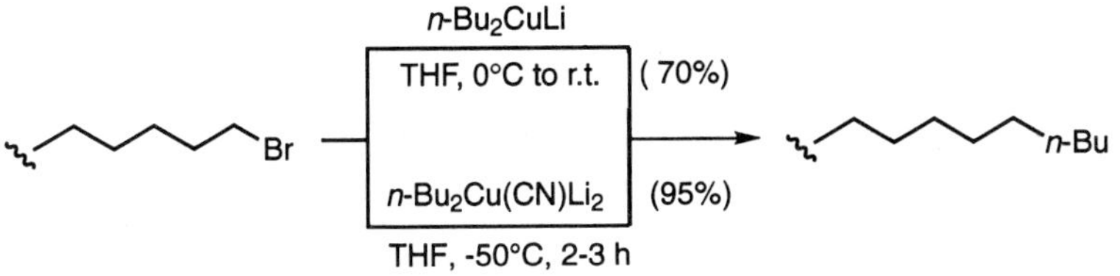

Scheme 5.1

and epoxides.[10,11] Thus, while even simple alkyl bromides (e.g. 1-bromopentane), upon treatment with n-Bu$_2$CuLi, are reported to take several hours between 0°C and room temperature,[12] the HO species n-Bu$_2$Cu(CN)Li$_2$ effects the same displacement at −50°C in far less time (Scheme 5.1; see also Protocol 2).[13]

Sulfonate leaving groups are also susceptible to cuprate attack, and while in simpler systems R$_2$CuLi is likely to lead to useful couplings,[14] more highly functionalized (especially oxygenated) substrates may benefit from the use of a HO cyanocuprate. For example, the C-6 tosylate of D-glucal **5**, upon treatment with n-Bu$_2$CuLi, afforded product **6** in low yield, accompanied by substantial amounts of by-products. Acceptable results were realized, however, by switching to n-Bu$_2$Cu(CN)Li$_2$ (eqn. 3).[15]

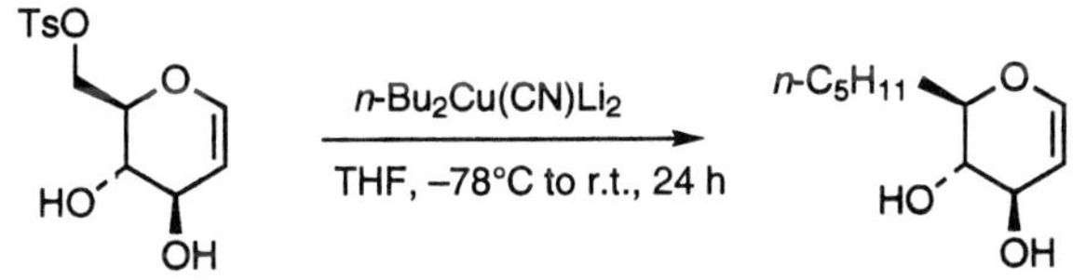

(3)

Differences between cuprates towards ring-opening reactions with epoxides can also be dramatic.[10,11] For example, the unactivated, trisubstituted cyclopentene oxide **7** is not particularly responsive to a LO cyanocuprate **8**, even in the less Lewis basic solvent Et$_2$O. However, the presence of an 'extra' equivalent of RLi, i.e. so as to form the HO cyanocuprate **9**, allows for a clean conversion to product **10** with the expected regio- (least-substituted carbon) and stereochemical (inversion at carbon) biases (Scheme 5.2; see also Protocols 3 and 6).

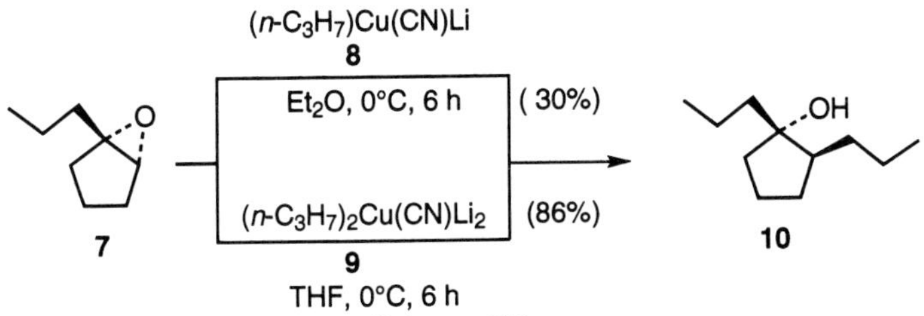

Scheme 5.2

Protocol 1.
Conjugate addition of a higher order divinylcyanocuprate to isophorone[a]

Caution! Carry out all procedures in a well-ventilated hood, and wear disposable vinyl or latex gloves and chemical-resistant safety goggles.

(vinyl)$_2$Cu(CN)Li_2 **12**; Et_2O, THF, – 50°C, 3.5 h

11 → **13** (88%)

Equipment

- One round-bottomed flask (10 mL) containing a side-arm, septum, and stirrer bar
- Argon supply and inlet
- Dry ice–acetone cooling bath
- Dry ice–cyclohexanone cooling bath
- Two glass syringes (2 mL)
- One glass syringe (100 μL)
- Three-way stopcock argon inlet/vacuum adapter
- One disposable plastic syringe (5 mL)

Materials

- CuCN[b] (FW 89.6) 0.087 g, 0.98 mmol — **highly toxic, irritant**
- Vinyllithium[c] in THF (FW 33.9) 2.07 M, 0.94 mL, 1.95 mmol — **flammable, moisture sensitive**
- Dry, distilled Et_2O — **flammable, irritant**
- Isophorone **11** (FW 138.2) 0.075 mL, 69 mg, 0.50 mmol — **cancer suspect agent, lachrymator**

1. Place a septum into the side-arm of the 10 mL flask, tighten with a ring of copper wire, and wrap the septum–flask interface with Parafilm®. Gently flame dry the flask containing the CuCN under vacuum, and allow it to cool under a positive pressure of argon.
2. Add Et_2O (1 mL) using a 2 mL syringe through the septum; stir to form a slurry.
3. Cool the mixture to −78°C and add the vinyllithium[c] using the second 2 mL syringe, dropwise over 1 min.
4. Remove the cooling bath and allow the mixture to warm until all of the CuCN has dissolved (*c.* −20°C), thereby forming cuprate **12**.
5. Recool the solution to −78°C and add the isophorone **11** via the 100 μL syringe dropwise.
6. After addition is complete, warm the solution to *c.* −50°C (dry ice–cyclohexanone bath) and stir at this temperature for 3.5 h.
7. Quench the reaction mixture by addition of 4 mL of an aqueous 10% concentrated NH_4OH–90% saturated NH_4Cl solution using the 5 mL plastic syringe

and allow the mixture to warm to ambient temperature with vigorous stirring until all solids have dissolved.

8. Pour the contents of the flask into a separatory funnel and extract with Et_2O (3 × 10 mL). Wash the combined extracts with H_2O and then brine. Dry them over anhydrous Na_2SO_4, then filter and rotary evaporate the solvents *in vacuo*.
9. Chromatograph the oil obtained on SiO_2 (ICN Biomedicals, 32–63, 60 Å) using 15% Et_2O–85% pentane[d] to obtain 68.3 mg (88%) of product **13**, and characterize the product by standard spectroscopic means as well as by high resolution mass spectral analysis.

[a] Based on Lipshutz, B. H.; Wilhelm, R. S.; Kozlowski, J. *Tetrahedron Lett.* **1982**, *23*, 3755–3758.
[b] *General comments on Protocols 1–6.* CuCN (from Aldrich) was stored in an Abderhalden over solid KOH at 56°C (refluxing acetone), although it is not hygroscopic. No formal purification is needed. Note that if obtained from other vendors, e.g. Fluka or Mallinckrodt, it is likely to appear as a green or tan powder, respectively, which are other forms of this salt. Nonetheless, it can be handled and used in exactly the same manner as described for the white form. Dry all syringes in an Abderhalden under vacuum (1 torr) at 56°C overnight, or alternatively store in an oven at 120°C overnight prior to use. Methyl-, *n*-butyl-, and vinyllithium were titrated using 1,10-phenanthroline as indicator.[e]
[c] Vinyllithium in THF (*c.* 2 M) can be purchased from Organometallics Inc., PO Box 287, East Hampstead, NH, USA.
[d] R_f Et_2O–pentane, 15:85 = 0.32.
[e] Watson, S. C.; Eastham, J. F. *J. Organomet. Chem.* **1967**, *9*, 165 (see Chapter 2, Protocol 6).

Protocol 2.
Displacement of a primary bromide by a higher order cyanocuprate[13]

Caution! Carry out all procedures in a well-ventilated hood, and wear disposable vinyl or latex gloves and chemical-resistant safety goggles.

14 —(15, $(\text{2-pyridyl-CH}_2\text{CH}_2)_2Cu(CN)Li_2$; THF, −50°C, 3 h)→ **16** (93%)

Equipment

- One round-bottomed flask (10 mL) equipped with a septum and a magnetic stirrer bar
- Argon supply
- Two dry ice–acetone cooling baths
- One dry ice–cyclohexanone cooling bath
- One round-bottomed flask (25 mL) with side-arm, septum, and stirrer bar
- Three-way stopcock argon inlet/vacuum adapter
- One stainless steel cannula (18 gauge, 6–12 in long)
- One glass syringe (2 mL)
- One glass syringe (0.5 mL)
- One glass syringe (1.0 mL)
- One glass syringe (100 mL)
- One plastic syringe (10 mL)

Protocol 2. *Continued*

Materials

- 2-Picoline (FW 93.0) 0.098 mL, 93 mg, 1.0 mmol — **flammable, toxic**
- *n*-BuLi in hexanes (FW 64.1) 2.33 M, 0.43 mL, 1.0 mmol — **flammable, moisture sensitive**
- CuCN (FW 89.6) 0.045 g, 0.50 mmol — **highly toxic, irritant**
- Dry, distilled THF — **flammable, irritant**
- (*S*)-(+)-Citronellyl bromide[a] **14** (FW 218.9) 0.074 mL, 76.71 mg, 0.35 mmol — **irritant**

1. Gently flame dry the 10 mL flask which is capped with a septum through which there is a needle connected to an argon line, then cool to room temperature.
2. Add THF (1.0 mL) using the 2 mL syringe, and then cool the flask to 0°C in an ice–water bath.
3. Add the 2-picoline via the 0.5 mL syringe, followed by the dropwise addition of *n*-BuLi using the 1.0 mL syringe. Stir at 0°C for 30 min to produce a red solution of the anion.
4. Place CuCN in the 25 mL flask mounted with the three-way stopcock and gently flame dry under a positive argon flow, then allow to cool to ambient temperature.
5. Add 0.5 mL THF using the 1.0 mL syringe and cool the stirred slurry to −78°C with a dry ice–acetone bath.
6. Transfer via the cannula the lithiated picoline, cooled to −78°C as well, to the CuCN–THF slurry, and then warm the mixture to 0°C to give a red-orange solution of cuprate **15**.
7. Recool to −50°C in the dry ice-cyclohexanone bath and add bromide **14** using the 100 μL syringe. Continue stirring at this temperature for 3 h.
8. Quench the reaction mixture using *c.* 5 mL of an aqueous solution made up of 10% concentrated NH_4OH and 90% saturated NH_4Cl using the 10 mL plastic syringe and allow the mixture to warm to ambient temperature with vigorous stirring until all solids have dissolved.
9. Chromatograph the resulting material on silica gel (ICN Biomedicals, 32–63, 60 Å) using pentane–Et_2O[b] (1:1) to afford 75.4 mg (93%) of a clear oil. The product **16** displays the appropriate spectral features and satisfactory HREIMS data.

[a] Purchased from Aldrich and distilled under vacuum prior to use (b.p. 111°C/12 mm Hg).
[b] R_f Et_2O–pentane, 1:1 = 0.65.

3. Reactions of mixed ligand cyanocuprates

In situations where a key carbon–carbon bond to be made calls for transfer of a non-trivial ligand, the 2RLi:CuCN ratio is no longer attractive, since only one of the two R groups on copper is 'active'.[16] To circumvent this issue, several lithiated, non-transferable ('dummy') ligands (R_r) have been developed,[1a] which combine with a valuable organolithium (R_tLi) and CuCN to form a mixed HO cyanocuprate $R_tR_rCu(CN)Li_2$ **17** (eqn 4). Representative non-transferable or 'dummy' ligands include species **18–22** (see also Chapter 2, Table 1.3).[17]

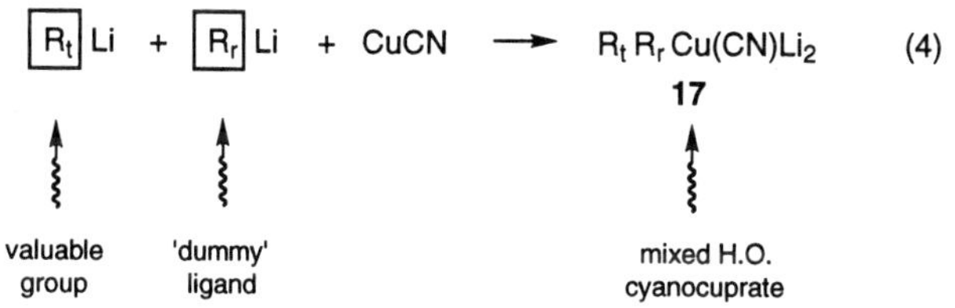

$$\boxed{R_t}Li + \boxed{R_r}Li + CuCN \longrightarrow R_tR_rCu(CN)Li_2 \quad (4)$$

17

valuable group — 'dummy' ligand — mixed H.O. cyanocuprate

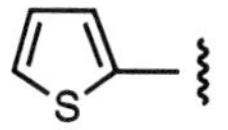

18  **19**  **20**

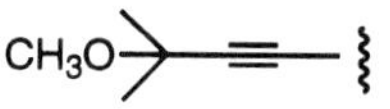

CH_3O **21**

$CH_3-S(=O)-CH_2-$ **22**

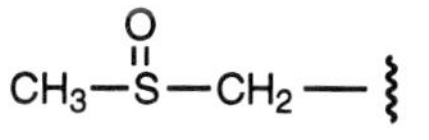

Implied in the use of mixed cuprates **17** is the requirement for two different organolithium precursors, which adds an element of experimental complexity to cuprate chemistry in general. In an effort to streamline this 1:1:1 formulation (*i.e.* eqn 4), it has been found that 2-lithiothiophene combines with CuCN to form a stable 'cuprate in a bottle', (2-Th)Cu(CN)Li **23** (2-Th = 2-thienyl),[18] which is a commercially available precursor to the mixed cuprate **24** (Scheme 5.3; see also Protocol 3).[19] One representative case in which the use of **24** was found advantageous is illustrated in eqn (5). Epoxide **25** could be transformed into the ring-opened acetoxy derivative **26** in good overall yield, providing a key intermediate *en route* to onnamide A, a marine natural product possessing antiviral and antitumour activity.[20]

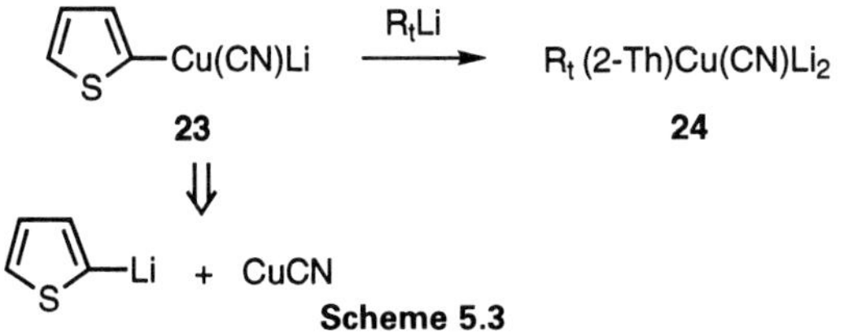

$$\text{(2-Th)}Cu(CN)Li \; \mathbf{23} \xrightarrow{R_tLi} R_t\text{(2-Th)}Cu(CN)Li_2 \; \mathbf{24}$$

$$\mathbf{23} \Downarrow \text{(2-Th)}Li + CuCN$$

Scheme 5.3

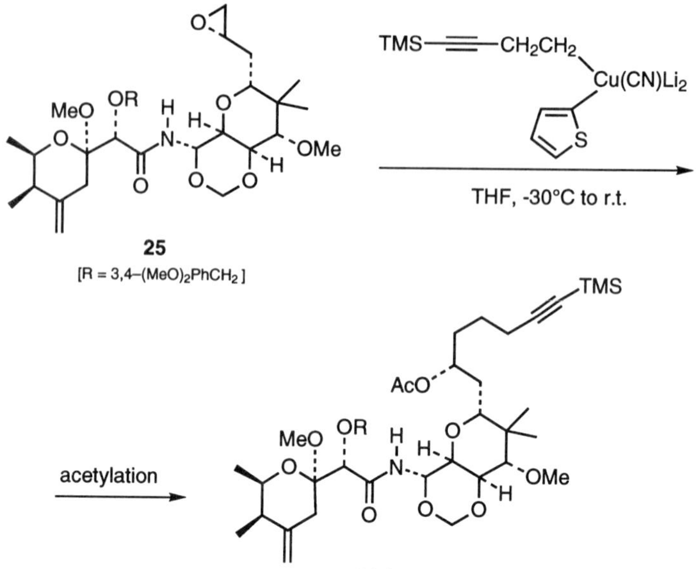

MeO OR H H O OMe H O O O

25

[R = 3,4–$(MeO)_2PhCH_2$]

TMS—≡—CH_2CH_2 $Cu(CN)Li_2$ S

THF, -30°C to r.t.

acetylation

TMS AcO MeO OR H H O OMe H O O O

26 (78%)

(5)

As with the 2-thienyl system, both *N*-lithiopyrrole and *N*-lithioimidazole form stable LO cyanocuprates **27** and **28**, respectively, which can be converted to their HO cuprates **29** and **30**, respectively, in the usual fashion, as shown below (Scheme 5.4; see also Protocol 4).[17b] Preliminary tests on both **27** and **28** have shown them to retain fully their integrity in THF solutions at 0.2 M when kept in a refrigerator (*c*. 4°C) for periods of over one week. They do not, however, appear to be quite as stable as (2-Th)Cu(CN)Li, which can be stored at ambient temperatures for over one week, thereby facilitating commercialization.

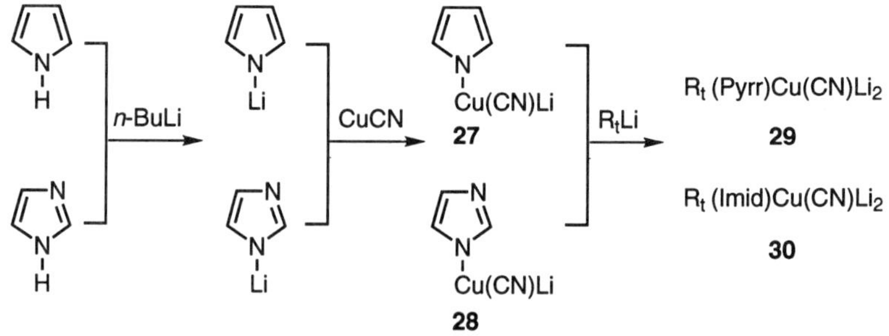

N H — *n*-BuLi → N Li — CuCN → N Cu(CN)Li **27** — R_tLi → R_t (Pyrr)Cu(CN)Li_2 **29**

N N H — N N Li — N N Cu(CN)Li **28** — R_t (Imid)Cu(CN)Li_2 **30**

Scheme 5.4

Protocol 3.
Opening of an epoxide by a mixed higher order vinylthienylcuprate[a]

Caution! Carry out all procedures in a well-ventilated hood, and wear disposable vinyl or latex gloves and chemical-resistant safety goggles.

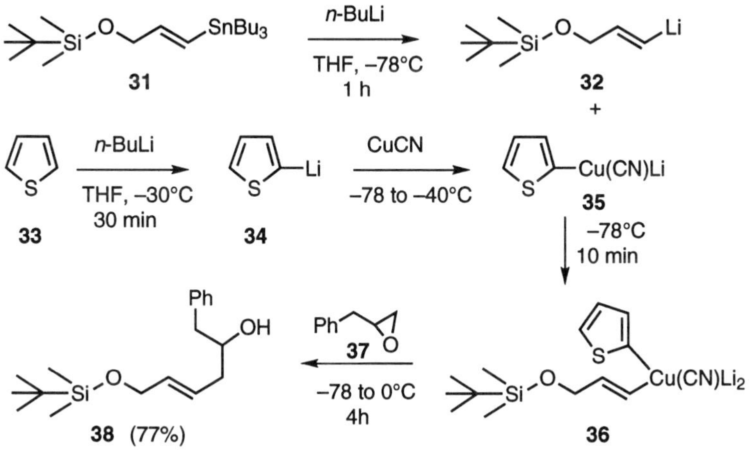

31 → (n-BuLi; THF, –78°C, 1 h) → 32

33 → (n-BuLi; THF, –30°C, 30 min) → 34 → (CuCN; –78 to –40°C) → 35

32 + 35 → (–78°C, 10 min) → 36 → (37; –78 to 0°C, 4h) → 38 (77%)

Equipment

- Four round-bottomed flasks (5 mL) each equipped with a septum and stirrer bar
- Argon supply and inlet
- Dry ice–acetone cooling bath
- Two syringes (100 mL)
- Two syringes (250 mL)
- Two syringes (0.5 mL)
- Three cannulae (18 gauge, 12 in long)

Materials

- Vinylstannane[b] **31** (FW 461.4) 183.4 mg, 0.398 mmol — **irritant**
- Two equal portions of *n*-BuLi (FW 64.1) 2.42 M in hexanes, 0.165 mL, 0.40 mmol — **flammable, moisture sensitive**
- Thiophene[c] **33** (FW 84.1) 33 μL, 34.50 mg, 0.41 mmol — **flammable, toxic**
- Allylbenzene epoxide[d] **37** (FW 134.0) 42 μL, 42.5 mg, 0.32 mmol — **irritant**
- CuCN (FW 89.6) 36.8 mg, 0.41 mmol — **highly toxic, irritant**

1. Gently flame dry a 5 mL round-bottomed flask under argon. After cooling to room temperature, add via the first 0.5 mL syringe dry THF (0.4 mL) and then cool the reaction vessel to −30°C (dry ice–CCl_4 bath).
2. Add thiophene **33** (33 μL, 0.41 mmol) to the flask using the first 100 μL syringe, followed by *n*-BuLi (0.165 mL, 0.40 mmol) via the second 0.5 mL syringe. Allow this solution to stir at −30°C for 30 min to produce a light-yellow solution of **34**.

Protocol 3. *Continued*

3. Flame dry a 5 mL round-bottomed flask under argon. After cooling to room temperature, add the CuCN to the flask, purge with argon, and insert a septum through which is placed a needle connected to the argon line.
4. Transfer 0.4 mL of THF to this flask to form a stirred slurry and then cool the flask to −78°C in a dry ice–acetone bath.
5. Transfer the lithiated thiophene solution into the stirred CuCN–THF slurry using a double-needle-tipped cannula. Warm the resulting mixture to room temperature to yield a clear tan solution of (2-Th)Cu(CN)Li **35**.
6. Flame dry another 5 mL round-bottomed flask under argon. After cooling to room temperature, add vinylstannane[e] **31** as a neat liquid via the first 250 μL syringe which had been previously dried in an Abderhalden overnight.
7. Azeotropically dry vinylstannane **31** using two 1 mL portions of toluene under a high vacuum pump at ambient temperature. Dissolve **31** in 0.6 mL of THF and cool this solution to −78°C
8. Add *n*-BuLi (0.165 mL, 0.40 mmol) to this cold solution using the second dry 0.5 mL syringe and stir this mixture at −78°C for 1 h.
9. Cannulate the resulting vinyl anion **32** into the lower order cuprate **35** and stir at −78°C for an additional 10 min, thereby forming **36**.
10. Flame dry another 5 mL round-bottomed flask under argon. After cooling to room temperature, add in to this flask epoxide **37** via the second dry 100 μL syringe.
11. Azeotropically dry epoxide **37** with two 1 mL portions of dry toluene and then dissolve the dried epoxide in 0.3 mL of dry THF.
12. Cannulate the THF solution of epoxide **37** into the higher order cuprate **36** previously formed and wash all remaining epoxide into the reaction flask using 0.2 mL of THF.
13. Once the addition is complete, remove the cooling bath and allow the reaction mixture to warm slowly to 0°C and stir at this temperature for 4 h.
14. Quench the reaction mixture with 20 mL of an aqueous 10% concentrated NH_4OH–90% saturated NH_4Cl solution. Transfer the contents of the reaction flask to a separatory funnel and extract the aqueous layer with diethyl ether (3 × 20 mL). Combine the extracts and wash them with brine (1 × 10 mL), dry them over anhydrous Na_2SO_4, then filter and evaporate all solvents *in vacuo*.
15. Purify the crude product by column chromatography on silica gel (ICN Biomedicals, 32–63, 60 Å) using ether–hexanes (1:5) to obtain alcohol **38**[f] (74.7 mg, 77%), along with 4.8 mg (11.3%) of the recovered epoxide. Characterize the product by spectral means as well as by high resolution mass spectral analysis.

[a] Based on Lipshutz, B. H.; Parker, D. A. unpublished work; *cf.* Parker, D. A. PhD Thesis, University of California, Santa Barbara, **1987**, 172–173.
[b] Prepared via hydrostannylation of propargyl alcohol: Corey, E. J.; Wollenberg, R. H. *J. Org. Chem.* **1975**, *40*, 2265; Jung, M. E.; Light, L. A. *Tetrahedron Lett.* **1982**, *23*, 3851.
[c] Distilled from CaH_2 under water aspirator pressures to give a clear, colourless liquid which is stored over 4 Å molecular sieves.
[d] Available from Aldrich.
[e] The quantity of stannane used was determined by weight using a pre-tared 250 μL syringe.
[f] TLC: R_f Et_2O–hexanes, 1:5 = 0.16

Protocol 4.
Conjugate addition of *n*-Bu(Imid)Cu(CN)Li_2 to 4-isopropyl-2-cyclohexenone[17b]

Caution! Carry out all procedures in a well-ventilated hood, and wear disposable vinyl or latex gloves and chemical-resistant safety goggles.

39 → (*n*-BuLi, THF, –78°C, 5 min) → 40 → (CuCN•2LiCl, THF, –78 to –20°C) → 41 (N–Cu(CN)Li)

→ (*n*-BuLi, –78 to –20°C) → 42 (Cu(CN)Li_2, Bu) → (1. BF_3•Et_2O, –78°C; 2. 43; THF, –78 to 0°C, 2h) → **44** (87%)

Equipment

- Two round-bottomed flasks (5 mL), each equipped with a magnetic stirrer bar and septum
- Argon supply and inlet
- Dry ice–acetone cooling bath
- One syringe (100 μL)
- One syringe (250 μL)
- Two syringes (0.5 μL)
- One cannula (18 gauge, 12 in long) cannula

Materials

• Imidazole **39** (FW 68.0) 68.1 mg, 1.0 mmol	**corrosive**
• CuCN (FW 89.6) 89.6 mg, 1.0 mmol	**highly toxic, irritant**
• LiCl[a] (FW 42.4) 84.8 mg 2.0 mmol	**irritant, hygroscopic**
• *n*-BuLi (FW 64.1) 2.22 M in hexanes, 0.90 mL, 2.0 mmol	**flammable, moisture sensitive**
• $BF_3{\cdot}Et_2O$[b] (FW 141.9) 98 μL, 114 mg, 0.80 mmol	**corrosive, moisture sensitive**
• 4-Isopropyl-2-cyclohexenone[e] **43** (FW 138.2) 110 μL, 0.80 mmol	**toxic**
• Dry, distilled THF	**flammable, irritant**

1. Gently flame dry one 5 mL round-bottomed flask containing a stirrer bar under a positive pressure of argon. After allowing the flask to cool in air to room temperature, dissolve imidazole **39** (previously dried via azeotropic

Protocol 4. *Continued*

removal from two 1 mL portions of toluene under high vacuum) in 1 mL of dry THF. Stir and cool the resulting solution to −78 °C with a dry ice–acetone bath.

2. Add titrated *n*-BuLi (0.45 mL) using the first 0.5 mL syringe, thereby generating **40**.
3. Flame dry the second 5 mL round-bottomed flask under argon. After cooling to room temperature, add the CuCN and LiCl, cap the flask with a septum, purge the flask with argon, and then syringe in 1.5 mL of dry THF to form a homogeneous solution at room temperature.
4. Cannulate the room temperature CuCN·2LiCl solution into the cold, lithiated imidazole **40**. Allow the mixture to warm to homogeneity (which occurs at approximately −20 °C), thereby giving rise to a yellow solution of cuprate **41**.
5. Recool the reaction mixture to −78 °C and then add *n*-BuLi (0.44 mL) via the second dry 0.5 mL syringe (no colour change).
6. Allow the yellow reaction mixture to warm over *c.* 5 min until it becomes homogeneous at *c.* −20 °C, at which point **42** is formed.
7. Recool the reaction mixture to −78 °C and then add $BF_3 \cdot Et_2O$ using the dry 100 μL syringe.
8. Add the 4-isopropyl-2-cyclohexenone **43** as a neat liquid via the dry 250 μL syringe.
9. Allow the reaction mixture to stir for 1 h at −78 °C. Remove the cooling bath and allow gradual warming to 0 °C, then stir for an additional 1 h at 0 °C, during which the mixture turns from yellow to green.
10. Quench the reaction mixture with 2 mL of 10% concentrated NH_4OH in saturated NH_4Cl. Extract with ether (3 × 20 mL), wash the extracts with brine (1 × 20 mL) and dried over anhydrous Na_2SO_4.
11. Isolate the product by filtration of the dried extracts, solvent removal *in vacuo*, and column chromatography of the resulting yellow oil on silica gel (ICN Biomedicals, 32–63, 60 Å) using 15% EtOAc in hexanes. The purified product **44** is obtained as a clear oil (136 mg, 87%) which displays the appropriate spectral characteristics and high resolution mass spectral data.[d]

[a] Obtained from Aldrich and stored in a vial in an Abderhalden over KOH. It was weighed out quickly in the air, and then transferred to the reaction vessel.
[b] Distilled at atmospheric pressure to a water-white liquid.
[c] Racemic cryptone **43** is apparently no longer commercially available. It can be prepared from 4-isopropylphenol following a three-step procedure: Nakamura, H.; Ye, B.; Murai, A. *Tetrahedron Lett.* **1992**, *33*, 8113 and references therein.
[d] Capillary GLC indicated the presence of only one isomer, assumed to be *trans*.

4. Cyanocuprates via transmetallations

During the past few years, alternative routes to certain types of HO cyanocuprates have been developed which no longer rely on organolithium precursors. Rather, an increasingly large pool of other organometallic species have been found to readily exchange a particular ligand of value for a trivial ligand (e.g. Me) present initially in a HO cyanocuprate. This exchange phenomenon is schematically represented in Scheme 5.5. Thus far, R_tML_n has been examined where R_t is a vinyl or substituted vinyl group, with $ML_n = SnBu_3$,[21] TeBu,[22] AlR_2,[23] or $ZrCp_2R$.[24] Likewise, both allylic stannanes[25] and zirconocenes[26] are prone to such transmetallations. Related exchanges with $Me_3SnSnMe_3$,[27] Bu_3SnH,[28a] Me_3SnH,[28b] and $R_3SnSiR'_3$[29] are also attractive routes to stannyl- or silylcuprates. These are stoichiometric processes (i.e. they rely on a full equivalent of a HO cyanocuprate), although some sequences involving catalytic copper(I) transmetallations of alanes,[30] zirconocenes,[31] and titanates[32] have very recently begun to show great promise.

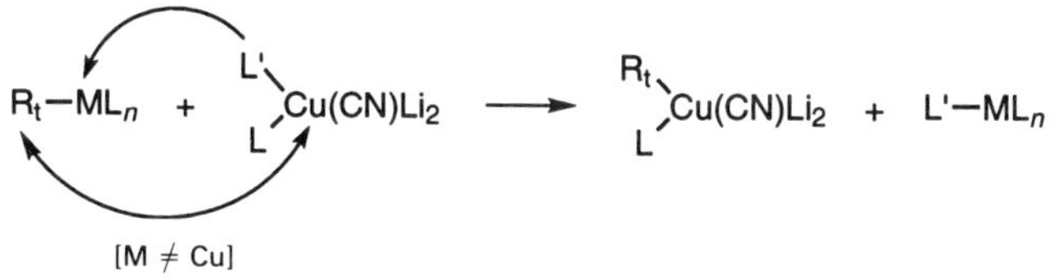

$$R_t{-}ML_n + L'L\,Cu(CN)Li_2 \longrightarrow R_tL\,Cu(CN)Li_2 + L'{-}ML_n \quad [M \neq Cu]$$

Scheme 5.5

One example which demonstrates the attractiveness of such a concept can be found in the prostaglandin area, where the lower side-chain of the potent, co-administered antisecretory agent Misoprostol is to be introduced into a cyclopentenone, *e.g.* **49**. Prior methodology had called for employment of either the (*E*)-vinylstannane or iodide as precursor to the vinyllithium **47**, itself a precursor to cuprate **48** (Scheme 5.6). Moreover, use of either stannane **45** or iodide **46** (which are patented intermediates) ultimately derives from their preparation via the corresponding acetylene, often relying on hydrostannylation or hydrozirconation–iodination. To bypass these series of reactions and associated intermediates which first lead up to cuprate **48**, a transmetallation protocol now exists which directly converts a terminal alkyne, e.g. **50** in Scheme 5.7, to an (*E*)-vinylic cuprate **52** (eqn 6) in one pot via a vinylzirconocene intermediate **51** (Scheme 5.7).[24] The method is highly tolerant of functional groups present in the 1-alkyne, including electrophilic centres such as TIPS esters, nitriles, and chlorides (*e.g.* as in Scheme 5.8).[33]

It should be noted at this point that mixed cuprates such as **52** which result from a ligand exchange phenomenon between zirconium and copper selectively transfer the vinylic rather than alkyl group to an α,β-unsaturated ketone. Hence, competitive delivery of the methyl group from, for example,

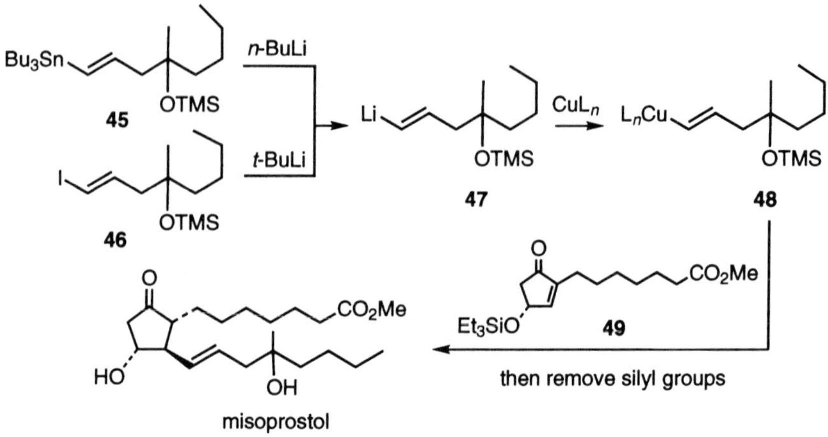

Bu_3Sn ... OTMS **45**

I ... OTMS **46**

n-BuLi

t-BuLi

Li ... OTMS **47**

CuL_n

L_nCu ... OTMS **48**

Et_3SiO ... CO_2Me **49**

then remove silyl groups

HO ... OH ... CO_2Me

misoprostol

Scheme 5.6

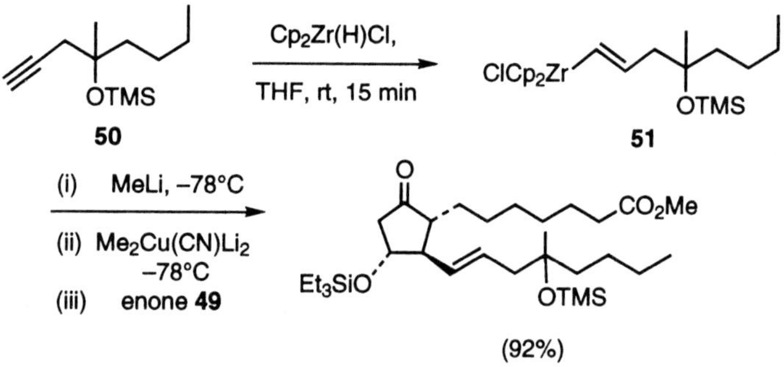

OTMS **50**

$Cp_2Zr(H)Cl$,

THF, rt, 15 min

$ClCp_2Zr$ ... OTMS **51**

(i) MeLi, –78°C

(ii) $Me_2Cu(CN)Li_2$ –78°C

(iii) enone **49**

Et_3SiO ... OTMS ... CO_2Me

(92%)

Scheme 5.7

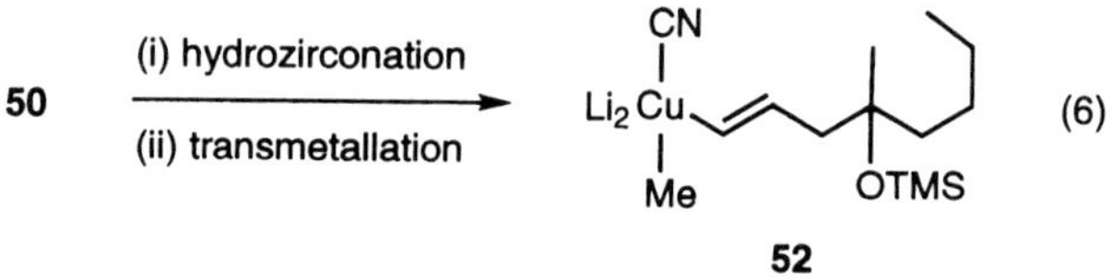

50

(i) hydrozirconation

(ii) transmetallation

CN, Li_2Cu, Me ... OTMS **52** (6) (6)

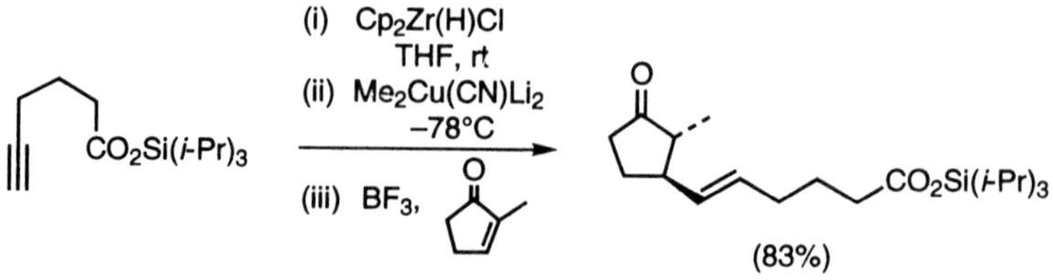

$CO_2Si(i\text{-}Pr)_3$

(i) $Cp_2Zr(H)Cl$ THF, rt

(ii) $Me_2Cu(CN)Li_2$ –78°C

(iii) BF_3,

$CO_2Si(i\text{-}Pr)_3$

(83%)

Scheme 5.8

52, is not a concern. However, when the nature of the substrate is altered from that of a Michael acceptor to an electrophile such as an alkyl halide or epoxide, mixed alkylvinylcuprates cannot be used effectively owing to the preference for alkyl ligand transfer (Scheme 5.9).[10]

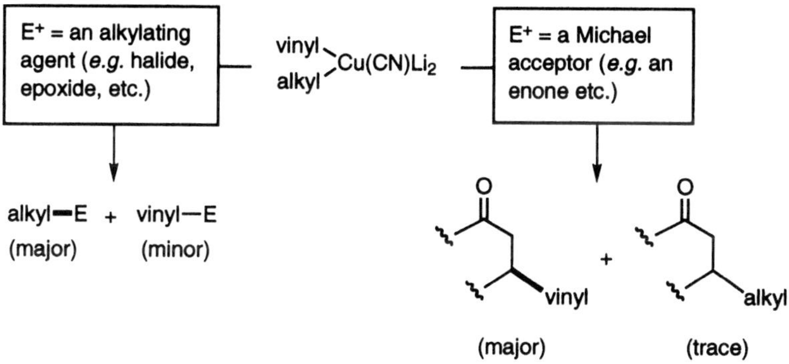

E+ = an alkylating agent (*e.g.* halide, epoxide, etc.)

vinyl, alkyl $Cu(CN)Li_2$

E+ = a Michael acceptor (*e.g.* an enone etc.)

alkyl—E (major) + vinyl—E (minor)

vinyl (major) + alkyl (trace)

Scheme 5.9

To obviate this issue, vinylzirconocenes like **51** can be treated with the mixed cyanocuprate Me(2-Th)Cu(CN)Li_2 (as opposed to $Me_2Cu(CN)Li_2$), thereby arriving after transmetallation at a vinylthienylcyanocuprate **53** (Scheme 5.10).[34] Selective release of the vinylic moiety over the 'dummy' thienyl ligand from **53** then ensues upon introduction of an alkylating agent (e.g. as in Scheme 5.11; see also Protocol 5).

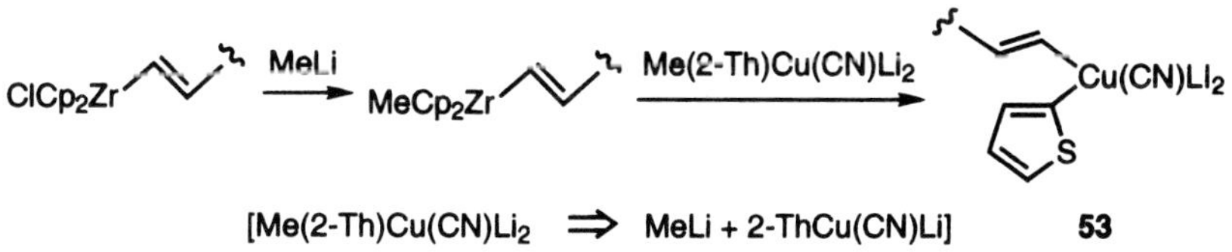

$ClCp_2Zr$ —MeLi→ $MeCp_2Zr$ —Me(2-Th)Cu(CN)Li_2→ Cu(CN)Li_2 **53**

[Me(2-Th)Cu(CN)Li_2 ⇒ MeLi + 2-ThCu(CN)Li]

Scheme 5.10

(i) $Cp_2Zr(H)Cl$, rt (ii) 2MeLi, −78°C (iii) 2-ThCu(CN)Li; Cl, −78°C to r.t. (75%)

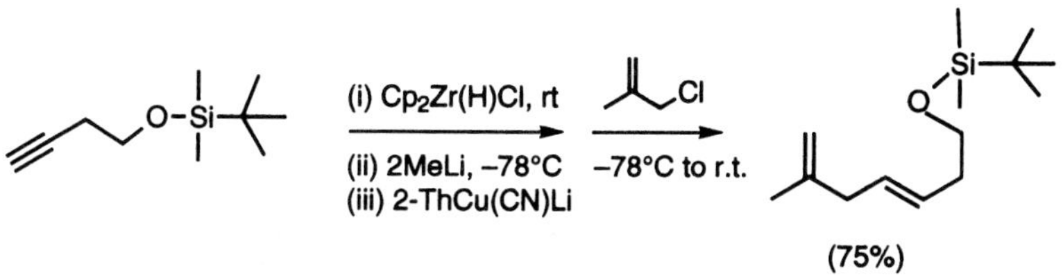

Scheme 5.11

Hydrozirconation of a 1-tributylstannylacetylene **54**, rather than a 1-alkyne, generates a 1,1-dimetallo species **55**.[35] Aqueous quench of **55** provides a trivial route to the difficult to prepare (*Z*)-vinylstannanes **56**. Intermediates **55** also allow for selective transmetallation to cuprates **57** involving exclusively the vinylzirconocene portion such that vinylstannane-containing coupling products of, for example, types **58** and **59** result, well suited for further manipulations (Scheme 5.12).[36]

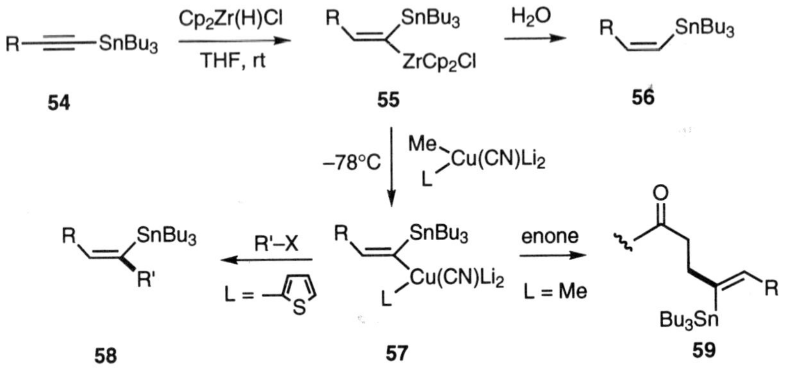

Scheme 5.12

Allylic stannanes are also highly prone to ligand exchange with $Me_2Cu(CN)Li_2$.[25] Tributyltin derivatives containing simple allyl, crotyl, methallyl, and prenyl ligands, i.e. **60–63**, respectively, react at −35 °C in THF to afford the mixed HO species **64**. Alkylations of **64**, however, are not especially useful since methyl rather than allyl ligand transfer to the electrophile is surprisingly competitive. Hence, a double transmetallation to arrive at diallylic cuprates **65** is required, the second transmetallation taking place at 0°C (Scheme 5.13; see also Protocol 6).

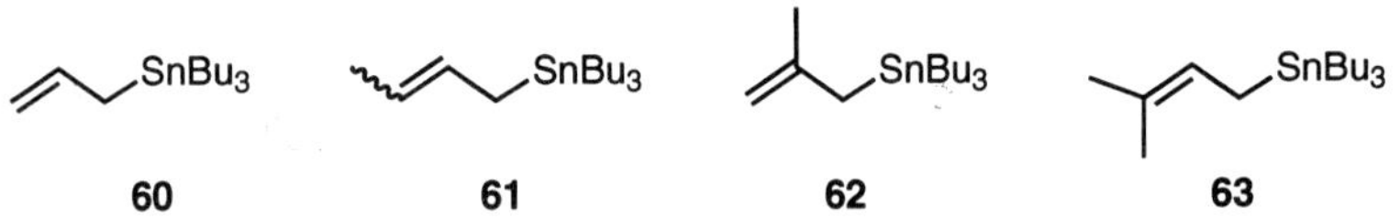

The reactivity of diallylic cuprates **65** is extraordinary.[25] Indeed, they are the most reactive cuprates known, reacting virtually upon mixing at −78°C with vinyl triflates (eqn 7).[37] Notwithstanding the rapidity of the coupling, strict retention of double-bond geometry is maintained. Although electrophilic partners such as primary, unactivated chlorides are rarely considered viable towards cuprates,[1a,38] they react at −78°C in minutes (eqn 8).[25]

Thus, it is easy to understand why allylic cuprates, in general, are ill-behaved in the presence of α,β-unsaturated ketones (i.e. they add pre-

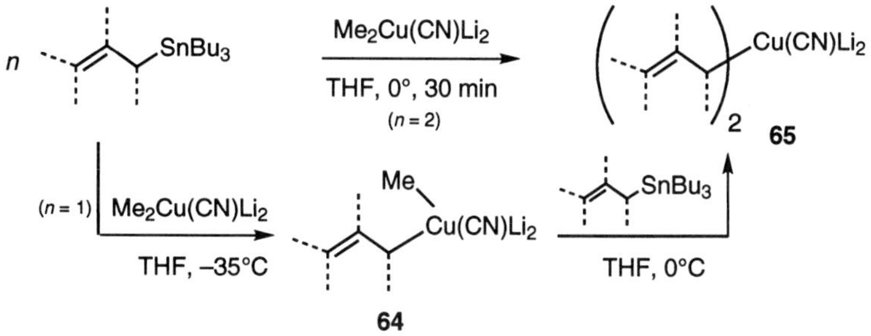

n SnBu$_3$ — $Me_2Cu(CN)Li_2$, THF, 0°, 30 min (n = 2) → Cu(CN)Li$_2$ (2) **65**

(n = 1) $Me_2Cu(CN)Li_2$, THF, −35°C → Me Cu(CN)Li$_2$ **64** — SnBu$_3$, THF, 0°C → **65**

Scheme 5.13

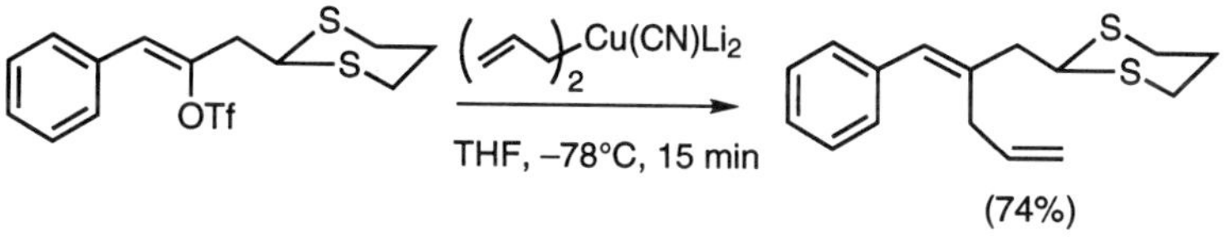

OTf — Cu(CN)Li$_2$ (2), THF, −78°C, 15 min → (74%) (7)

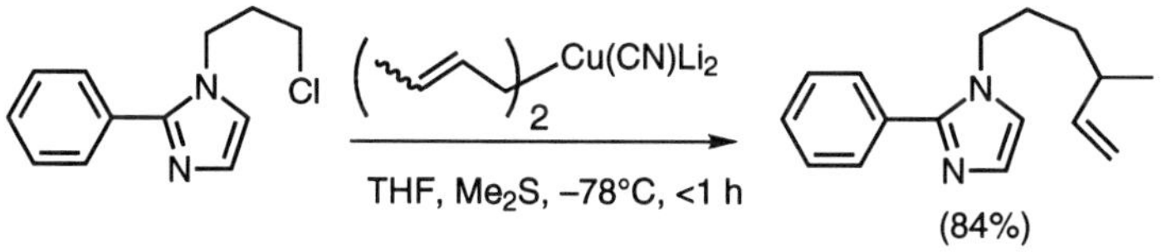

Cl — Cu(CN)Li$_2$ (2), THF, Me_2S, −78°C, <1 h → (84%) (8)

dominantly in a 1,2-sense).[39] Fortunately, their reactivity can be tamed such that these valuable ligands can be successfully introduced into the β-position of various conjugated enones.[40] The 'trick' is to use what is routinely referred to as a neutral organocopper complex, written as 'RCu'.[1a,38,41] NMR and chemical studies, however, have provided strong evidence that the lithium salts (LiX) present in the generation of these materials are part of the cluster, i.e. forming halocuprates RCu(X)Li.[42] These reagents, together with an equivalent of TMS-Cl,[43] provide an effective means of delivering allylic ligands in a Michael sense, initially affording the corresponding trimethylsilyl enol ethers and ultimately the corresponding ketones (eqn 9).

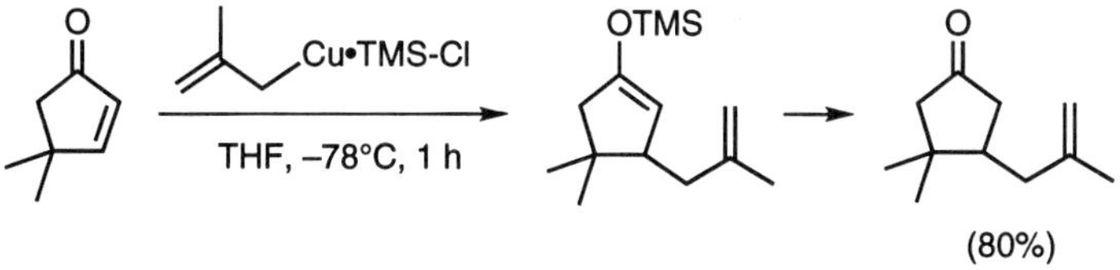

O — Cu•TMS-Cl, THF, −78°C, 1 h → OTMS → O (80%) (9)

Additional examples based on a hydrozirconation–transmetallation theme for acetylenic stannanes (see Protocol 5) have recently been described,[36] as has a new method for three-component couplings which relies on cyano-cuprate technology.[44]

Protocol 5.
Hydrozirconation–transmetallation–alkylation of a stannylacetylene[36]

Caution! Carry out all procedures in a well-ventilated hood, and wear disposable vinyl or latex gloves and chemical-resistant safety goggles.

Equipment

- One round-bottomed flask (10 mL) with a magnetic stirrer bar and septum
- Argon supply and inlet
- Dry ice—acetone cooling bath
- Two glass syringes (1 mL)
- One glass syringe (5 mL)
- Three stainless steel cannulae (18 gauge, each 5–10 in long)

Materials

- 1-Tributylstannylpropyne[a] **66** (FW 329.1) 50 mg, 0.15 mmol — **toxic**
- Zirconocene chloridehydride[b] **67** (FW 265.7) 45 mg, 0.17 mmol — **moisture sensitive, light sensitive**
- Thienylcyanocuprate[c] **69** (FW 179.6) 0.27 M in THF, 0.63 mL, 0.17 mmol — **highly toxic, flammable**
- Methyllithium[d] (FW 22.0) 1.45 M in ether, 0.23 mL. 0.34 mmol — **pyrophoric, moisture sensitive**
- Methallyl chloride[e] **71** (FW 78.6) 23 μL, 21 mg, 0.23 mmol — **flammable, cancer suspect agent**

1. Remove the reaction vessel from an oven (heated to 120°C, dried for 24 h or flame dried immediately prior to use) and fit with a Teflon-coated magnetic stirrer bar. Cool the flask under argon and cap with a septum. Argon flow into the vessel is provided via an 18-gauge needle which pierces the septum.
2. Concurrently remove three glass syringes (a 5 mL and two 1 mL syringes, Luer locking) and three 18-gauge needles from the oven and assemble while still hot. Tighten the Luer locks and cover with Teflon tape; fill the syringes with argon by placing the tips of the needles in an argon-flushed vial (fitted with a septum, argon inlet, and KOH as a desiccant). Allow the syringes to cool in this argon atmosphere.
3. Remove the flask from the argon inlet and weigh with the septum attached. Gently remove the septum, add the zirconocene chloridehydride **67**, and re-attach the septum.[f] Flush the reaction vessel and the parent vial with a

stream of argon and shield from light with aluminum foil. Reinsert the argon inlet and use the 5 mL syringe to add 3 mL of distilled THF to the reaction vessel. Wrap the septum with a layer of Parafilm® and use the magnetic stirrer to produce a white suspension.

4. Add 50 mg of the tributylstannylpropyne **66** to the flask using a 100 μL syringe. Stir the solution for 15 min: the suspension will become clear and yellow.[g] Cool the solution to −78°C with a dry ice–acetone bath.
5. Add 0.63 mL of the thienylcuprate **69** in THF dropwise (use a 1 mL syringe; total addition time about 3 min). Withdraw 0.23 mL of methyllithium using a second 1 mL syringe, and add it dropwise to the stirred solution at −78°C (the order of addition of **69** and MeLi to **68** is not significant in this case). Allow the solution to stir for about 10 min to ensure transmetallation to cuprate **70**.
6. Use a 100 μL syringe to add 23 μL (1.3 equivalents) of methallyl chloride **71** to the solution, and allow to stir at −78°C for 2 h.[h]
7. Remove the −78°C bath and remove the septum from the reaction vessel. Pour the THF solution into a beaker containing 15 mL of 10% sodium bicarbonate and 15 mL of diethyl ether. Transfer the cloudy bilayer to a separatory funnel and remove the organic layer. Extract the aqueous layer and emulsion four times with 5–10 mL of fresh ether. Wash the combined organic layers with brine, dry over magnesium sulfate, filter through a Celite® bed (on a fritted glass funnel), and concentrate *in vacuo*.[i]
8. Chromatograph[j] the resulting oil using reverse-phase silica gel[k] (40% MeCN–60% CH_2Cl_2) to yield 52 mg (87%) of (2*Z*)-3-tributylstannyl-5-methyl-hexa-2,5-diene **72**, which displays the appropriate spectral characteristics and high resolution mass spectral data.

[a] Tributylstannylpropyne **66** was prepared by treating excess propyne, condensed at −78°C in THF (*c.* 0.2 M), with less than one equivalent of *n*-BuLi in hexanes at this temperature. The solution is allowed to warm to ambient temperature and is then recooled to 0°C, at which point tributyltin chloride (one equivalent to *n*-BuLi) is then added. The solution is allowed to warm to room temperature and stirred overnight to afford a cloudy mixture. Filtration through a bed of Celite® gives a clear solution which is concentrated *in vacuo* to an oil. Kugelrohr distillation at 90°C (0.1 mm Hg) yields the pure product, typically in 75–80% yield. The stannane was stored in a glass vial under an inert atmosphere. The syringe used in measuring out **66** was weighed before and after withdrawing the stannane to gauge the proper amount required.

[b] The zirconocene chloridehydride **67** (purchased from Boulder Scientific, Mead, CO, or Aldrich) was typically stored in a polyethylene-stoppered glass vial sealed with Parafilm, covered with aluminum foil, and placed in a desiccator over Drierite.

[c] This can be obtained from Aldrich or prepared according to our earlier procedure: Lipshutz, B. H.; Koerner, M.; Parker, D. A. *Tetrahedron Lett.* **1987**, *28*, 945 (see Protocol 3).

[d] Available as a solution in diethyl ether from Aldrich.

[e] Distilled at atmospheric pressure from CaH_2 at 71°C under argon.

[f] If care is taken not to tip the flask, the amount of argon lost in the weighing process is minimal. However, the time allowed for exposure of the reagents to atmospheric moisture should be minimized.

[g] Occasionally, brownish or even green colours can be seen. This is *not* a reproducible phenomenon nor does it appear to affect the subsequent chemistry.

[h] Following the progress of the reaction via TLC proved difficult as the reactants and products are too non-polar for standard silica plates (which provided little if any separation), while the use of C_{18}

Protocol 5. *Continued*

reverse-phase silica plates was hampered by the polar metallic by-products (which caused streaking on the plates).

[i]Often the aqueous work-up failed to remove all traces of metals and polar by-products from the crude sample. If desired, the resulting oil can be 'cleaned up' by dissolving in 5% ethyl acetate–hexanes and forcing it through a 1 in long silica gel plug (fashioned from a Pasteur pipette and a bit of glass wool) with a stream of nitrogen, and then re-concentrating the solution.

[j]Subjecting the oil to high vacuum (i.e. 0.1 mm Hg) overnight tended to remove excess electrophile and any volatile by-products. When combined with the procedure in footnote i above, this was found on occasion to give pure products without resorting to column chromatography.

[k]Purchased from Aldrich, labelled as 'Octadecyl-functionalized Silica Gel'. See Farina, V. *J. Org. Chem.* **1991**, *56*, 4985.

Protocol 6. An allylic cuprate-mediated opening of an epoxide[25]

Caution! Carry out all procedures in a well-ventilated hood, and wear disposable vinyl or latex gloves and chemical-resistant safety goggles.

$SnBu_3$ **73** —($Me_2Cu(CN)Li_2$, THF, 0°C, 30 min)→ ( )$_2Cu(CN)Li_2$ **74** —(**75**, 0°C, 1 h)→ OH **76** (90%)

Equipment

- One round-bottomed, two-necked flask (10 mL) containing a magnetic stirrer bar, with a side-arm bearing a Teflon stopcock
- Three-way stopcock with argon inlet and vacuum adapter, placed onto the two-necked flask
- One pear-bottomed flask (5 mL) with stirrer bar and septum
- One stainless steel cannula (20 gauge)
- Two glass syringes (2 mL)
- One glass syringe (1 mL)
- One glass syringe (100 mL)
- Mineral oil bubbler
- Argon supply and inlet
- Dry ice–acetone cooling bath

Materials

- Methallyltri-*n*-butylstannane[a–f] **73** (FW 345.1) 0.63 mL, 0.674 g, 1.95 mmol — **toxic**
- Cyclohexene oxide **75** (FW 98.1) 0.040 mL, 0.046 g, 0.47 mmol — **flammable, toxic**
- CuCN (FW 89.6) 0.0873 g, 0.975 mmol — **lightly toxic, irritant**
- MeLi (FW 22.0) 1.6 M in Et_2O, 1.22 mL, 1.95 mmol — **pyrophoric, moisture sensitive**
- Dry, distilled THF — **flammable, irritant**

1. Gently flame dry the 10 mL flask containing the CuCN under vacuum and purge with argon through both the three-way stopcock and side-arm stopcock, the latter bearing a septum through which is placed a syringe needle.

2. Remove the needle and then close the side-arm stopcock under a positive pressure of argon, which continues to enter through the three-way stopcock.
3. Introduce THF (1.15 mL) via the 2.0 mL syringe through the side-arm and stir, thereby forming a slurry.
4. Cool the slurry to −78°C and add the MeLi in Et_2O through the side-arm via the 2.0 mL syringe dropwise.
5. Remove the −78°C cooling bath and place the vessel in an ice–water bath to allow the dimethylcuprate to form and equilibrate at 0°C with stirring.
6. Add the methallyltri-*n*-butylstannane **73** as a neat liquid via a 1.0 mL syringe in one portion rapidly at 0°C, whereupon the solution takes on a yellow colour. Stir for an additional 30 min at this temperature, after which transmetallation is complete, thereby forming **74**.
7. Remove the vacuum line from the three-way stopcock and replace it with a line connected to the mineral oil bubbler.
8. Flame dry the 5 mL pear-bottomed flask under vacuum and purge several times with argon to ensure dryness. Add THF (0.5 mL) through the septum using the 2.0 mL syringe followed by cyclohexene oxide **75** via the 100 μL syringe.
9. Recool the cuprate to −78°C and add the cyclohexene oxide solution in THF dropwise via the cannula under a positive pressure of argon, venting the cuprate flask through the bubbler accessed through the three-way stopcock. Re-establish a positive pressure of argon after addition is complete.
10. Remove the low temperature cooling bath and bring the resulting mixture to 0°C in an ice–water bath; stir the mixture for an additional 1 h at this temperature.
11. Quench the reaction mixture by transferring it to an Erlenmeyer flask cooled to 0°C containing a rapidly stirred mix of *c*. 5 mL of Et_2O and *c*. 10 mL of an aqueous solution made up of 10% concentrated NH_4OH and 90% saturated NH_4Cl. Allow the mixture to warm to ambient temperature with stirring until all solids dissolve.
12. Transfer the mixture to a separatory funnel and extract it with Et_2O (3 × 10 mL). Wash the combined organic extracts with water and then brine. Dry the organic phase over anhydrous Na_2SO_4, filter the solvents, and evaporate *in vacuo*.
13. Chromatograph the resulting oil on silica gel (ICN Biomedicals, 32–63, 60 Å) using distilled petroleum ether (30–60°C)–ethyl acetate (9:1) to yield *trans*-2-(2-methyl-1-propen-3-yl)cyclohexan-1-ol **76** (0.104 g, 90%) as a colourless oil. The product displays characteristic spectroscopic and high resolution mass spectral data.

Protocol 6. *Continued*

[a] Prepared from methallyl Grignard and n-Bu_3SnCl in THF: Keck, G. E.; Enholm, E. J.; Yates, J. B.; Wiley, M. R. *Tetrahedron* **1985**, *41*, 4079; Seyferth, D.; Weiner, M. A. *J. Org. Chem.* **1961**, *26*, 4797.
[b] A very high purity of the allylic stannanes is crucial to the success of the *in situ* transmetallation.
[c] Exposure of an allylic stannane to septa should be totally avoided as such treatment results in inhibition of the transmetallation process.
[d] The material should be stored in small lots under glass or polyethylene at freezer temperatures when not in use. The stannane should not be exposed to screw-cap vials without a Teflon liner.
[e] After repeated use, or in the event that the stannane develops a yellow tint, it should be redistilled and repackaged in small lots.
[f] Stannanes possess an objectionable odour and are mildly toxic. Work should be performed in a fume hood whenever possible and gloves should be worn during the work-up procedure. Contaminated glassware may be cleaned with household bleach to remove any residual stannane present.

References

1. (a) Lipshutz, B. H., Sengupta, S. *Org. React.* **1992**, *41*, 135–631. (b) Lipshutz, B. H. *Synlett* **1990**, *3*, 119–128. (c) Lipshutz, B. H. *Synthesis* **1987**, 325–341. (d) Lipshutz, B. H.; Wilhelm, R. S.; Kozlowski, J. A. *Tetrahedron* **1984**, *40*, 5005–5038.
2. Gilman, H.; Jones, R. G.; Woods, L. A. *J. Org. Chem.* **1952**, *17*, 1630–1634.
3. House, H. O.; Koepsell, D. G.; Campbell, W. J. *J. Org. Chem.* **1972**, *37*, 1003–1011. Ashby, E. C.; Watkins, J. J. *J. Am. Chem. Soc.* **1977**, *99*, 5312–5317. Bertz, S. H.; Dabbagh, G. *J. Am. Chem. Soc.* **1988**, *110*, 3668–3670.
4. (a) Macdonald, T. L.; Still, W. C. *J. Am. Chem. Soc.* **1975**, *97*, 5280–5281. (b) Ashby, E. C.; Lin, J. J. *J. Org. Chem.* **1977**, *42*, 2805–2808. (c) Bertz, S. H. *Tetrahedron Lett.* **1980**, *21*, 3151–3154. (d) Palmisano, G.; Pellagata, R. *J. Chem. Soc., Chem. Commun.* **1975**, 892–893. (e) Bertz, S. H.; Dabbagh, G. *Tetrahedron* **1989**, *45*, 425–434.
5. Lipshutz, B. H. In *Advances in Metal-Organic Chemistry*; Liebeskind, L., ed.; JAI Press: London, in press.
6. Lipshutz, B. H.; Siegmann, K.; Garcia, E. *J. Am. Chem. Soc.* **1991**, *113*, 8161–8162. Lipshutz, B.H.; Siegmann, K.; Garcia, E.; Kayser, F. *J. Am. Chem. Soc.* **1993**, *115*, 9276–9282. Lipshutz, B. H.; Siegmann, K.; Garcia, E. *Tetrahedron* **1992**, *48*, 2579–2588.
7. Lipshutz, B. H.; Sharma, S.; Ellsworth, E. L. *J. Am. Chem. Soc.* **1990**, *112*, 4032–4034. Bertz, S. H. *J. Am. Chem. Soc.* **1990**, *112*, 4031–4032. Stemmler, T.; Penner-Hahn, J. E.; Knochel, P. *J. Am. Chem. Soc.* **1993**, *115*, 348–350.
8. Stolze, D. A.; Perron-Sierra, F.; Heeg, M. J.; Albizati, K. F. *Tetrahedron Lett.* **1991**, *32*, 4081–4084.
9. Lipshutz, B. H.; Wilhelm, R. S.; Floyd, D. M. *J. Am. Chem. Soc.* **1981**, *103*, 7672–7674.
10. Lipshutz, B. H.; Wilhelm, R. S.; Kozlowski, J. A.; Parker, D. A. *J. Org. Chem.* **1984**, *49*, 3928–3938.
11. Lipshutz, B. H.; Kozlowski, J. A.; Wilhelm, R. S. *J. Am. Chem. Soc.* **1982**, *104*, 2305–2307.
12. Whitesides, G. M.; Fischer, W. F.; San Filippo, J.; Bashe, R. W.; House, H. O. *J. Am. Chem. Soc.* **1969**, *91*, 4871–4882.

13. Lipshutz, B. H.; Parker, D. A.; Kozlowski, J. A.; Miller, R. D. *J. Org. Chem.* **1983**, *48*, 3334–3336.
14. Johnson, C. R.; Dutra, G. A. *J. Am. Chem. Soc.* **1973**, *95*, 7777–7782. Johnson, C. R.; Dutra, G. A. *J. Am. Chem. Soc.* **1973**, *95*, 7783–7788.
15. Quinton, P.; Le Gall, T. *Tetrahedron Lett.* **1991**, *32*, 4909–4912. Smith, A. B.; Salvatore, B. A.; Hull, K. G.; Duan, J. J.-W. *Tetrahedron Lett.* **1991**, *32*, 4859–4862. Hanessian, S.; Thavonekham, B.; DeHoff, B. *J. Org. Chem.* **1989**, *54*, 5831–5833.
16. Lipshutz, B. H.; Kozlowski, J. A.; Wilhelm, R. S. *J. Org. Chem.* **1984**, *49*, 3943–3949.
17. (a) 2-Thienyl: Lipshutz, B. H.; Kozlowski, J. A.; Parker, D. A.; Nguyen, S. L.; McCarthy, K. E. *J. Organomet. Chem.* **1985**, *285*, 437–447; Malmberg, H.; Nilsson, M.; Ullenius, C. *Tetrahedron Lett.* **1982**, *23*, 3823–3826. (b) *N*-Pyrrolyl/*N*-imidazoyl: Lipshutz, B. H.; Fatheree, P.; Hagen, W.; Stevens, K. L. *Tetrahedron Lett.* **1992**, *33*, 1041–1044. (c) 3-Methyl-3-methoxy-1-butynyl: Corey, E. J.; Floyd, D.; Lipshutz, B. H. *J. Org. Chem.* **1978**, *43*, 3418–3420. (d) Dimsyl: Johnson, C. R.; Dhanoa, D. S. *J. Org. Chem.* **1987**, *52*, 1885–1888.
18. Lipshutz, B. H.; Koerner, M.; Parker, D. A. *Tetrahedron Lett.* **1987**, *28*, 945–948.
19. Sold by Aldrich under lithium 2-thienylcyanocuprate.
20. Hong, C. Y.; Kishi, Y. *J. Am. Chem. Soc.* **1991**, *113*, 9693–9694.
21. Behling, J. R.; Babiak, K. A.; Ng, J.S.; Campbell, A. L.; Moretti, R.; Koerner, M.; Lipshutz, B. H. *J. Am. Chem. Soc.* **1988**, *110*, 2641–2643.
22. Comasseto, J. V.; Berriel, J. N. *Synth. Commun.* **1990**, *20*, 1681–1685.
23. Ireland, R. E.; Wipf, P. *J. Org. Chem.* **1990**, *55*, 1425–1426.
24. Lipshutz, B. H., Ellsworth, E. L. *J. Am. Chem. Soc.* **1990**, *112*, 7440–7441. Babiak, K. A., Behling, J. R.; Dygos, J. H.; McLaughlin, K. T., Ng, J. S., Kalish, V. J.; Kramer, S. W.; Shone, R. L. *J. Am. Chem. Soc.* **1990**, *112*, 7441–7442.
25. Lipshutz, B. H.; Crow, R.; Dimock, S. H.; Ellsworth, E. L.; Smith, R. A. J.; Behling, J. R. J. *Am. Chem. Soc.* **1990**, *112*, 4063–4064.
26. Lipshutz, B. H.; Wood., M. Unpublished results.
27. Oehlschlager, A. C.; Hutzinger, M. W.; Aksela, R.; Sharma, S.; Singh, S. M. *Tetrahedron Lett.* **1990**, *31*, 165–168.
28. (a) Lipshutz, B. H.; Ellsworth, E. L.; Dimock, S. M.; Reuter, D. C. *Tetrahedron Lett.* **1989**, *30*, 4617–4620.
29. Lipshutz, B. H.; Reuter, D. C.; Ellsworth, E. L. *J. Org. Chem.* **1989**, *54*, 4975–4977. Lipshutz, B.H.; Sharma, S.; Reuter, D. C. *Tetrahedron Lett.* **1990**, *31*, 7253–7256.
30. Lipshutz, B. H.; Dimock, S. H. *J. Org. Chem.* **1991**, *56*, 5761–5763. Wipf, P.; Smitrovich, J. H.; Moon, C.-W. *J. Org. Chem.* **1992**, *57*, 3178–3186.
31. Wipf, P.; Smitrovich, J. H. *J. Org. Chem.* **1991**, *56*, 6494–6496. Wipf, P.; Xu, W. *Synlett* **1992**, 718–721. Venanzi, L. M.; Lehmann, R.; Keil, R.; Lipshutz, B. H. *Tetrahedron Lett.* **1992**, *33*, 5857–5860.
32. Arai, M.; Lipshutz, B. H.; Nakamura, E. *Tetrahedron* **1992**, *48*, 5709–5718. Arai, M.; Nakamura, E.; Lipshutz, B. H. *J. Org. Chem.* **1991**, *56*, 5489–5491.
33. Lipshutz, B. H.; Keil, R. *J. Am. Chem. Soc.* **1992**, *114*, 7919–7920.
34. Lipshutz, B. H.; Kato, K. *Tetrahedron Lett.* **1991**, *32*, 5647–5650.
35. Lipshutz, B. H., Keil, R.; Barton, J. C. *Tetrahedron Lett.* **1992**, *33*, 5861–5864.

36. Lipshutz, B. H.; Keil, R. *Inorg. Chim. Acta.* **1994**, *220*, 41–44.
37. Lipshutz, B. H.; Elworthy, T. R. *J. Org. Chem.* **1990**, *55*, 1695–1696.
38. Posner, G. H. *Org. React.* **1975**, *22*, 253–400
39. Some representative examples include: Majetich, G.; Cesares, A.; Chapman, D.; Behnke, M. *J. Org. Chem.* **1986**, *51*, 1745–1753; Hutchinson, D. K.; Fuchs, P. L. *Tetrahedron Lett.* **1986**, *27*, 1429–1432. See also reference 4e.
40. Lipshutz, B. H.; Ellsworth, E. L.; Dimock, S. H.; Smith, R. A. J. *J. Am. Chem. Soc.* **1990**, *112*, 4404–4410.
41. Posner, G. H. *Org. React.* **1972**, *19*, 1–113. Posner, G. H. *An Introduction to Synthesis Using Organocopper Reagents.* **1980**. Wiley: New York.
42. Lipshutz, B. H.; Ellsworth, E. L.; Dimock, S. H. *J. Am. Chem. Soc.* **1990**, *112*, 5869–5871.
43. Johnson, C. R.; Marren, T. J. *Tetrahedron Lett.* **1987**, *28*, 27–30.
44. Lipshutz, B. H.; Wood, M. R. *J. Am. Chem. Soc.* **1993**, *115*, 12625–12626.

6

Me_3SiCl-accelerated conjugate addition reactions of organocopper reagents

EIICHI NAKAMURA

1. Introduction

Conjugate addition to α,β-unsaturated carbonyl compounds represents an archetypical reaction of organocopper reagents (eqn 1).[1] Various reagents including the classical Grignard-based catalytic copper reagent, the Gilman reagent (R_2CuLi), and the '$RCu{\cdot}BF_3$' reagent have been developed to improve the efficiency of the process by enhancing the reactivities of the reagents and by suppressing side-reactions such as 1,2-addition. Because of the paramount importance of the conjugate addition in organic synthesis, research efforts continue to make the reaction more versatile and more reliable. This chapter describes the Me_3SiCl-accelerated reactions of organocopper reagents—a methodology which has proven its great synthetic utility in the past several years.

R-Cu + [enone] ⟶ [enolate, R, O⁻] —H⁺⟶ [ketone, R]

1

2. Scope and limitations

Me_3SiCl has long been used as a reagent to 'trap' the enolate anion **1** generated by the conjugate addition reaction of organocopper reagents. Since chlorosilane neutralizes the highly basic enolate anion which may undergo further reactions, it was found to be particularly useful for additions involving α,β-unsaturated aldehydes, wherein an unstable aldehyde enolate can be trapped *in situ* to produce a neutral enol silyl ether as the final product.[2] In 1984, however, it was recognized that Me_3SiCl is not simply a 'trapping agent' for the enolate anion, but acts as an active participant in such reactions and accelerates the conjugate addition itself. The initial discovery was made for the conjugate addition of a functionalized organozinc reagent that a mixture of Me_3SiCl and HMPA (or DMPU) accelerates the copper-catalysed conjugate addition of zinc homoenolate **2**, which is otherwise inert as a Michael donor (eqn 2). Thus, the chlorosilane–HMPA mixture effected quantitative addition of **2** to α,β-unsaturated ketones, aldehydes, and esters, as well as to acetylenic ketones and esters.[3] The details of this procedure have been described in the literature.[4] It was subsequently demonstrated that this neutral and readily available reagent, Me_3SiCl, as well as other related silylating agents, dramatically improves the performance of a variety of catalytic and stoichiometric reagents based[5] on Grignard,[5c,5e] organozinc,[6] titanium,[7] manganese,[8] and sodium[9] reagents as well as standard stoichiometric reagents including $RCu \cdot LiX$, $R_2CuLi \cdot LiX$, and higher order species.

$Et_2O \rightarrow Zn(CH_2CH_2C(OEt){=}O)(CH_2CH_2COOEt)$ (**2**) + cyclohexenone $\xrightarrow[\text{cat. CuBr·Me}_2\text{S, HMPA/THF}]{\text{Me}_3\text{SiCl}}$ 1-($OSiMe_3$)-3-(CH_2CH_2COOEt)cyclohexene 100% (2)[4]

Recently, the accelerating effects of chlorosilane were also observed for 1,4-additions of alkynylzinc reagents[10] and tin(II) enolates,[11] and for 1,2-carbonyl additions of organocopper[12] and zinc reagents.[13]

The nature of the Me_3SiCl-mediated reactions may be illustrated first by the experimental details of some representative reactions. Conjugate addition of an *o*-tolyl Grignard reagent to 3-methylcyclohexenone[5c,5e] is a difficult reaction to achieve, because of the steric hindrance at the reacting centres on both reactants. Thus, they are quite unreactive with each other at −70 °C both in the absence and in the presence of 5% $CuBr \cdot Me_2S$ in THF (or HMPA–THF) and produce mainly a 1,2-addition product in low yield (eqn 3). However, in the presence of Me_3SiCl and HMPA (one or two equivalents) the reaction is complete within several minutes. Other additives such as TMEDA, triethylamine, and *N*,*N*-dimethylimidazolidone are ineffective in this reaction. The accelerating effects may be observed only in THF and not in ether: the same was found for the reactions of Gilman reagents.[5a,5d,5e] As Me_3SiCl by itself is inactive at low temperatures either to carbonyl compounds, to Grignard reagents, or to organocopper reagents, the mixing order of reagents is not crucial insofar as the reagents are mixed below −70 °C. Since chlorosilane may often contain HCl even after distillation, the silylating reagent may be pre-mixed with triethylamine and used as it is. It has also been reported that Me_3SiCl makes copper(II) complex-catalysed conjugate additions possible.[14] The primary product of the chlorosilane-accelerated addition is an enol silyl ether having its double bond regiospecifically placed at the mechanistically expected position, and this may be isolated by careful neutral aqueous work-up.[4,5e] Alternatively, the silyl group can be removed under acidic or basic conditions either during work-up or after isolation of the enol silyl ether. While HMPA has proven to be a versatile co-additive, it was recently found that LiBr can replace this cancer suspect agent.[15] With the chlorosilane modifications, the conventional copper-catalysed conjugate addition of Grignard reagents, once deemed obsolete, has acquired versatility conforming to the standard of modern organic synthesis.

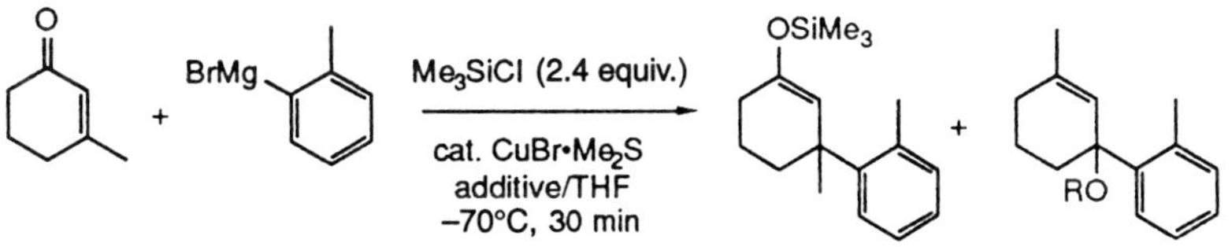

BrMg; Me_3SiCl (2.4 equiv.); cat. $CuBr{\cdot}Me_2S$, additive/THF, –70°C, 30 min; $OSiMe_3$; RO; R = H or Me_3Si (3)[5e]

additive	1,4/1,2	yield (%)
none	18:82	13
HMPA	500:1	100
TMEDA	36:64	16

Protocol 1. Me_3SiCl–copper-mediated conjugate addition of a Grignard reagent[5e]

Caution! Carry out all procedures in a well-ventilated hood, and wear disposable vinyl or latex gloves and chemical-resistant safety goggles.

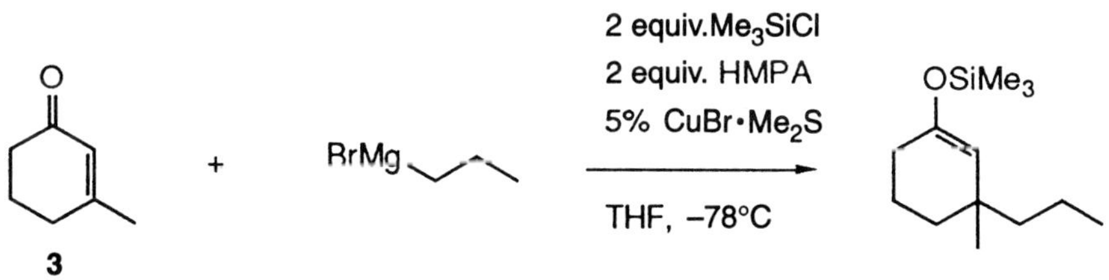

3 + BrMg; 2 equiv. Me_3SiCl, 2 equiv. HMPA, 5% $CuBr{\cdot}Me_2S$, THF, –78°C → $OSiMe_3$ **4**

Equipment

- Three-necked, round-bottomed flask (100 mL) fitted with a condenser, a dropping funnel, and a magnetic stirring bar. A three-way stopcock is fitted to the top of the condenser and connected to a vacuum/nitrogen source
- Vacuum/inert gas source (nitrogen source may be a nitrogen balloon)
- Heating mantle

Materials

- Distilled 3-methylcyclohex-2-en-1-one **3** (FW 110.2) 0.971 g, 8.81 mmol **toxic**
- Distilled propyl bromide (FW 123.0) 1.51 g, 12.3 mmol **flammable, irritant**

Protocol 1. *Continued*

- Magnesium turnings (FW 24.3) 0.32 g, 13.2 mmol — **flammable solid**
- Copper(I) bromide–dimethyl sulfide complex[a] (FW 205.6) 0.091 g, 0.44 mmol — **moisture sensitive**
- Freshly distilled (CaH_2) chlorotrimethylsilane (FW 108.6) 3.15 mL, 24.8 mmol — **flammable, corrosive**
- Distilled hexamethylphosphoric triamide (FW 179.2) 4.29 mL, 24.7 mmol — **cancer suspect agent**
- Distilled triethylamine (FW 101.2) 2.5 mL, 17.9 mmol — **flammable, corrosive**
- Dry, distilled THF — **flammable, irritant**

1. Flame dry the reaction vessel under dry nitrogen. Quickly add the magnesium turnings and a stirring bar while the flask is hot, and then refill the flask with nitrogen through several vacuum cycles. Maintain a slightly positive nitrogen pressure throughout this reaction.
2. When the flask has cooled to room temperature, add a small amount of iodine (*c.* 10 mg). Add THF (2 mL) and a portion of propyl bromide (*c.* 0.2 mL) and let the mixture stand until the iodine colour disappears. When a vigorous reaction starts, initiate stirring, which is maintained throughout the reaction. Dilute the mixture with 10 mL of THF. Add dropwise the remainder of the propyl bromide in 6 mL of THF over a 30 min period while maintaining gentle heat production. Heat the resulting mixture to reflux for and additional 1 h. Cool the mixture to −78°C with a dry ice–acetone bath.
3. Remove the condenser and add copper(I) bromide–dimethyl sulfide complex (while nitrogen is gently blown from the nitrogen inlet). Close the side-arm with a stopper. Add hexamethylphosphoric triamide through the dropping funnel with care to drop it directly into the stirred solution so as to avoid freezing this additive.
4. Add a mixture of 3-methylcyclohex-2-en-1-one and chlorotrimethylsilane in 8 mL of THF during 20 min. After 30 min, warm the mixture to −40°C (dry ice–MeCN bath).
5. After stirring for 30 min, add triethylamine. Add 30 mL of a 1:1 ether–hexane mixture and 1 mL of water to the reaction mixture. Warm the mixture to room temperature and filter the suspension through a pad of Celite®, and wash the filter cake five times with hexane. Wash the filtrate 10 times with 5 mL portions of water. Extract the water layer once with 5 mL of hexane and wash the combined extracts with 5 mL of brine.
6. Concentrate the crude product under vacuum and distil the residue at 7.6 mm Hg to collect 1.82 g (91%) of the product **4** boiling at 87–91°C. Characterize the product by elemental analysis. This sample may contain several per cent of hexamethylphosphoric triamide as determined by ^{1}H NMR, and less than 1% of 1,2-addition product as determined by capillary GLC analysis (capillary column coated with OV-1).

[a] From Aldrich, used as received.

Protocol 2. Me_3SiCl–LiBr-accelerated conjugate addition[15]

Caution! Carry out all procedures in a well-ventilated hood, and wear disposable vinyl or latex gloves and chemical-resistant safety goggles.

methyl vinyl ketone + t-BuMgCl → (2 equiv. Me_3SiCl, 2 equiv. LiBr, 5 mol% CuBr; ether, −40°C) → Me_3SiO-enol ether

(*E*):(*Z*) = 2:1

Equipment

- Two-necked, round-bottomed flask (300 mL) fitted with a dropping funnel and a magnetic stirring bar. A three-way stopcock is connected to a vacuum/nitrogen source
- Vacuum/inert gas source (nitrogen source may be a nitrogen balloon)

Materials

- Freshly distilled methyl vinyl ketone (FW 70.1) 7.01 g, 0.10 mol — **flammable, highly toxic**
- *t*-Butylmagnesium chloride (FW 116.9) 2 M solution in diethyl ether, 60 mL, 0.12 mol — **flammable, moisture, sensitive**
- Dry LiBr (FW 86.85) 17.4 g, 0.20 mol — **hygroscopic**
- Dry, finely powdered copper(I) bromide (FW 143.5) 7.18 g, 0.05 mol — **irritant, hygroscopic**
- Chlorotrimethylsilane[a] (FW 108.6) 21.7 g, 0.20 mol — **flammable, corrosive**
- Distilled triethylamine (FW 101.2) 30 mL, 0.21 mol — **flammable, corrosive**
- Dry, distilled ether — **flammable, irritant**

1. Flame dry the reaction vessel under dry nitrogen. Quickly add LiBr, finely powdered CuBr, and a stirring bar while the flask is hot, and then refill the flask with nitrogen through several vacuum cycles. Maintain a slightly positive nitrogen pressure throughout this reaction.
2. Add 10 mL of dry THF to the flask through the dropping funnel. Add a solution of *t*-butylmagnesium chloride in 100 mL of diethyl ether to the light-green solution while keeping the temperature of the mixture below −40°C with a dry ice–MeCN bath.
3. Add dropwise a mixture of Me_3SiCl and freshly distilled methyl vinyl ketone in 50 mL of diethyl ether over 45 min with efficient stirring. Keep the temperature between −40 and −50°C. After each drop the grey suspension turns yellow for a few seconds.
4. After 15 min at −40°C, warm the reaction mixture to 0°C. Add 30 mL of triethylamine, and pour the mixture into 200 mL of an ice-cold solution of ammonium chloride (30 g).

Protocol 2. *Continued*

5. Extract with ether, and then wash the combined ethereal layers several times with aqueous ammonium chloride until the aqueous phase reaches a pH of *c.* 7. Dry the extract over $MgSO_4$ and concentrate under reduced pressure. Distil the residual liquid at 10–15 mm Hg to obtain the enol silyl ether in 80% yield. The product is >96% pure by ^1H NMR, ^{13}C NMR, GLC analysis, and may be characterized further by elemental analysis.

[a] Freshly distilled from a small amount of *N,N,*-diethylaniline.

In a like manner, Me_3SiCl accelerates the conjugate addition of Gilman reagents (R_2CuLi). While the standard Gilman reagent ($Bu_2CuLi{\cdot}LiI$) itself is virtually inert to 3-methylcyclohexenone at −70°C in THF, Me_3SiCl brings the reaction to instantaneous completion to afford exclusively the 1,4-adduct in quantitative yield after several minutes (eqn 4). The simplicity of the experimental procedure is particularly appealing: to a solution of a Gilman reagent prepared in the usual manner in THF at −78°C is added one or two equivalents of Me_3SiCl and an acceptor olefin either at the same time or successively, and the reaction proceeds instantaneously. *Since the modification of the conventional procedure is so slight, detailed experimental instructions are not given in this chapter.* The reaction may be worked up by following the examples of the copper-catalysed conjugate additions. *t*-$BuMe_2SiCl$ may also achieve considerable rate acceleration,[5a,5d] especially in the presence of HMPA or DMAP. This observation is rather surprising from a mechanistic viewpoint, since *t*-$BuMe_2SiCl$ is unable to silylate enolate or alkoxide anions under the low temperature conditions employed in the chlorosilane-assisted

3-methylcyclohexenone + $Bu_2CuLi{\cdot}LiI$ or $BuCu{\cdot}LiI$ —(R_3SiCl/additive, THF, −70°C, 1 h)→ 3-butyl-3-methyl-1-($OSiR_3$)cyclohexene (4)[5e]

	R_3SiCl/additive	yield (%)
$Bu_2CuLi{\cdot}LiI$	none	0
	Me_3SiCl	99
	t-$BuMe_2SiCl$	0
	t-$BuMe_2SiCl$/HMPA	85
	t-$BuMe_2SiCl$/DMAP	86
$BuCu{\cdot}LiI$	none	0
	Me_3SiCl	~30
	Me_3SiCl/HMPA	89

conjugate addition, suggesting the mechanistic complexity of the chlorosilane acceleration. RCu reagents have an obvious advantage over the R_2CuLi reagents in terms of the R group economy.[16] Me_3SiCl–HMPA or Me_3SiCl–TMEDA[17] assists the conjugate addition of RCu reagents, which are unreactive alkyl donors by themselves. Higher order cuprates, $R_2Cu(CN)Li_2$, also benefit from the presence of Me_3SiCl. The Me_3SiCl acceleration has been recognized for zinc-based stoichiometric reagents, e.g. RCu(CN)ZnI.[6] Me_3SiI,[18] Me_3SiOTf,[10] and Me_3SiCl–$BF_3{\cdot}Et_2O$[19] may be used as more powerful accelerators for the less reactive substrates, including unsaturated esters.

Protocol 3.
Me_3SiCl–copper-mediated conjugate addition of a functionalized organozinc reagent[20]

Caution! Carry out all procedures in a well-ventilated hood, and wear disposable vinyl or latex gloves and chemical-resistant safety goggles.

I, COOMe → [Zn/Me_3SiCl, $BrCH_2CH_2Br$, THF] → ZnI, COOMe → [(1) CuCN•2LiCl/Me_3SiCl; (2) $CH_3COCH{=}C(CH_3)_2$; (3) H_3O^+] → O, COOMe

Equipment

- A three-necked, round-bottomed flask (200 mL) fitted with a condenser, a dropping funnel, a glass stopper, and a magnetic stirring bar. A three-way stopcock is fitted to the top of the condenser and connected to a vacuum/nitrogen source
- A two-necked, round bottomed flask (200 mL) fitted with a septum, a nitrogen inlet, and a magnetic stirring bar
- Vacuum/inert gas source (nitrogen source may be a nitrogen balloon)
- Syringes
- Schlenk flask (50 mL)
- Schlenk flask (100 mL)

Materials

• Activated zinc powder[a] (FW 55.4) 6.13 g, 93.7 mmol	**flammable solid, irritant**
• Methyl 6-iodohexanoate[21] (FW 256.1) 12.0 g, 46.9 mmol	**irritant**
• 1,2-Dibromoethane (FW 187.9) 0.2 mL, 2.32 mmol	**toxic, cancer suspect agent**
• Freshly distilled (CaH_2) chlorotrimethylsilane (FW 108.6) 0.2 mL, 1.6 mmol and 4.57 mL, 36 mmol	**flammable, corrosive**
• Distilled 4-methylpent-3-en-2-one (FW 98.1) 2.29 mL, 20 mmol	**flammable, lachrymator**
• 1 M CuCN·2LiCl solution[b] (FW 174.4) 43.2 mL, 43.2 mmol	**flammable, highly toxic**
• Copper(I) cyanide (FW 89.6) 3.87 g, 43.2 mmol	**highly toxic, irritant**
• Lithium chloride (FW 42.4) 3.69 g, 87 mmol	**irritant**
• Dry, distilled THF	**flammable, irritant**

1. Flame dry the three-necked flask under dry nitrogen. Quickly add zinc powder while the flask is hot, and then refill the flask with nitrogen through

Protocol 3. *Continued*

several vacuum cycles. Maintain a slightly positive nitrogen pressure throughout the reaction.

2. When the flask has cooled to room temperature, add THF (4 mL) and 1,2-dibromoethane and stir the mixture at 60°C for a few minutes. Cool the mixture to room temperature and add chlorotrimethylsilane (0.2 mL). Add methyl 6-iodohexanoate and then THF (44 mL) and heat the mixture to reflux. Stop stirring and let the mixture stand overnight. Monitor the disappearance of the iodide by TLC (visualized with UV). This procedure will make a *c.* 0.9 M solution of the desired zinc reagent.
3. Flame dry and fill the two-necked 200 mL flask with nitrogen as above. Add a 40 mL THF solution of the above reagent (36 mmol) with the aid of a syringe through the septum, and cool the flask to −10°C. Add a 1 M THF solution of CuCN·2LiCl (43.2 mL) and stir the mixture at −10°C for 30 min. Add 4.57 mL (36 mmol) of chlorotrimethylsilane, and then 4-methylpent-3-en-2-one. Remove the cooling bath, and stir the mixture at room temperature for 6 h.
4. Add saturated aqueous ammonium chloride solution (100 mL). Separate the organic layer with the aid of a separating funnel. Extract the aqueous layer with hexane (30 mL). Wash the combined organic layers with brine (30 mL), then dry them over anhydrous magnesium sulfate. Remove the desiccant and concentrate the filtrate. Dissolve the crude product in methanol (50 mL) and add concentrated sulfuric acid (0.5 mL). After stirring for 30 min at room temperature, add water (50 mL) and ether (50 mL). Separate the organic layer and extract the aqueous layer with ether (50 mL). Wash the combined organic layers with saturated sodium bicarbonate and then with water. After drying over anhydrous magnesium sulfate and filtration, the filtrate is concentrated and distilled. Distil the desired product (3.83 g, 84%) as a fraction boiling at 124°C at 0.3 mm Hg and characterize by IR, ^{1}H NMR, and elemental analysis.

[a] Wash 10 g of commercial zinc powder (~100 mesh) several times with 30 mL of 1 M HCl. Then wash successively with 30 mL each of ethanol and ether. Transfer the powder to a 50 mL Schlenk flask and dry at 100°C for 5 h; store under argon.

[b] Flame dry a 100 mL Schlenk flask under dry argon. Quickly add copper(I) cyanide (3.87 g, 43.2 mmol), lithium chloride (3.69 g, 87 mmol) (previously dried under vacuum at 150°C for 2 h), and a stirring bar while the flask is hot, and then refill the flask with argon through several vacuum cycles. Maintain a slightly positive argon pressure throughout this reaction. Add 43 mL of THF and stir the mixture for 3 h at room temperature under argon. Let the clear solution stand overnight.

Table 6.1 illustrates various examples of halosilane-accelerated conjugate additions. The first few examples illustrate the successful use of various unsaturated aldehydes both in catalytic and stoichiometric chemistry. The high (*E*)-stereoselectivity and the high yield are of particular note, providing a unique entry to stereodefined aldehyde enolates (entry 1). Conjugate addition to an enal in entry 2 was used as a key step in the synthesis of (+)-olivin.[22] Conjugate addition to a lactam led to a total synthesis of the alkaloid geissoschizine (entry 3).[23] The chlorosilane-mediated reaction overcomes considerable steric hindrance at the electrophilic centre of the substrate olefin: thus, sterically hindered isophorone reacted smoothly with an *o*-methoxyphenyl Grignard reagent to provide a starting material for the synthesis of (±)-membranolide (entry 4).[24] Entry 5 illustrates the use of Me_3SiOTf in a grandisol synthesis, wherein Me_2CuLi was added to a hindered ester.[25] Addition to γ-amino-α,β-unsaturated esters proceeded smoothly only in the presence of Me_3SiCl to give the desired adducts with very high stereoselectivity (entry 6).[26] The chemoselectivity of the reaction may be illustrated for an intermediate in the cortisone synthesis (entry 7), wherein various other vinylcopper reagents gave very capricious yields.[27] Enantioselectivity of the conjugate addition in the presence of a chiral diamine has been found to improve dramatically in the presence of a chlorosilane, the structure of which also significantly affects the enantioselectivity (entry 8).[28]

The chlorosilane methodology is also effective for organocopper reagents prepared by non-conventional methods. The reaction in entry 9 illustrates the use of an organocopper reagent prepared directly from the corresponding alkyl halide.[29] Entries 10 and 11 illustrate the use of organozinc and organocadmium reagents for conjugate additions in the presence of chlorosilanes. In entry 10, an amino acid derivative has been employed successfully without racemization of the chiral centre.[30] Entry 11 shows the conjugate addition of a functionalized cadmium–copper complex reagent.[31] Conjugate additions to acetylenic carbonyl compounds are also accelerated by chlorosilanes (entry 12).[4]

A change in the stereochemical course of the conjugate addition under the influence of a chlorosilane has been noted in many instances.[28,32] For instance, the addition to the 4-alkoxy spiroenone in eqn (5) showed complete reversal of the selectivity in the presence and the absence of Me_3SiCl.[5a] Also notable is the enhancement of enantioselectivity in chiral ligand-induced conjugate addition reactions in the presence of chlorosilanes.[29]

Me_2CuLi/Me_3SiCl, THF, −78°C → >99% diastereoselectivity

Me_2CuLi, THF, −78°C–0°C → 92% diastereoselectivity

(5)[5a]

Table 6.1. Chlorosilane-accelerated conjugate additions of organocopper reagents[a]

Entry	Carbonyl compound	Copper reagent	Product
1	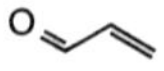	BuMgBr/cat. CuBr•Me_2S/Me_3SiCl or Bu_2CuLi/Me_3SiCl	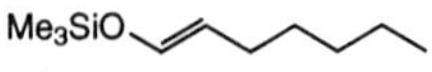 Me_3SiO; 95%, >95% (*E*)
2	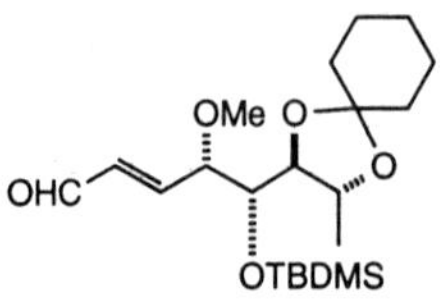 OHC, OMe, O, O, OTBDMS	$(CH_2{=}CH)_2CuLi/Me_3SiCl$	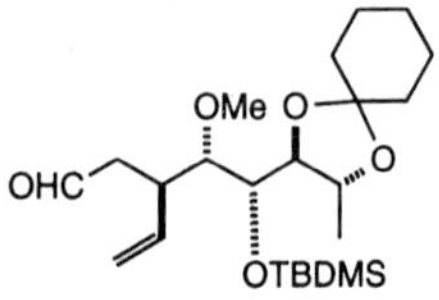 OHC, OMe, O, O, OTBDMS; 91%
3	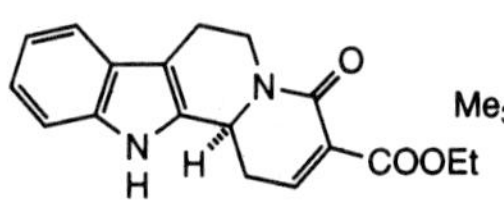 N, H, H, N, O, COOEt	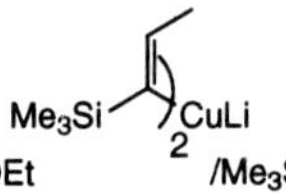 Me_3Si, CuLi, 2, $/Me_3SiCl$	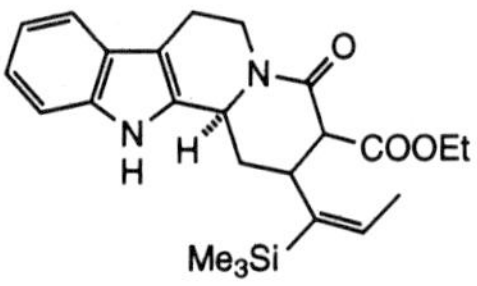 N, H, H, N, O, COOEt, Me_3Si; 75%
4	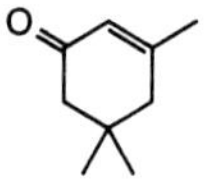 O	*o*-$MeOC_6H_4MgBr/Me_3SiCl$	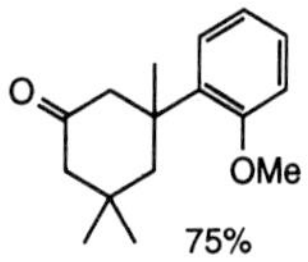 O, OMe; 75%
5	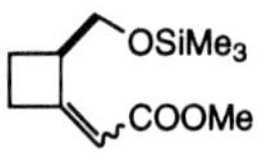 $OSiMe_3$, COOMe	Me_2CuLi/Me_3SiOTf	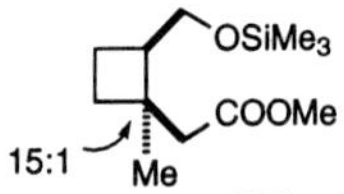 $OSiMe_3$, COOMe, 15:1, Me; >66%
6	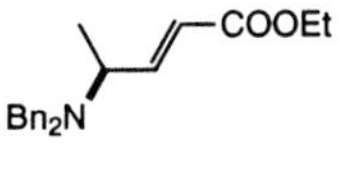 Bn_2N, COOEt	Ph_2CuLi/Me_3SiCl	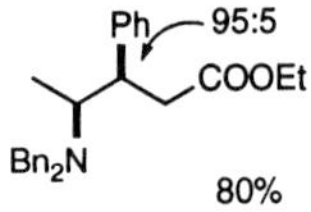 Ph, 95:5, COOEt, Bn_2N; 80%

[a] Conjugate addition products form initially as enol silyl ethers, which may be hydrolysed during work-up either intentionally or unintentionally. Cat* in entry 8 is

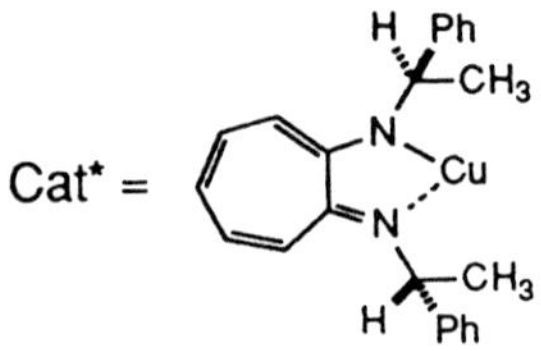

Cat* = (H, Ph, CH_3, N, Cu, N, CH_3, H, Ph)

Entry	Carbonyl compound	Copper reagent	Product
7	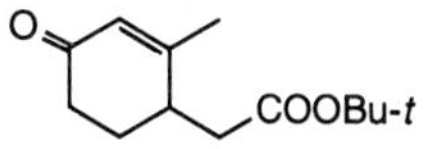	$Me_2C{=}CHMgBr$/ Me_3SiCl	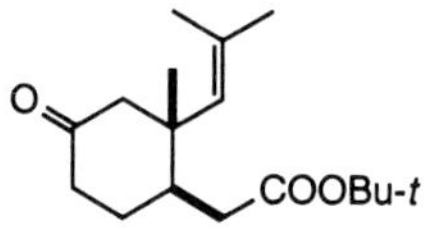 >81%
8	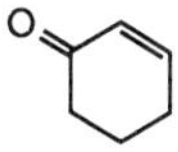	BuMgCl/Ph_2(*t*-Bu)SiCl/HMPA Cat*	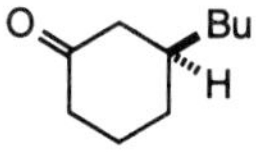 97% (74-78%ee)
9	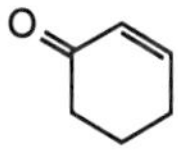	$Br(CH_2)_3COOEt$ $CuI{\bullet}PBu_3$/Li/Naphthalene	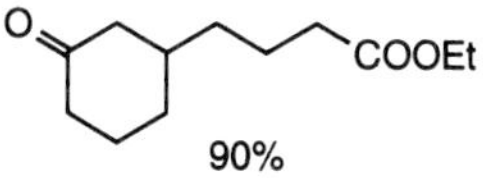 90%
10	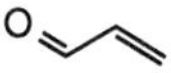	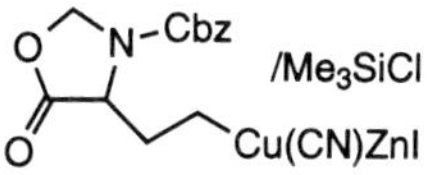 N-Cbz /Me_3SiCl Cu(CN)ZnI	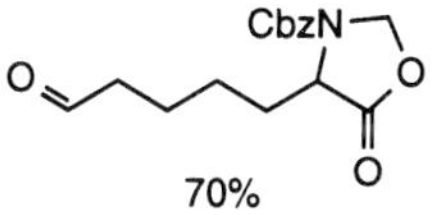 70%
11	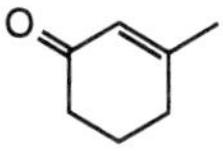	$Me_3COOCH_2Cu(CN)CdI$/ Me_3SiCl	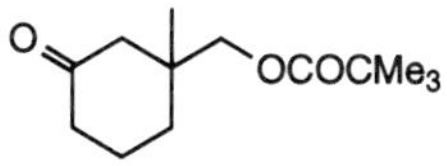 71%
12	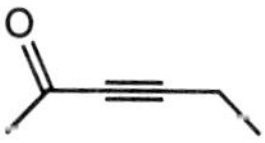	$Zn(CH_2CH_2COOEt)_2$	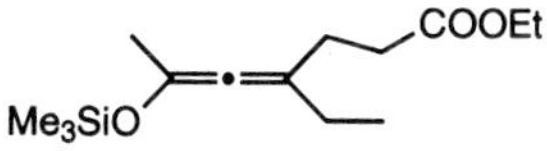 73%

Little mechanistic information on the Me_3SiCl acceleration is available. At a glance, chlorosilane appears simply to trap the enolate anion formed by the standard mechanism of the conjugate addition. However, this simplistic 'enolate trapping' rationale does not account for the fact that chlorosilane-assisted conjugate addition reactions always take place much faster than the reaction of an enolate anion with chlorosilane. In addition, the stereochemistry with respect to the enolate or to the chiral centre which is formed by the conjugate addition often changes in the presence of Me_3SiCl. Such evidence clearly indicates that all three reactants—chlorosilane, copper reagent, and substrate—cooperate in the transition state to effect dramatic rate enhancement of the conjugate addition.

3. Chlorosilane-assisted 1,2-additions

Me_3SiCl also accelerates 1,2-addition reactions of Gilman reagents to aldehydes and changes the stereoselectivity of the Cram-type diastereoselective carbonyl additions.[12] Similarly, the stereoselectivity of a higher order cuprate may also be significantly improved.[33] It was recently reported that Gilman reagents undergo chlorosilane-assisted 1,2-addition to certain α,β-unsaturated aldehydes possessing a γ-chiral centre, wherein a high level of 1,4-asymmetric induction was observed. Thus, the 1,2-addition of an alkylcuprate to the aldehyde carbon of a γ-amino-[34] or γ-alkoxy-α,β-unsaturated aldehyde[35] (eqn 6) was found to take place with good diastereoselectivity under the stereochemical influence of the remote γ-position. The reaction represents the first case of the vinylogous Cram's rule.

H OMOM ... O, H —[Bu_2CuLi/R^3Me_2SiCl ; R^3 = Me or *t*-Bu]→ H OMOM ... Bu, R^3Me_2SiO H (6)[35]

The chorosilane acceleration is being accepted as a standard procedure in organocopper chemistry, and further applications have appeared in the recent literature.[36,37]

References

1. See Chapters 1, 7, 8, and 9 for pertinent references.
2. Chuit, C,; Foulon, J. P.; Normant, F. J. *Tetrahedron* **1980**, *36*, 2305–2310.
3. Nakamura, E.; Kuwajima, I. *J. Am. Chem. Soc.* **1984**, *106*, 3368–3370. Enda, J.; Kuwajima, I. *J. Am. Chem. Soc.* **1985**, *107*, 5495.
4. Nakamura, E.; Kuwajima, I. *Org. Synth.* **1987**, *66*, 43. Nakamura, E.; Aoki, S.; Sekiya, K.; Oshino, H.; Kuwajima, I. *J. Am. Chem. Soc.* **1987**, *109*, 8056–8066.

5. (a) Corey, E. J.; Boaz, N. W. *Tetrahedron Lett.* **1985**, *26*, 6015–6018, 6019–6021. (b) Alexakis, A.; Berlan, J.; Besace, Y. *Tetrahedron Lett.* **1986**, *27*, 1047–1050. (c) Horiguchi, Y.; Matsuzawa, S.; Nakamura, E.; Kuwajima, I. *Tetrahedron Lett.* **1986**, *27*, 4025–4028. (d) Nakamura, E.; Matsuzawa, S.; Horiguchi, Y.; Kuwajima, I. *Tetrahedron Lett.* **1986**, *27*, 4029–4032. (e) Matsuzawa, S.; Horiguchi, Y.; Nakamura, E.; Kuwajima, I. *Tetrahedron* **1989**, *45*, 349–362.
6. Tamaru, Y.; Ochiai, H.; Nakamura, T.; Yoshida, Z.-i. *Angew. Chem., Int. Ed. Engl.* **1987**, *26*, 1157.
7. Arai, M.; Nakamura, E.; Lipshutz, B. H. *J. Org. Chem.* **1991**, *56*, 5489–5491.
8. Cahiez, G.; Alami, M. *Tetrahedron Lett.* **1990**, *31*, 7423–7424.
9. Bertz, S. H.; Gibson, C. P.; Dabbagh, G. *Organometallics* **1988**, *7*, 227–232.
10. Kim, S.; Lee, J. M. *Tetrahedron Lett.* **1990**, *31*, 7627–7630.
11. Mukaiyama, T.; Iwasawa, N.; Yura, T.; Clark, R. S. J. *Tetrahedron* **1987**, *43*, 5003–5017.
12. Matsuzawa, S.; Isaka, M.; Nakamura, E.; Kuwajima, I. *Tetrahedron Lett.* **1989**, *30*, 1975–1978.
13. Oshino, H.; Nakamura, E.; Kuwajima, I. *J. Org. Chem.* **1985**, *50*, 2802–2804.
14. Sakata, H.; Aoki, Y.; Kuwajima, I. *Tetrahedron Lett.* **1990**, *31*, 1161–1164.
15. Andringa, H.; Oosterveld, I.; Brandsma, L. *Synth. Commun.* **1991**, *21*, 1393–1396.
16. Bertz, S. H.; Dabbagh, G. *Tetrahedron* **1989**, *45*, 425–434.
17. Johnson, C. R.; Marren, T. J. *Tetrahedron Lett.* **1987**, *28*, 27–30.
18. Bergdahl, M.; Lindstedt, E.-L.; Olsson, T. *J. Organomet. Chem.* **1989**, *365*, C11–C14. Bergdahl, M.; Lindstedt, E.-L.; Nilsson, M.; Olsson, T. *Tetrahedron* **1989**, *45*, 535–543.
19. Sakata, H.; Kuwajima, I. *Tetrahedron Lett.* **1987**, *28*, 5719–5722.
20. Sato, F.; Yamamoto, K.; Imamoto, T. (eds). *Jikken Yukikinzoku Kagaku (Experimental Organometallic Chemistry)*; Kodansha: Tokyo, 1992; pp. 53–55.
21. Olah, G. A.; Narang, S. C.; Gupta, B. G. B.; Malhotra, R. *J. Org. Chem.* **1979**, *44*, 1247–1251.
22. Roush, W. R.; Michaelides, M. R.; Tai, D. F.; Lesur, B. M.; Chong, W. K. M.; Harris, D. J. *J. Am. Chem. Soc.* **1989**, *111*, 2984–2995.
23. Overman, L. E.; Robichaud, A. J. *J. Am. Chem. Soc.* **1989**, *111*, 300–308.
24. Yoo, S.-e.; Yi, K. Y. *Synlett* **1990**, 697–699.
25. Narasaka, K.; Kusama, H.; Hayashi, Y. *Bull. Chem. Soc. Jpn.* **1991**, *64*, 1471–1478.
26. Reetz, M. T.; Röhrig, D. *Angew. Chem., Int. Ed. Engl.* **1989**, *28*, 1706–1709. Jako, I.; Uiber, P.; Mann, A.; Wermuth, C.-G. *J. Org. Chem.* **1991**, *56*, 5729–5733. Scolastico, C. *Pure Appl. Chem.* **1988**, *60*, 1689–1698.
27. Horiguchi, Y.; Nakamura, E., Kuwajima, I. *J. Am. Chem. Soc.* **1989**, *111*, 6257–6265.
28. Ahn, K.-H.; Klassen, R. B.; Lippard, S. J. *Organometallics* **1990**, *9*, 3178–3181. Bert, S. H.; Dabbagh, G.; Sundararajan, G. *J. Org. Chem.* **1986**, *51*, 4953–4959.
29. Wehmeyer, R. M.; Rieke, R. D. *J. Org. Chem.* **1987**, *52*, 5056–5057.
30. Tamaru, Y.; Tanigawa, H.; Yamamoto, T.; Yoshida, Z. *Angew. Chem., Int. Ed. Engl.* **1989**, *28*, 351.
31. Knochel, P.; Chou, T.-S.; Chem, H. G.; Yeh, M. C. P.; Rozema, M. J. *J. Org.*

Chem. **1989**, *54*, 5202–5204. Linderman, R. J.; Griedel, B. D. *J. Org. Chem.* **1990**, *55*, 5428–5430.

32. Alexakis, A.; Sedrani, R.; Mangeney, P. *Tetrahedron Lett.* **1990**, *31*, 345–348. Zhao, S.-K.; Helquist, P. *Tetrahedron Lett.* **1991**, *32*, 447–448.
33. Lipshutz, B. H.; Ellsworth, E. L.; Siahaan, T. J.; Shirazi, A. *Tetrahedron Lett.* **1988**, *29*, 6677–6680.
34. Reetz, M.; Wang, F.; Harms, K. *J. Chem. Soc., Chem. Commun.* **1991**, 1309–1311.
35. Arai, M.; Nemoto, T.; Ohashi, Y.; Nakamura, E. *Synlett* **1992**, 309–310.
36. Lipshutz, B. H.; Dimock, S. H.; James, B. *J. Am. Chem Soc.* **1993**, *115*, 9283–9284. Chounan, Y.; Yamamoto, Y. *Chemtracts* **1993**, *6*, 310–314.
37. Lipshutz, B. H.; James, B. *Tetrahedron Lett.* **1993**, *34*, 6689–6692. Bergdahl, M.; Eriksson, M.; Nilsson, M.; Olsson, T. *J. Am. Chem. Soc.* **1993**, *115*, 9283–9284.

7

Boron trifluoride/aluminium trichloride-mediated conjugate addition and substitution reactions

TOSHIRO IBUKA and YOSHINORI YAMAMOTO

1. Introduction

The conjugate addition and the substitution reactions of ordinary organocopper reagents are the most important carbon–carbon bond-forming processes in synthetic organic chemistry today.[1,2] However, owing to either moderate reactivity of the conventional organocopper reagents or severe steric hindrance in the substrates, the attempted conjugate addition to certain α,β-unsaturated carbonyl compounds or substitution to allylic and propargylic systems often gives none of the desired products. Failure to react with ordinary organocuprates has undoubtedly detracted from the application of organocopper reactions in synthesis.

In 1977 Yamamoto noticed that when RCu was treated with boron trifluoride etherate, a reagent more reactive than RCu or a Gilman-type reagent was formed corresponding to the stoichiometry $RCu{\cdot}BF_3$ (eqn 1).[3–5] A few years later, Yamamoto and Ibuka discovered that a mixture composed of RCu and $AlCl_3$ adds to 'unreactive' α,β-enones in a stereoselective manner and showed the synthetic potential of these reagents as Michael donors of an 'R^-' group (eqn 2).[6–9]

$$RCu + BF_3{\cdot}Et_2O \rightarrow \text{`}RCu{\cdot}BF_3\text{'} \quad (1)$$

$$RCu + AlCl_3 \rightarrow \text{`}RCu{\cdot}AlCl_3\text{'} \quad (2)$$

$$RLi + TiCl_4 \rightarrow RTiCl_3 + LiCl \quad (3)$$

$$2RMgX + ZnCl_2 \rightarrow R_2Zn + 2Mg(X)Cl \quad (4)$$

$$3PhMgBr + AlCl_3 \rightarrow Ph_3Al + 3Mg(Br)Cl \quad (5)$$

Organocopper compounds do coexist with BF_3 or $AlCl_3$ at low temperatures (eqns 1 and 2), and the two species are not so easily transmetallated as can be seen in the reaction of organometallics with $TiCl_4$, $ZnCl_2$, or $AlCl_3$ (eqns 3–5).

The admixture of $BF_3 \cdot Et_2O$ or $AlCl_3$ with certain Gilman reagents and higher order cyanocuprates has become an effective technique for enhancing carbon–carbon bond formation.[2,10] Other Lewis acids ($ZnCl_2$, $TiCl_4$, $SnCl_4$) are incompatible with higher order cyanocuprates even at −78°C, and form intractable gums, suggesting a lack of compatibility.[10] Note that the $RCu \cdot BF_3$ and $RCu \cdot AlCl_3$ notations are written merely to indicate the molar ratio of RCu to $BF_3 \cdot Et_2O$ and $AlCl_3$, though this notation is frequently employed in publications. Although the actual nature of the species responsible for the synthetic contribution is not clear, organocopper(I)–BF_3 or –$AlCl_3$ reagents have been acclaimed as some of the most useful of known organocopper reagents.

The effects of $BF_3 \cdot Et_2O$ on RCu reactions in THF have recently been investigated by Lipshutz. From NMR spectral and chemical studies, it was concluded that both BF_3 and lithium salts such as lithium iodide are essential ingredients for the generation of a reactive species.[11] More importantly, the $BF_3 \cdot Et_2O$ serves to modify the RCu,[11] R_2CuLi,[12] and $R_2Cu(CN)Li_2$,[13] the result of which affords a more reactive, different alkylcopper–BF_3 combination.

The choice of solvent and copper(I) salt (CuX, where X = Br, I, CN, etc.) and the need for additives all contribute to the complexity of determining which reagent is best suited for the substrate in question. In certain situations, reagents formed from CuI + 2RLi + $BF_3 \cdot Et_2O$ behave in a remarkably different manner from those prepared from CuCN + 2RLi + $BF_3 \cdot Et_2O$.[14,15] The reactivity of organocopper–Lewis acid complexes can be altered by changing several parameters: (1) the choice of Lewis acid ($BF_3 \cdot Et_2O$, $AlCl_3$, R_3SiCl, etc.); (2) the choice of solvent (tetrahydrofuran, ether, HMPA, etc.); (3) the source of copper(I) salt (CuX, where X = Cl, Br, I, CN, SCN, etc.); (4) the ratio of organometallic (R—M, where M = Li or MgX) to CuX; (5) the inorganic salt involved (LiX, MgX_2, etc.); and (6) the presence of an additive (trialkylphosphines, dialkyl sulfides, etc.).

2. Addition reactions

2.1 Additions to α,β-unsaturated acids and esters

In spite of the fact that the conjugate addition reaction of an organocopper reagent to an α,β-unsaturated acid or ester is an indispensable synthetic operation, the attempted conjugate additions to these substrates with conventional dialkylcuprates are reported to lead to poor results.[1,2,10] α,β-Enoate **1** and α,β-enoic acid **3** are challenging examples. Compounds **1** and **3** gave 1,4-alkylated ester **2** and acid **4**, respectively, in acceptable yields by treatment with *n*-$BuCu \cdot BF_3$.[4,5]

n-Butylcopper–triethylborane adds to dimethyl acetylenedicarboxylate **5** with a high stereospecificity at −70°C, a reaction which cannot be achieved with reagents such as *n*-Bu_2CuLi and *n*-BuCu.[18]

n-BuCu.BF_3 (52%) / n-Bu_2CuLi (<1%): 1 → 2

n-BuCu.BF_3 (81%) / n-Bu_2CuLi (<1%): 3 → 4

$MeO_2C{\equiv}C$-CO_2Me (5) → 6 + 7

		6		7
n-BuCu.BEt_3	(85%)	>99	:	<1
n-Bu_2CuLi	(98%)	60	:	40

Protocol 1. Synthesis of 3-methylheptanoic acid. Reaction of an α,β-unsaturated carboxylic acid with RCu·BF_3

The following procedure for the synthesis of 3-methylheptanoic acid **4** by reaction of (*E*)-2-butenoic acid 3 with *n*-BuCu·BF_3 is representative.[4]

Caution! Carry out all procedures in a well-ventilated hood, and wear disposable vinyl or latex gloves and chemical-resistant safety goggles. Because of their high reactivity and relatively low thermal stability, organocopper–Lewis acid reagents should be prepared *in situ* and used immediately.

Equipment

- Magnetic stirrer
- Three-necked, round-bottomed flask[a] (200 mL)
- Two-way and three-way stopcocks
- Septum (Aldrich Z10,075–7 or Z10,074–9)
- Teflon-coated magnetic stirring bar (octagonal or egg-shaped bar: 2.5 × 0.8 cm)
- All-glass syringe with a needle-lock Luer (volume appropriate for quantity of solution to be transferred)
- Medium-gauge needle (20 or 22 gauge, 6 in long)
- Source of dry argon or nitrogen (preferably from an argon or a nitrogen line: see Chapter 2, Section 6.2 for a suitable arrangement)
- Water-jacketed, semi-micro, short-path distillation apparatus

Materials

- Dry ether[b] 65 mL — **flammable, irritant**
- Copper(I) iodide[c] (FW 190.4) 5.7 g, 30 mmol — **irritant**
- (*E*)-2-Butenoic acid[d] (FW 86.1) 0.86 g, 10 mmol — **corrosive, toxic**
- Boron trifluoride etherate[e] (FW 141.9) 3.9 mL, 30 mmol — **corrosive, moisture sensitive**
- *n*-Butyllithium in *n*-hexane[f] (FW 64.1) 1.3 M solution, 23.1 mL, 30 mmol — **flammable, moisture sensitive**
- Technical ether for extraction 120 mL — **flammable, irritant**

Protocol 1. *Continued*

1. Clean all glassware, syringes, needles, and stirring bar and dry for at least 2 h in a 105°C electric oven before use.
2. Assemble the flask, stirrer bar, and stopcocks under argon while the apparatus is still hot.
3. Support the assembled flask using a clamp and a stand with a heavy base.
4. Put purified copper(I) iodide (5.7 g, 30 mmol) into the flask, and equip the neck of the flask with a septum as shown in Fig. 7.1.
5. Dry the apparatus with an electric heat gun under vacuum (0.5–1 mm Hg) for 5 min, then back-fill the flask with argon. Repeat to a total of three times. Do not allow the apparatus containing copper(I) iodide to remain in direct heat for a long time. Overheating under reduced pressure often leads to decomposition of copper(I) iodide.
6. Cool the flask with a dry ice–ethanol or dry ice–acetone bath to −40°C (bath temperature) under a slow stream of argon.
7. Assemble the syringe and needle while hot and allow the assembled syringe to cool to room temperature in a desiccator. Flush the syringe with argon or nitrogen (see Fig. 7.2).
8. Charge the flask with dry ether (60 mL) using a syringe by piercing the septum on the reaction flask.
9. Support the bottle containing *n*-butyllithium in hexane using a ring or clamp and a stand with a heavy base.
10. Fill a syringe with *n*-butyllithium in hexane from the bottle containing *n*-butyllithium using argon pressure. Apply the argon or nitrogen pressure to fill the syringe slowly with the required volume. The argon or nitrogen pressure pushes the plunger back as the *n*-butyllithium in hexane enters the syringe (see Fig. 7.3). Do not pull back the plunger: this tends to cause leaks and generate gas bubbles in the syringe. Transfer the reagent in the syringe to the reaction flask by piercing septum on the flask.
11. Keep the flask temperature at −40°C (dry ice–ethanol bath temperature) and add *n*-butyllithium dropwise from the syringe, through the septum on the reaction flask over a 5 min period with stirring using a magnetic stirrer. Stir the mixture at 40 to −30°C for 10 min. During the reaction, keep the reaction apparatus under a slight pressure of argon or nitrogen. Temperature variation can affect both the rate of formation of organocopper reagents and their stability. Optimum conditions for organocopper formation require careful control of reaction temperature. During this period, the slurry turns dark brownish-black.
12. Cool the reaction flask with a dry ice–acetone bath to −70°C. Stir the mixture vigorously and add boron trifluoride etherate dropwise by means of a syringe through the septum on the reaction flask.

13. Add (*E*)-2-butenoic acid in 5 mL of dry ether dropwise via a syringe at −70 °C with stirring. Allow the mixture to warm to ambient temperature with stirring and stir for an additional 15 min.
14. Recool the reaction flask to −70 °C in a dry ice–acetone bath, and then quench the mixture with distilled water (25 mL) with vigorous stirring. Remove the dry ice–acetone bath and allow the temperature to rise to ambient temperature. The colour of the ethereal layer turns gradually through reddish-brown to grey to nearly colourless.
15. Filter the mixture through Celite® with suction to remove any inorganic precipitate, and rinse the flask and the filter cake with three 30 mL portions of ether.
16. Transfer the filtrate to a separatory funnel and separate the two layers. Extract the water layer with ether (30 mL). Combine the ether layers and wash with two 20 mL portions of saturated brine.
17. Transfer the ethereal layer to a 200 mL flask. Dry the layer over anhydrous magnesium sulfate, and filter through a filter paper. Concentrate the filtrate under reduced pressure using a rotary evaporator (25 °C/40 mm Hg).
18. Transfer the oily residue to a water-jacketed, semi-micro, short-path distillation apparatus equipped with a thermometer. Distil the crude product under reduced pressure to obtain 3-methylheptanoic acid (b.p. 75–76 °C/1 mm Hg; 1.16 g, 81% yield) as a colourless oil which displays the appropriate ^1H NMR (in $CDCl_3$) and combustion analytical data.

[a] For volumes smaller than 20 mL, it is convenient to use the apparatus shown in *Fig. 7.4.*
[b] Distil ether from lithium aluminum hydride or benzophenone ketyl under an inert atmosphere (nitrogen or argon) and use immediately (see Chapter 2, Section 4).
[c] Commercially available copper(I) iodide retains moisture and air-dried copper(I) iodide contains *c.* 4% water. Purify technical-grade CuI by published methods (see Chapter 2, Protocol 2).[16,17] Dry the purified CuI for 24 h in a phosphorus pentoxide drying pistol at 0.5 mm Hg at 100 °C. Extra-pure CuI (99.999% purity) can be purchased from Aldrich. Generally, since the copper(I) iodide used to prepare the organocopper–Lewis acid reagents is not too hygroscopic, the use of a glove-box or a dry-box procedure is not necessary.
[d] Commercially available (*E*)-2-butenoic acid (crotonic acid) can be used without purification.
[e] Commercially available boron trifluoride etherate darkens rapidly because of air oxidation. Boron trifluoride etherate can be purified by addition of ether (1 mL) to boron trifluoride etherate (50 mL) followed by distillation from calcium hydride (0.2 g) in an all-glass apparatus at 46 °C/10 mm Hg and should be stored in a brown glass bottle. The hydride reduces bumping and removes volatile acids. Purified and redistilled boron trifluoride etherate can be purchased from Aldrich.
[f] Titrate *n*-butyllithium from Kanto with a 1.00 M solution of *s*-butanol in xylene using 1,10-phenanthroline as the indicator just prior to use (see Chapter 2, Protocol 6).

Although it has been reported that reactions of electron-deficient alkenes with some organometallic reagents yield a complex mixture of products, the reaction of diethyl fumarate **8** in Et_2O with the reagent *n*-$Bu_2CuLi{\cdot}AlCl_3$ gives predominantly a conjugate adduct, diethyl butylsuccinate **10**. Somewhat surprisingly, diethyl maleate **9** yields predominantly the reduction product **11**.

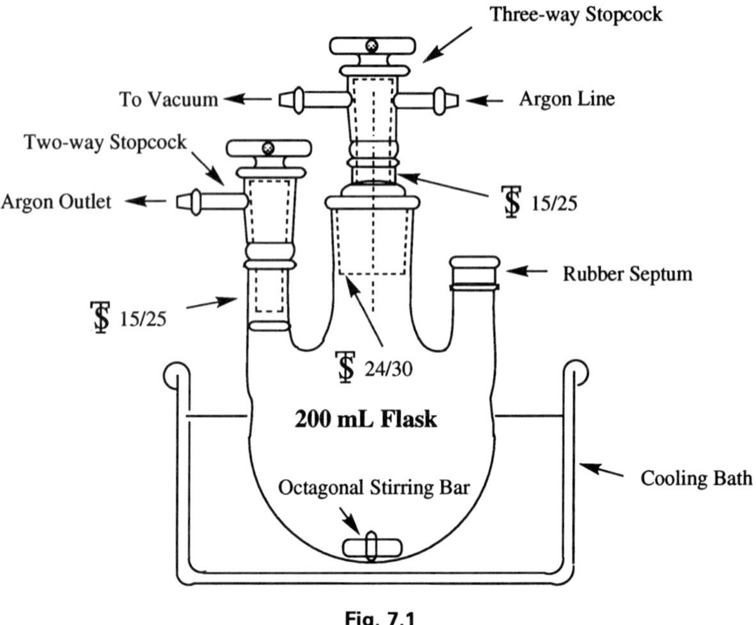

Fig. 7.1

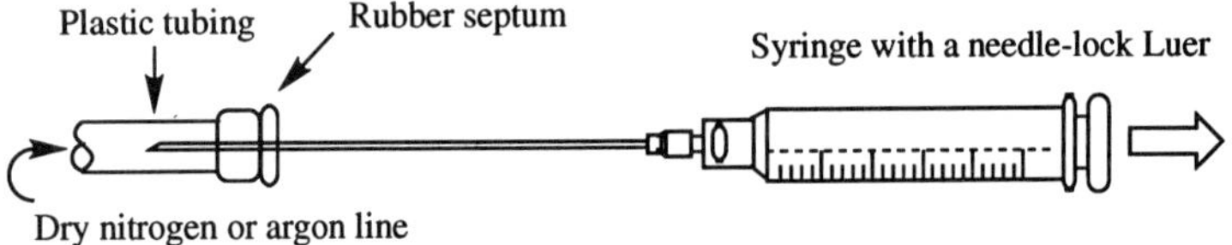

Fig. 7.2

Apparently, the direction of the reaction path is dependent upon the geometry of the double bond. In these reactions, Et_2O is the solvent of choice, because reaction in THF or in a mixture of Et_2O and THF is too sluggish to be practicable.[19,20]

Substrate	Reagent	**10** (EtO_2C, H / *n*-Bu, CO_2Et)	**11** (EtO_2C, CO_2Et / H, H)
8 (EtO_2C, H / H, CO_2Et)	$n\text{–}Bu_2CuLi.AlCl_3$	94%	trace
9 (EtO_2C, CO_2Et / H, H)	$n\text{–}Bu_2CuLi.AlCl_3$	2%	92%

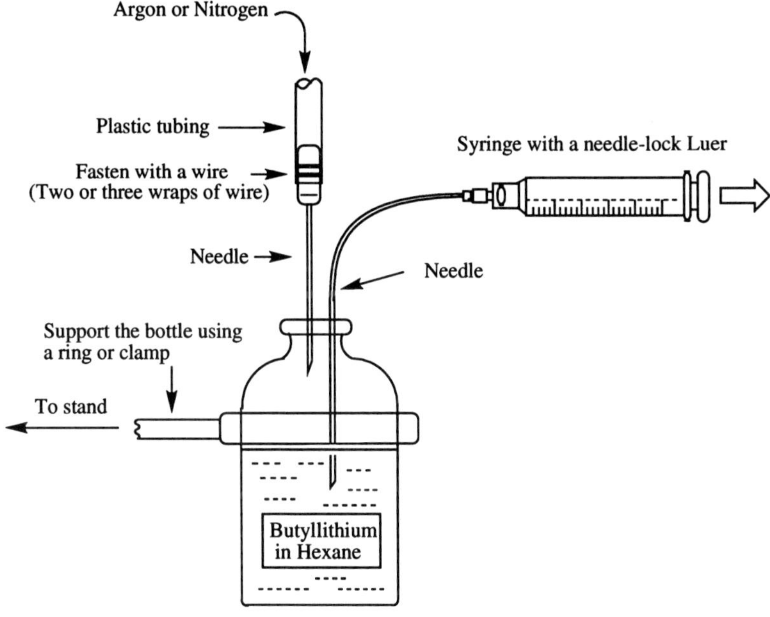

Fig. 7.3

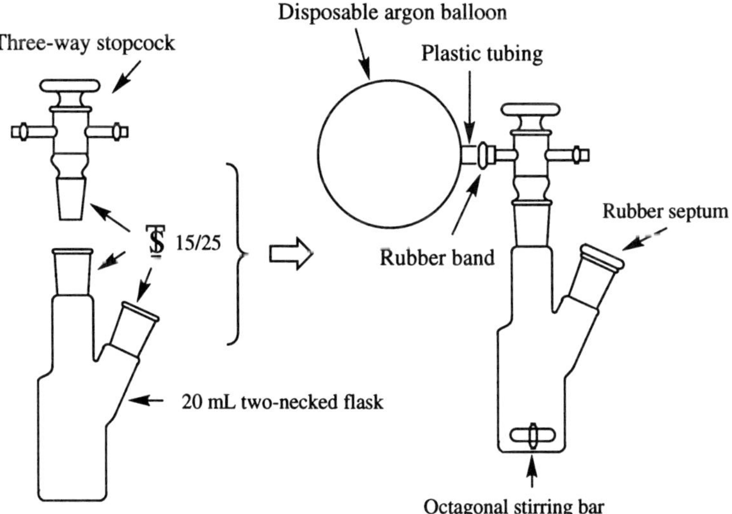

Fig. 7.4

Boron trifluoride-promoted conjugate addition of 4-methyl-3-pentenyl-copper to the crotonate **12** in the presence of tributylphosphine yields the adduct **13** in high chemical and optical yields. Both boron trifluoride and tributylphosphine are essential additives for a clean reaction.[21,22]

2.2 Additions to α,β-unsaturated ketones

The addition of conventional ordinary organocuprates to sterically hindered α,β-unsaturated ketones has often met with limited success. Even less reactive electrophiles (hindered α,β-unsaturated ketones) can be used provided that either $BF_3{\cdot}Et_2O$ or $AlCl_3$ is present to enhance reactivity. In this way, weakly electrophilic **14** affords the 1,4-addition product **15** by treatment with either $MeCu{\cdot}BF_3$[23,24] or $Me_2CuLi{\cdot}1.8BF_3$.[25,26] Noteworthy is the successful boron trifluoride-mediated conjugate addition to sterically crowded α,β-enone **16**. Without $BF_3{\cdot}Et_2O$, none of the desired material **17** is formed.[27]

Differences in reactivity between classical Gilman reagents and $RCu{\cdot}BF_3$ or $RCu{\cdot}AlCl_3$ are manifested in the reactions of R_2CuLi and $RCu{\cdot}BF_3$ or $RCu{\cdot}AlCl_3$ with the α,β-enone **18**.[6,9,28,29] The reaction between the α,β-unsaturated ketone **18** and $BuCu{\cdot}AlCl_3$ or $BuCu{\cdot}BF_3$ (see Table 7.1) proceeds in a synthetically acceptable yield to give the expected 1,4-adduct **21**, which can be transformed into perhydrohistrionicotoxin. The ability of $RCu{\cdot}BF_3$ or $RCu{\cdot}AlCl_3$ to undergo conjugate additions with γ-oxygenated α,β-unsaturated carbonyl compounds appears to exceed that of any other ordinary organocopper reagent.

CO-Me
OTBDMS
18
TBDMS = *t*-Butyldimethylsilyl

TBDMSO ... COMe, R: **19** R = Me; **21** R = *n*-Bu; **23** R = Ph

\+

TBDMSO ... COMe, R: **20** R = Me; **22** R = *n*-Bu; **24** R = Ph

Table 7.1. Effect of Lewis acids on reactions of the enone **18** with organocopper reagents

Reagent	Yield (%)	Product ratios			References
MeCu(CN)Li	0	**19**	0:0	**20**	6, 9, 28
Ph_2CuLi	0	**23**	0:0	**24**	6, 9, 28
$MeCu{\cdot}AlCl_3$	84	**19**	91:9	**20**	6, 9, 28
n-$BuCu{\cdot}AlCl_3$	84	**21**	92:8	**22**	6, 9, 28
n-$BuCu{\cdot}BF_3$	75	**21**	100:0	**22**	6
$PhCu{\cdot}AlCl_3$	69	**23**	100:0	**24**	6, 9, 28

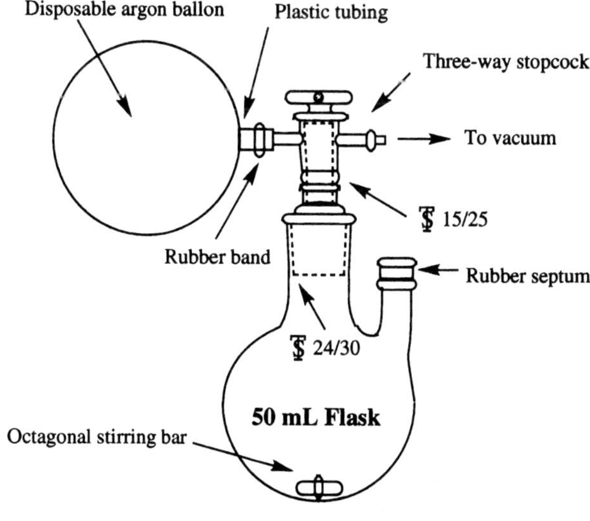

Fig. 7.5

Protocol 2.
Synthesis of 1-[(1*R**,2*S**,3*S**)-2-butyl-3-*t*-butyldimethylsiloxycyclohexyl]ethanone 21 and its (1*S**)-isomer 22. Reaction of α,β-unsaturated ketone 18 with BuCu·$AlCl_3$[6,9,28]

Caution! Carry out all procedures in a well-ventilated hood, and wear disposable vinyl or latex gloves and chemical-resistant safety goggles.

Equipment

- Magnetic stirrer
- Two-necked, round-bottomed flask[a] (50 mL) (see Fig. 7.5)
- Two-necked flask (20 mL) (see Fig. 7.4)
- Two three-way stopcocks (see Fig. 7.4)
- Two septa (Aldrich Z10,075–7 or Z10,074–9)
- Two Teflon-coated magnetic stirring bars (octagonal or egg-shaped bar: 2.0 × 0.8 cm)
- All-glass syringe with a needle-lock Luer (volume appropriate for quantity of solution to be transferred)
- Medium-gauge needle (20 or 22 gauge, 6 in long)
- Disposable argon balloon
- Separatory funnel (100 mL)
- Flask (100 mL)
- Column for flash chromatography

Materials

- Dry ether[b] 28 mL — **flammable, irritant**
- Copper(I) iodide[c] (FW 190.4) 0.8 g, 4.2 mmol — **irritant**
- Aluminum trichloride[d] (FW 133.3) 0.533 g, 4.0 mmol — **corrosive, moisture sensitive**
- *n*-Butyllithium in *n*-hexane[e] (FW 64.1) 1.56 M solution in hexane, 2.5 mL, 4.0 mmol — **flammable, moisture sensitive**
- α,β-Unsaturated ketone **18**[6,9,28] 0.2 g, 0.79 mmol — **harmful**
- Technical ether for extraction 50 mL — **flammable, irritant**
- Saturated aqueous ammonium chloride solution 5 mL — **corrosive, toxic**
- 28% aqueous ammonium hydroxide 8 mL — **corrosive, toxic**
- Silica gel for flash chromatography 35 g, Merck 9385 — **irritant dust**
- *n*-Hexane for flash chromatography — **flammable, irritant**
- Chloroform for flash chromatography — **highly toxic, cancer suspect agent**

1. Clean all glassware, syringes, needles, and stirring bars and dry for at least 2 h in a 105°C electric oven before use.
2. Assemble the flask and the stopcock under argon while the apparatus is still hot.
3. Support the assembled flask using a clamp and a stand with a heavy base.
4. Put purified copper(I) iodide (0.8 g, 4.2 mmol) into the flask, and equip the neck of the flask with a septum as shown in Fig. 7.5.
5. Follow steps 5–7 of Protocol 1.
6. Charge the flask with dry ether (15 mL) using a syringe by piercing the septum on the reaction flask.

7. Follow steps 9–11 of Protocol 1.
8. Cool the reaction flask with a dry ice–acetone bath to −70°C. Stir the mixture vigorously and add dropwise an ethereal solution of aluminum trichloride[f] by means of a syringe through the septum on the reaction flask.
9. Add enone **18** in 3 mL of dry ether dropwise via a syringe at −70°C with stirring. Allow the mixture to warm to −25°C with stirring and stir for a further 30 min at −25°C.
10. Recool the reaction flask to −70°C in a dry ice–acetone bath, and then quench the mixture with 13 mL of a 5:8 mixture of saturated ammonium chloride and 28% ammonium hydroxide with vigorous stirring. Remove the dry ice–acetone bath, allow the temperature to rise to ambient temperature, and stir for a further 30 min.
11. Filter the mixture through Celite® with suction to remove any inorganic precipitate, and rinse the flask and the filter cake with three 10 mL portions of technical ether.
12. Transfer the mixture to a separatory funnel and separate the two layers. Extract the water layer with ether (50 mL). Combine the ether layers and wash with two 15 mL portions of saturated brine.
13. Transfer the ethereal layer to a 100 mL flask. Dry the layer over anhydrous magnesium sulfate, and filter through a filter paper. Concentrate the filtrate under reduced pressure by means of a rotary evaporator (25°C/40 mm Hg).
14. Apply the colourless, oily residue to a flash column packed with silica gel. Elute with a mixed solvent of *n*-hexane–chloroform (1:4) to obtain successively 17 mg (7% yield, Kugelrohr distillation, 120°C/1 mm Hg) of **22** and 190 mg (77% yield, Kugelrohr distillation, 120°C/1 mm Hg) of **21**. Compounds **21** and **22** display the appropriate ^1H NMR (in $CDCl_3$), IR (in $CHCl_3$), and combustion analytical and mass spectral data.

[a] For volumes smaller than 20 mL, it is convenient to use the apparatus shown in Fig. 7.4.
[b] For the preparation of dry ether, see Protocol 1 and Chapter 2, Section 4.
[c] For the purification of dry copper(I) iodide, see Protocol 1 and Chapter 2, Protocol 2.
[d] Commercially available crystalline aluminum trichloride can be used without purification. Do not use aged (white powder) aluminum trichloride.
[e] For titration of *n*-butyllithium, see Protocol 1 and Chapter 2, Protocol 6.
[f] An ethereal aluminum trichloride solution can be prepared as follows. Charge aluminum trichloride (0.533 g, 4.0 mmol) into a hot, two-necked flask equipped with a stirrer bar, an argon balloon, and a septum (see Fig. 7.4). Evacuate and dry the flask by using a rotary pump and an electric heat gun, then flush the flask with argon. Repeat the process three times. Cool the flask to 10°C with a water bath and add dry ether (10 mL) with stirring via a syringe to obtain a light-yellow ethereal solution of aluminum trichloride. Note that aluminum trichloride is insoluble in ether at around 0°C. In addition, aluminum trichloride is insoluble in tetrahydrofuran and dioxane. **Caution!** Dissolution of aluminum trichloride in ether is exothermic.

3. Substitution reactions

The regio- and stereochemical outcomes of the reactions of organocopper reagents with allylic esters,[30] halides,[31] sulfonates,[14,15] phosphonates,[32] carbamates,[33] alcohols and ethers,[34] ammonium salts,[35] and oxiranes,[36] as well as with propargylic[37] and allenic[38] substrates, are well documented.[39] However, the lack of regioselectivity (S_N2 or S_N2') has prevented wider application of the method. Organocopper–Lewis acid complexes are one of the reagents of choice in the (*E*)-stereoselective and S_N2'-selective reactions.

As illustrated for the pair of allylic chlorides **25** and **26**, the butyl group in *n*-BuCu·BF_3 transfers to the γ-position with very high regioselectivity and in high yield.[3]

Me-CH=CH-CH(Cl) (**25**) —*n*–BuCu.BF_3 (90%)→ Me-CH(*n*-Bu)-CH=CH_2 (**27**) >98 : <2 Me-CH=CH-CH_2-Bu-*n* (**28**)

CH_2=CH-CH(Cl)-Me (**26**) —*n*–BuCu.BF_3 (96%)→ *n*-Bu-CH_2-CH=CH-Me (**29**) 4 : 96 CH_2=CH-CH(*n*-Bu)-Me (**30**)

Chiral δ-oxygenated α-alkyl-β,γ-enoates such as **32** and **34** bearing an (*E*)-double bond are promising intermediates for the synthesis of natural products. Exposure of **31** and **32** to organocyanocopper–boron trifluoride complexes yields α-alkyl-(*E*)-β,γ-enoates **32** and **34** in a very high chemical yield.[40,41]

31 (MsO, TBDMSO, CO_2Et, CO_2Et) —MeCu(CN)Li.BF_3; 90% yield; >98% de→ **32** (TBDMSO, EtO_2C, CO_2Et, Me)

33 (EtO_2C, BnO, MsO, S, S) —MeCu(CN)Li.BF_3; 75%→ **34** (EtO_2C, BnO, Me, S, S)

The reaction between the mesylate **35** and organocyanocopper–BF_3 reagents yields only the desired *trans*-alkene isostere **36**.[42,43] The present reaction of γ-methanesulfonyloxy-α,β-unsaturated esters with organocyanocopper–BF_3 reagents in THF or mixed solvents involving THF is significantly faster and usually attains completion within a few minutes, even at −78°C. The reaction conditions used do not cause any epimerization at the chiral carbon centre.

$$\text{R}^1\text{CH(NR}^2\text{-Boc)CH(OMs)CH=CHCO}_2\text{Me}\ (\mathbf{35}) \xrightarrow[\text{>90\% yield, >98\% de}]{\text{RCu(CN)M.BF}_3} \text{R}^1\text{CH(NR}^2\text{-Boc)CH=CHCH(R)CO}_2\text{Me}\ (\mathbf{36})$$

R^1 = alkyl, benzyl, etc; R^2 = H or ring residue;
R = alkyl or benzyl; M = Li or MgX

The synthetic utility of these reagents has been further demonstrated in recent studies.[44]

Protocol 3.
Synthesis of diethyl (3*E*,6*E*,5*R*,2*S*)-3-*t*-butyldimethylsiloxy-2-methyl-3,6-octadienedioate 32. Reaction of γ-methanesulfonyloxy-α,β-unsaturated ester 31 with MeCu(CN)Li·BF_3[40,41]

Caution! Carry out all procedures in a well-ventilated hood, and wear disposable vinyl or latex gloves and chemical-resistant safety goggles.

Equipment

- Magnetic stirrer
- Two-necked, round-bottomed flask (50 mL) (see Fig. 7.5)
- Three-way stopcock (see Fig. 7.4)
- Two septa (Aldrich Z10,075–7 or Z10,074–9)
- Two Teflon-coated magnetic stirring bars (octagonal or egg-shaped bar: 2.5 × 0.8 cm)
- All-glass syringe with a needle-lock Luer (volume appropriate for quantity of solution to be transferred)
- Medium-gauge needle (20 or 22 gauge, 6 in long)
- Disposable argon balloon
- Separatory funnel (100 mL)
- Flask (100 mL)
- Column for flash chromatography
- Argon supply

Materials

- Dry tetrahydrofuran[a] 18 mL — **flammable, irritant**
- Copper(I) cyanide[b] (FW 89.6) 600 mg, 6.6 mmol — **highly toxic, irritant**
- γ-Methanesulfonyloxy-α,β-unsaturated ester (FW 450.6) **31**[40] 1.0 g, 2.2 mmol — **harmful**
- Boron trifluoride etherate[c] (FW 141.9) 0.79 mL, 6.6 mmol — **corrosive, moisture sensitive**
- Methyllithium (complexed with lithium bromide in ether)[d] 1.5 M solution in ether, 4.4 mL, 6.6 mmol — **flammable, moisture sensitive**
- Technical ether for extraction 60 mL — **flammable, irritant**
- Saturated aqueous ammonium chloride solution 5 mL — **corrosive, toxic**
- 28% aqueous ammonium hydroxide 5 mL — **corrosive, toxic**
- Silica gel for flash chromatography 35 g, Merck 9385 — **irritant dust**
- *n*-Hexane for flash chromatography — **flammable, irritant**
- Ethyl acetate for flash chromatography — **flammable, irritant**

1. Clean all glassware, syringes, needles, and stirring bar and dry for at least 2 h in a 105°C electric oven before use.

Protocol 3. *Continued*

2. Assemble the flask and stopcock under argon while the apparatus is still hot.
3. Support the assembled flask using a clamp and stand with a heavy base.
4. Put purified copper(I) cyanide (600 mg, 6.6 mmol) into the flask and equip the neck of the flask with a septum as shown in Fig. 7.5.
5. Dry the apparatus with an electric heat gun under vacuum (0.5–1 mm Hg) for 5 min, then back-fill the flask with argon. Repeat to a total of three times. **Caution!** Do not flame the apparatus containing copper(I) cyanide with a Bunsen burner. Overheating under reduced pressure can lead to decomposition of copper(I) cyanide.
6. Cool the flask with a dry ice–acetone bath to −78°C (bath temperature) under a slightly positive pressure from an argon balloon.
7. Assemble the syringe and needle while hot and allow the assembled syringe to cool to room temperature in a desiccator. Flush the syringe with argon (see Fig. 7.2).
8. Charge the flask with dry tetrahydrofuran (15 mL) using a syringe by piercing the septum on the reaction flask.
9. Support the bottle containing methyllithium in ether using a ring or clamp and stand with a heavy base.
10. Fill a syringe with methyllithium in ether from the bottle containing methyllithium using argon pressure, following the procedure described in Protocol 1, step 10 (Fig. 7.3).
11. Keep the flask temperature at −78°C (dry ice–acetone bath temperature) and add methyllithium dropwise by means of the syringe through the septum on the reaction flask over a 5 min period with stirring using a magnetic stirrer. Remove the dry ice–acetone bath and allow the temperature to rise to 0°C. Stir the mixture at 0°C for 10 min. During the reaction, keep the reaction apparatus under a slight pressure of argon.
12. Recool the flask to −78°C in a dry ice–acetone bath. Stir the mixture vigorously and add boron trifluoride etherate dropwise by means of a syringe through the septum on the reaction flask. Stir at −78°C for 10 min.
13. Add enoate **31** in 3 mL of dry tetrahydrofuran dropwise via a syringe at −78°C under stirring and stir for 30 min.
14. Quench the mixture with 10 mL of a 1:1 saturated ammonium chloride–28% ammonium hydroxide solution under vigorous stirring. Remove the dry ice–acetone bath and allow the temperature to rise to ambient temperature.
15. Remove both the stopcock and the septum from the flask and stir the mixture for 30 min. The colour of the mixture turns gradually from grey to clear sky blue.

16. Transfer the solution to a separatory funnel (100 mL), add 40 mL of ether, and shake the funnel. Extract the water layer with 20 mL of ether. Combine the ether layers and wash with 30 mL of ice-cooled, saturated brine. **Caution!** Do not discharge the blue water layer and the washings to a sink (copper salts and the cyanide ion are hazardous).

17. Transfer the ethereal layer to a 200 mL flask. Dry the layer over anhydrous magnesium sulfate, and filter through a filter paper. Concentrate the filtrate under reduced pressure using a rotary evaporator (25°C/40 mm Hg).

18. Apply the residual oil to a flash silica gel column (3 × 30 cm) using a 4:1 mixture of *n*-hexane and ethyl acetate. Elute the column with a 4:1 mixture of *n*-hexane and ethyl acetate under a nitrogen pressure of 1.5 kg cm^{-2} to obtain pure product **32** (740 mg, 90.3% yield) as a colourless oil which displays the appropriate ^1H NMR (in $CDCl_3$), IR (in $CHCl_3$), circular dichroism (in *n*-octane), and mass spectral data.

[a] Distil THF from lithium aluminum hydride or the benzophenone, ketyl under an inert atmosphere (nitrogen or argon) and use immediately (see Chapter 2, Section 4). The use of THF or solvents involving THF is essential for a clean reaction.
[b] Reagent-grade CuCN (from Mitsuwa) was dried at 56°C (0.5 mm Hg) for 24 h before use. **Caution!** Copper(I) cyanide cannot be replaced with other copper(I) salts such as CuI and $CuBr{\cdot}Me_2S$.
[c] For purification of boron trifluoride etherate, see Protocol 1.
[d] Titrate methyllithium in ether from Kanto with a 1.00 M solution of *s*-butanol in xylene using 1,10-phenanthroline as the indicator just prior to use (see Chapter 2, Protocol 6).

References

1. Posner, G. H. *Org. React.* **1972**, *19*, 1–113.
2. Lipshutz, B. H.; Sengupta, S. *Org. React.* **1992**, *41*, 135–631.
3. Maruyama, K.; Yamamoto, Y. *J. Am. Chem. Soc.* **1977**, *99*, 8068–8070.
4. Yamamoto, Y.; Maruyama, K. *J. Am. Chem. Soc.* **1978**, *100*, 3240–3241.
5. Yamamoto, Y. *Angew. Chem., Int. Ed. Engl.* **1986**, *25*, 947–959.
6. Yamamoto, Y.; Yamamoto, S.; Yatagai, H.; Ishihara, Y.; Maruyama, K. *J. Org. Chem.* **1982**, *47*, 119–126.
7. Ibuka, T.; Minakata, H. *Synth. Commun.* **1980**, 119–125.
8. Ibuka, T.; Minakata, H.; Mitsui, Y.; Kinoshita, K.; Kawami, Y. *J. Chem. Soc., Chem. Commun.* **1980**, 1193–1194.
9. Ibuka, T.; Minakata, H.; Mitsui, Y.; Kinoshita, K.; Kawami, Y.; Kimura, N. *Tetrahedron Lett.* **1980**, *21*, 4073–4076.
10. Lipshutz, B. H. *Synthesis* **1987**, 325–341.
11. Lipshutz, B. H.; Ellsworth, E. L.; Dimock, S. H. *J. Am. Chem. Soc.* **1990**, *112*, 5869–5871.
12. Lipshutz, B. H.; Ellsworth, E. L.; Siahaan, T. J. *J. Am. Chem. Soc.* **1989**, *111*, 1351–1358.
13. Lipshutz, B. H.; Ellsworth, E. L.; Siahaan, T. J. *J. Am. Chem. Soc.* **1988**, *110*, 4834–4835.

14. Ibuka, T.; Tanaka, M.; Nishii, S.; Yamamoto, Y. *J. Am. Chem. Soc.* **1989**, *111*, 4864–4872.
15. Ibuka, T.; Akimoto, N.; Tanaka, M.; Nishii, S.; Yamamoto, Y. *J. Org. Chem.* **1989**, *54*, 4055–4061.
16. Kauffmann, G. B.; Teter, L. A. *Inorg. Synth.* **1963**, *7*, 9–12.
17. Posner, G. H.; Whitten, C. E.; Sterling, J. J. *J. Am. Chem. Soc.* **1973**, *95*, 7788–7800.
18. Yamamoto, Y.; Yatagai, H.; Maruyama, K. *J. Org. Chem.* **1979**, *44*, 1744–1746.
19. Ibuka, T.; Aoyagi, T.; Yamamoto, Y. *Chem. Pharm. Bull.* **1986**, *34*, 2417–2427.
20. Ibuka, T.; Aoyagi, T.; Kitada, K.; Yoneda, F.; Yamamoto, Y. *J. Organomet. Chem.* **1985**, *287*, C18–C22.
21. Oppolzer, W. *Tetrahedron* **1987**, *43*, 1969–2004.
22. Oppolzer, W.; Moretti, R.; Godel, T.; Meunier, A.; Löher, H. *Tetrahedron Lett.* **1983**, *24*, 4971–4974.
23. Karpf, M.; Dreiding, A. S. *Tetrahedron Lett.* **1980**, *21*, 4569–4570.
24. Karpf, M.; Dreiding, A. S. *Helv. Chim. Acta* **1981**, *64*, 1123–1133.
25. Smith, III, A. B.; Jerris, P. J. *J. Am. Chem. Soc.* **1981**, *103*, 194–195.
26. Smith, III, A. B.; Jerris, P. J. *J. Org. Chem.* **1982**, *47*, 1845–1855.
27. Lipshutz, B. H.; Parker, D. A.; Kozlowski, J. A.; Nguyen, S. L. *Tetrahedron Lett.* **1984**, *25*, 5959–5962.
28. Ibuka, T.; Minakata, H.; Mitsui, Y.; Hayashi, K.; Taga, T.; Inubushi, Y. *Chem. Pharm. Bull.* **1982**, *30*, 2840–2859.
29. Ibuka, T.; Minakata, H.; Mitsui, Y.; Tabushi, Y.; Taga, T.; Inubushi, Y. *Chem. Lett.* **1981**, 1409–1412.
30. Underiner, T. L.; Goering, H. L. *J. Org. Chem.* **1991**, *56*, 2563–2572.
31. Arai, M.; Nakamura, E.; Lipshutz, B. H. *J. Org. Chem.* **1991**, *56*, 5489–5491.
32. Yanagisawa, A.; Noritake, Y.; Nomura, N.; Yamamoto, H. *Synlett* **1991**, 251–253.
33. Denmark, S. E.; Marble, L. K. *J. Org. Chem.* **1990**, *55*, 1984–1986.
34. Goering, H. L.; Kantner, S. S. *J. Org. Chem.* **1981**, *46*, 2144–2148.
35. Pan, Y.; Hutchinson, D. K.; Nantz, M. H.; Fuchs, P. L. *Tetrahedron* **1989**, *45*, 467–478.
36. Marino, J. P.; Kelly, M. G. *J. Org. Chem.* **1981**, *46*, 4389–4393.
37. Alexakis, A.; Marek, I.; Mangeney, P.; Normant, J. F. *Tetrahedron Lett.* **1989**, *30*, 2387–2390.
38. Corey, E. J.; Boaz, N. W. *Tetrahedron Lett.* **1984**, *25*, 3059–3062.
39. Marshall, J. A. *Chem. Rev.* **1989**, *89*, 1503–1511.
40. Ibuka, T.; Tanaka, M.; Yamamoto, Y. *J. Chem. Soc., Chem. Commun.* **1989**, 967–969.
41. Chen, S.-H.; Horvath, R. F.; Joglar, J.; Fisher, M. J.; Danishefsky, S. J. *J. Org. Chem.* **1991**, *56*, 5834–5845.
42. Ibuka, T.; Habashita, H.; Funakoshi, S.; Fujii, N.; Oguchi, Y.; Uyehara, T.; Yamamoto, Y. *Angew. Chem., Int. Ed. Engl.* **1990**, *29*, 801–803.
43. Ibuka, T.; Habashita, H.; Otaka, A.; Fujii, N.; Oguchi, Y.; Uyehara, T.; Yamamoto, Y. *J. Org. Chem.* **1991**, *56*, 4370–4382.
44. Kempf, D. J.; Wang, X. C.; Spanton, S. G. *Int. J. Pept. Protein Res.* **1991**, *38*, 237–241. Ibuka, T.; Taga, T.; Habashita, H.; Nakai, K.; Tamamura, H.; Fujii, N.; Chounan, Y.; Nemoto, H.; Yamamoto, Y. *J. Org. Chem.* **1993**, *58*, 1207–1214.

8

Asymmetric conjugate addition

A. ALEXAKIS

1. Introduction

The conjugate addition, the most popular reaction in organocopper chemistry, is very well suited for the formation of new stereogenic centres. Several examples of very high levels of diastereoselectivity[1] show that organocopper reagents are very sensitive to the steric and electronic environment of the prochiral sp^2 carbon β to the carbonyl. Owing to the importance of this synthetic transformation it is not surprising that the asymmetric version of this reaction was attempted at the very early stages of the organocopper era.[2] There are several ways to attain this goal. The most obvious one is to attach covalently a chiral auxiliary, do a diastereoselective conjugate addition, and then remove this chiral auxiliary, the overall result being an enantioselective transformation. This chiral auxiliary may be placed either on the electrophile or on the R group of the organocopper reagent itself.

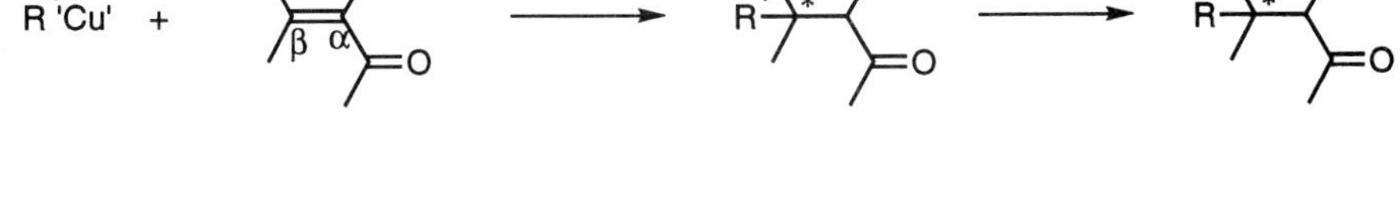

R'Cu' + [β,α-unsaturated carbonyl bearing R*] → [R, R* adduct] → [R adduct]

R*Cu' + [β,α-unsaturated carbonyl] → [R* adduct] → [R adduct]

The other most studied way to effect asymmetric conjugate addition is to use a chirally transformed organocopper complex. This is achieved with chiral heterocuprates (see Chapter 1 for the definition of heterocuprates) or, alternatively, by using chiral non-covalently bound ligands. In both cases the chiral 1,4-adduct is obtained directly.

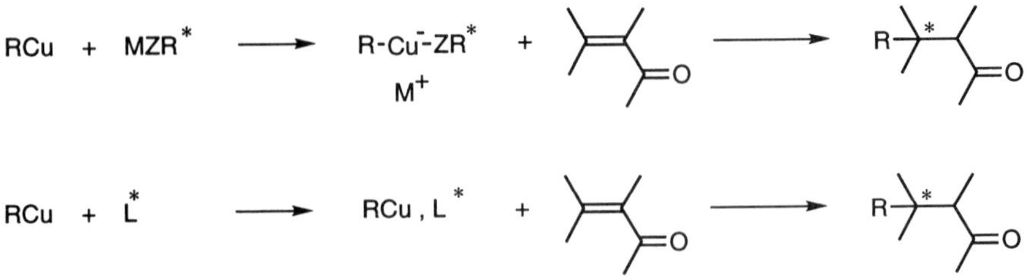

2. Covalent chiral auxiliary

Among the various α,β-ethylenic carbonyl compounds, the easiest way to attach a chiral auxiliary is on to esters or amides. One has just to choose the most adequate chiral alcohol or amine.

With ketones and aldehydes (or even with esters) the chiral auxiliary has to be placed in the neighbourhood of the prochiral sp^2 carbon. This can be achieved with, for example, chiral sulfoxides, or chiral aminals, such as

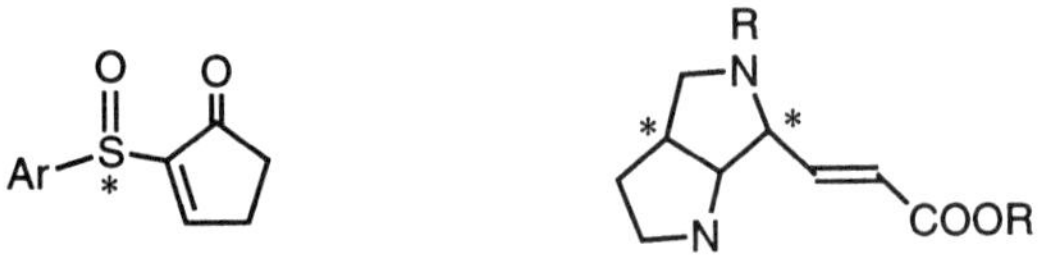

Alternatively, the carbonyl functionality itself may be transformed into another, equivalent but chiral one, and then regenerated

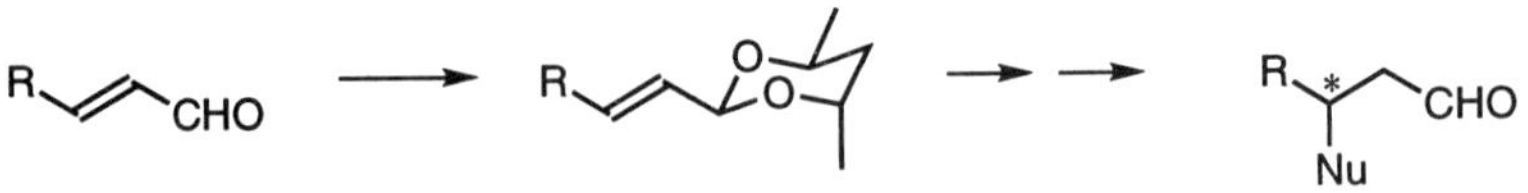

2.1 Chiral esters

This approach is one of the most versatile, since a very large variety of chiral alcohols are available from natural and unnatural sources. However, the early attempts were not very successful. In fact, two problems have to be dealt with. One is the control of the reactive conformer: *s-cis* and *s-trans* conformers give opposite stereochemical results. The second is the steric or chelation control by the auxiliary itself, allowing the organometallic reagent to approach from one side of the molecule or from the other.

The first effective control of these two parameters was achieved with ester **1**, having phenylmenthol as auxiliary.[3] The π-stacking between the phenyl group and the enoate is highest in the *s-trans* conformer, and this same phenyl group masks completely the *si* face of the prochiral sp^2 carbon.

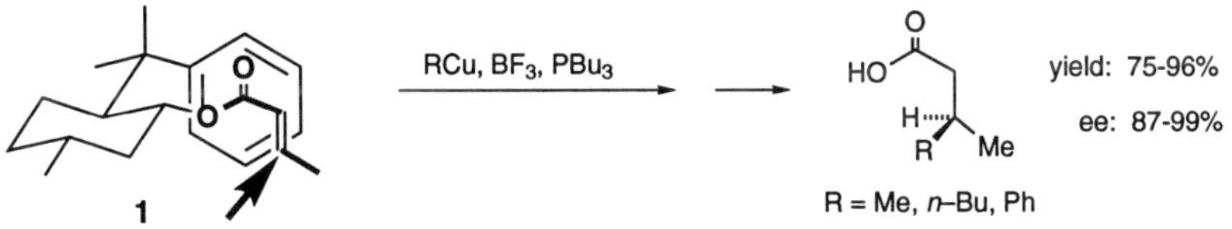

1

RCu, BF_3, PBu_3

yield: 75-96%

ee: 87-99%

R = Me, *n*-Bu, Ph

In fact, esters are not among the most reactive substrates for conjugate addition reaction, and quite often modified organocopper reagents have to be used. This is the case in the above reaction, where activation by $BF_3 \cdot OEt_2$ and solubilization by PBu_3 are necessary to get a good chemical yield.

Several other chiral alcohols, most derived from camphor, have, since then, been used successfully. All of them are based on the same principle. A bulky group masks one face of the prochiral centre, and the reactive conformer is the *s-trans* one. The fact that a Lewis acid (such as $BF_3 \cdot OEt_2$) is often necessary also helps in fixing the *s-trans* conformer.

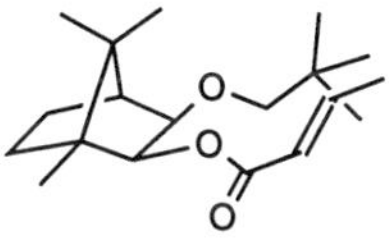

2 Oppolzer *et al.*[4] 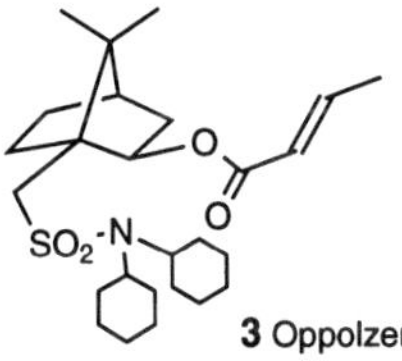 **3** Oppolzer *et al.*[5] 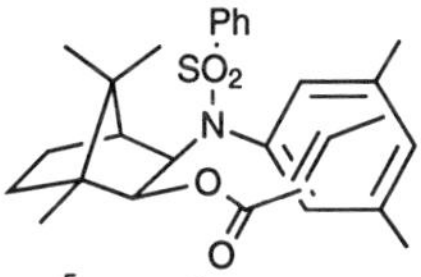 **4** Helmchen and Wegner[6]

With these chiral esters, several skeletal modifications are allowed on the enoate system and several organic groups could be efficiently transferred. There are also several applications of these reactions in natural product synthesis.[2] The obtention of the opposite enantiomer can be achieved by using the other enantiomer of camphor. However, this unnatural enantiomer is very expensive and a much simpler way to do that is to use a different diastereomer of the chiral auxiliary **2** or **4**.[4a,6]

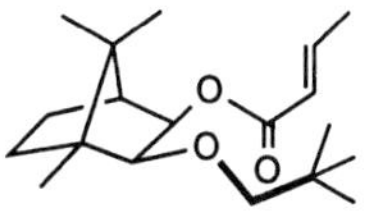

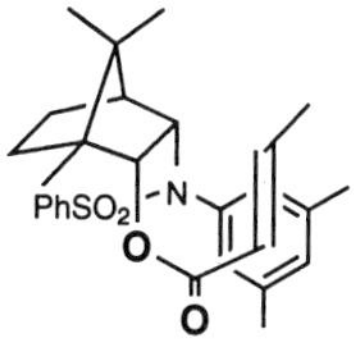

It has also been found that the nature of the organocopper reagent itself may be of great importance. The stereochemistry of the adduct on ester **5** is sensitive to this fact. The basic higher order reagent formed from three equivalents of RLi and one equivalent of CuI[7a] (eqn 1) seems to react with the *s-cis* conformer (Protocol 1), whereas the copper-rich reagent (3RLi + 2CuI) (eqn 2) gives the opposite result by reacting with the *s-trans* conformer. The activation of the reactivity of organocopper species may also be achieved by addition of trimethylsilyl derivatives.[7b] In the present case, the same result is obtained with the assistance of trimethylsilyl iodide[7a] (Protocol 2).

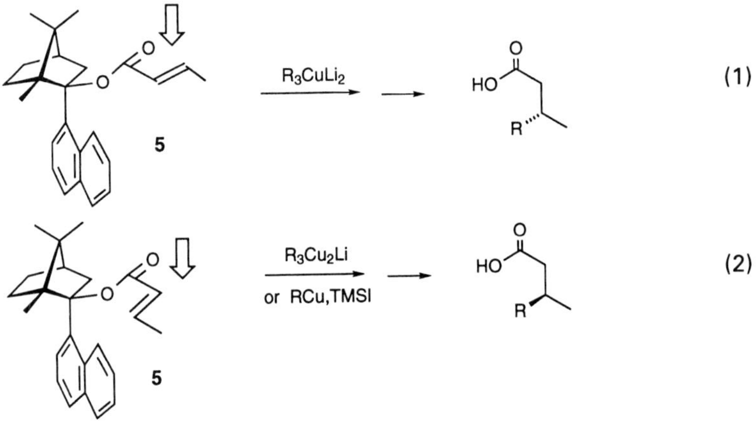

R_3CuLi_2 (1)

R_3Cu_2Li or RCu,TMSI (2)

The steric control exerted by the chiral auxiliary may be replaced by a chelation control. Chiral diols have been used to that purpose, as shown by ester **6.**[8] Addition of the organocopper reagent produces an *in situ* heterocuprate which then directs the conjugate addition from the same face.

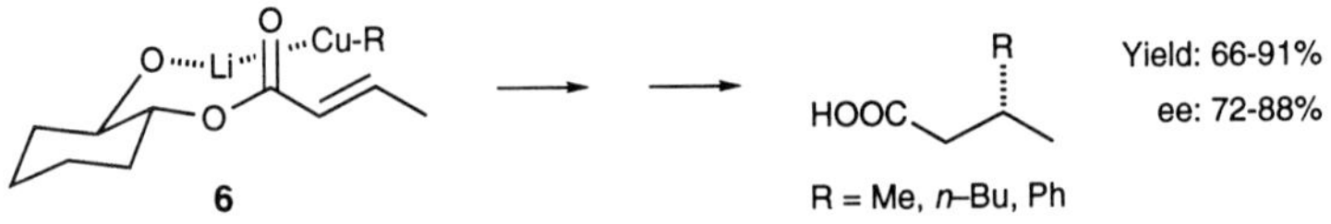

R = Me, *n*-Bu, Ph

Yield: 66-91%

ee: 72-88%

Protocol 1.
Higher order cuprate addition to naphthylcamphor ester 5[7a]

Caution! Carry out all procedures in a well-ventilated hood, and wear disposable vinyl or latex gloves and chemical-resistant safety goggles.

5 → (Bu_3CuLi_2)

Np =

Equipment

- Three-necked, round-bottomed flask (250 mL) with magnetic stirring bar, addition funnel, and septa
- Inert gas supply and inlet
- Dry, gas-tight syringes

Materials

- Copper(I) iodide (FW 190.4) 950 mg, 5 mmol **irritant, light sensitive**
- *n*-Butyllithium (FW 64.1) 1.6 M solution in hexane, 9.2 mL, 15 mmol **flammable, moisture sensitive**
- Crotonic ester[a] **5** (FW 348.5) 348 mg, 1 mmol **harmful**

1. Flame dry the reaction vessel and accessories under nitrogen. After cooling to room temperature, place into the flask the copper iodide and ether (10 mL).
2. Cool the stirred slurry to −78 °C (dry ice–acetone bath). Add the *n*-butyllithium dropwise. Warm to −45 °C by removing the cooling bath, and stir until a homogeneous solution is obtained (30 min).
3. Cool again to −78 °C and add slowly (2 mL min^{-1}) the chiral ester **5** dissolved in ether (20 mL). Stir the mixture at −60 °C until no starting material is left, as shown by TLC of a hydrolysed aliquot (1 h).
4. Hydrolyse the reaction mixture by addition of a mixture of aqueous NH_4OH (5 mL) and NH_4Cl (20 mL). When the contents of the flask reach room temperature, filter the salts on Celite®, and extract the aqueous phase with ether (3 × 50 mL). Wash the combined organic phases with brine (25 mL) and dry over sodium sulfate. Remove the solvents *in vacuo* and purify the adduct by column chromatography on silica gel using ether–pentane (5:95). The yield is 390 mg (>95%) and the diastereomeric excess amounts to 86% by ^1H NMR spectroscopy.

[a] Prepared according to a published procedure.[7a]

Protocol 2. TMSI-assisted addition to naphthylcamphor ester 5[7a]

Caution! Carry out all procedures in a well-ventilated hood, and wear disposable vinyl or latex gloves and chemical-resistant safety goggles.

RCu,TMSI

Np =

Equipment

- Three-necked, round-bottomed flask (250 mL) with magnetic stirring bar, addition funnel, and septa
- Inert gas supply and inlet
- Dry, gas-tight syringes

Materials

- Copper(I) iodide (FW 190.4) 1.05 g, 5.5 mmol — **irritant, light sensitive**
- *n*-Butyllithium (FW 64.1) 1.6 M solution in hexane, 3.1 mL, 5 mmol — **flammable, moisture sensitive**
- Crotonic ester[a] **5** (FW 348.5) 348 mg, 1 mmol — **harmful**
- Freshly distilled (from copper powder), colourless trimethylsilyl iodide (FW 200.1) 0.71 mL, 5 mmol — **flammable, corrosive**

1. Flame dry the reaction vessel and accessories under nitrogen. After cooling to room temperature, place into the flask the copper iodide and ether (10 mL).
2. Cool the stirred slurry to −78 °C (dry ice–acetone bath). Add dropwise the *n*-butyllithium. Warm to −60 °C by removing the cooling bath, and stir the mixture for 40 min.
3. Add trimethylsilyl iodide dropwise and stir the slurry at −60 °C for 5 min.
4. Cool again to −78 °C and add slowly (2 mL min^{-1}) a solution of ester **5** in Et_2O (15 mL). Stir the mixture for 20 h at −60 °C (acetone bath with a cryocool).
5. Quench with pyridine–ether (1:9, 10 mL) and stir the suspension (2 h) at −60 °C. Then hydrolyse and work up as in Protocol 1. The yield is 378 mg (93%) and the ^1H NMR spectrum shows a single diastereomer (>98%) of reverse stereochemistry from the one obtained in Protocol 1. Obtain the pure diastereomer by recrystallization (three times) from aqueous ethanol to give white, cotton-like needles, m.p. 118–120 °C, $[\alpha]_D^{24} = -116°$ ($c = 0.63$, CH_2Cl_2). Characterize the product by ^1H and ^{13}C NMR spectroscopy.

[a] Prepared according to a published procedure.[7a]

2.2 Chiral amides

Chiral amides have essentially been used in the same way as chiral esters, although they are usually less reactive. For example, amide **7** does not react with organocopper reagents, but it does so with Grignard reagents at reflux in THF.[9] In contrast, the doubly activated amide **8** reacts very easily.[10]

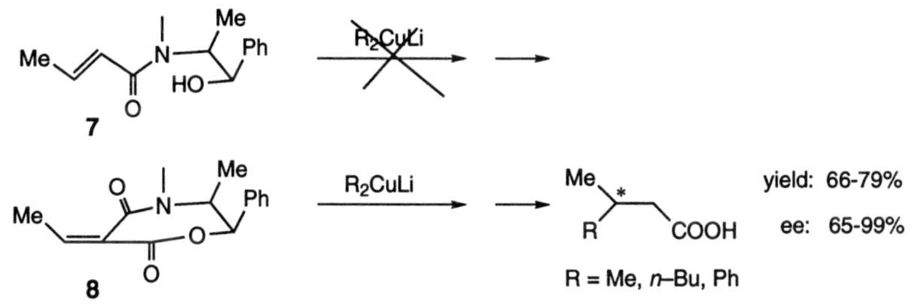

N-Enoyl sultams of type **9** are very attractive compounds and more reactive than simple amides. Moreover, they are advantageous because they readily crystallize and allow an easy improvement of the optical purity of the adduct.[11] Another interesting point is that tertiary amides are fixed in the *s-cis* conformation. Sultams **9a** (R = Me)[12a] and **9b** (R = $SiPhMe_2$)[12b] react diastereoselectively with organocopper reagents in the presence of magnesium or aluminum salts, which chelate both the SO_2 and C=O systems. In contrast, the diastereoselectivity may be reversed with $BF_3 \cdot OEt_2$ as activator; only the C=O system is chelated and this different conformation leads to the opposite diastereomer.[12b]

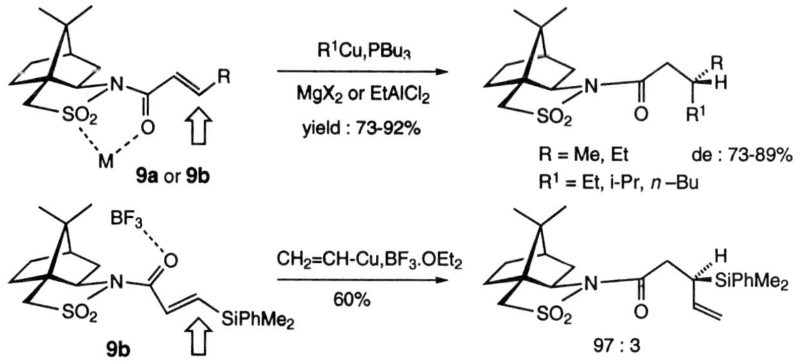

This method has been very efficiently used, not only for the conjugate addition of a variety of R groups, but also for the *in situ* trapping of the enolate.[13] The diastereoselectivity of both steps is in the 75–90% range and can be easily boosted to >95% after one or two recrystallizations.

Another very efficient chiral amide is based on the very sterically hindered γ-butyrolactam **10**[14] (Protocol 3). The bulky triphenylmethyl group masks completely the *re* face, whereas the magnesium salts tightly bind the two carbonyl groups.

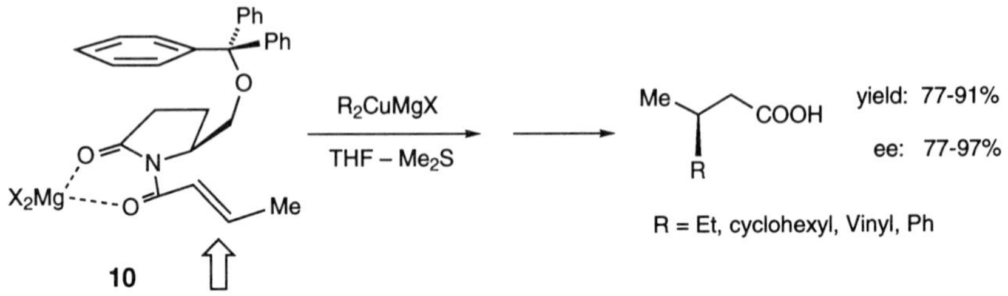

Ph, Ph, O, O, N, X_2Mg, O, Me, **10**, R_2CuMgX, THF – Me_2S, Me, COOH, R, yield: 77-91%, ee: 77-97%

R = Et, cyclohexyl, Vinyl, Ph

Protocol 3.
Addition to chiral amide 10[14]

Caution! Carry out all procedures in a well-ventilated hood, and wear disposable vinyl or latex gloves and chemical-resistant safety goggles.

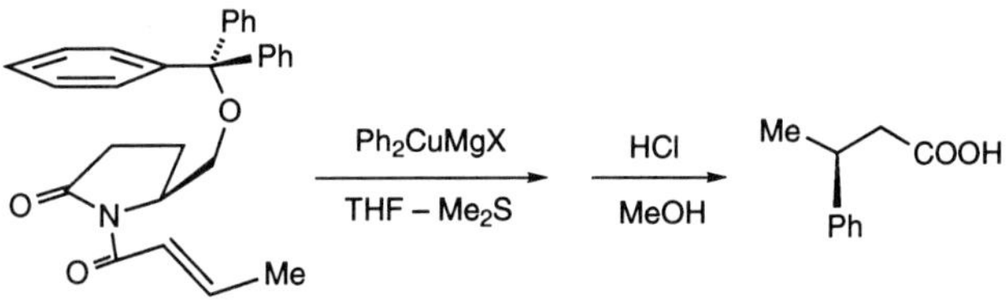

Ph, Ph, O, O, N, O, Me, **10**, Ph_2CuMgX, THF – Me_2S, HCl, MeOH, Me, COOH, Ph

Equipment

- Three-necked, round-bottomed flask (100 mL) with magnetic stirring bar, addition funnel, and septa
- Round-bottomed flask (50 mL) equipped with a reflux condenser
- Inert gas supply and inlet
- Dry, gas-tight syringes

Materials

- Phenylmagnesium chloride (FW 136.9) 1.2 M solution in THF, 3.75 mL, 4.5 mmol — **flammable, moisture sensitive**
- Copper(I) bromide–dimethyl sulfide complex (FW 205.6) 462 mg, 2.25 mmol — **moisture sensitive**
- Dimethyl sulfide (FW 62.1) 6 mL — **flammable, stench**
- Chiral amide[a] **10** (FW 425.5) 636 mg, 1.5 mmol — **harmful**

1. Flame dry the reaction vessel and accessories under nitrogen. After cooling to room temperature, dissolve, in the three-necked flask, $CuBr{\cdot}Me_2S$ in Me_2S (6 mL), then slowly add THF (15 mL) at room temperature.

2. Cool this solution to −48 °C (acetone bath with some dry ice) and add dropwise the THF solution of PhMgCl. Warm this mixture and stir for 10 min at −23 °C (dry ice–CCl_4 bath).
3. Add the chiral amide **10** in solution in THF (3 mL) and stir for 1.5 h at −23 °C.
4. Hydrolyse with saturated aqueous NH_4Cl solution (30 mL) and let the mixture warm to room temperature. Add ether (50 mL), filter the salts, and separate the layers. Extract the aqueous phase with ether (2 × 30 mL) and wash the combined organic phases with aqueous NH_4Cl (2 × 30 mL). Dry over Na_2SO_4, concentrate the solution, and purify by silica gel chromatography (hexane–EtOAc, 7:1).
5. Place the purified product (737 mg) in MeOH (10 mL) containing concentrated HCl (0.4 mL) and heat to reflux for 12 h in the round-bottomed flask equipped with a reflux condenser.
6. Cool to room temperature and add a solution of KOH (0.5 g) in water (2 mL). Heat again to reflux for 5 h.
7. After cooling to room temperature add distilled water (20 mL) and evaporate the methanol under vacuum. Add ether (50 mL) and separate the layers. Take the aqueous layer and add 1M HCl (20 mL), then extract it with ether (2 × 50 mL). Dry the organic layers over Na_2SO_4, evaporate the solvents at reduced pressure, and purify by bulb-to-bulb distillation (100 °C/20 mm Hg) to obtain (*S*)-3-phenylbutyric acid (214 mg, 87%, 94% ee), for which $[\alpha]_D^{20} = +53.5°$ (c = 10, benzene).

[a] Prepared according to a published procedure.[14]

2.3 Remote chiral auxiliary

The chiral auxiliary can be placed in other parts of the α,β-ethylenic carbonyl compound. Chiral sulfoxides placed α to the carbonyl, such as in compound **11**, have effectively been used and widely exploited for the synthesis of complex natural products.[15]

Ar S(O) 11 —R'Cu'→ Ar S(O) R * —Al – Hg→ R *

Although organocopper reagents were first used, it was later shown that many other organometallic reagents could undergo the same conjugate addition, with much higher diastereomeric discrimination. Both diastereoselectivities could be obtained according to the nature of the metal.[15]

Chiral acetals have been placed in the neighbourhood of the prochiral sp^2 carbon, for example

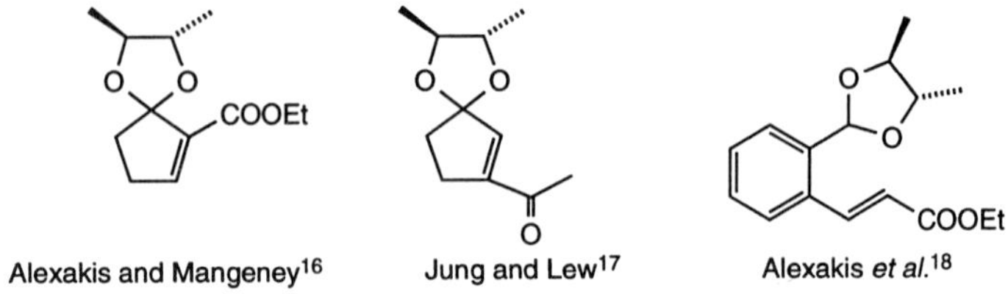

Alexakis and Mangeney[16] Jung and Lew[17] Alexakis *et al.*[18]

However, poor diastereoselectivities were observed. More successful were the attempts with chiral oxazolidines and aminals.

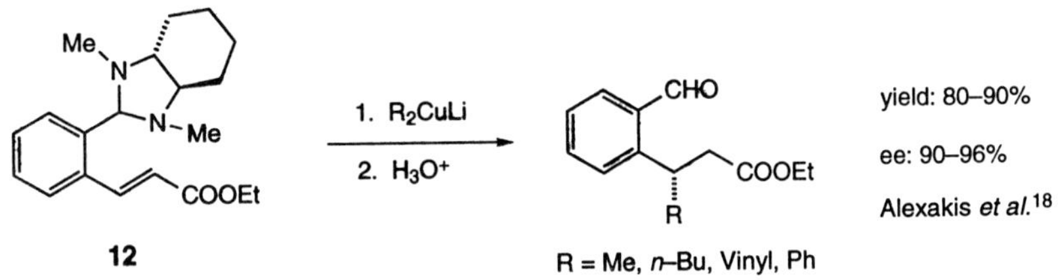

1. R_2CuLi 2. H_3O^+

12

R = Me, *n*-Bu, Vinyl, Ph

yield: 80–90%
ee: 90–96%
Alexakis *et al.*[18]

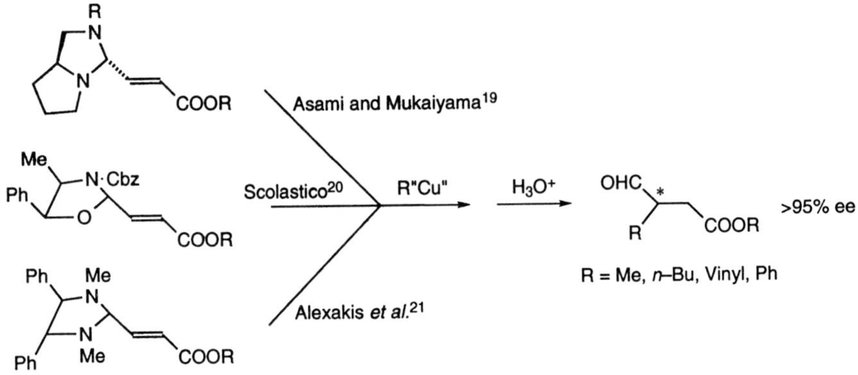

Asami and Mukaiyama[19]
Scolastico[20]
Alexakis *et al.*[21]

R"Cu" H_3O^+

>95% ee

R = Me, *n*-Bu, Vinyl, Ph

Steric control is assumed to operate when the chiral auxiliary is next to the prochiral centre, as well as with cinnamate **12**[18] (Protocol 4). However, in the latter case either steric or chelation control operates when a different diamine is used as the auxiliary.[7b]

Protocol 4.
Addition to an ester having a remote chiral auxiliary[18]

Caution! Carry out all procedures in a well-ventilated hood, and wear disposable vinyl or latex gloves and chemical-resistant safety goggles.

Me N N–Me COOEt **12** 1. Me_2CuLi 2. H_3O^+ CHO COOEt Me

Equipment

- Three-necked, round-bottomed flask (250 mL) with magnetic stirring bar, addition funnel, and septa
- Inert gas supply and inlet
- Dry, gas-tight syringes

Materials

- Methyllithium–lithium bromide complex (FW 22.0) 1 M solution in ether, 8 mL, 8 mmol — **flammable, moisture sensitive**
- Copper(I) bromide–dimethyl sulfide complex (FW 205.6) 830 mg, 4.04 mmol — **moisture sensitive**
- Aminal[a] **12** (FW 328.5) 656 mg, 2 mmol — **harmful**

1. Flame dry the reaction vessel and accessories under nitrogen. After cooling to room temperature, place into the flask the copper(I) bromide–dimethyl sulfide complex and ether (50 mL).
2. Add MeLi·LiBr solution to this slurry at −50 °C (acetone bath with some dry ice) under N_2, then warm to −10 °C and stir until a homogeneous solution is obtained (10 min).
3. Cool to −30 °C and add the chiral enoate **12** in solution in ether (10 mL). Warm to −10 °C (ice–salt bath) and stir for 2 h. A yellow precipitate is formed.
4. Hydrolyse with 1 M HCl (50 mL), remove the stoppers and the inert gas inlet, and continue stirring for 1 h at room temperature. Filter the precipitate off, and rinse with 50 mL ether. Separate the layers and wash the organic phase with saturated aqueous NH_4Cl solution (2 × 30 mL). Dry over Na_2SO_4, concentrate *in vacuo*, and purify the residue by silica gel chromatography with cyclohexane–ethyl acetate (90:10) as eluent to yield 374 mg of pure (*S*)-aldehyde (85% yield, 94% ee), for which $[\alpha]_D^{20} = -1.5°$ (c = 2.56, chloroform). Characterize the product by ^{1}H and ^{13}C NMR spectroscopy. The enantiomeric excess is determined through formation of the diastereomeric acetals with (2*S*,3*S*)-1,4-dimethoxy-2,3-butanediol.

[a] Prepared according to a published procedure.[18]

Finally, related to the conjugate addition reaction is the addition, in position 4, to pyridinium salts. This could be done in an asymmetric version by using the chiral aminal **13**.[22a]

Application of this concept to more complex acid chlorides (Protocol 5) resulted in an efficient and very short synthesis of the indoloquinolizine class of alkaloids.[22b]

Protocol 5.
Addition to activated chiral pyridine 13[2b]

Caution! Carry out all procedures in a well-ventilated hood, and wear disposable vinyl or latex gloves and chemical-resistant safety goggles.

Equipment

- Three-necked, round-bottomed flask (250 mL) with magnetic stirring bar, addition funnel, and septa
- Inert gas supply and inlet
- Dry, gas-tight syringes

Materials

• Copper(I) bromide–dimethyl sulfide complex (FW 205.6) 205 mg, 1 mmol	**moisture sensitive**
• Anhydrous lithium bromide (FW 86.9) 174 mg, 2 mmol	**hygroscopic**
• Ethyl magnesium bromide (FW 133.3) 1 M solution in THF, 1 mL, 1 mmol	**flammable, moisture sensitive**
• Pyridine aminal[a] **13** (FW 329.4) 330 mg, 1 mmol	**harmful**
• Indolylacetic acid chloride[b] (FW 193.6) 0.58 g, 3 mmol	**corrosive, moisture sensitive**

1. Flame dry the reaction vessel and accessories under nitrogen. After cooling to room temperature, place into this flask, under N_2, copper bromide–dimethyl sulfide complex (1 mmol) and anhydrous lithium bromide (2 mmol). Add anhydrous THF (50 mL) and stir at room temperature until a clear solution is obtained (5–10 min).
2. Cool to −40 °C (acetone bath with some dry ice) and add the THF solution of ethylmagnesium bromide (1 mmol). Stir for 10 min.
3. Cool to −78 °C (acetone bath saturated with dry ice) and add the pyridine aminal **13** (1 mmol) dissolved in THF (5 mL).
4. Immediately after, add dropwise a solution of indolylacetic acid chloride (3 mmol) in THF (20 mL). After the addition is completed, warm to −60 °C (by lowering the cooling bath) and stir for 3 h.
5. Hydrolyse the reaction mixture by addition of a saturated aqueous solution of NH_4Cl (40 mL) mixed with a concentrated aqueous solution of NH_4OH (10 mL). Remove the cooling bath and allow the reaction vessel to warm up to room temperature.
6. Separate the phases. Extract the aqueous phase twice with ether (2 × 100 mL). Combine the organic phases and wash them once with aqueous NH_4Cl solution (50 mL). Dry the organic phase over K_2CO_3, and remove the solvents *in vacuo*.
7. Purify the crude product by column chromatography on silica gel using cyclohexane–ethyl acetate (70:30). The chiral dihydropyridine adduct is obtained as an oily product (439 mg, 85%, 95% > de) characterized by ^1H and ^{13}C NMR spectroscopy.
8. The aminal functionality may be hydrolysed as follows. Pour the crude, non-dried aminal adduct (step 6) into Et_2O (50 mL) and add, at room temperature and with stirring, 1 M aqueous HCl solution (50 mL). After 1 h, separate the phases and extract the aqueous phase with ether (3 × 50 mL). Combine the organic phases and wash them once with aqueous NH_4Cl solution (50 mL).

Protocol 5. *Continued*

Dry the organic phase over K_2CO_3, and remove the solvents *in vacuo*. Purify the crude product by column chromatography on silica gel using cyclohexane–ether (50:50). The chiral dihydropyridine carboxaldehyde adduct is obtained as an oily product (232 mg, 79% overall yield), for which $[\alpha]_D^{20} = -160°$ (c = 0.9, $CHCl_3$). The product is characterized by ^1H and ^{13}C NMR spectroscopy.

[a] Prepared according to a published procedure.[22]
[b] Prepared from indolylacetic acid and oxalyl chloride.

2.4 Transformation of the carbonyl group

The carbonyl group can be transformed itself into a chiral derivative, which, after reaction with the organocopper reagent, allows an easy regeneration of the carbonyl group. This is effectively done with α,β-ethylenic aldehydes by transforming them into chiral oxazolidines,[23] or, much more efficiently, into chiral acetals **14**[16,24a] (Protocol 6).

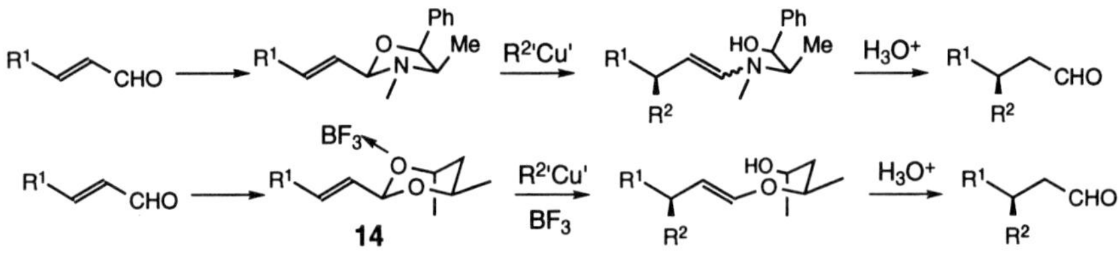

Whereas oxazolidines react easily with organocuprate reagents, chiral acetals need the assistance of a strong Lewis acid such as $BF_3 \cdot OEt_2$. The reaction is stereoelectronically controlled, since the incoming nucleophile (RCu) attacks, by S_N', *anti* to the C—O bond that is cleaved. The selection of this C—O bond is done by the Lewis acid (which coordinates on the oxygen next to the axial Me group) in the most stable conformation, i.e. the one having the *s-trans* vinyl group in the equatorial position. With chiral ketals these conditions are not met, since both equatorial and axial positions are allowed.

as stable as

Protocol 6.
Asymmetric cleavage of an α,β-ethylenic chiral acetal[24a]

Caution! Carry out all procedures in a well-ventilated hood, and wear disposable vinyl or latex gloves and chemical-resistant safety goggles.

Me … **14** —PhCu,2LiBr,PBu₃ / BF₃.Et₂O→ Me … Ph —HCOOH→ Me … CHO, Ph

Equipment

- Three-necked, round-bottomed flask (250 mL) with magnetic stirring bar, addition funnel, and septa
- Round-bottomed flask (50 mL) equipped with a reflux condenser
- Inert gas supply and inlet
- Dry, gas-tight syringes

Materials

- Phenyllithium–lithium bromide complex (FW 170.9) 1 M solution in ether, 5 mL, 5 mmol — **flammable, moisture sensitive**
- Copper(I) bromide–dimethyl sulfide complex (FW 205.6) 1.025 g, 5 mmol — **moisture sensitive**
- Freshly redistilled tributylphosphine (FW 202.3) 1.43 mL, 5 mmol — **flammable, corrosive**
- Chiral acetal[a] **14** (FW 156.2) 312 mg, 2 mmol — **harmful**
- Boron trifluoride diethyl etherate (FW 141.9) 0.615 mL, 5 mmol — **corrosive, moisture sensitive**
- Formic acid 10 mL — **corrosive**

1. Flame dry the reaction vessel and accessories under nitrogen. After cooling to room temperature, place into this flask, under N_2, copper bromide–dimethyl sulfide complex (5 mmol) and Et_2O (50 mL). Add PBu_3 to this slurry at 0°C (ice–water bath) under N_2. When a homogeneous solution is obtained (10 min), cool to −30 °C (acetone bath with some dry ice) and add the solution of PhLi·LiBr in Et_2O. Stir at 0°C for 30 min.
2. Cool to −60 °C (acetone bath with some dry ice), and add the chiral acetal **14** dissolved in ether (5 mL). After 10 min, add dropwise and slowly (10 min) a solution of $BF_3{\cdot}Et_2O$ in Et_2O (10 mL). After the addition is complete, warm to −40 °C and stir for 30 min.
3. Hydrolyse, at −50°C (acetone bath with some dry ice), with a mixture of concentrated aqueous NH_4OH (10 mL) and saturated aqueous NH_4Cl (20 mL). Remove the cooling bath and let the mixture warm to room temperature. Remove the stoppers and the inert gas inlet and continue stirring in air for 1 h.
4. Separate the layers and extract the aqueous phase with ether (2 × 50 mL). Combine the organic phases and evaporate the solvents.

Protocol 6. *Continued*

5. Put the residue into the round-bottomed flask (50 mL) equipped with a reflux condenser and add pure formic acid (10 mL). Heat, under nitrogen, to reflux for 2 h. After cooling to room temperature, add pentane (50 mL) and separate the layers. Extract the lower formic acid layer once with pentane (50 mL), combine the organic phases, wash them with saturated aqueous $NaHCO_3$ (20 mL), dry over $MgSO_4$, and evaporate the solvents *in vacuo*. Purify the residue by silica gel chromatography with cyclohexane–ethyl acetate (90:10) as eluent. (*S*)-3-Phenylbutyraldehyde is obtained (222 mg, 75% yield) in > 95% ee (as determined by diastereomeric aminal formation[24b]) and characterized by ^1H and ^{13}C NMR spectroscopy.

[a] Prepared according to a published procedure.[24a]

2.5 Chiral organocopper reagents

This approach has not been well explored, although several examples of diastereoselective conjugate addition exist (without a removable chiral auxiliary). Moderate success has been achieved with an arylcopper bearing a chiral aminal.[21]

R, R¹, N, 'Cu' + Me, COOEt, COOEt → R, R¹, N, *, COOEt, COOEt, Me — 56% de

Chiral copper azaenolates are much more efficient, and synthetically useful levels of overall enantioselectivity have been attained with imine **15**[25] (Protocol 7).

15 (Ph, Ph, N, OMe) → 1. BuLi; 2. alkynylcopper → (Ph, Ph, N, OMe, Cu, Li+, OMe) → 1. cyclopentenone; 2. NH_4Cl–NH_3 → yield: 78%, ee: 78%

Protocol 7.
Use of a chiral organocopper azaenolate[25a]

Caution! Carry out all procedures in a well-ventilated hood, and wear disposable vinyl or latex gloves and chemical-resistant safety goggles.

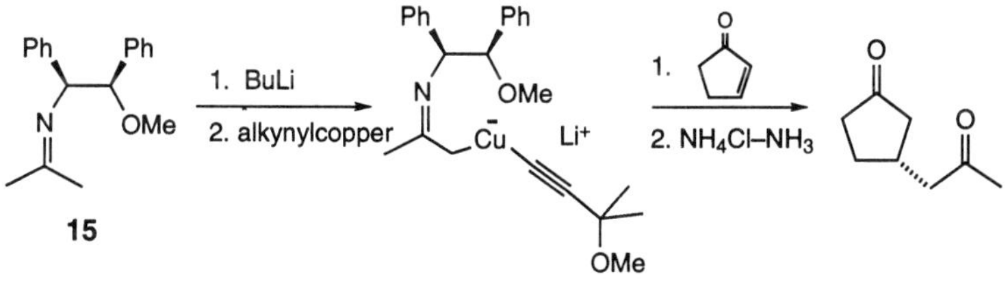

Equipment

- Two three-necked, round-bottomed flasks (100 mL) with magnetic stirring bars, addition funnels, and septa
- Inert gas supply and inlet
- Dry, gas-tight syringes
- Double-ended cannula

Materials

- Imine[a] **15** (FW 267.3) 614 mg, 2.3 mmol — **harmful**
- *n*-Butyllithium (FW 64.1) 1.6 M solution in hexane, 2 × 1.5 mL, 2 × 2.4 mmol — **flammable, moisture sensitive**
- 3-Methoxy-3-methyl-1-butyne[b] (FW 98.1) 225 mg, 2.5 mmol — **harmful**
- Copper(I) bromide–dimethyl sulfide complex (FW 205.6) 514 mg, 2.5 mmol — **moisture sensitive**
- 2-Cyclopentenone (FW 82.1) 164 mg, 2 mmol — **harmful**

1. Flame dry the two reaction vessels and accessories under nitrogen.
2. After cooling to room temperature, add imine **15** and THF (10 mL) into one of these flasks. Cool to −40 °C (acetone bath with some dry ice) and add *n*-BuLi (2.4 mmol) dropwise to the stirred solution. Stir at this temperature for 30 min.
3. In a separate flask, add *n*-BuLi (2.4 mmol) to a stirred solution of 3-methoxy-3-methyl-1-butyne (2.5 mmol) in THF (10 mL) at −40 °C (acetone bath with some dry ice). Remove the cooling bath and let the mixture warm to room temperature. After 10 min, cool again to −20 °C and add at once $CuBr{\cdot}Me_2S$ (2.5 mmol). Warm to room temperature until a homogeneous solution is obtained (10–20 min).
4. Cool the lithiated imine (first flask) to −78 °C (acetone bath saturated with dry ice) and add the solution of the copper acetylide (second flask) via the double-ended cannula. Stir for 30 min at this temperature.
5. Add a solution of 2-cyclopentenone (2 mmol) in THF (2 mL) at −78 °C, stir for 1 h at −60 °C, then warm to −20 °C.

Protocol 7. *Continued*

6. Hydrolyse with a mixture of aqueous concentrated NH_4OH (5 mL) and saturated NH_4Cl solution (10 mL). Remove the stoppers and the inert gas inlet and stir for 1–2 h at room temperature.
7. Add Et_2O (30 mL), separate the layers, and extract the aqueous phase with Et_2O (2 × 30 mL). Wash the combined organic phases with saturated aqueous NH_4Cl (30 mL). After drying over Na_2SO_4, concentrate the solution *in vacuo* and purify the product by column chromatography on silica gel using cyclohexane–ethyl acetate (70:30). The yield is 218 mg (78%) with a 78% ee ($[\alpha]_D^{20} = -59.8°$ in benzene). Characterize the product by ^1H and ^{13}C NMR spectroscopy.

[a] Prepared according to a published procedure.[25a]
[b] Prepared according to a published procedure.[25c]

---

3. Non-covalent chiral auxiliaries

3.1 Heterocuprates

The discovery of heterocuprates as new organocopper reagents[26] brought new possibilities in the attempts to effect asymmetric conjugate addition. This way seemed very attractive, since one has just to deprotonate a chiral alcohol, amine, or thiol to bring a chiral environment near the copper atom itself

$$R^*ZH \xrightarrow{RM} R^*ZM \xrightarrow{RCu} R^*Z\bar{Cu}R\ M^+ \rightleftarrows 0.5\ R_2^*CuM + 0.5\ R_2CuM$$

However, this concept has to deal with the problem of disproportionation and the formation, in equilibrium, of a true homocuprate (optically inactive) and a diheterocuprate.

Several dozens of alcohols, thiols, and amines were tried by several authors with more or less success. From all these attempts, the most successful ones involve auxiliaries bearing at least one extra heteroatom, which serves to stabilize the heterocuprate by an intramolecular chelation, thus avoiding the above disproportionation.

R–Cu—X
M----Y

The substrates in all these reactions are α,β-ethylenic ketones, usually cyclic ones (the most synthetically useful), which avoid the problem of *s-cis* and *s-trans* conformations. In the acyclic series the most common enone is chalcone (1,3-diphenylpropenone). Quite often, the results obtained with the *s-trans* cyclic enones cannot be reproduced with chalcone, the representative *s-cis* enone. Nevertheless, quite high enantiomeric excesses have been attained in recent years using such heterocuprates. The most representative are the following.

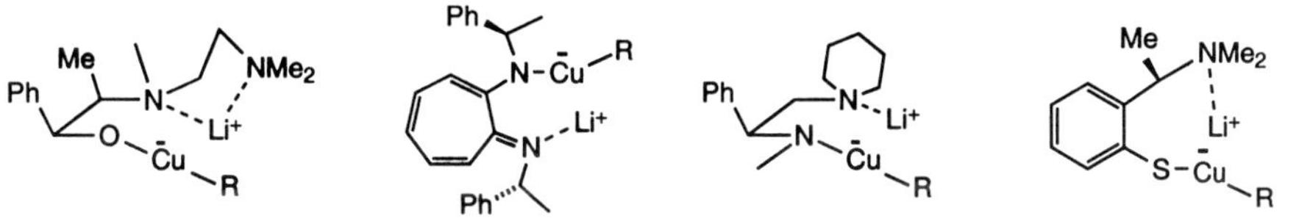

16 Corey *et al.*[27] **17** Lippard *et al.*[28] **18** Rossiter *et al.*[29] **19** van Koten *et al.*[30]

These reactions are not as simple as one would expect. The quality of the copper salt and its nature (CuI, CuBr, or CuCN) are quite important. Several authors have reported a strange reversal of enantioselectivity just by changing the solvent![2] In most cases the heterocuprate is prepared from an organolithium reagent, but in some other cases Grignard reagents seem more efficient. It has also been reported that the quality of the organolithium reagent plays a crucial role,[27] since aged bottles may contain variable amounts of lithium alkoxides which also form heterocuprates. Therefore, the experimental procedure has to be followed exactly as small changes dramatically lower the enantiomeric excess (Protocol 8).

The ideal situation would be the use of catalytic amounts of the chiral auxiliary. Although nobody has yet reported the miracle auxiliary, two examples are noteworthy.

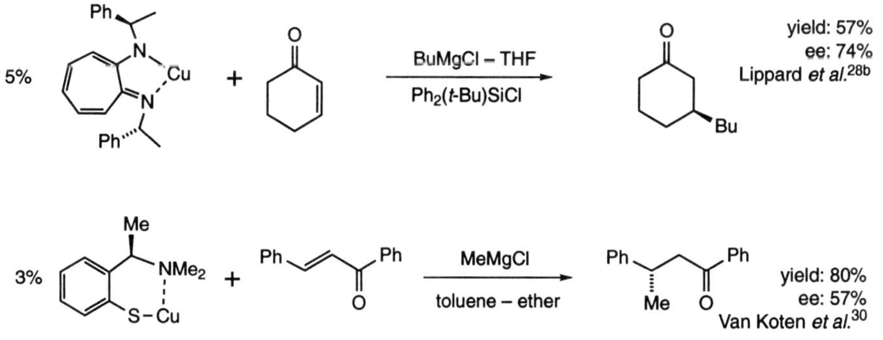

5% + BuMgCl – THF / $Ph_2(t\text{-Bu})SiCl$ → yield: 57% ee: 74% Lippard *et al.*[28b]

3% + MeMgCl / toluene – ether → yield: 80% ee: 57% Van Koten *et al.*[30]

Protocol 8.
Use of a chiral heterocuprate[29]

Caution! Carry out all procedures in a well-ventilated hood, and wear disposable vinyl or latex gloves and chemical-resistant safety goggles.

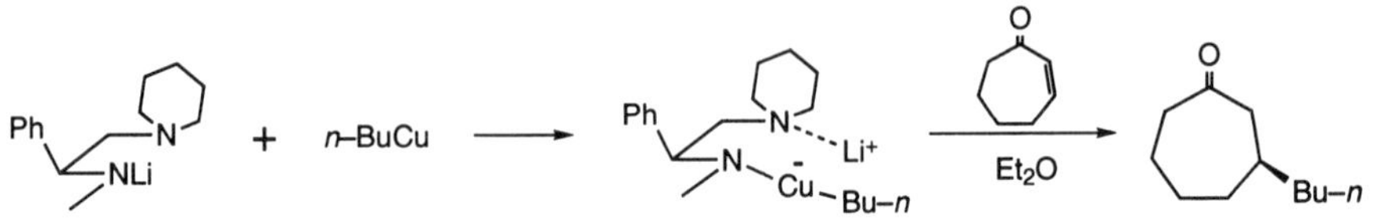

Equipment

- Two three-necked, round-bottomed flasks (100 mL) with magnetic stirring bars, addition funnels, and septa
- Inert gas supply and inlet
- Dry, gas-tight syringes
- Double-ended cannula

Materials

- (*S*)-*N*-Methyl-1-phenyl-2-(1-piperidinyl)ethanamine[a] (MAPP) **18** (FW 218.3) 348 mg, 1.6 mmol — **harmful**
- *n*-Butyllithium (FW 64.1) 2.5 M solution in hexanes, 0.61 mL, 1.53 mmol and 0.51 mL, 1.28 mmol — **flammable, moisture sensitive**
- Copper(I) iodide (FW 190.4) 254 mg, 1.33 mmol — **irritant, light sensitive**
- 2-Cycloheptenone (FW 110.2) 141 mg, 1.28 mmol — **harmful**

1. Flame dry the two reaction vessels and accessories under nitrogen.
2. After cooling to room temperature, add (*S*)-*N*-methyl-1-phenyl-2-(1-piperidinyl) ethanamine (MAPP) **18** and ether (12 mL) into one of these flasks. Cool to −65°C (acetone bath with some dry ice) and add *n*-BuLi (0.61 mL) dropwise to this solution under N_2. Stir the resulting solution for 5 min at this temperature, then remove the cooling bath and let it warm gradually to 0°C.
3. In a separate flask, place the copper iodide and Et_2O (10 mL). Cool to −78°C (acetone bath saturated with dry ice) and add *n*-BuLi (0.51 mL) to the stirred slurry, then warm to −50°C (by lowering the cooling bath) for 30 min.
4. Add, via the double-ended cannula, the cooled (−35°C) lithium amide solution to the suspension of *n*-BuCu, also at −35°C. After 10–15 min, cool the resulting solution to −78°C and stir for 30 min, then add dropwise a solution of 2-cycloheptenone in Et_2O (2 mL).
5. After 1 h at −78°C, hydrolyse the reaction mixture by adding saturated aqueous NH_4Cl solution (15 mL) and let the mixture warm to room temperature. Separate the phases and extract the aqueous phase with ether (2 × 20 mL). Wash the combined organic phases with 1 M HCl (15 mL), dry over Na_2SO_4, and concentrate *in vacuo*. Purify the residue by silica gel chromatography with hexane–ethyl acetate (10:1) as eluent to obtain (*S*)-3-*n*-

butylcycloheptanone (136 mg, 63% yield, 96% enantiomeric excess). Characterize the product by ^1H and ^{13}C NMR spectroscopy. The enantiomeric purity is determined by formation of the diastereomeric ketals with (*R*,*R*)-2,3-butanediol[31a] (^{13}C NMR).

[a] Prepared according to a published procedure.[29]

3.2 Chiral ligands

As with all organometallics, organocopper reagents need some ligands to complete their coordination spheres. Such non-covalently bound ligands may be chiral compounds, generating a chiral environment around the organometallic reagent. Earlier attempts with (−)-sparteine[32] or chiral amino ethers[33] (derived from tartaric acid) gave low enantiomeric excesses with a variety of enones. The first breakthrough came with the tridentate ligand **20** prepared from hydroxyproline.[34]

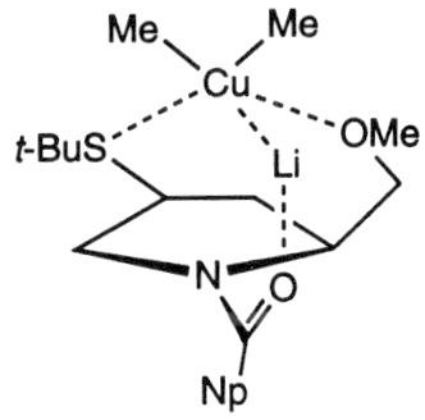

It has been found that the bulky naphthoyl group plays an important role in fixing the geometry of the complex. This ligand is very effective on an *s-cis* enone, such as chalcone. With an *s-trans* enone, such as cyclohexenone, poor results are obtained. However, α-phenylselenocyclohexenone is a good substrate, the α-phenylseleno group and the double bond playing the role of an *s-cis* enone.[35]

The best results are achieved with lithium diorganocuprates and with two equivalents of the ligand.

For our part, we have introduced a new class of ligand, based on the idea that phosphorous ligands are excellent ones for transition metals such as copper. From a variety of phosphorous ligands tried, ligand **21** prepared from inexpensive norephedrine is the most efficient.[31b]

Ph, Me, O, P, N, NMe_2 **21**

With two equivalents of this ligand, the best conditions for the efficient transfer of a butyl group onto cyclohexenone were screened. It was found that addition of LiBr in THF is essential for high enantiomeric excesses, and that cuprates of the stoichiometry $R_5Cu_3Li_2$ are the best. With these cuprates, any *n*-butyllithium (hydrocarbon solution or ethereal solution, new or aged bottle) works just as well (Protocol 9).[31d]

$$5BuLi + 3(CuI,2L^*,4LiBr) \xrightarrow{THF} Bu_5Cu_3Li_2,L^*_6 + 2 \text{ cyclohexenone} \longrightarrow 2 \text{ 3-butylcyclohexanone}$$

yield: 80%
ee: 95%

Protocol 9.
Use of a non-covalent chiral phosphorous ligand[31d]

Caution! Carry out all procedures in a well-ventilated hood, and wear disposable vinyl or latex gloves and chemical-resistant safety goggles.

$$5BuLi + 3(CuI,2L^*,4LiBr) \xrightarrow{THF} Bu_5Cu_3Li_2,L^*_6 \xrightarrow{2 \text{ cyclohexenone}} 2 \text{ 3-butylcyclohexanone}$$

L* = **21** (Ph, Me, O, P, N, NMe_2)

Equipment

- Three-necked, round-bottomed flask (250 mL) with magnetic stirring bar and septa
- Inert gas supply and inlet
- Dry, gas-tight syringes

Materials

Material	Hazard
• Copper(I) iodide (FW 190.4) 476 mg, 2.5 mmol	**irritant, light sensitive**
• Phosphorous ligand[a] **21** (FW 266.3) 1 M solution in toluene, 5 mL, 5 mmol	**harmful**
• Anhydrous lithium bromide (FW 86.8) 870 mg, 10 mmol	**hygroscopic**
• *n*-Butyllithium (FW 64.1) 1.6 M solution in hexane, 2.59 mL, 4.15 mmol	**flammable, moisture sensitive**
• 2-Cyclohexenone (FW 96.1) 193.6 μl, 192 mg, 2 mmol	**highly toxic**

1. Flame dry the reaction vessel and accessories under nitrogen. After cooling to room temperature add copper(I) iodide (476 mg, 2.5 mmol) and THF (40 mL).
2. Cool to −30 °C (acetone bath with some dry ice) and add the toluene solution of the phosphorous ligand **21**. Remove the cooling bath and observe the solubilization of the copper salt. Then add, at once, the solid lithium bromide.
3. When everything is solubilized, cool to −40 °C (acetone bath with some dry ice) and add dropwise the hexane solution of *n*-BuLi. Let the mixture warm to −5 °C in 10 min.
4. Cool again to −78 °C (acetone bath saturated with dry ice) and add 2-cyclohexenone in solution in THF (10 mL). After stirring for 1 h, remove an aliquot and check that the reaction is over by TLC.
5. Hydrolyse the reaction by adding 5 M aqueous HCl (20 mL) and remove the cooling bath. Remove the inert gas inlet and the stopper, and continue stirring at room temperature in air for 1 h to hydrolyse the phosphorous ligand completely.
6. Separate the phases and extract the aqueous phase twice with ether (2 × 100 mL). Wash the combined organic phases twice with aqueous NH_4OH (2 × 50 mL) to remove the copper salts.
7. Dry over $MgSO_4$, then filter and remove the solvents *in vacuo*. Purify butyl-3-cyclohexanone by silica gel chromatography (cyclohexane-ethyl acetate, 90:10). The yield is 247 mg (80%). Characterize the product by ^{1}H and ^{13}C NMR spectroscopy. The enantiomeric purity is determined by formation of the diastereomeric aminals with (*R*, *R*)-1,2-diphenyl ethylene diamine[31e] (^{13}C NMR) and is >95% ee.

[a] Prepared according to a published procedure.[31c]

4. Conclusion

This short overview (for an exhaustive review see elsewhere[2]) shows that the asymmetric conjugate addition with organocopper reagents is becoming a relatively reliable synthetic tool. Recent publications[36] further illustrate that this research field is a very active one. Many syntheses of natural compounds have included such a chiral step; in most cases it is a covalent chiral auxiliary which is chosen. Although no efficient catalytic system has yet been found, fast progress is being made in this field which gives much hope for the future.

References

1. For reviews covering this subject see: (a) Lipshutz, B. H.; Sengupta, S. *Org. React.* **1992**, *41*, 135–631; (b) Posner, G. H. *Org. React.* **1972**, *19*, 1–113.
2. For recent reviews see: (a) Rossiter, B. E.; Swingle, N. E. *Chem. Rev.* **1992**, *92*, 771–806; (b) Perlmutter, P. *Conjugate Addition Reactions in Organic Synthesis*; Tetrahedron Organic Chemistry Series, No. 9; Pergamon Press: Oxford, **1992**.
3. Oppolzer, W.; Löher, H. *Helv. Chim. Acta* **1981**, *64*, 2808–2811.
4. (a) Oppolzer, W.; Moretti, R.; Godel, T.; Meunier, A.; Löher, H. *Tetrahedron Lett.* **1983**, *24*, 4971–4974. (b) Oppolzer, W.; Stevenson, T. *Tetrahedron Lett.* **1986**, *27*, 1139–1140.
5. Oppolzer, W.; Dudfield, P.; Stevenson, T.; Godel, T. *Helv. Chim. Acta* **1985**, *68*, 212–215.
6. Helmchen, G.; Wegner, G. *Tetrahedron Lett.* **1985**, *26*, 6051–6054.
7. (a) Bergdahl, M.; Nilsson, M.; Olsson, T.; Stern, K. *Tetrahedron* **1991**, *47*, 9691–9702. (b) Alexakis, A.; Sedrani, R.; Mangeney, P. *Tetrahedron Lett.* **1990**, *31*, 345–348.
8. (a) Fang, C.; Suemune, H.; Sakai, K. *Tetrahedron Lett.* **1990**, *31*, 4751–4756. (b) Fang, C.; Ogawa, T.; Suemune, H.; Sakai, K. *Tetrahedron Asymm.* **1991**, *2*, 389–398.
9. Mukaiyama, T.; Iwasawa, N. *Chem. Lett.* **1981**, 913–916.
10. Mukaiyama, T.; Takeda, T.; Fujimoto, K. *Bull. Chem. Soc. Jpn.* **1978**, *51*, 3368–3372.
11. Oppolzer, W. *Tetrahedron* **1987**, *43*, 1969–2004.
12. (a) Oppolzer, W.; Poli, G.; Kingma, A. J.; Starkemann, C.; Bernardinelli, G. *Helv. Chim. Acta* **1987**, *70*, 2201–2214. (b) Oppolzer, W.; Mills, R. J.; Pachinger, W.; Stevenson, T. *Helv. Chim. Acta* **1986**, *63*, 1542–1545.
13. Oppolzer, W.; Kingma, A.; Poli, G. *Tetrahedron* **1989**, *45*, 479–488.
14. Tomioka, K.; Suenaga, T.; Koga, K. *Tetrahedron Lett.* **1986**, *27*, 369–372.
15. Posner, G. H. *Acc. Chem. Res.* **1987**, *20*, 72–78.
16. Alexakis, A.; Mangeney, P. *Tetrahedron Asymm.* **1990**, *1*, 477–511.
17. Jung, M. F.; Lew, W. *Tetrahedron Lett.* **1990**, *31*, 623–626.
18. Alexakis, A.; Sedrani, R.; Mangeney, P.; Normant, J. F. *Tetrahedron Lett.* **1988**, *29*, 4411–4414.
19. Asami, M.; Mukaiyama, T. *Chem. Lett.* **1979**, 569–572.
20. Scolastico, C. *Pure Appl. Chem.* **1988**, *60*, 1689–1698.
21. Alexakis, A.; Sedrani, R.; Christenson, B.; Mangeney, P. Unpublished results.
22. (a) Gosmini, R.; Mangeney, P.; Alexakis, A.; Commerçon, M.; Normant, J. F. *Synlett* **1991**, 111–113. (b) Mangeney, P.; Gosmini, R.; Alexakis, A. *Tetrahedron Lett.* **1991**, *32*, 3981–3984. (c) Mangeney, P.; Gosmini, R.; Raussou, S.; Commerçon, M.; Alexakis, A. *J. Org. Chem.* **1994**, *59*, 1877–1888.
23. (a) Mangeney, P.; Alexakis, A.; Normant, J. F. *Tetrahedron* **1984**, *40*, 1803. (b) Berlan, J.; Besace, Y.; Pourcelot, G.; Cresson, P. *Tetrahedron* **1986** *42*, 4757–4765. (c) Berlan, J.; Besace, Y. *Tetrahedron* **1986**, *42*, 4767–4776.
24. (a) Alexakis, A.; Mangeney, P.; Ghribi, A.; Marek, I.; Sedrani, R.; Guir, C.; Normant, J. F. *Pure Appl. Chem.* **1988**, *60*, 49–56. (b) Cuvinot, D.; Mangeney, P.; Alexakis, A.; Normant, J. F. *J. Org. Chem.* **1989**, *54*, 2420–2425.
25. (a) Yamamoto, K.; Kanoh, M.; Yamamoto, N.; Tsuji, J. *Tetrahedron Lett.* **1987**,

28, 6347–6350. (b) Yamamoto, K.; Iijima, M.; Ogimura, Y. *Tetrahedron Lett.* **1982**, *23*, 3711–3714. (c) Corey, E. J.; Floyd, D.; Lipshutz, B. H. *J. Org. Chem.* **1978**, *43*, 3418–3420.

26. Posner, G. H.; Whitten, C. E.; Sterling, J. J. *J. Am. Chem. Soc.* **1973**, *95*, 7788–7800.
27. Corey, E. J.; Naef, R.; Hannon, F. J. *J. Am. Chem. Soc.* **1986**, *108*, 7114–7116.
28. (a) Villacorta, G. M.; Rao, C. P.; Lippard, S. J. *J. Am. Chem. Soc.* **1988**, *110*, 3175–3182. (b) Ahn, K. H.; Klassen, R. B.; Lippard, S. J. *Organometallics* **1990**, *9*, 3178–3181.
29. (a) Rossiter, B. E.; Eguchi, M. *Tetrahedron Lett.* **1990**, *31*, 965–968. (b) Rossiter, B. E.; Eguchi, M.; Hernandez, A. E.; Vickers, D. *Tetrahedron Lett.* **1991**, *32*, 3973–3976.
30. (a) Lambert, F.; Knotter, D. M.; Janssen, M. D.; van Klaveren, M.; Boersma, P.; van Koten, G. *Tetrahedron Asymm.* **1991**, *2*, 1097–1100. (b) Knotter, D. M.; Grove, D. M.; Smeets, W. J. J.; Speck, A. L.; van Koten, G. *J. Am. Chem. Soc.* **1992**, *114*, 3400–3410.
31. (a) Hiemstra, H.; Wynberg, H. *Tetrahedron Lett.* **1977**, 2183–2186. (b) Alexakis, A.; Mutti, S.; Normant, J. F. *J. Am. Chem. Soc.* **1991**, *113*, 6332–6334. (c) Bernard, D.; Burgada, R. *Phosphorus* **1974**, *3*, 187–195. (d) Alexakis, A.; Frutos, J. C.; Mangeney, P. *Tetrahedron: Asymmetry* **1993**, *4*, 2427–2430. (e) Alexakis, A.; Frutos, J. C.; Mangeney, P. *Tetrahedron: Asymmetry* **1993**, *4*, 2431–2434.
32. Kretchmer, R. A. *J. Org. Chem.* **1972**, *37*, 2744–2747.
33. Langer, W.; Seebach, D. *Helv. Chim. Acta* **1979**, *62*, 1710–1722.
34. Leyendecker, F.; Laucher, D. *New J. Chem.* **1985**, *9*, 13–19.
35. Laucher, D. Ph.D. thesis, Strasbourg, France, 1985.
36. (a) Kanai, M.; Koga, K.; Tomioka, K. *Tetrahedron Lett.* **1992**, *33*, 7193–7196. (b) Melnyk, O.; Stephan, E.; Pourcelot, G.; Cresson, P. *Tetrahedron* **1992**, *48*, 841–850. (c) Tanaka, K.; Matsui, J.; Suzuki, H. *J. Chem. Soc., Perkin Trans. 1* **1993**, 153–157. (d) Rossiter, B. E.; Eguchi, M.; Miao, G.; Swingle, N. E.; Hernandez, A. E.; Vickers, D.; Fluckinger, E.; Patterson, R. G.; Reddy, V. *Tetrahedron* **1993**, *49*, 965–986. (e) Spescha, M.; Rihs, G. *Helv. Chim. Acta* **1993**, *76*, 1219–1230. (f) Zhou, Q.-L.; Pfaltz, A. *Tetrahedron Lett.* **1993**, *34*, 7725–7728. (g) Kanai. M.; Tomioka, K. *Tetrahedron Lett.* **1994**, *35*, 895–898.

9

Conjugate addition–enolate-trapping reactions

M. SUZUKI and R. NOYORI

1. Introduction

Enolate anions are one of the most widely used nucleophilic species in organic synthesis.[1] Organocopper conjugate addition to α,β-unsaturated carbonyl compounds generates *in situ* metal enolates under mild conditions with the incorporation of an organic group to the β-position.[2] Therefore, the organocopper-mediated reaction followed by the regioselective enolate trapping provides an extremely powerful tool for controlled construction of organic frameworks (see Scheme 9.1).[3] The organocopper reagents commonly used for this tandem reaction can be divided into four general classes: (1) Grignard reagents plus catalytic amounts of copper salts (RMgX + cat. CuX; X = Cl, Br, or I); (2) complexed mono-organocopper compounds (RCu complexed with phosphine, phosphite, or sulfide ligand); (3) homocuprates or mixed cuprates [$MCuR_2$, $MCuRR^1$, or MCuR(Z); M = Li or MgX, Z = C≡CPr or CN]; and (4) higher order cuprates [$Li_2CuR_2(CN)$ or $Li_2CuRR^1(CN)$].[2,3] Homocuprates and higher order cuprates also include silylcuprates [$Li_2Cu(SiR_3)_2(CN)$],[4] stannylcuprates [$Li_2CuSnMe_3$(2-thienyl)(CN)],[4c,5] and aminocuprates {$LiCu[NSiMe_3(CH_2Ph)]_2$ and $Li_2Cu[NSiMe_3(CH_2Ph)]_2(CN)$}.[6] Although the combination of Grignard reagents and copper catalysts is often the first choice, lithium diorganocuprates and higher order cuprates[2g,7] have been used more widely in view of the higher efficiency, selectivity, and reproducibility of the conjugate addition reactions[2] and the enhanced reactivity of the resulting enolate intermediates.[4–7] The nature of the enolates contaminated with copper species is complicated. Indeed, the enolates generated by organocuprate conjugate addition to enones are considerably less reactive towards electrophiles than pure lithium enolates.[8] However, such detrimental effects may be improved by proper choice of organocopper reagents, solvents, additives, and reaction conditions (see the next section).[3] The necessity to use valuable building blocks has prompted the development of methods using mixed cuprates comprising cheap, anionic dummy ligands.[9] An efficient conjugate addition method based on the 1:1 use of an organometallic reagent and a carbonyl substrate has been elaborated from the need for control of the enolate reactivity, particularly in prostaglandin synthesis.[10,11] Electrophiles such as alkyl halides, aldehydes, ketones, esters, Mannich bases, acyl halides, formates, chloroformates, phosphorochloridates, chlorosilanes, *N*-phenyltrifluoromethanesulfonimide, disulfides, sulfinyl chloride, selenyl bromides, α-silylated methyl vinyl ketones, nitroalkenes, ketene bis(methylthio)acetal monoxides, and so on are employable as agents for enolate trapping.[3] Metal enolates are ambident nucleophiles reacting at either the carbon or oxygen terminus. In general, hard electrophiles such as chlorosilanes or phosphorochloridates tend to give *O*-trapping products, while soft electrophiles such as alkyl halides produce *C*-trapping products selectively. Notably, the organocopper conjugate addition reaction is facilitated by the presence of chlorosilanes, i.e. hard electrophiles.[12]

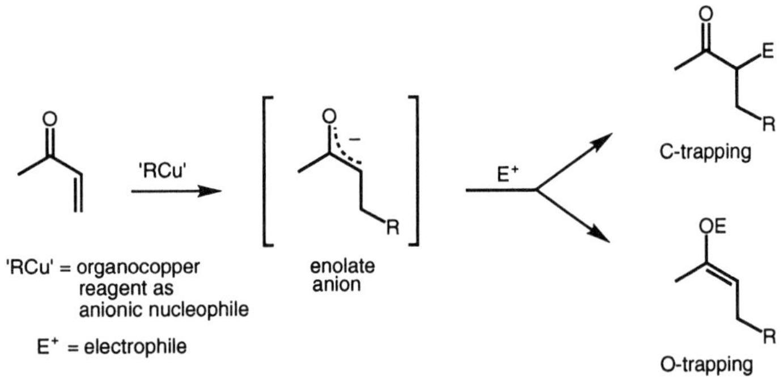

'RCu'

enolate anion

E^+

C-trapping

O-trapping

'RCu' = organocopper reagent as anionic nucleophile

E^+ = electrophile

Scheme 9.1

The utility of the organocopper conjugate addition–enolate-trapping protocol has been amply proved by synthesis of a number of biologically active complex natural products and related compounds including steroids,[8a,13] terpenoids,[2b,14] prostaglandins,[3a,3c,6] lactones,[15] and β-lactams.[6,16]

2. Methods and procedures—general

This section gives a brief overview of the reaction types, some selected synthetic methods, including detailed experimental procedures, and their utility.

2.1 Trapping of the enolates generated by organocopper conjugate addition to α,β-unsaturated ketones

The importance of this process stems from the possibility of controlled construction of molecular structures based on the regiospecific generation of the

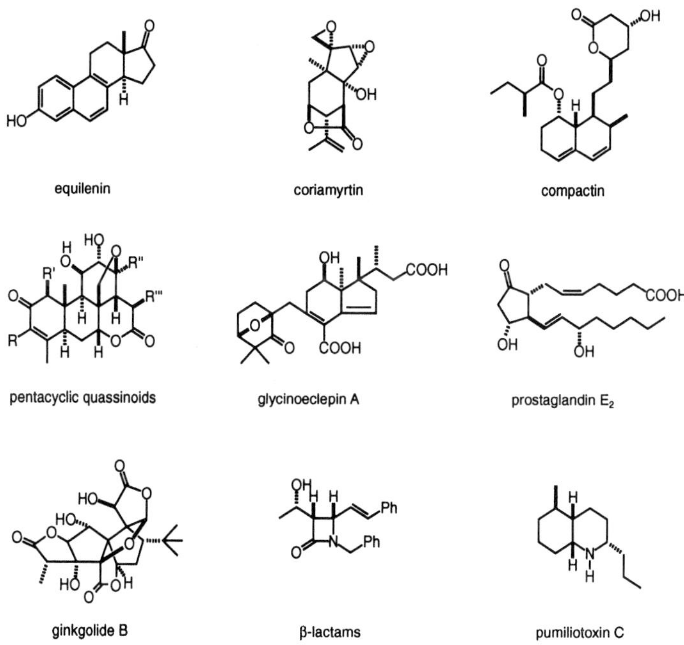

enolate anions on unsymmetrical ketone structures, which is difficult by conventional methods.[1,3] In this view, the organocopper conjugate addition–enolate-trapping method for enones has been studied more extensively than for enals, enoates, or ynoates.[2,3]

2.1.1 *C*-Trapping

i. Alkylation

Alkylation of the ketone enolate generated by conjugate addition of organocopper reagents such as homocuprates or mixed cuprates is restricted to highly reactive alkylating agents such as methyl, allyl, propargyl, benzyl, and α-carbonylalkyl halides. Even with these agents, disappointing results are often obtained,[3] particularly with cyclohexanone and cyclopentanone enolates.[1b,3] This is mainly due to proton exchange between the enolates and alkylated products, leading to polyalkylation or other side-reactions.[1b,3,8b,8d,17,18]. This problem is solved partially by using dimethoxyethane (DME)[16a,19] as solvent, or more suitably by using a THF–hexamethylphosphoric triamide (HMPA) reaction medium (Protocol 1).[20] Changing the countercation of the cuprate from lithium to magnesium is also effective, often doubling the yield (Protocol 1).[13a,13b] Sufficiently active lithium enolates can be generated via a two-step procedure by silylation of organocopper-generated enolates (see Section 2.1.2) followed by treatment with lithium amide in liquid ammonia.[17] By this procedure, relatively unreactive alkylating agents such as iodobutane give satisfactory results.[17] The undesired proton exchange in the reaction of organocopper-generated enolates and alkyl halides is completely suppressed by addition of a triorganostannane and HMPA (see Section 3).[11]

Protocol 1.
Vicinal dialkylation of 2-methyl-2-cyclopentenone[13a]

Caution! Carry out all procedures in a well-ventilated hood, and wear disposable vinyl or latex gloves and chemical-resistant safety goggles.

a. Ar(PrC≡C)CuMgBr
b. $BrCH_2COOEt$

1 → 2

Ar = 6-methoxy-2-naphthyl

Equipment

- A round-bottomed flask (5 mL) with magnetic stirring bar and septum
- A two-necked, round-bottomed flask (30 mL) with magnetic stirring bar and septa
- Inert gas supply and inlet
- Outlet equipped with a liquid paraffin (or mercury) trap
- Dry syringes with stainless steel needles

Materials

- 1-Pentynylcopper[21] (FW 130.7) 65 mg, 0.5 mmol (see Chapter 2, Protocol 4) — **flammable, moisture sensitive**
- 6-Methoxy-2-naphthylmagnesium bromide[13a] (FW 261.4) 0.82 M solution in THF, 0.61 mL, 0.5 mmol — **flammable, moisture sensitive**
- 2-Methyl-2-cyclopentenone[a] **1** (FW 96.1) 0.05 mL, 48 mg, 0.5 mmol — **harmful**
- HMPA[b] (FW 179.2) 10 mL — **highly toxic, cancer suspect agent**
- Ethyl iodoacetate[a] (FW 214.0) 0.66 mL, 1.19 g, 5.6 mmol — **highly toxic, corrosive**
- Dry, distilled THF[c] 2.5 mL — **flammable, irritant**

Dry the reaction apparatus either (1) by flame under a stream of nitrogen or argon or (2) by heating under high vacuum and then introducing nitrogen or argon after the temperature has dropped to room temperature. Set up the reaction apparatus under a stream of inert gas and maintain an inert gas atmosphere by slow and continuous supply of fresh nitrogen or argon throughout the reactions.[d]

1. Place 1-pentynylcopper in a dry, argon-purged, round-bottomed flask.
2. Add a solution of 6-methoxy-2-naphthylmagnesium bromide in THF via

syringe and then stir the mixture rapidly for 1 h at room temperature. During this time the solution becomes dark green and homogeneous.

3. Add 2-methyl-2-cyclopentenone **1** to the THF solution of (6-methoxy-2-naphthyl)(1-pentynyl)copper magnesium bromide generated in the fashion described above. During the course of stirring for 3 h the solution turns black but remains homogeneous.
4. Place dry HMPA and ethyl iodoacetate in a separate, dry, argon-purged, two-necked, round-bottomed flask.
5. Transfer the enolate solution diluted with dry THF (2.5 mL) via syringe (or through a stainless steel cannula) to the room temperature HMPA solution and then continue the stirring for 16 h. The dark green-black solution becomes a faint yellow over this period.
6. Dilute the reaction mixture with ether (10 mL) and saturated aqueous ammonium chloride solution (see Section 6.4 in Chapter 2), and then separate the organic phase.
7. Evaporate the solvent to obtain the alkylated product **2** in >95% yield,[e] which is characterized by IR, ^{1}H NMR, and mass spectra. The product is identical to the authentic material after saponification of the ester function.

[a] The best commercial grade of material is freshly distilled under reduced pressure and the purity is tested before use.
[b] Commercial HMPA is dried over CaH_2 under argon followed by distillation from the same drying agent under reduced pressure, and kept in a sealed ampoule or in a tightly capped bottle with dried molecular sieves (4A) under argon.
[c] Freshly distilled from $LiAlH_4$ or sodium benzophenone ketyl under argon.
[d] For the apparatus set-up, see also Chapter 2, Section 6.3 and Fig. 2.4.
[e] HPLC analysis indicates that unalkylated material is not formed and ^{1}H NMR indicates a stereochemical purity of >99%.

---

ii. Michael reaction

The reaction of vinyl alkyl ketones with the enolates generated by copper-catalysed reaction of Grignard reagents or homocuprates with α,β-unsaturated ketones is representative.[3b] This type of reaction became synthetically viable with the use of α-silylated vinyl ketones as trapping agents (see Protocol 2).[22,23] This method is useful for steroid synthesis.[8a,13c]

iii. Aldol condensation

The enolates produced by copper-catalysed conjugate addition of Grignard reagents to α,β-unsaturated ketones react with aldehydes.[27] The enolates generated by use of homocuprates undergo intramolecular aldol condensation.[28] However, intermolecular trapping of such enolates with aldehydes requires the aid of $ZnCl_2$ (see Protocol 3),[29a,29b] providing a stereoisomeric mixture of aldol products.[29a,29b,30]

Protocol 2.
Organocopper conjugate addition-enolate trapping with an α-silylated vinyl ketone[23]

Caution! Carry out all procedures in a well-ventilated hood, and wear disposable vinyl or latex gloves and chemical-resistant safety goggles.

Equipment

- Three-necked, round-bottomed flask (100 mL) with a magnetic stirring bar, septa, and pressure-equalizing dropping funnel (25 mL)
- Inert gas supply and inlet
- Outlet equipped with a liquid paraffin (or mercury) trap
- Dry glass syringes with stainless steel needles

Materials

- Copper(I) iodide[a] (FW 190.4) 1.9 g, 10 mmol — **irritant, light sensitive**
- Methyllithium[b] (FW 22.0) 2 M solution in ether, 10 mL, 20 mmol — **flammable, moisture sensitive**
- 2-Methyl-2-cyclohexenone[24] **3** (FW 110.2) 1.1 g, 10 mmol — **harmful**
- 3-Trimethylsilyl-3-buten-2-one[22d] (FW 142.3) 2.13 g, 15 mmol — **harmful**
- Dry, distilled ether[c] 60 mL — **flammable, irritant**

Dry the reaction apparatus and maintain under an inert gas atmosphere as described in Protocol 1. Increase the argon flow rate when the flask is immersed in the 78°C bath.

1. Put purified CuI and anhydrous ether (40 mL) into a three-necked, round-bottomed flask.
2. Cool the mixture with stirring in an ice bath while an ether solution of methyllithium is injected through the septum into the flask.
3. Cool the resulting straw-yellow solution of lithium dimethylcuprate to −78°C in a dry ice–acetone bath, and then inject a solution of 2-methyl-2-cyclohexenone **3** in dry ether (10 mL) into the flask with stirring over a 2–3 min period. Warm the cooling bath slowly to about −20°C over an interval of *c.* 1 h.
4. Add a solution of 3-trimethylsilyl-3-buten-2-one in dry ether (10 mL) dropwise during 5 min at −20°C. Stir the mixture and cool it to between −20 and −30°C for another 1 h by occasional addition of dry ice to the cooling bath.

5. Pour the contents of the flask into an ammonium chloride–ammonium hydroxide buffer solution (100 mL, pH 8) (see Section 6.4 in Chapter 2) that has been cooled to 0°C, and then extract the ether layer with two or three additional 100 mL portions of the buffer solution. Extract the combined aqueous solutions with two 75 mL portions of ether, and then wash with saturated NaCl, dry over anhydrous $MgSO_4$, and evaporate the solvent under reduced pressure.
6. Dissolve the residual yellow liquid (2.7–3.4 g) in a mixture of methanol (40 mL) and 4% KOH in H_2O (5 mL), and then heat the resulting solution at reflux under an argon atmosphere for 4 h. Evaporate the methanol from the cooled solution under reduced pressure and then dissolve the residue in ether (50 mL). Wash the ether solution with water, dry over anhydrous $MgSO_4$, and then evaporate the solvent.
7. Distil the residual liquid with a Kugelrohr apparatus (0.5 mm Hg) to afford, after separation of a 0.1–0.2 g sample collected at an oven temperature of 50°C, the octalone[d] **4** (0.76–1.02 g, 43–57%) at an oven temperature of 85–90°C, which is characterized by IR and ^{1}H NMR spectra.

[a] CuI is purified by precipitation from a concentrated aqueous solution of potassium iodide and dried at 100°C over phosphorus pentoxide at high vacuum,[25] or commercial ultrapure CuI is dried at high vacuum before use (see Protocol 2 in Chapter 2).
[b] The commercial reagent (Ventron) is used directly from the bottle after standardization by titration[26] (see also Section 5.4 in Chapter 2).
[c] Freshly distilled over $LiAlH_4$ or sodium benzophenone ketyl under argon.
[d] GLC analysis using a 50 ft capillary column coated with Carbowax 20 M at 140°C indicates a stereochemical purity of >95% *cis*.

Protocol 3. Trapping of a homocuprate-generated enolate with an aldehyde in the presence of $ZnCl_2$[29a,29b]

Caution! Carry out all procedures in a well-ventilated hood, and wear disposable vinyl or latex gloves and chemical-resistant safety goggles.

a. $LiCuMe_2$
b. PhCHO
c. H_3O^+

5 → **6**

Equipment

- Three-necked, round-bottomed flask (50 mL) with magnetic stirring bar and septa
- Inert gas supply and inlet
- Outlet equipped with a liquid paraffin (or mercury) trap
- Dry glass syringes with stainless steel needles

Protocol 3. *Continued*

Materials

- Mesityl oxide[a] **5** (FW 98.2) 98 mg, 1 mmol — **flammable, lachrymator**
- Lithium dimethylcuprate[b] (FW 100.7) 0.1 M solution in ether, 20 mL, 2 mmol — **flammable, moisture sensitive**
- Freshly fused $ZnCl_2$ (FW 136.3) dissolved in ether[c] 0.69 M, 3 mL, 2.1 mmol — **flammable, moisture sensitive**
- Benzaldehyde[a] (FW 106.1) 1.06 g, 10 mmol — **toxic, irritant**
- Dry, distilled ether[d] 1 mL — **flammable, irritant**

Dry the apparatus and maintain under an inert gas atmosphere as described in Protocol 1. Increase the inert gas flow rate when the flask is immersed in the −78 °C bath.

1. Add a solution of mesityl oxide **5** in dry ether (1 mL) to the solution of lithium dimethylcuprate in ether via a syringe at 0 °C under N_2 (or argon). A yellow precipitate develops immediately. Stir the mixture for 30 min at 0 °C.
2. Add a saturated ether solution of freshly fused $ZnCl_2$ and then add benzaldehyde (1.06 g, 10 mmol) via a syringe to the mixture at −78 °C (dry ice–acetone bath). Stir the resulting mixture at −78 °C for 5 min.
3. Pour the mixture into 10% aqueous ammonium chloride (100 mL) to give a blue aqueous phase. Extract the mixture with ether, and then evaporate the solvent.
4. Conduct preparative layer chromatography on silica gel (20 × 20 cm with a layer thickness of 1.25 mm) using a 2:3 mixture of ether and hexane as solvent to afford *erythro*-3-(1′-hydroxybenzyl)-4, 4-dimethyl-2-pentanone[31] **6** (97 mg, 44% yield; m.p. 72 °C) and the *threo*-aldol[31] (48 mg, 22% yield), whose structures are characterized by IR and ^{1}H NMR spectra and elemental analysis, and definitively by X-ray crystallographic analysis[29b] of *erythro*-**6**.

[a] The best commercial grade of material is freshly distilled before use.
[b] See Protocol 2 for its preparation.
[c] Commercially available (Aldrich, 1.0 M solution).
[d] Freshly distilled over $LiAlH_4$ or sodium benzophenone ketyl under argon.

2.1.2 *O*-Trapping

Silylation, phosphorylation, trifluoromethanesulfonylation, acylation, and so on belong to this category.[3]

i. Silylation

In the presence of triethylamine, chlorotrimethylsilane reacts with enolates, which are generated by using Grignard reagents containing catalytic amounts of copper salts,[32] homocuprates,[17] or higher order cuprates[15,33] in ethereal solvents, to give the corresponding trimethylsilyl enol ethers (see Protocol 4). Acceleration of the conjugate addition of organocopper reagents to enones, enals, or enoates by a chlorosilane opened a new dimension in the regio-, stereo-, and chemoselective reactions, enhancing the value of the organocopper method (see Chapter 6).[12] The resulting silyl enol ether is not only a precursor of a lithium enolate[17,34] or a naked enolate,[31,35] but also a useful nucleophile in the Lewis acid-promoted reaction (see Scheme 9.2).[12f,36]

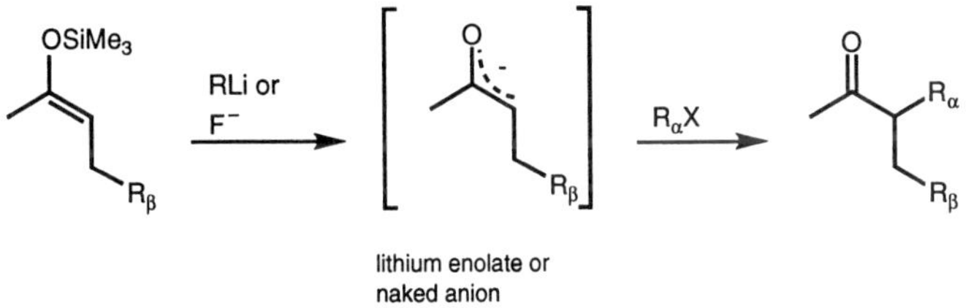

Scheme 9.2

Protocol 4.
Silyl trapping of an organocopper-generated enolate[17]

Caution! Carry out all procedures in a well-ventilated hood, and wear disposable vinyl or latex gloves and chemical-resistant safety goggles.

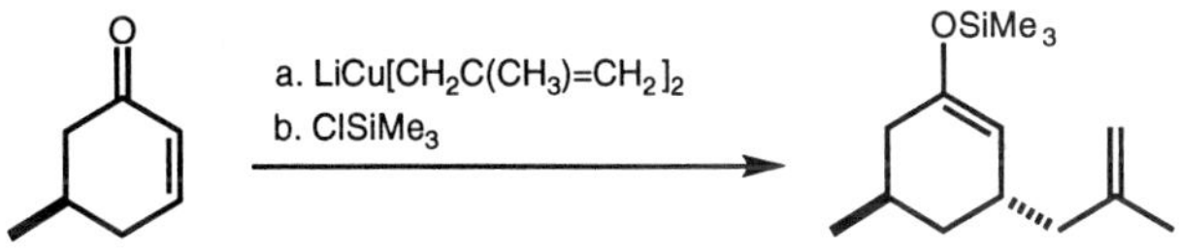

Equipment

- Two-necked, round-bottomed flask (300 mL) with magnetic stirring bar and septa
- Inert gas supply and inlet
- Outlet equipped with a liquid paraffin (or mercury) trap
- Dry syringes

Materials

- Purified CuI[a] (FW 190.4) 9.5 g, 50 mmol — **irritant, light sensitive**
- Freshly distilled di-*n*-butyl sulfide[b] (FW 146.3) 14.6 g, 100 mmol — **irritant, stench**
- Methallyllithium[37] (FW 62.0) 0.5 M solution in ether, 80 mL, 40 mmol — **flammable, moisture sensitive**
- 5-Methyl-2-cyclohexenone[38] **7** (FW 110.2) 1.84 g, 16.7 mmol — **harmful**
- Chlorotrimethylsilane[c] 6.1 mL — **flammable, corrosive**
- Triethylamine[d] 7.6 mL — **flammable, corrosive**
- HMPA[e] 3.8 mL — **highly toxic, cancer suspect agent**
- Dry, distilled ether[f] (20 mL) — **flammable irritant**

Dry the apparatus and maintain under an inert gas atmosphere as described in Protocol 1. Increase the inert gas flow rate when the flask is immersed in the −78°C bath.

Protocol 4. *Continued*

1. Add purified CuI to freshly distilled di-*n*-butyl sulfide under N_2 (or argon) with stirring. Filter the resulting clear-orange liquid complex and store it under N_2 (or argon).
2. Cool a solution of this CuI complex (9.65 g, 20 mmol) in dry ether (10 mL) to −78°C (dry ice–acetone bath) under nitrogen in a two-necked, round-bottomed flask.
3. Add an ether solution of methallyllithium slowly via a syringe. After one equivalent of methallyllithium in ether solution has been introduced, the reaction mixture becomes a bright-red slurry, which changes to a clear, pale-yellow solution upon addition of the second equivalent.
4. Stir this resulting lithium dimethallylcuprate solution for an additional 15 min at −78°C, and then add dropwise a solution of 5-methyl-2-cyclohexenone **7** in dry ether (10 mL) via a syringe. Stir the mixture for an additional 15 min at −78°C, and then warm it to 0°C.
5. Add chlorotrimethylsilane rapidly and then immediately add freshly distilled triethylamine and HMPA via syringes. Stir the resulting reaction mixture at room temperature for 1 h.
6. Dilute the reaction mixture with pentane (150 mL) and then decant the liquid layer from the insoluble copper salts. Wash the resulting solid materials with pentane. Combine the organic portions and wash successively with two 50 mL portions of 5% HCl and 5% $NaHCO_3$, and then dry over $MgSO_4$.
7. Remove the solvent and distil the residue to yield the silylated product **8** (3.41 g, 86% yield; b.p. 74°C/0.9 mm Hg), which is characterized by ^1H NMR and elemental analysis.

[a] Commercial ultrapure CuI can be used directly after drying at high vacuum or after continuous extraction with THF followed by drying at high vacuum (see Protocol 2 in Chapter 2).
[b] The best commercial grade of material is distilled under reduced pressure before use.
[c] The best commercial grade is distilled immediately before use.
[d] Freshly distilled from CaH_2 before use.
[e] Commercial HMPA is dried over CaH_2 under argon followed by distillation from the same drying agent under reduced pressure, and kept in a sealed ampoule or in a tightly capped bottle with dried molecular sieves (4A) under argon.
[f] Freshly distilled over $LiAlH_4$ or sodium benzophenone ketyl under argon.

---

ii. Phosphorylation

The enolates generated by the conjugate addition of homocuprates to enones can be trapped with diethyl phosphorochloridate to give the enol phosphates in good yield (see Protocol 5).[39] The C—O bonds of the resulting enol phosphates can be cleaved by dissolving metals to give the corresponding alkenes without double-bond migration.[39]

Protocol 5.
Synthesis of an enol phosphate[39a]

Caution! Carry out all procedures in a well-ventilated hood, and wear disposable vinyl or latex gloves and chemical-resistant safety goggles.

a. $LiCuMe_2$
b. $ClPO(OEt)_2$

9 → 10 ($(EtO)_2OPO$)

Equipment

- Three-necked, round-bottomed flask (100 mL) with a magnetic stirring bar, septa, and a pressure-equalizing dropping funnel
- Inert gas supply and inlet
- Outlet equipped with a liquid paraffin (or mercury) trap
- Dry glass syringe with stainless steel needle
- Dry dropping funnel

$iMaterials

- Purified CuI [a] (FW 190.4) 384 mg, 2 mmol — **irritant, light sensitive**
- Methyllithium [b] (FW 22.0) 1.64 M solution in ether, 2.44 mL, 4 mmol — **flammable, moisture sensitive**
- 4-Cholesten-3-one [c] **9** (FW 384.7) 576 mg, 1.5 mmol
- Diethyl phosphorochloridate [d] (FW 172.6) 2 g, 11.5 mmol — **highly toxic, corrosive**
- Triethylamine [e] 4 mL — **flammable, corrosive**
- Dry, distilled ether [f] 40 mL — **flammable, irritant**

Dry the apparatus and maintain under an inert gas atmosphere as described in Protocol 1. Replace the first dropping funnel by the second one under a vigorous stream of inert gas.

1. Add purified CuI and anhydrous ether (20 mL) to the dry, three-necked, round-bottomed flask fitted with the pressure-equalizing dropping funnel.
2. Cool the mixture in an ice bath and then add an ether solution of methyllithium via syringe, dropwise and with stirring. As the methyllithium is added, the initial yellow precipitate of polymeric methylcopper(I) redissolves to form a colourless to pale-yellow solution of lithium dimethylcuprate.

Protocol 5. *Continued*

3. Add a solution of 4-cholesten-3-one **9** in anhydrous ether (20 mL) to the resulting cold solution, dropwise and with stirring over 20 min via the dropping funnel. During the addition of the enone, a yellow precipitate of polymeric methylcopper(I) separates from the reaction solution.
4. Remove the cooling bath and then stir the reaction mixture for 2 h while it warms to room temperature.
5. Replace the dropping funnel in the apparatus by a second dry dropping funnel which contains a loose plug of glass wool above the stopcock.
6. Cool the reaction mixture again in an ice bath, and then add a mixture of triethylamine and diethyl phosphorochloridate from the dropping funnel to the reaction mixture rapidly and with stirring.
7. Remove the cooling bath and then stir the mixture for 1 h.
8. Add saturated aqueous $NaHCO_3$ to hydrolyse any remaining organometallic reagents, and then transfer the mixture to a separatory funnel and wash successively with two 50 mL portions each of cold (0°C), aqueous 1 M ammonium hydroxide and water.
9. Extract the aqueous washes with ether (30 mL) and combine the extract with the ether solution. Dry over anhydrous Na_2SO_4, and then concentrate the solution by using a rotary evaporator.
10. Submit the residual liquid in ether (3 mL) to column chromatography on a 2.5×15 cm column packed with a slurry of silica gel (50 g), using ether as eluent, to yield the phosphate ester **10** (420–480 mg, 52–60% yield) as a colourless liquid which is characterized by ^{1}H NMR.

[a] Purified-grade commercial material can be used without further purification.
[b] The commercial reagent should be titrated[26] immediately before use (see Section 5.4 in Chapter 2).
[c] The commercial sample can be used without further purification.
[d] The commercial material is freshly distilled under reduced pressure before use.
[e] Freshly distilled from CaH_2 before use.
[f] Freshly distilled over $LiAlH_4$ or sodium benzophenone ketyl under argon.

iii. Trifluoromethanesulfonylation

The organocopper-generated enolates can be trapped with *N*-phenyltrifluoromethanesulfonimide [PhN(Tf)$_2$] in ethereal solvents to give the enol triflates (see Protocol 6) in good yield.[16c,40]

The demand for enol triflates has now increased as partners in regiospecific cross-coupling reactions with organocopper reagents[42] or organostannanes in the presence of palladium catalysts (see Scheme 9.3).[43]

Acylation of organocopper-generated enolates tends to occur at both carbon and oxygen termini,[3b] but an α-substituted enone gives the *O*-trapping product with high selectivity.[44] For other variations on *O*-trapping, refer to an excellent review.[3b]

Protocol 6.
Synthesis of an enol triflate[16c]

Caution! Carry out all procedures in a well-ventilated hood, and wear disposable vinyl or latex gloves and chemical-resistant safety goggles.

a. $LiCuMe_2$
b. $PhN(Tf)_2$

11 → **12**

Equipment

- Two-necked, round-bottomed flask (50 mL) with magnetic stirring bar and septa
- Inert gas supply and inlet
- Outlet equipped with a liquid paraffin (or mercury) trap
- Dry glass syringes (1 mL) with stainless steel needles

Materials

- Enone[16c] **11** (FW 281.4) 87 mg, 0.31 mmol — **harmful**
- Lithium dimethylcuprate[a] (FW 100.7) 0.052 M solution in ether, 9 mL, 0.47 mmol — **flammable, moisture sensitive**
- *N*-Phenyltrifluoromethanesulfonimide[b] (FW 357.3) 114 mg, 0.32 mmol — **irritant**
- Dry, distilled THF[c] 12 mL — **flammable, irritant**

Dry the apparatus and maintain under an inert gas atmosphere as described in Protocol 1. Increase the inert gas flow rate when the flask is immersed in the −78 °C bath.

1. Add a solution of enone **11** in THF (7 mL) via a syringe to an ether solution of lithium dimethylcuprate cooled to −78 °C in a dry ice–acetone bath, and then stir the mixture for 45 min at the same temperature.
2. Warm the mixture to 0 °C. Stir it for 35 min at 0 °C, and then add *N*-phenyltrifluoromethanesulfonimide in THF (5 mL) via a syringe. Stir the mixture for 2 h at 0 °C, and then warm it to ambient temperature and stir for 1 h.
3. Filter the mixture through a glass frit, and then concentrate the filtrate to give a brown oil. Submit the oil to flash chromatography on silica gel (Kieselgel 60, 230–400 mesh) using a gradient of hexane to 1:9 ethyl acetate–hexane as eluent to afford the enol triflate **12** (61 mg, 46% yield) as an oil which is characterized by IR and ^{1}H NMR spectra and elemental analysis.

[a] Generated *in situ* at −20 to 0 °C by a procedure similar to that described in Protocol 5.
[b] The commercial reagent is purified by recrystallization from hexane.[41]
[c] Freshly distilled from sodium benzophenone ketyl under argon.

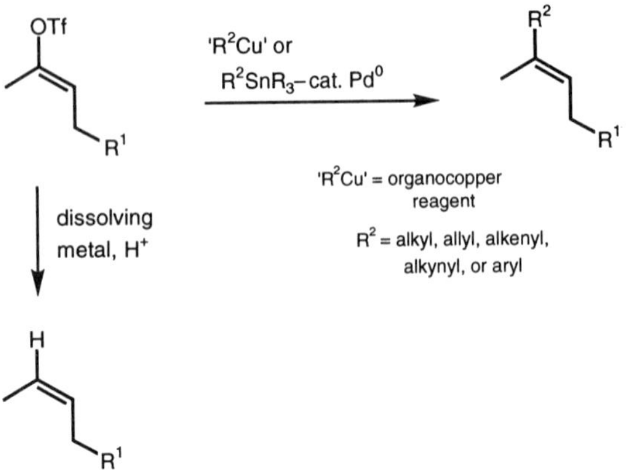

'R^2Cu' or
R^2SnR_3–cat. Pd^0

'R^2Cu' = organocopper reagent
R^2 = alkyl, allyl, alkenyl, alkynyl, or aryl

dissolving metal, H^+

Scheme 9.3

2.2 Trapping of the enolates generated by organocopper conjugate addition to α,β-unsaturated aldehydes

Conjugated ethylenic aldehydes undergo conjugate addition of homocuprates, mixed cuprates, and aggregated cuprates,[45] and the resulting enolates can be trapped with a chlorosilane to give the corresponding enol silyl ethers.[46] The efficiency of this transformation is enhanced by conducting the copper-catalysed conjugate addition reaction in the presence of chlorotrimethylsilane and HMPA.[12c] However, organocopper-generated aldehyde enolates in turn give poor results in the trapping reaction with an alkyl halide, presumably because of their tendency to undergo facile aldol self-condensation.[1e] Careful work-up is necessary for termination of the reaction at the enolate stage because excess organometallic reagents remaining in the reaction medium react with an exposed aldehyde product under aqueous conditions. It is better to isolate the resulting aldehyde as the corresponding enol silyl ether and then regenerate an aldehyde under mild aqueous acid conditions.[47] The enol silyl ether of a hindered aldehyde (a secondary aldehyde) generated by organocopper reaction can be alkylated after converting it to the potassium enolate by treatment with $(Me_3Si)_2NK$ in THF.[1e,47,48] Indirect methods using naphthyloxazolines[49a] or aldimines,[49b] masked forms of enals, are recommended for vicinal dialkylation of enals. In contrast, the enolate generated by reaction of a silylcuprate with an enal can be trapped with methyl iodide without any trouble.[50]

2.3 Trapping of the enolates generated by organocopper conjugate addition to α,β-unsaturated esters

Conjugated enoates, including allenic esters, undergo conjugate addition of organocopper reagents.[2a,2f,2h] The limitation to the use of homocuprates owing to the competitive 1,2- and 1,4-addition reactions can be removed by employment of higher order cuprates[7,51] or an organocopper–BF_3 system (see Chapters 5 and 7 and also Protocol 7).[2h,12e] The conjugate addition reaction proceeds for short reaction times at low temperatures with the use of only a slight excess of reagent. These methods realize high yields even for tri- or tetrasubstituted enoates.[7,12e] The enolates generated by copper-catalysed conjugate addition of a Grignard reagent react with carbonyl compounds[14a] and methanesulfinyl chloride.[52] Trapping of the ester enolates is achieved in higher yield in comparison with the reactions of ketone and aldehyde enolates because of their higher reactivity[53] and lesser tendency to undergo proton exchange or self-condensation.[1e,3d] Carbocupration of acetylenic esters gives vinylcopper species[2] which are equilibrated with the corresponding allenolates, as judged by observation of the loss of stereointegrity.[3d,54] Ketones,[54] esters,[55] and epoxides[54,56] react with either vinylcopper species or allenolates to give the condensed product at the carbon terminus. The use of a higher order cuprate is particularly advantageous for the reaction with an epoxide.[56]

2.4 Trapping of the enolates generated by conjugate addition of silylcuprates, stannylcuprates, or aminocuprates to α,β-unsaturated carbonyl compounds

Silyllithium reagents (two equivalents) mixed with CuI or CuCN react with α,β-unsaturated carbonyl compounds such as enones, enals, enoates, and ynoates to give the metal enolates with the incorporation of the silyl group into the β-position of the starting carbonyl compound (see also Chapter 12).[4,50,57] Such enolates can efficiently be trapped by methyl iodide[4,50,57] to give the *threo*-derivatives[31] for acyclic substrates[50] or the *trans*-adducts for cyclic compounds[4,57] as the major products. (*Z*)-Enolate intermediates generated by the reaction of (*Z*)-β-monosubstituted enoates with higher order disilylcyanocuprates react with aldehydes to give the 1,2-*threo*-2,3-*erythro*-aldols[31] as the major products (see Protocol 7; the numbering is indicated in the aldol **14**).[58] In contrast, condensation of aldehydes with the lithium enolates generated by the reaction of the corresponding saturated ester having a silyl group at the C-3 position with lithium diisopropylamide (LDA) gives the stereoisomeric 1,2-*threo*-2,3-*threo*-aldols[31] predominantly (numbering is indicated in **14**).[58] The resulting Si—C bond can be oxidatively cleaved to give the hydroxyl compound with retention of stereochemistry,[59] thus enhancing the synthetic value of this method.

The enolates generated by the reaction of conjugated acetylenic esters with higher order stannylcuprates react with highly reactive alkylating agents such as methyl iodide, allyl iodide, and propargyl bromide to give the corresponding *C*-alkylated products.[5] However, the enolate intermediates generated by the use of other conventional stannylcopper(I) reagents cannot be trapped with electrophiles other than proton donors.

An efficient chiral synthesis of β-lactams has been elaborated on the basis of the conjugate addition of higher order diaminocyanocuprates to enoates followed by enolate trapping with aldehydes.[6a]

Protocol 7.
Stereoselective three-component coupling of (*E*)-methyl crotonate, a silylcuprate, and crotonaldehyde[58b,60]

Caution! Carry out all procedures in a well-ventilated hood, and wear disposable vinyl or latex gloves and chemical-resistant safety goggles.

COOMe **13** → a. $Li_2Cu(SiMe_2Ph)_2(CN)$ b. (*E*)-MeCH=CHCHO → $PhMe_2Si$, COOMe, OH **14**

Equipment

- Three-necked, round-bottomed flask (500 mL) with magnetic stirring bar and septa
- Inert gas supply and inlet
- Outlet equipped with a liquid paraffin (or mercury) trap
- Dry syringes with stainless steel needles

Materials

- Dimethyl(phenyl)silyllithium[4a] (FW 142.2) 0.428 M solution in THF, 152 mL, 65.1 mmol (see Chapter 12, Protocol 1) **$vcflammable, moisture sensitive**
- Copper(I) cyanide[a] (FW 89.6) 2.87 g, 32 mmol **highly toxic, irritant**
- Dry, distilled THF[b] 131 mL **flammable, irritant**
- (*E*)-Methyl crotonate[c] **13** (FW 100.1) 2.9 g, 29 mmol **flammable, irritant**
- Crotonaldehyde[c] (FW 70.1) 3.4 g, 48 mmol **highly toxic, flammable**

Dry the apparatus and maintain under an inert gas atmosphere as described in Protocol 1. Increase the inert gas flow rate when the flask is immersed in the −78°C bath.[d]

1. Add a THF solution of dimethyl(phenyl)silyllithium under argon at 0°C via a syringe (or a cannula) to a suspension of dry CuCN in dry THF (106 mL) placed in the three-necked, round-bottomed flask with magnetic stirring bar and septa, and then stir the mixture for 30 min at the same temperature.[60] This higher order silylcuprate solution in THF, after being cooled to −78°C in a dry ice–acetone bath, is used for the following conjugate addition reaction.

2. Add dropwise a solution of (*E*)-methyl crotonate **13** in THF (20 mL) via a syringe to a stirred solution of bis[dimethyl(phenyl)silyl]cyanocuprate, prepared as above, under argon (or N_2) at −78°C, and then stir the mixture for 2 h at the same temperature.
3. Add a solution of freshly distilled crotonaldehyde in dry THF (5 mL) via a syringe and stir the mixture for 1 h at −78°C.
4. Quench the reaction mixture with basic aqueous ammonium chloride solution (20 mL, pH 8–9) (see also Section 6.4 in Chapter 2) and then extract it with ether (3 × 50 mL). Wash the combined ether layers with basic aqueous ammonium chloride solution (2 × 10 mL, pH 8–9) and then brine (10 mL). Dry the resulting solution over $MgSO_4$ and then evaporate the solvent.
5. Submit the residue to flash chromatography on silica gel using a 5:1 mixture of hexane and ethyl acetate as eluent to yield the aldol product **14** (6.0 g, 68% yield) having the 1,2-*threo*-2,3-*erythro*-configuration[31] and its stereoisomer (1.5 g, 16% yield) having the 1,2-*threo*-2,3-*threo*-configuration,[31] whose structures are characterized by IR, ^{1}H NMR, and high resolution mass spectrometry.

[a] The commercial material is dried in a desiccator (see also Table 2.1).
[b] Freshly distilled over $LiAlH_4$ or sodium benzophenone ketyl under argon.
[c] The best commercial grade of material is distilled immediately before use.
[d] For related reactions see Chapter 12.

---

3. Prostaglandin synthesis via three-component coupling

The tandem enone conjugate addition–enolate trapping is particularly useful for prostaglandin (PG) synthesis.

3.1 Single-step synthesis of PGE_2 and 5,6-didehydro-PGE_2 derivatives

As illustrated by Scheme 9.4 (M = metal, X = halogen), the one-pot construction of the whole PG framework is accomplished via organometallic-aided conjugate addition of the eight-carbon ω side-chain unit to 4-oxygenated 2-cyclopentenones followed by the trapping of the regiochemically defined enolate species by seven-carbon halides having the α side-chain structure.[3a,3c,61] Organocopper conjugate addition–enolate trapping realizes this ideal PG synthesis. Enolates generated by using a Gilman lithium diorganocuprate or related mixed cuprate react only with highly reactive formaldehyde[62] or 1,1-bis(methylthio)ethene *S*-oxide (a strong Michael acceptor).[8d] However, conjugate addition of a phosphine-complexed mono-organocopper reagent generates an enolate capable of reacting with an actual α side-chain alkyl halide in the presence of a triorganostannane and HMPA (see Protocol 8).[3a,8d,11a] The organocopper reagent generated by mixing equimolar amounts of CuI and an ω side-chain vinyllithium with two to three equivalents of tributylphosphine in THF or ether undergoes conjugate addition to the chiral siloxyenone under the use of a 1:1 reagent : substrate molar ratio (see Protocols 8–10).[10,11] The addition of one equivalent of chlorotriphenylstannane and excess HMPA to the resulting enolate suppresses proton exchange in the alkylation stage.[63] Additives such as chlorotributylstannane and trimethylaluminum are found to be less effective in this case.[64] Thus, a one-pot, sequential treatment of the organocopper reagent with the enone, HMPA, chlorotriphenylstannane, and an α side-chain (*Z*)-allylic iodide leads to the PGE_2 or 5,6-didehydro-PGE_2 derivative in *c*. 80% yield (see Protocol 8).[11a] PGE_2 and prostacyclin (PGI_2) can be synthesized via only three and five steps with 61 and 57% overall yields, respectively, from a chiral 2-cyclopentenone.[3c,11a] The above 5,6-didehydro-PGE_2 derivative is a useful intermediate in the synthesis of isocarbacyclin,[65] a chemically stable PGI_2 analogue, and an azido-functionalized isocarbacyclin analogue as a probe for PGI_2 receptor proteins.[66] This phosphine-complexed mono-organocopper reagent also serves as a key tool for another convergent PG synthesis using a protected 4,5-dihydroxy-2-cyclopentenone as the enone substrate.[67,68] Conjugate addition of a higher order organocuprate[7] to the *O*-methyl derivative of the oxime of a 4-siloxy-2-cyclopentenone followed by trapping of the resulting stabilized metal carbanion with an α side-chain alkyl halide has been exploited for the short-step synthesis of PGE_2.[69]

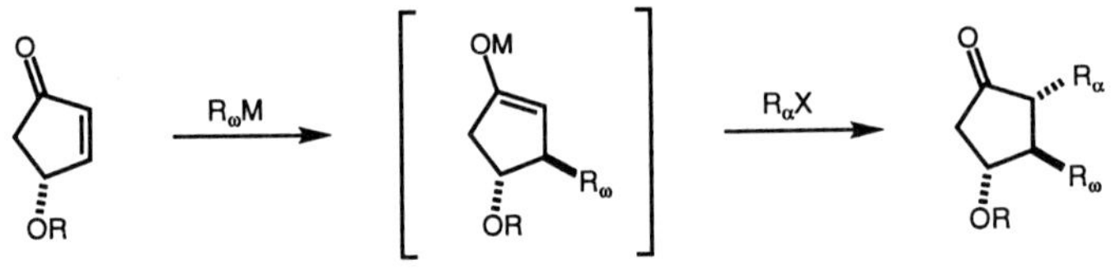

Scheme 9.4

The apparatus used in these reactions is shown in Fig. 9.1. A 1 L Pyrex ampoule **F** equipped with a spiral tube **D** is used as the reaction vessel. Inlets **A** and **J** and outlet **C** are capped with septa (Aldrich) and sealed tightly by Parafilm. Outlet **C** of three-way stopcock **B** is connected to a paraffin bubbler. The use of the ampoule not only avoids air contamination but also, by using a narrow, long Dewar bottle **H** as the cooling bath which accommodates the whole vessel, effects efficient cooling of the reactants. The set of the spiral side-arm ensures precooling of the enone substrate to the same temperature as the organocopper reagent, and therefore the reaction vessel **F** is dropped into the Dewar bottle **H** until the spiral tube **D** is entirely covered by a cold medium **I**. The slow and continuous addition of a solution of the enone substrate in a glass syringe **K** is conducted through inlet **A** by a syringe pump. Prior to the introduction of solvent and materials, dry reaction vessel **F** *in vacuo* by heating with a heat gun and then fill with argon after the temperature falls to room temperature. Dry the round-bottomed flask by a similar procedure. Conduct these operations using vacuum lines, and maintain the reaction vessels under an inert gas atmosphere by a slow and continuous supply of fresh argon during the reactions. Increase the argon flow rate when the flask is immersed in the −78 °C bath. Use methanol sherbet (−95 °C), dry ice–methanol (−78 °C), and cold methanol (−30 °C), for cooling.

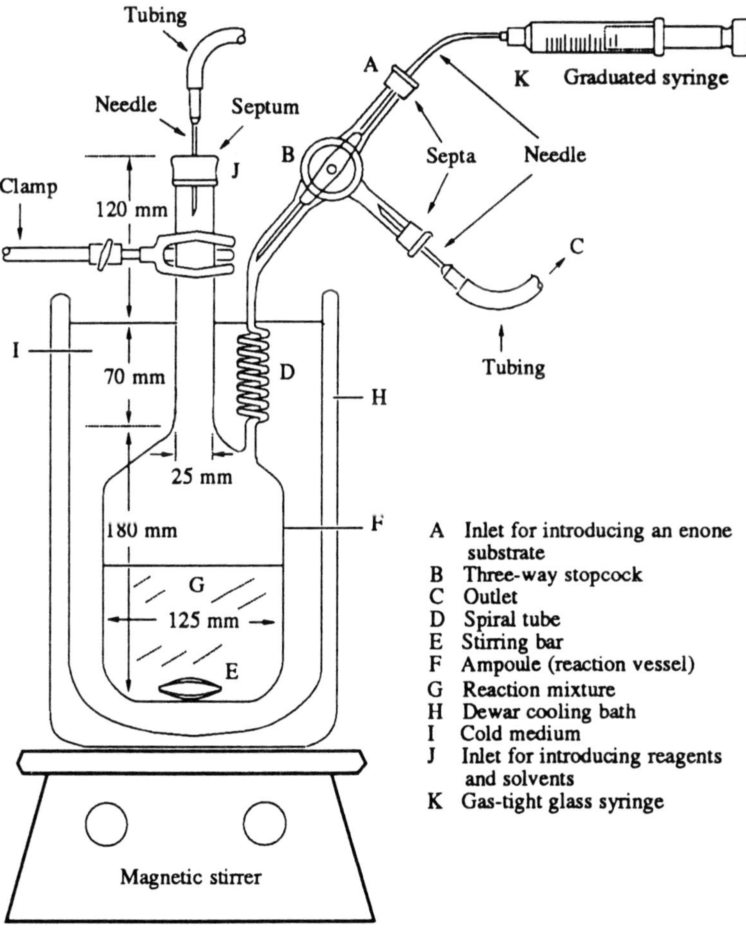

Fig. 9.1

Protocol 8.
Single-step synthesis of a 5,6-didehydro-PGE_2 derivative, an intermediate for the general synthesis of PGs[11a]

Caution! Carry out all procedures in a well-ventilated hood, and wear disposable vinyl or latex gloves and chemical-resistant safety goggles.

a. $R_\omega Li$ + CuI + 2.6 n-Bu_3P
b. Ph_3SnCl, HMPA (9 equiv)
c. $R_\alpha I$ (5 equiv)

16 → 18

$R_\omega Li$ = 15

$R_\alpha I$ = 17

TBDMS = $SiMe_2Bu$–t

Equipment

- An ampoule (l L) with a spiral tube, magnetic stirring bar, and septa (Fig. 9.1)
- A round-bottomed flask (200 mL) with magnetic stirring bar and septum
- Inert gas supply and inlet
- Outlet equipped with a liquid paraffin trap
- Dry glass syringes (50 and 300 mL) with stainless steel needles
- Syringe pump
- Large Dewar bottle
- Temperature-controlled bath
- Cannula

Materials

- (S,E)-3-(t-Butyldimethylsiloxy)-1-iodo-1-octene[11a,70] (FW 368.4) 10.149 g, 27.6 mmol — **harmful**
- t-Butyllithium[a] (FW 64.1) 1.77 M solution in pentane, 31.13 mL, 55.1 mmol — **flammable, moisture sensitive**
- Purified CuI[b] (FW 190.4) 5.247 g, 27.6 mmol — **irritant, light sensitive**
- Tributylphosphine[c] (FW 202.3) 17.85 mL, 71.6 mmol — **flammable, corrosive**
- (R)-4-(t-Butyldimethylsiloxy)-2-cyclopentenone[11a,11c] **16** (FW 212.4) 5.793 g, 27.3 mmol — **harmful**
- Chlorotriphenylstannane[d] (FW 385.5) 10.741 g, 27.6 mmol — **highly toxic, moisture sensitive**
- Methyl 7-iodo-5-heptynoate[11a,71] **17** (FW 266.1) 36.29 g, 136 mmol — **harmful**
- HMPA[e] 47.46 mL — **highly toxic, cancer suspect agent**
- Dry, distilled ether[f] 100 mL — **flammable, irritant**
- Dry, distilled THF[f] 410 mL — **flammable, irritant**

1. Put (S,E)-3-(t-butyldimethylsiloxy)-1-iodo-1-octene into the 1 L Pyrex ampoule **F** under an argon atmosphere, and then dissolve it in anhydrous ether (100 mL).
2. Cool the solution to −95 °C, and then add a pentane solution of t-butyllithium

through inlet **J** (do not use an aged *t*-butyllithium solution contaminated with white precipitates). Stir the mixture for 3 h at −78°C to give a suspension containing the lithiated ω side-chain component **15**.

3. Put purified CuI into a separate 200 mL round-bottomed flask under argon, and then add anhydrous THF (100 mL) and freshly distilled tributylphosphine at room temperature. Stir the mixture until the suspension becomes a clear solution (5 min).

4. Cool the mixture to −78°C, and then add it to the lithium derivative in ampoule **F** via stainless steel cannula through inlet **J** under an argon stream at −78°C. Rinse the flask with anhydrous THF (100 mL) and add to the mixture in ampoule **F**. Stir the mixture for 10 min at −78°C.

5. Add a solution of (*R*)-4-(*t*-butyldimethylsiloxy)-2-cyclopentenone **16** in anhydrous THF (200 mL) to this organocopper reagent from inlet **A** at −78°C through the cooled (−78°C) spiral tube **D** over a period of 3.5 h by using a glass syringe **K** with a syringe pump (slow addition is crucial for a high yield). Rinse the syringe with anhydrous THF (10 mL) and add to the mixture in ampoule **F** with cooling at −78°C. Stir the mixture for 10 min at −78°C.

6. Add HMPA (25 mL) to this mixture from inlet **J** and then stir for 30 min at −78°C.

7. Add a solution of chlorotriphenylstannane in anhydrous THF (30 mL) to the mixture from inlet **J** at −78°C.

8. Warm the resulting mixture to −30°C and then add a solution of methyl 7-iodo-5-heptynoate **17** in HMPA (22.46 mL) from inlet **J** at this temperature. Replace quickly the cold bath by a cryocool-controlled −30°C bath and then stir the mixture for 39 h at −30°C under argon.

9. Pour the mixture onto saturated aqueous ammonium chloride (300 mL) and then separate the organic layer. Wash the resulting layer with saturated aqueous NaCl (300 mL), dry it over anhydrous Na_2SO_4, and evaporate the solvent. At this stage, TLC (5:1 hexane–ethyl acetate as solvent) shows seven spots having R_f values of 0.16 (triphenyltin derivatives, tailing spot), 0.45 (iodide **17**), 0.50 (desired product **18**), 0.58 (unknown product in a very small quantity), 0.64 (organocopper–phosphine complex), 0.67 (1,4-adduct), and 0.82 (tributylphosphine). Isolate the desired product **18** from the mixture as follows. First, submit the crude mixture to chromatography on a short column filled with silica gel (50 g) using a 1:5 mixture of ethyl acetate and hexane as eluent (500 mL). By this rough chromatographic operation the organotin derivatives, organocopper–phosphine complex, simple 1,4-adduct, and tributylphosphine are mostly removed. Then submit the fractions containing **17** and **18** to silica gel (600 g) column chromatography using a 1:60 and then 1:20 mixture of ethyl acetate and hexane as eluent (9.0 and 6.3 L each) to obtain the pure **18** (12.31 g, 76%; a colourless oil),[g] characterized by IR, ^{1}H NMR, ^{13}C NMR, and high resolution mass spectrometry, and

Protocol 8. *Continued*

then **17** (23.24 g, 80% recovery).[h] A small-scale reaction using **16** (318.5 mg, 1.5 mmol) under similar conditions gives the product **18** in 82% yield.

[a] The commercial reagent (Aldrich, Kanto) is used directly from the bottle after standardization by titration (see Section 5.4 in Chapter 2).[26]
[b] The commercial material is purified by continuous extraction using a Soxhlet extractor and then dried in vacuum at room temperature for several hours (see Protocol 2 in Chapter 2).
[c] The commercial material is freshly distilled under reduced pressure before use.
[d] The commercial reagent (Aldrich) is used without further purification.
[e] The commercial material is dried over CaH_2 under argon followed by distillation from the same drying agent under reduced pressure, and kept in a sealed ampoule or a tightly capped bottle with molecular sieves (4A) under argon.
[f] Freshly distilled over sodium benzophenone ketyl under argon.
[g] Homogeneous as judged by ^{13}C NMR measurements.
[h] The product **18** is most conveniently separated from iodide **17** by preparative HPLC using a Japan Analytical Industry model LC-09 chromatograph (column, JAIGEL AJ2H; eluent, $CHCl_3$).

3.2 Synthesis of functionalized prostaglandins

The three-component method based on the use of phosphine-complexed mono-organocopper reagents allows wide variations in nucleophilic/electrophilic vicinal carba-condensation.[3a,3c] The reaction of a regiodefined enolate with a saturated seven-carbon aldehyde gives the 7-hydroxy-PGE_1 derivative (stereomixtures at C-7) in 83% yield (see Protocol 9).[11c,72] This aldol product is convertible to PGE_1 in four steps with 66% overall yield. The double dehydration of the same aldol product produces the tumour-suppressing compound Δ^7-PGA_1.[11c,72,73] This aldol process also allows a large-scale preparation of Mexiprostil, a drug for the inhibition of gastric acid secretion.[74] Trapping of the enolate with α,β-unsaturated aldehydes is facilitated by the addition of BF_3 etherate,[11c] and this process has been used in the synthesis of 7-fluoro-PGI_2.[75] Reaction of an enolate with a seven-carbon nitroalkene followed by exposure of the nitronate intermediate to aqueous titanium(III) chloride ($TiCl_3$) gives a precursor of the anti-ulcer compound 6-oxo-PGE_1 in excellent yield (see Protocol 10).[11b]

Recent publications further emphasize the synthetic importance of the conjugate addition–enolate-trapping process.[77]

Protocol 9.
Three-component synthesis of a 7-hydroxy-PGE$_1$ derivative[11c,72]

Caution! Carry out all procedures in a well-ventilated hood, and wear disposable vinyl or latex gloves and chemical-resistant safety goggles.

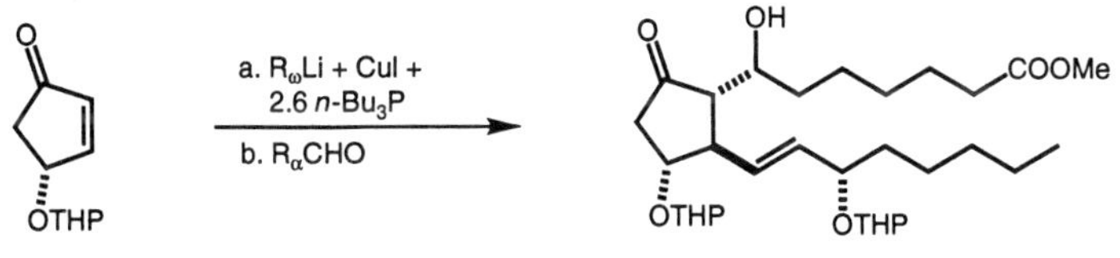

a. $R_\omega Li$ + CuI + 2.6 n-Bu_3P
b. $R_\alpha CHO$

20 → 22

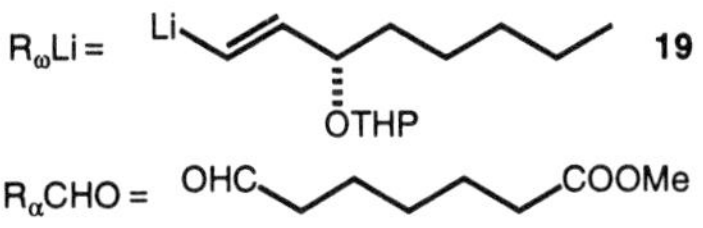

$R_\omega Li$ = 19

$R_\alpha CHO$ = 21

Equipment

- A test-tube (20 mL) with magnetic stirring bar and septum
- An ampoule (150 mL) with magnetic stirring bar and septum[a]
- Inert gas supply and inlet
- Outlet equipped with a liquid paraffin trap
- Dry glass syringes (10 and 30 mL) with stainless steel needles
- Cannula

Materials

- (*E*,3*S*)-3-(Tetrahydropyran-2-yloxy)-1-iodo-1-octene[11c,70] (FW 338.2) 744 mg, 2.20 mmol — **harmful**
- *t*-Butyllithium[b] (FW 64.1) 1.38 M solution in pentane, 3.19 mL, 4.40 mmol — **flammable, moisture sensitive**
- Purified CuI[c] (FW 190.4) 419 mg, 2.20 mmol — **irritant, light sensitive**
- Tributylphosphine[d] (FW 202.3) 1.43 mL, 5.72 mmol — **flammable, corrosive**
- 4-(Tetrahydropyran-2-yloxy)-2-cyclopentenone[11c] **20** (FW 182.1) 364 mg, 2.0 mmol — **harmful**
- Methyl 6-formylhexanoate[11c,76] **21** (FW 158.2) 348 mg, 2.20 mmol — **harmful**
- Dry, distilled ether[e] 85 mL — **flammable, irritant**

Dry the apparatus and maintain under an inert gas atmosphere as described in Protocol 1. Increase the inert gas flow rate when the flask is immersed in the −78°C bath.

1. Prepare the ω side-chain alkenyllithium **19** in ether (white suspension) by a procedure similar to that described in Protocol 8. Mix a solution of (*E*,3*S*)-3-(tetrahydropyran-2-yloxy)-1-iodo-1-octene in anhydrous ether (10 mL) with a solution of *t*-butyllithium in pentane in the 20 mL test-tube and stir at −78°C for 3 h.

Protocol 9. *Continued*

2. Generate the enolate by slow addition of a solution of 4-(tetrahydropyran-2-yloxy)-2-cyclopentenone **20** in dry ether (20 mL) via a syringe to the organocopper reagent (generated *in situ*[f]) at −78 °C along the cooled (−78 °C) wall of the reaction vessel[a] over a period of 35 min under stirring. Stir the resulting mixture at −78 °C for 15 min.
3. Add a solution of methyl 6-formylhexanoate **21** in anhydrous ether (or THF[e]) (5 mL) to the enolate solution at −78 °C with stirring via a syringe, and then stir the mixture for 15 min at this temperature.
4. Quench the mixture with saturated aqueous ammonium chloride solution (40 mL) and shake it vigorously. Separate the organic layer and extract the aqueous layer twice with ether (25 mL each). Dry the combined organic extracts with $MgSO_4$ and evaporate the solvent.
5. Submit the residual material to column chromatography on silica gel (30 g) using a 3:1 to 1:1 mixture of hexane and ethyl acetate as eluent to yield the aldol product **22** (919 mg, 83% yield; a mixture of two stereoisomers) as a colourless oil, which is characterized by IR, ^{1}H NMR, and high resolution mass spectrometry.

[a] For large-scale reactions, use the ampoule equipped with the spiral side-arm (see Fig. 9.1) in view of the high reproducibility of the reaction.
[b] The commercial reagent (Aldrich, Kanto) is used directly from the bottle after standardization by titration (see Section 5.4 in Chapter 2).[26]
[c] The commercial material is purified by continuous extraction using a Soxhlet extractor and then dried in vacuum at room temperature for several hours (see Protocol 2 in Chapter 2).
[d] The commercial material is freshly distilled under reduced pressure before use.
[e] Freshly distilled over sodium benzophenone ketyl under argon.
[f] Generate the organocopper reagent *in situ* in a 150 mL ampoule by mixing a solution of CuI–tributylphosphine in anhydrous ether (40 mL) with the ω side-chain alkenyllithium **19** in ether, prepared in step 1, via a cannula at −78 °C with stirring; rinse the test-tube with dry ether (10 mL); then stir the mixture at −78 °C for 10 min.

Protocol 10.
Three-component synthesis of a 6-oxo-PGE_1 derivative from a siloxycyclopentenone, a phosphine-complexed ω side-chain alkenylcopper, and an α side-chain nitroalkene[11b]

Caution! Carry out all procedures in a well-ventilated hood, and wear disposable vinyl or latex gloves and chemical-resistant safety goggles.

16: OTBDMS

a. $R_\omega Li$ + CuI + 2.6 n-Bu_3P
b. nitroalkene

nitronate intermediate: ^-O, N^+, O^-, COOMe, OTBDMS, OTBDMS

aq. $TiCl_3$

24: O, O, COOMe, OTBDMS, OTBDMS

$R_\omega Li$ = Li ... OTBDMS **15**

nitroalkene = ... COOMe, NO_2 **23**

Equipment

- An ampoule (50 mL) with magnetic stirring bar and septum
- An ampoule (100 mL) with magnetic stirring bar and septum[a]
- Inert gas supply and inlet
- Outlet equipped with a liquid paraffin trap
- Dry glass syringes with stainless steel needles
- Cannula

Materials

- (R)-4-(t-Butyldimethylsiloxy)-2-cyclopentenone[11a,11c] **16** (FW 212.4) 424 mg, 2.0 mmol — **harmful**
- (S,E)-3-(t-Butyldimethylsiloxy)-1-iodo-1-octene[11a,70] (FW 368.4) 883 mg, 2.4 mmol — **harmful**
- t-Butyllithium[b] (FW 64.1) 2.0 M solution in ether, 2.4 mL, 4.8 mmol — **flammable, moisture sensitive**
- Purified CuI[c] (FW 190.4) 457 mg, 2.4 mmol — **irritant, light sensitive**
- Tributylphosphine[d] (FW 202.3) 1.2 mL, 4.8 mmol — **flammable, corrosive**
- Methyl 6-nitro-6-heptenoate[11b] **23** (FW 187.2) 449 mg, 2.4 mmol — **harmful**
- 25% aqueous $TiCl_3$ solution[e] 15 mL, 24 mmol — **corrosive**
- Dry, distilled ether[f] 36 mL — **flammable, irritant**

Dry the apparatus and maintain under an inert gas atmosphere as described in Protocol 1. Increase the inert gas flow rate when the flask is immersed in the −78°C bath.

Protocol 10. *Continued*

1. Prepare the ω side-chain vinyllithium **15** by a procedure similar to that described in Protocol 8. In a 50 mL ampoule under argon mix a solution of (*S*,*E*)-3-(*t*-butyldimethylsiloxy)-1-iodo-1-octene in dry ether (20 mL) with a solution of *t*-butyllithium in pentane and stir at −78°C for 3 h.
2. Generate the enolate by addition of a solution of (*R*)-4-(*t*-butyldimethylsiloxy)-2-cyclopentenone **16** in dry ether (5 mL) via a syringe to the phosphine-complexed mono-organocopper reagent (generated *in situ*[g]) at −78°C along the cooled (−78°C) wall of the reaction vessel[a] over a period of 10 min under stirring. Stir the resulting mixture at −78°C for 15 min.
3. Add a solution of methyl 6-nitro-6-heptenoate **23** in dry ether (5 mL) to the enolate at −78°C with stirring via a syringe, and then stir the mixture for 20 min at −78°C, 10 min at −30°C, then 5 min at 0°C.
4. Pour the resulting mixture into a mixture of 25% aqueous $TiCl_3$ solution, 18% aqueous ammonium acetate (50 mL, 144 mmol), and THF (100 mL), and then stir the mixture for 18 h at room temperature.
5. Neutralize the mixture with aqueous $NaHCO_3$ and take up in hexane (300 mL). Separate the organic layer and extract the aqueous layer twice with hexane (500 mL). Wash the combined organic extracts with brine, dry over $MgSO_4$, and evaporate the solvent.
6. Submit the residual material to column chromatography on silica gel (100 g) using a 19:1 to 1:1 (gradient) mixture of hexane and ethyl acetate as eluent to obtain the desired oxo compound **24** (806 mg, 66% yield; more polar) and its C-8 epimer (118 mg, 10%; less polar),[h] whose structures are characterized by IR, ^{1}H NMR, and high resolution mass spectrometry.

[a] For large-scale reactions, use the ampoule equipped with the spiral side-arm (see Fig. 9.1) in view of the high reproducibility of the reaction.
[b] The commercial reagent (Aldrich, Kanto) is used directly from the bottle after standardization by titration (see Section 5.4 in Chapter 2).[26]
[c] The commercial material is purified by continuous extraction using a Soxhlet extractor and then dried in vacuum at room temperature for several hours (see Protocol 2 in Chapter 2).
[d] The commercial material is freshly distilled under reduced pressure before use.
[e] The best commercial grade of $TiCl_3$ is used without further purification.
[f] Freshly distilled over sodium benzophenone ketyl under argon.
[g] Generate the reagent *in situ* in a 100 mL ampoule by mixing a solution of CuI–tributylphosphine in dry ether (6 mL) with the ω side-chain alkenyllithium **15** in ether, prepared in step 1, via a cannula at −78°C and then stir the mixture at −78°C for 10 min.
[h] The less polar C-8 epimer readily epimerizes to **24** by treatment with aqueous potassium acetate or with silica gel on a TLC plate.

References

1. Reviews: (a) House, H. O. *Modern Synthetic Reaction*, 2nd edn; Benjamin: Menlo Park, CA, 1972; Chapters 9–11; (b) Augustine, R. L. *Carbon–Carbon Bond Formation*; Marcel Dekker: New York, 1979; Vol. 1; (c) Evans, D. A. In *Asymmetric Synthesis*; Morrison, J. D., ed.; Academic Press: London, 1984; Vol. 3, Chapter 1, pp. 1–110; (d) Heathcock, C. H. In *Asymmetric Synthesis*; Morrison, J. D., ed.; Academic Press: London, 1984; Vol. 3, Chapter 2, pp. 111–212; (e) Caine, D. In *Comprehensive Organic Synthesis*; Trost, B. M., Fleming, I., eds; Pergamon: Oxford, 1991; Vol. 3, Chapter 1.1, pp. 1–63.
2. Reviews: (a) Posner, G. H. *Org. React.* **1972**, *19*, 1–113; (b) Posner, G. H. *An Introduction to Synthesis Using Organocopper Reagents*; Wiley: New York, 1980; (c) Norman, J. F. *Synthesis* **1972**, 63–80; (d) Kozlowski, J. A. In *Comprehensive Organic Synthesis*; Trost, B. M., Fleming, I., eds; Pergamon: Oxford, 1991; Vol. 4, Chapter 1.4, pp. 169–198; (e) Klunder, J. M.; Posner, G. H. In *Comprehensive Organic Synthesis*; Trost, B. M., Fleming, I., eds; Pergamon; Oxford, 1991; Vol. 3, Chapter 1.5, pp. 207–239; (f) Schmalz, H.-G. In *Comprehensive Organic Synthesis*; Trost, B. M., Fleming, I., eds; Pergamon: Oxford, 1991; Vol. 4, Chapter 1.5, pp. 199–236; (g) Lipshutz, B. H.; Sengupta, S. *Org. React.* **1992**, *41*, 135–631; (h) Rossiter, B. E.; Swingle, N. M. *Chem. Rev.* **1992**, *92*, 771–806.
3. Reviews: (a) Noyori, R.; Suzuki, M. *Angew. Chem., Int. Ed. Engl.* **1984**, *23*, 847–876; (b) Taylor, R. J. K. *Synthesis* **1985**, 364–392; (c) Noyori, R.; Suzuki, M. *Chemtracts Org. Chem.* **1990**, *3*, 173–197; (d) Hulce, M.; Chapdelaine, M. J. In *Comprehensive Organic Synthesis*; Trost, B. M., Fleming, I., eds; Pergamon: Oxford, 1991; Vol. 4, Chapter 1.6, pp. 237–268. See also references 1 and 2.
4. (a) Ager, D. J.; Fleming, I.; Patel, S. K. *J. Chem. Soc., Perkin Trans. 1* **1981**, 2520–2526. (b) Fleming, I.; Newton, T. W. *J. Chem. Soc., Perkin Trans. 1* **1984**, 1805–1808. For structural studies in THF see: (c) Sharma, S.; Oehlschlager, A. C. *J. Org. Chem.* **1991**, *56*, 770–776.
5. Piers, E.; Tillyer, R. D. *J. Org. Chem.* **1988**, *53*, 5366–5369.
6. (a) Yamamoto, Y.; Asao, N.; Uyehara, T *J. Am. Chem. Soc.* **1992**, *114*, 5427–5429. (b) Shida, N.; Uyehara, T.; Yamamoto, Y. *J. Org. Chem.* **1992**, *57*, 5049–5051.
7. Reviews: (a) Lipshutz, B. H.; Wilhelm, R. S.; Kozlowski, J. A. *Tetrahedron* **1984**, *40*, 5005–5038; (b) Lipshutz, B. H. *Synthesis* **1987**, 325–341; (c) Lipshutz, B. H. *Synlett* **1990**, 119–128. For structural studies on higher order cuprates in THF or ether see: (d) Lipshutz, B. H.; Sharma, S.; Ellsworth, E. L. *J. Am. Chem. Soc.* **1990**, *112*, 4032–4034; (e) Bertz, S. H. *J. Am. Chem. Soc.* **1991**, *113*, 5470–5471; Bertz, S. H.; Dabbagh, G.; He, X.; Power, P. P. *J. Am. Chem. Soc.* **1993**, *115*, 11640–11641. (f) Stemmler, T.; Penner-Hahn, J. E.; Knochel, P. *J. Am. Chem. Soc.* **1993**, *115*, 348–350.
8. (a) Boeckman, Jr., R. K. *J. Am. Chem. Soc.* **1974**, *96*, 6179–6180. (b) Posner, G. H.; Lentz, C. M. *J. Am. Chem. Soc.* **1979**, *101*, 934–956. (c) House, H. O.; Wilkins, J. M. *J. Org. Chem.* **1976**, *41*, 4031–4033. (d) Davis, R.; Untch, K. G. *J. Org. Chem.* **1979**, *44*, 3755–3759.
9. Corey, E. J.; Beames, D. J. *J. Am. Chem. Soc.* **1972**, *94*, 7210–7211. For other variations see references cited in reference 2g.
10. Suzuki, M.; Suzuki, T.; Kawagishi, T.; Noyori, R. *Tetrahedron Lett.* **1980**, *21*,

1247–1250. Suzuki, M.; Suzuki, T.; Kawagishi, T.; Morita, Y.; Noyori, R. *Isr. J. Chem.* **1984**, *24*, 118–124.

11. (a) Suzuki, M.; Yanagisawa, A.; Noyori, R. *J. Am. Chem. Soc.*, **1985**, *107*, 3348–3349. Suzuki, M.; Yanagisawa, A.; Noyori, R. *J. Am. Chem. Soc.*, **1988**, *110*, 4718–4726. (b) Tanaka, T.; Hazato, A.; Bannai, K.; Okamura, N.; Sugiura, S.; Manabe, K.; Toru, T.; Kurozumi, S.; Suzuki, M.; Kawagishi, T.; Noyori, R. *Tetrahedron*, **1987**, *43*, 813–824. (c) Suzuki, M.; Kawagishi, T.; Yanagisawa, A.; Suzuki, T.; Okamura, N.; Noyori, R. *Bull. Chem. Soc. Jpn.* **1988**, *61*, 1299–1312.
12. (a) Corey, E. J.; Boaz, N. W. *Tetrahedron Lett.* **1985**, *26*, 6015–6018, 6019–6022. (b) Alexakis, A.; Berlan, J.; Besace, Y. *Tetrahedron Lett.*, **1986**, *27*, 1047–1050. (c) Horiguchi, Y.; Matsuzawa, S.; Nakamura, E.; Kuwajima, I. *Tetrahedron Lett.* **1986**, *27*, 4025–4028. Nakamura, E.; Matsuzawa, S.; Horiguchi, Y.; Kuwajima, I. *Tetrahedron Lett.* **1986**, *27*, 4029–4032. Matsuzawa, S.; Horiguchi, Y.; Nakamura, E.; Kuwajima, I. *Tetrahedron*, **1989**, *45*, 349–362. Review: Nakamura, E. *Synlett* **1991**, 539–547. (d) Johnson, C. R.; Marren, T. J. *Tetrahedron Lett.* **1987**, *28*, 27–30. Likewise, BF_3 etherate accelerates organocopper conjugate addition: (e) (review) Yamamoto, Y. *Angew. Chem., Int. Ed. Engl.* **1986**, *25*, 947–959; (f) Lee, V. J. In *Comprehensive Organic Synthesis*; Trost, B. M., Fleming, I., eds; Pergamon: Oxford, 1991; Vol. 4, Chapter 1.3, pp. 139–168.
13. (a) Posner, G. H.; Chapdelaine, M. J.; Lentz, C. M. *J. Org. Chem.* **1979**, *44*, 3661–3665. (b) Funk, R. L.; Vollhardt, K. P. C. *J. Am. Chem. Soc.* **1980**, *102*, 5253–5261. (c) Takahashi, T.; Naito, Y.; Tsuji, J. *J. Am. Chem. Soc.* **1981**, *103*, 5261–5263. Doi, T.; Shimizu, K.; Takahashi, T.; Tsuji, J.; Yamamoto, K. *Tetrahedron Lett.* **1990**, *31*, 3313–3316.
14. Coriamyrtin: (a) Tanaka, K.; Uchiyama, F.; Sakamoto, K.; Inubushi, Y. *J. Am. Chem. Soc.* **1982**, *104*, 4965–4967. Pentacyclic quassinoids: (b) Batt, D. G.; Takamura, N.; Ganem, B. *J. Am. Chem. Soc.* **1984**, *106*, 3353–3354.
15. Ginkgolide B: Corey, E. J.; Kang, M.; Desai, M. C.; Ghosh, A. K.; Houpis, I. N. *J. Am. Chem. Soc.* **1988**, *110*, 649–651.
16. Compactin: (a) Hsu, C.-T.; Wang, N.-Y.; Latimer, L. H.; Sih, C. J. *J. Am. Chem. Soc.* **1983**, *105*, 593–601. Glycinoeclepin A: (b) Murai, A.; Tanimoto, N.; Sakamoto, N.; Masamune, T. *J. Am. Chem. Soc.* **1988**, *110*, 1985–1986. Pumiliotoxin C: (c) Polniaszek, R. P.; Dillard, L. W. *J. Org. Chem.* **1992**, *57*, 4103–4110.
17. (a) Binkley, E. S.; Heathcock, C. H. *J. Org. Chem.* **1975**, *40*, 2156–2160. (b) Snider, B. B.; Yang, K. *J. Org. Chem.* **1992**, *57*, 3615–3626.
18. (a) Posner, G. H.; Sterling, J. J.; Whitten, C. E.; Lentz, C. M.; Brunelle, D. J. *J. Am. Chem. Soc.* **1975**, *97*, 107–118. (b) Borowitz, I. J.; Casper, E. W. R.; Crouch, R. K.; Yee, K. C. *J. Org. Chem.* **1972**, *37*, 3873–3878.
19. Coates, R. M.; Sandefur, L. O. *J. Org. Chem.* **1974**, *39*, 275–277.
20. (a) Stork, G. *Pure Appl. Chem.*, **1968**, *17*, 383–401. (b) Boeckman, Jr., R. K. *J. Org. Chem.* **1973**, *38*, 4450–4452.
21. Stephens, R. D.; Castro, C. E. *J. Org. Chem.* **1963**, *28*, 3313–3315.
22. (a) Stork, G.; Ganem, B. *J. Am. Chem. Soc.* **1973**, *95*, 6152–6153. (b) Boeckman, Jr., R. K. *J. Am. Chem. Soc.* **1973**, *95*, 6867–6869. (c) Stork, G.; Singh, J. *J. Am. Chem. Soc.* **1974**, *96*, 6181–6182. (d) Boeckman, Jr., R. K.; Blum, D. M.; Ganem, B.; Halvey, N. *Org. Synth.* **1978**, *58*, 152–157. Boeckman, Jr., R. K.; Blum, D. M.; Ganem, B.; Halvey, N. *Org. Synth. Coll. Vol.* **1988**, *6*, 1033–1036.

(e) (review) Jung, M. E. *In Comprehensive Organic Synthesis*; Trost, B. M., Fleming, I., eds; Pergamon: Oxford, 1991; Vol. 4, Chapter 1.1, pp. 1–67.

23. Boeckman, Jr., R. K.; Blum, D. M.; Ganem, B. *Org. Synth.* **1978**, *58*, 158–162. Boeckman, Jr., R. K.; Blum, D. M.; Ganem B. *Org. Synth. Coll. Vol.* **1988**, *6*, 666–669.
24. Warnhoff, E. W.; Martin, D. G.; Johnson, W. S. *Org. Synth. Coll. Vol.* **1963**, *4*, 163–166.
25. Kauffman, G. B.; Teter, L. A. *Inorg. Synth.* **1963**, *7*, 9–12.
26. (a) Watson, S. C.; Eastham, J. F. *J. Organomet. Chem.* **1967**, *9*, 165–168. (b) Gall, M.; House, H. O. *Org. Synth. Coll. Vol.* **1988**, *6*, 121–130. (c) Kofron, W. G.; Baclawski, L. M. *J. Org. Chem.* **1976**, *41*, 1879–1880. (d) Lipton, M. F.; Sorensen, C. M.; Sadler, A. C. *J. Organomet. Chem.* **1980**, *186*, 155–158.
27. Näf, F.; Decorzant, R. *Helv. Chim. Acta* **1974**, *57*, 1317–1327.
28. (a) Näf, F.; Decorzant, R.; Thommen, W. *Helv. Chim. Acta* **1975**, *58*, 1808–1812. For the intramolecular reaction of enolates (generated by conjugate addition of a Grignard reagent containing a copper catalyst) with the ester function see: (b) Pearson, A. J. *Tetrahedron Lett.* **1980**, *21*, 3929–3932.
29. (a) Heng, K. K.; Smith, R. A. J. *Tetrahedron* **1979**, *35*, 425–435. (b) Heng, K. K.; Simpson, J.; Smith, R. A. J. *J. Org. Chem.* **1981**, *46*, 2932–2934. (c) Stevens, R. V.; Albizati, K. F. *J. Org. Chem.* **1985**, *50*, 632–644.
30. For the reaction of lithium enolates with aldehydes see: (a) Stork, G.; d'Angelo, J. *J. Am. Chem. Soc.* **1974**, *96*, 7114–7116; Stork, G.; Kraus, G. A.; Garcia, G. A. *J. Org. Chem.* **1974**, *39*, 3459–3460; Stork, G.; Kraus, G. *J. Am. Chem. Soc.* **1976**, *98*, 6747–6748. For the $ZnCl_2$ effect in the aldol condensation of lithium enolates see: (b) House, H. O.; Crumrine, D. S.; Teranishi, A. Y.; Olmstead, H. D. *J. Am. Chem. Soc.* **1973**, *95*, 3310–3324.
31. For the nomenclature of *threo* and *erythro* see: Noyori, R.; Nishida, I.; Sakata, J. *J. Am. Chem. Soc.* **1983**, *105*, 1598–1608.
32. (a) Stork, G.; Hudrlik, P. F. *J. Am. Chem. Soc.* **1968**, *90*, 4462–4464. (b) Bornack, W. K.; Bhagwat, S. S.; Ponton, J.; Helquist, P. *J. Am. Chem. Soc.* **1981**, 103, 4647–4648.
33. Luo, F.-T.; Negishi, E. *J. Org. Chem.* **1985**, *50*, 4762–4766.
34. Stork, G.; Hurdlik, P. F. *J. Am. Chem. Soc.* **1968**, *90*, 4464–4465.
35. Kuwajima, I.; Nakamura, E.; Shimizu, M. *J. Am. Chem. Soc.* **1982**, *104*, 1025–1030.
36. Review: (a) Mukaiyama, T. *Org. React.* **1982**, *28*, 203–331. For recent advances see: (b) Hollis, T. K.; Robinson, N. P.; Bosnich, B. *Tetrahedron Lett.* **1992**, *33*, 6423–6426; (c) Corey, E. J.; Cywin, C. L.; Roper, T. D. *Tetrahedron Lett.* **1992**, *33*, 6907–6910 and references therein. See also: (d) Murata, S.; Suzuki, M.; Noyori, R. *Tetrahedron* **1988**, *44*, 4259–4275.
37. Seyferth, D.; Weiner, M. A. *J. Org. Chem.* **1961**, *26*, 4797–4800.
38. House, H. O.; Fischer, Jr., W. F. *J. Org. Chem.* **1968**, *33*, 949–956.
39. (a) Muchmore, D. C. *Org. Synth.* **1972**, *52*, 109–114. (b) Heathcock, C. H.; DelMar, E. G.; Graham, S. L. *J. Am. Chem. Soc.* **1982**, *104*, 1907–1917. (c) Plata, D. J.; Kallmerten, J. *J. Am. Chem. Soc.* **1988**, *110*, 4041–4042. (d) Heathcock, C. H.; Davidsen, S. K.; Mills, S. G.; Sanner, M. A. *J. Org. Chem.* **1992**, *57*, 2531–2544.
40. McMurray, J. E.; Scott, W. J. *Tetrahedron Lett.* **1983**, *24*, 979–982.

41. Crisp, G. T.; Scott, W. J.; Stille, J. K. *J. Am. Chem. Soc.* **1984**, *106*, 7500–7506.
42. (a) McMurray, J. E.; Scott, W. J. *Tetrahedron Lett.* **1980**, *21*, 4313–4316. (b) Lipshutz, B. H.; Elworthy, T. R. *J. Org. Chem.* **1990**, *55*, 1695–1696. Reviews: (c) Knight, D. W. In *Comprehensive Organic Synthesis*; Trost, B. M., Fleming, I., eds; Pergamon: Oxford, **1991**; Vol. 3, Chapter 2.3, pp. 481–520; (d) Sonogashira, K. In *Comprehensive Organic Synthesis*; Trost, B. M., Fleming, I., eds; Pergamon: Oxford, 1991; Vol. 3, Chapter 2.4, pp. 521–549.
43. (a) Scott, W. J.; Peña, M. R.; Swärd, K.; Stoessel, S. J.; Stille, *J. K. J. Org. Chem.* **1985**, *50*, 2302–2308. (b) Scott, W. J.; Crisp, G. T.; Stille, J. K. *Org. Synth.* **1989**, *68*, 116–129. Review: (c) Scott, W. J.; McMurry, J. E. *Acc. Chem. Res.* **1988**, *21*, 47–54; (d) Ritter, K. *Synthesis* **1993**, 735–762.
44. Kowalski, C. J.; Weber, A. E.; Fields, K. W. *J. Org. Chem.* **1982**, *47*, 5088–5093.
45. Clive, D. L. J.; Farina, V.; Beaulieu, P. L. *J. Org. Chem.* **1982**, *47*, 2572–2582.
46. Commercon, M. B.; Foulon, J. P.; Normant, J. F. *J. Organomet. Chem.* **1982**, *228*, 321–326. Rehnberg, N.; Frejd, T.; Magnusson, G. *Tetrahedron Lett.* **1987**, *28*, 3589–3592.
47. Marshall, J. A.; Audia, J. E.; Shearer, B. G. *J. Org. Chem.* **1986**, *51*, 1730–1735.
48. Ireland, R. E.; Mander, L. N. *J. Org. Chem.* **1967**, *32*, 689–696.
49. (a) Roth, G. P.; Rithner, C. D.; Meyers, A. I. *Tetrahedron* **1989**, *45*, 6949–6962. Rawson, D. R.; Meyers, A. I. *J. Org. Chem.* **1991**, *56*, 2292–2294. Robichaud, A. J.; Meyers, A. I. *J. Org. Chem.* **1991**, *56*, 2607–2609. (b) Tomioka, K.; Masumi, F.; Yamashita, T.; Koga, K. *Tetrahedron Lett.* **1984**, *25*, 333–336.
50. Bernhard, W.; Fleming, I.; Waterson, D. *J. Chem. Soc., Chem. Commun.* **1984**, 28–29; Crump, R. A. N. C.; Fleming, I.; Hill, J. H.M.; Parker, D.; Reddy, L.; Waterson, D. *J. Chem. Soc., Perkin Trans 1* **1992**, 3277–3294.
51. Lipshutz, B. H. *Tetrahedron Lett.* **1983**, *24*, 127–130.
52. Fujisawa, T.; Noda, A.; Kawara, T.; Sato, T. *Chem. Lett.* **1981**, 1159–1160.
53. (a) Bertrand, M.; Gil, G.; Viala, J. *Tetrahedron Lett.* **1977**, 1785–1788. (b) Fang, C.; Suemune, H.; Sakai, K. *Tetrahedron Lett.* **1990**, *31*, 4751–4754.
54. (a) Marino, J. P.; Linderman, R. J. *J. Org. Chem.* **1981**, *46*, 3696–3702. (b) Marino, J. P.; Linderman, R. J. *J. Org. Chem.* **1983**, *48*, 4621–4628.
55. (a) Nugent, W. A.; Hobbs, Jr., F. W. *J. Org. Chem.* **1983**, *48*, 5364–5366. (b) Crimmins, M. T.; DeLoach, J. A. *J. Org. Chem.* **1984**, *49*, 2076–2077. (c) Crimmins, M. T.; Mascarella, S. W.; DeLoach, J. A. *J. Org. Chem.* **1984**, *49*, 3033–3035.
56. Lewis, D. E.; Rigby, H. L. *Tetrahedron Lett.* **1985**, *26*, 3437–3440.
57. (a) Chow, H.-F.; Fleming, I. *J. Chem. Soc., Perkin Trans, 1* **1984**, 1815–1819. (b) Engel, W.; Fleming, I.; Smithers, R. H. *J. Chem. Soc., Perkin Trans. 1* **1986**, 1637–1641. (c) Fleming, I.; Reddy, N. L.; Takaki, K.; Ware, A. C. *J. Chem. Soc., Chem. Commun.* **1987**, 1472–1474. For efficient trapping with butyl iodide see: (d) Fleming, I.; Waterson, D. *J. Chem. Soc., Perkin Trans. 1* **1984**, 1809–1813.
58. (a) Fleming, I.; Kilburn, J. D. *J. Chem. Soc., Chem. Commun.* **1986**, 305–306. (b) Fleming, I.; Sarkar, A. K.; Doyle, M. J.; Raithby, P. R. *J. Chem. Soc., Perkin Trans. 1* **1989**, 2023–2030; (c) Fleming, I.; Kilburn, J. D. *J. Chem. Soc., Perkin Trans 1* **1992**, 3295–3302.
59. (a) Fleming, I.; Lawrence, N. J. *Tetrahedron Lett.* **1988**, *29*, 2077–2080. For a

review on the oxidative cleavage of Si—C bonds see: (b) Tamao, K. *J. Synth. Org. Chem. Jpn.* **1988**, *46*, 861–878.
60. Fleming, I.; Terrett, N. K. *J. Organomet. Chem.* **1984**, *264*, 99–118.
61. For an organozinc aid in enolate alkylation and its application to the three-component PG synthesis see: (a) Morita, Y.; Suzuki, M.; Noyori, R. *J. Org. Chem.* **1989**, *54*, 1785–1787; (b) Suzuki, M.; Morita, Y.; Koyano, H.; Koga, M.; Noyori, R. *Tetrahedron* **1990**, *46*, 4809–4822.
62. Stork, G.; Isobe, M. *J. Am. Chem. Soc.* **1975**, *97*, 6260–6261.
63. We see a similar stannyl effect in the alkylation of the enolate generated by conjugate addition of lithium dibutylcuprate to 2-cyclopentenone: Yamamoto, Y.; Yatagai, H.; Maruyama, K. *Silicon, Germanium, Tin, and Lead Compounds* **1986**, *9*, 25–40 (Freund Pub. House: Tel-Aviv).
64. For the effects of these compounds in the alkylation of pure lithium enolates see: (a) Tardella, P. A. *Tetrahedron Lett.* **1969**, 1117–1120; Nishiyama, H.; Sakuta, K.; Itoh, K. *Tetrahedron Lett.* **1984**, *25*, 223–226, 2487–2488. For the effect of chlorotriphenylstannane see: (b) Binns, M. R.; Haynes, R. K.; Lambert, D. E.; Schober, P. A. *Tetrahedron Lett.* **1985**, *26*, 3385–3388; Haynes, R. K.; Lambert, D. E.; Schober, P. A.; Turner, S. G. *Aust. J. Chem.* **1987**, *40*, 1211–1222.
65. (a) Suzuki, M.; Koyano, H.; Noyori, R. *J. Org. Chem.* **1987**, *52*, 5583–5588. (b) Tanaka, T.; Bannai, K.; Hazato, A.; Koga, M.; Kurozumi, S.; Kato, Y. *Tetrahedron* **1991**, *47*, 1861–1876.
66. (a) Suzuki, M.; Koyano, H.; Noyori, R.; Hashimoto, H.; Negishi, M.; Ichikawa, A.; Ito, S. *Tetrahedron* **1992**, *48*, 2635–2658. (b) Ito, S.; Hashimoto, H.; Negishi, M.; Suzuki, M.; Koyano, H.; Noyori, R.; Ichikawa, A. *J. Biol. Chem.* **1992**, *267*, 20326–20330; (c) Noyori, R.; Suzuki, M. *Science* **1993**, *259*, 44–45.
67. (a) Johnson, C. R.; Penning, T. D. *J. Am. Chem. Soc.* **1986**, *108*, 5655–5656. Johnson, C. R.; Penning, T. D. *J. Am. Chem. Soc.* **1988**, *110*, 4726–4735. (b) Johnson, C. R.; Chen, Y.-F. *J. Org. Chem.* **1991**, *56*, 3344–3351.
68. For the application of phosphine-complexed organocopper reagents to the syntheses of steroids, quassinoids, and glycinoeclepin see references 13c, 14b, and 16b, respectively.
69. Corey, E. J.; Niimura, K.; Konishi, Y.; Hashimoto, S.; Hamada, Y. *Tetrahedron Lett.* **1986**, *27*, 2199–2202.
70. Kluge, A. F.; Untch, K. G.; Fried, J. H. *J. Am. Chem. Soc.* **1972**, *94*, 7827–7832.
71. Martel, J. Japan Patent 46–5625, 1971.
72. Suzuki, M.; Kawagishi, T.; Suzuki, T.; Noyori, R. *Tetrahedron Lett.* **1982**, *23*, 4057–4060.
73. Kato, T.; Fukushima, M.; Kurozumi, S.; Noyori, R. *Cancer Res.* **1986**, *46*, 3538–3542.
74. Hijfte, L. V.; Kolb, M. *Tetrahedron* **1992**, *48*, 6393–6402.
75. Sugiura, S.; Toru, T.; Tanaka, T.; Okamura, N.; Hazato, A.; Bannai, K.; Manabe, K.; Kurozumi, S. *Chem. Pharm. Bull.* **1984**, *32*, 1248–1251. Holland, G. W.; Maag, H.; Rosen, P. Ger. Offen DE 3 208 880, 1982 (*Chem. Abstr.* **1983**, *98*, 89 048r). Toru, T.; Sugiura, S.; Kurozumi, S. Japan Patent 59–10 577. Yasuda, A.; Arai, T.; Kato, M.; Uchida, K.; Yamabe, M. Japan Patent 59–227 888 (WO 89/04917A1), 1984 (*Chem. Abstr.* **1985**, *103*, 22 360q).
76. (a) Wohl, R. A. *Synthesis* **1974**, 38–40. (b) Schreiber, S. L.; Claus, R. E.; Reagan, J. *Tetrahedron Lett.* **1982**, *23*, 3867–3870.

77. (a) Piers, E.; Renaud, J. *J. Org. Chem.* **1993**, *58*, 11–13. (b) Burke, S. D.; Piscopio, A. D.; Kort, M. E.; Matulenko, M. A.; Parker, M. H.; Armistead, D. M.; Shankaran, K. *J. Org. Chem.* **1994**, *59*, 332–347. Review and books: (c) Perlmutter, P. *Conjugate addition reactions in organic synthesis*. Pergamon Press, Oxford, **1992**, pp. 137–197. (d) Wipf, P. *Synthesis* **1993**, 537–557. (e) Noyori, R. *Asymmetric catalysis in organic synthesis*. Wiley, New York, **1994**, pp. 298–322.

10

The Sonogashira Cu–Pd-catalysed alkyne coupling reaction

I. B. CAMPBELL

1. Introduction

Carbon–carbon bond-forming reactions are of crucial importance to the practising organic chemist. In particular, those which are technically simple, efficient, high-yielding, and which tolerate a wide variety of functional groups are of special use. One such reaction is the Sonogashira copper–palladium-catalysed coupling of terminal alkynes with aromatic and vinyl halides.[1,2] The general forms of the reaction are given in eqns (1) and (2).

$$\text{Ar-X} + \text{RC}\equiv\text{CH} \xrightarrow{\text{Cu(I), Pd(II), Base}} \text{Ar-C}\equiv\text{C-R} \quad (1)$$

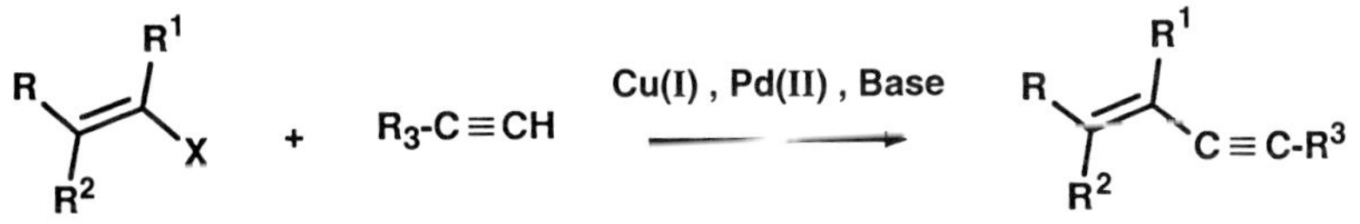

$$\text{RR}^2\text{C}=\text{CR}^1\text{X} + \text{R}_3\text{-C}\equiv\text{CH} \xrightarrow{\text{Cu(I), Pd(II), Base}} \text{RR}^2\text{C}=\text{CR}^1\text{-C}\equiv\text{C-R}^3 \quad (2)$$

X = I, Br, Cl, OTf

The reaction was developed in 1975 by Sonogashira[1] at the same time as both Heck[3] and Cassar[4] reported a similar process which did not involve copper catalysis but which required much more forcing conditions. The reaction can in fact be envisaged as an extension of the well-used Heck palladium-catalysed arylation of alkenes.[5] Prior to these investigations the only method available for coupling alkynes and iodoarenes was the Stephens–Castro reaction, involving a preformed copper acetylide reacting in pyridine at high temperatures.[6] The Sonogashira process can be seen as a major advance as it removes the requirement for the quite serious technical difficulties involved in the preparation and safe handling of copper acetylides and allows a huge range of substrates to couple under very mild conditions. The reaction has recently been reviewed by Sonogashira.[7]

The reaction is of particular use to chemists working in research laboratories who need to explore structure–activity relationships within a series of compounds. Thus, an important aryl system of interest may be readily substituted by a number of chains of varying length and functional complexity. Conversely, a wide variety of differently substituted aryl rings can be easily appended to a molecule of interest. The introduced triple bond can be removed by hydrogenation or converted into many other functionalities.[8] Similarly, the use of vinylic halides in the reaction allows the rapid construction of highly functionalized systems which are of particular use in the field of natural product chemistry. This tremendous versatility in terms of changing rings, chains, and functionality within a compound class means that the Sonogashira coupling has established itself as one of the most important carbon–carbon bond-forming reactions available to organic chemists.

2. Mechanism

The Sonogashira reaction certainly follows the normal oxidative addition–reductive elimination process common to palladium-catalysed carbon–carbon bond-forming reactions (Scheme 10.1). The exact mechanism of the reaction, however, is not known and in particular the role the copper catalyst plays remains unclear. The process may be envisaged as involving a palladium(0) species **1**, generated from the palladium(II) precatalyst, which inserts into the aryl halide bond to give the arylpalladium(II) intermediate **2**. Subsequent reaction with the alkyne, possibly via a transient copper acetylide species, leads to the alkynylpalladium(II) derivative **3** which collapses to give the required coupled product and to regenerate the active palladium(0) catalyst. The oxidative addition of the aryl halide to the palladium(0) species is the rate-determining step of the reaction. Substrates carrying electron-withdrawing groups *ortho* or *para* to the halide will therefore react more readily as the more electron-deficient aryl halides will undergo oxidative addition more rapidly.[9] The reaction does proceed without the copper co-catalyst but only under more forcing conditions and not for less active substrates.[3,4]

3. Scope of the reaction

3.1 Aryl iodides

The most commonly used form of the Sonogashira reaction is that in which an aromatic iodide is coupled with a terminal alkyne. The reaction takes place readily at room temperature and although generally left overnight is often

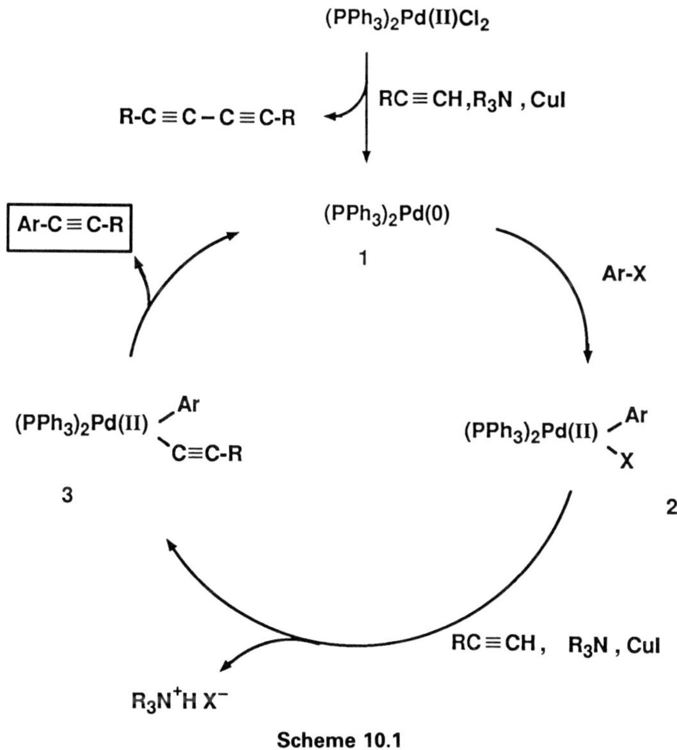

Scheme 10.1

complete in a few hours. An example is given in Protocol 1. Quite marked colour changes tend to occur during the course of the reaction, although not in a usefully predictable manner. A small amount of heat can occasionally be noted soon after the reaction commences.

The reaction of iodides is a very reliable one and will normally proceed readily without vigorous purification of substrates and without needing to dry or freshly distil solvents. It has been shown that the use of a modified phosphine ligand on the palladium catalyst allows the Sonogashira reaction to take place in aqueous media and that phase-transfer conditions can also be employed effectively.[10,11] The presence of a copper(I) species is essential for the reaction to proceed at room temperature. Copper(I) iodide is most often used but copper(I) bromide has also been used successfully.[12] Commercial-grade (Aldrich, 98% pure) copper(I) iodide is perfectly adequate in this procedure. The palladium catalyst originally employed by Sonogashira, $PdCl_2(PPh_3)_2$, is the catalyst most commonly used to effect the reaction. The reaction has proved so robust that little investigation of alternative palladium catalysts has been deemed necessary. Amongst other catalysts, both $Pd(PPh_3)_4$ and $Pd(OAc)_2$–PPh_3 have been used effectively,[13,14] and it is likely that any of the commonly used palladium(0) or palladium(II) species will function equally well in the coupling. An interesting recent development involves the preparation of polymer-bound palladium–phosphine catalysts.[15] Such systems may prove to be of great use in industrial processes where catalyst recovery and metal contamination of products are important issues. The choice of catalyst in the literature is usually determined simply by the availability of a particular palladium catalyst to the chemist performing the reaction. The palladium(II) catalysts, however, do generally exhibit greater long-term stability than the palladium(0) species. Samples of $PdCl_2(PPh_3)_2$ which have been stored under normal laboratory conditions for several years still prove effective in the reaction, whereas $Pd(PPh_3)_4$ rapidly deteriorates unless stored under carefully controlled conditions.

The amount of copper and palladium catalysts used is generally of the order of 2 mol% of each with respect to the halide and alkyne. This type of ratio is likely to lead to product formation in an acceptable time period but is not necessarily optimum. In one study, the most effective ratio of $Pd(PPh_3)_4$ to copper(I) bromide was found to be 2:3, and it was also shown that the reaction rate doubled on increasing catalyst amounts from 2 and 3% to 4 and 6%, respectively.[12] In large-scale preparations (>100 g), catalyst ratios as low as 0.5 mol% have still proved to be effective.

In the initially used procedure of Sonogashira, diethylamine was chosen as both base and solvent and this medium continues to be used effectively. There appears to be no particular advantage of diethylamine over many of the commonly used organic bases, several of which have also been used successfully in the reaction. Diethylamine is, of course, reactive towards alkylating agents, and should be avoided if these are present in favour of triethylamine or *N,N*-diisopropylethylamine. One modification which has proved effective in our laboratories is the use of one equivalent of dicyclohexylamine as the base in a co-solvent (THF or MeCN). An example is given in Protocol 3. The dicyclohexylamine hydriodide produced in the reaction is highly crystalline and precipitates from solution. The coupling reactions are often accelerated under these conditions and may be complete within an hour. The use of dicyclohexylamine is not recommended for the preparation of amine-containing products, as any excess material is difficult to remove other than by extraction with acid. Triethylamine at reflux is the medium of choice for unreactive substrates (see Protocol 2). The functional group tolerance inherent in the reaction means that any reasonable co-solvent may be used where required in order to aid solubility of reagents.

3.2 Aryl bromides

Aromatic bromides react much less readily than the corresponding iodides and will generally require solvents at reflux in order to effect reaction. Whereas with iodides no special care is needed to ensure useful conversions

Protocol 1.
Coupling of an aryl iodide with an alkyne under Sonogashira conditions[16]

Caution! Carry out all procedures in a well-ventilated hood, and wear disposable vinyl or latex gloves and chemical-resistant safety goggles.

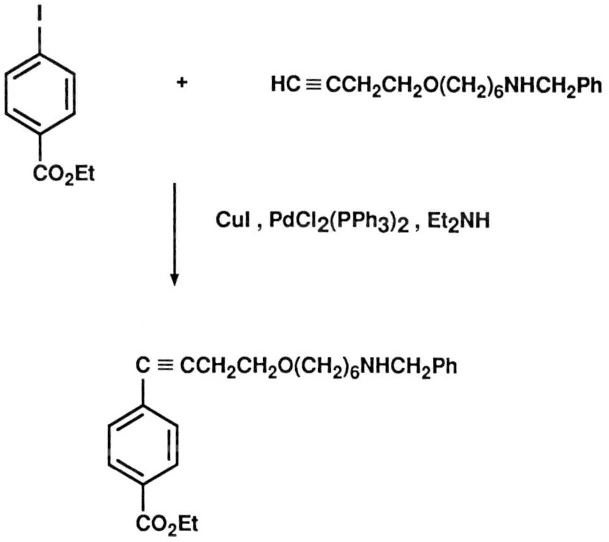

$HC{\equiv}CCH_2CH_2O(CH_2)_6NHCH_2Ph$

CO$_2$Et

CuI , $PdCl_2(PPh_3)_2$, Et_2NH

$C{\equiv}CCH_2CH_2O(CH_2)_6NHCH_2Ph$

CO$_2$Et

Equipment

- Two-necked, round-bottomed flask (250 mL) with magnetic stirring bar and septa
- Nitrogen source and Inlet

Materials

• Ethyl 4-iodobenzoate (FW 276.1) 4.65 g, 16.8 mmol	**harmful**
• *N*-[(3-Butynyloxy)hexyl]benzylamine[16] (FW 259.4) 4.12 g, 15.9 mmol	**harmful**
• Copper(I) iodide (FW 190.4) 70 mg, 0.37 mmol, 2 mol%	**irritant, light sensitive**
• Bis(triphenylphosphine)palladium(II) chloride (FW 701.9) 120 mg, 0.17 mmol, 1 mol%	**hygroscopic**
• Diethylamine 90 mL	**flammable, corrosive**
• Diethyl ether 100 mL	**flammable, irritant**

1. Add all reagents to the flask[a] and stir the mixture for 16 h at room temperature under a nitrogen atmosphere.
2. Dilute the resulting suspension with diethyl ether (100 mL) and filter.
3. Evaporate the filtrate under reduced pressure to give an oil.

Protocol 1. *Continued*

4. Purify the product by flash column chromatography on silica gel (Merck 9385), eluting with diethyl ether–triethylamine (99:1). Ethyl 4-(3-{6-[(phenylmethyl)amino]hexyl}oxy-1-butynyl)benzoate is obtained as an orange oil (6.0 g, 93%) displaying the appropriate spectroscopic and analytical data.

[a] There is no requirement to use specially dry apparatus or reagents in the Sonogashira reaction.

of starting materials to products, the more reticent nature of the bromides requires a more careful technical approach. It is advisable in this case to use a high quality solvent (triethylamine most commonly), although vigorous drying is unnecessary. An important requirement, however, is to deoxygenate the reaction mixture by bubbling nitrogen through for *c.*10 minutes prior to addition of the palladium and copper catalysts. This technique is useful to help to maintain the integrity of the catalytic cycle during the course of the reaction. It may still be the case that a further addition of copper and palladium catalysts will be effective if the reaction fails to proceed to completion. Although most substrates will react in 24 hours, it is possible for some reactions to take several days before high yields of products are obtained. An example of the use of aryl bromides in the reaction is given in Protocol 2, and in Protocol 3 the use of a sealed vessel, necessary for the reaction of volatile alkynes with less active substrates, is exemplified.

Protocol 2. Coupling of an aryl bromide with an alkyne under Sonogashira conditions[17]

Caution! Carry out all procedures in a well-ventilated hood, and wear disposable vinyl or latex gloves and chemical-resistant safety goggles.

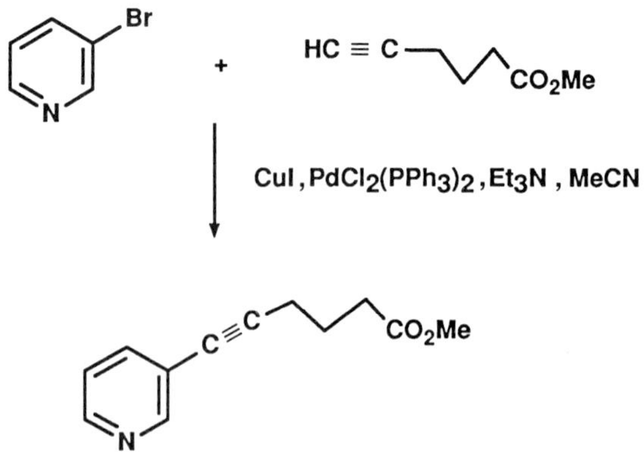

Br

N

+ $HC \equiv C$ … CO_2Me

$CuI, PdCl_2(PPh_3)_2, Et_3N$, MeCN

$C \equiv C$ … CO_2Me

N

Equipment

- Two-necked, round-bottomed flask (50 mL) with magnetic stirring bar and stopper
- Reflux condenser fitted with septum
- Magnetic stirrer–hotplate
- Nitrogen source and inlet

Materials

• 3-Bromopyridine (FW 158.0) 1.0 g, 6.33 mmol	**highly toxic, irritant**
• Methyl 5-hexynoate (FW 126.2) 882 mg, 7.0 mmol	**harmful**
• Copper(I) iodide (FW 190.4) 25 mg, 0.13 mmol, 2 mol%	**irritant, light sensitive**
• Bis(triphenylphosphine)palladium(II) chloride (FW 701.9) 90 mg, 0.13 mmol, 2 mol%	**hygroscopic**
• Triethylamine 10 mL	**flammable, corrosive**
• Acetonitrile 10 mL	**flammable, lachrymator**
• Dichloromethane 150 mL	**toxic, irritant**

1. Assemble the apparatus, omitting the septum from the reflux condenser.
2. Introduce the halide, alkyne, triethylamine, and acetonitrile to the flask and stir to produce a solution.
3. Place the needle from the nitrogen inlet below the surface of the solution and pass nitrogen gently through the solution for 10 min.
4. Add the palladium and copper catalysts, fit the septum and nitrogen inlet to the top of the condenser, and stopper the flask.
5. Heat the mixture under reflux for 4 h, cool to room temperature, and add to aqueous sodium bicarbonate (1 M, 100 mL).
6. Extract with dichloromethane (3 $\times$ 50 mL), dry ($MgSO_4$) the combined extracts, and evaporate under reduced pressure.
7. Purify the product by flash column chromatography on silica gel (Merck 9385) using hexane–diethyl ether (3:1). Methyl 6-(3-pyridinyl)-5-hexynoate is obtained as a colourless oil (900 mg, 70%) displaying the appropriate spectroscopic and analytical data.

3.3 Aryl chlorides

The reaction of aromatic chlorides with alkynes under Sonogashira conditions is much more restricted in the nature of substrate which will participate in the process. Only those benzenoid aromatic chlorides which also possess suitably sited electron-withdrawing groups—particularly nitro—are likely to react to any appreciable extent.[19] Many heteroaromatic systems will, however, react quite readily if one or more nitrogen atoms is in an activating position relative to the chloride.[20] Examples of the use of aromatic chlorides are given in eqns (3)[12] and (4),[20] and the reaction conditions for these would be very similar to those given in Protocol 3.

$$\text{2,4-}(O_2N)_2C_6H_3Cl \xrightarrow[\text{reflux, 10 min, 87\%}]{HC\equiv C\text{-}C_5H_{11} \quad Pd(PPh_3)_4,\ CuBr,\ Et_3N} \text{2,4-}(O_2N)_2C_6H_3C\equiv C\text{-}C_5H_{11} \quad (3)$$

$$\text{2-chloropyridine} \xrightarrow[\text{12 h, 120°C, 80\%}]{HC\equiv C-SiMe_3 \quad PdCl_2(PPh_3)_2,\ CuI,\ Et_3N} \text{2-}(C\equiv C-SiMe_3)\text{pyridine} \quad (4)$$

Protocol 3. Coupling of an aryl bromide with a volatile alkyne under Sonogashira conditions[17]

Caution! Carry out all procedures in a well-ventilated hood, and wear disposable vinyl or latex gloves and chemical-resistant safety goggles.

$$\text{2-Br-}C_6H_4CO_2Me \xrightarrow{HC\equiv C-SiMe_3,\ CuI,\ MeCN,\ Pd(PPh_3)_2Cl_2,\ (C_6H_{11})_2NH} \text{2-}(Me_3SiC\equiv C)C_6H_4CO_2Me$$

Equipment

- Autoclave[a] (Parr; Hasteloy, 100 mL) with magnetic stirring bar
- Nitrogen source and inlet
- Magnetic stirrer–hotplate

Materials

• Methyl 2-bromobenzoate (FW 215.1) 6.0 g, 28 mmol	**harmful**
• Trimethylsilylacetylene (FW 98.2) 5.5 g, 56 mmol	**flammable, irritant**
• Copper(I) iodide (FW 190.4) 200 mg, 1 mmol, 3.6 mol%	**irritant, light sensitive**
• Bis(triphenylphosphine)palladium(II) chloride (FW 701.9) 500 mg, 0.7 mmol, 2.5 mol%	**hygroscopic**
• Dicyclohexylamine (FW 181.3) 6.2 mL, 31 mmol	**corrosive, toxic**
• Acetonitrile 50 mL	**flammable, lachryator**
• Diethyl ether 100 mL	**flammable, irritant**

1. Add the bromide, dicyclohexylamine, and acetonitrile to the autoclave and stir to give a solution.

2. Place the needle of the nitrogen inlet below the surface of the solution and pass nitrogen gently through the solution for 10 min.
3. Add the alkyne and the palladium and copper catalysts, remove the nitrogen line, and seal the autoclave.
4. Place the reaction behind a safety shield and heat at *c.* 80°C for 3h.
5. Cool the autoclave to room temperature, open it, and add the mixture to diethyl ether (100 ml).
6. Filter the suspension and evaporate the filtrate under reduced pressure to give a black oil.
7. Purify the product by flash column chromatography on silica gel (Merck 9385), eluting with hexane–diethyl ether (98:2). Methyl 2-(trimethylsilyl-ethynyl)benzoate[18] is obtained as an orange oil (5.25 g, 81%) displaying the appropriate spectroscopic and analytical data.[b]

[a] For small-scale reactions a suitable thick-walled glass vial fitted with a screw cap may be used (e.g. Reacti-Vial, Pierce and Warriner).
[b] Silylalkynes are readily cleaved to give terminal alkynes using fluoride ion or mild basic (aq. Na_2CO_3) hydrolysis.

3.4 Vinyl halides and related compounds

Substitution of aromatic systems is perhaps the more common use of the Sonogashira coupling, but the reaction is equally applicable to substitution of vinylic halides.[7] Thus, reaction of vinyl iodides, bromides, and in this case chlorides and trifluoromethanesulfonates under the same conditions as those used for aryl halides (Protocols 1 and 2) leads to coupled products (eqn 5). An example is given in Protocol 4.

$$\mathrm{RR_2C{=}CR_1X} + \mathrm{R_3\text{-}C{\equiv}CH} \xrightarrow{\text{Cu(I), Pd(II), Base}} \mathrm{RR_2C{=}CR_1\text{-}C{\equiv}C\text{-}R_3} \quad (5)$$

X = I, Br, Cl, OTf

The resulting enynes produced are of defined stereochemistry and this simple methodology allows the rapid preparation of long-chain molecules containing high levels of unsaturation in the presence of multiple functional groups. The reaction has been extensively used in several areas of natural product synthesis. The highly unsaturated products from the arachidonic acid cascade provide important targets for potential drug therapies, and many groups have synthesized these compounds by making use of the alkyne coupling as the key step in their synthetic strategies (eqn 6).[21–23]

The use of vinyl triflates further extends the scope of the coupling as the

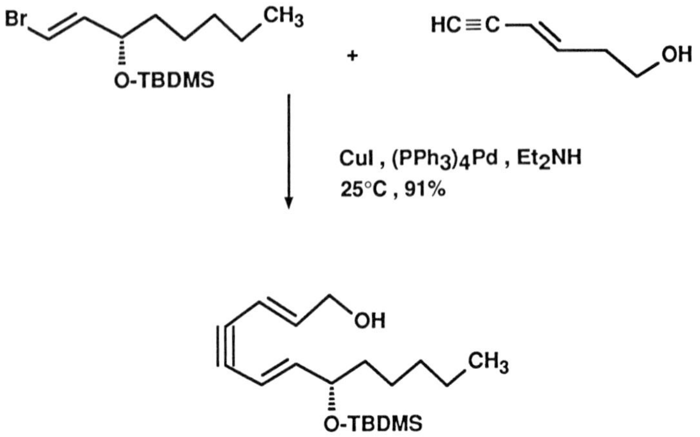

Br, CH$_3$, O-TBDMS + HC≡C—, OH (6)

CuI, $(PPh_3)_4Pd$, Et_2NH
25°C, 91%

OH, CH$_3$, O-TBDMS

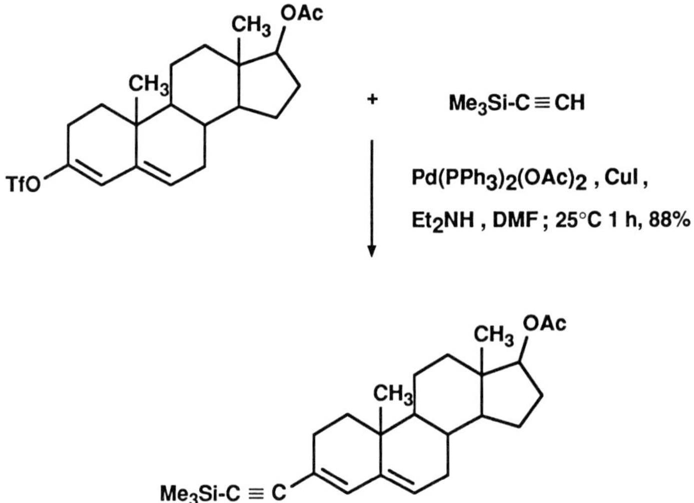

CH$_3$ OAc, CH$_3$, TfO + Me_3Si-C≡CH (7)

$Pd(PPh_3)_2(OAc)_2$, CuI,
Et_2NH, DMF; 25°C 1 h, 88%

CH$_3$ OAc, CH$_3$, Me_3Si-C≡C
4

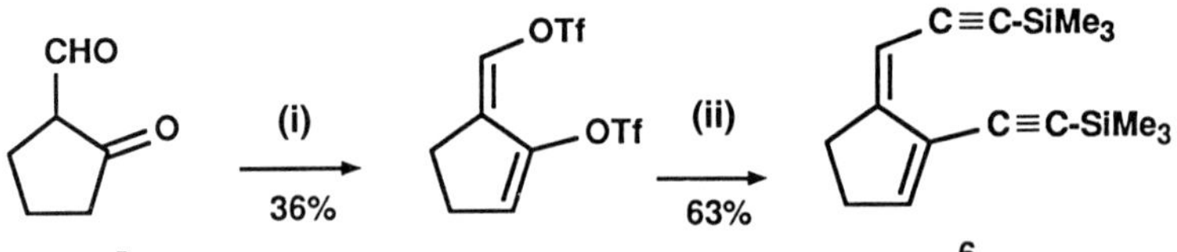

CHO, O
5
(i)
36%
OTf, OTf
(ii)
63%
C≡C-SiMe$_3$, C≡C-SiMe$_3$
6

(i): (a) *t*-BuLi, THF; (b) Tf_2O; (c) LiHMDS, $PhN(Tf)_2$
(ii): CuI, $PdCl_2(PPh_3)_2$, *i*-Pr_2NH, THF, HC≡C-SiMe$_3$ (8)

required triflates are readily prepared from available aldehydes and ketones. Examples of the usefulness of this approach are shown (eqns 7 and 8) in the preparation of the substituted steroid nucleus[24] **4** and in the conversion of the formylcyclopentanone **5** to the dienediyne **6**—a model for the construction of the dienediyne unit present in the antitumour agent neocarzinostatin.[25]

Several other groups have used this strategy to construct the complex enediyne unit of the esperamicin antitumour antibiotic family in an elegant and efficient manner,[26,27] further demonstrating the remarkable effectiveness of the Sonogashira coupling.

Although of more limited synthetic value, the preparation of diynes via the coupling of 1-iodoalkynes and terminal alkynes under Sonogashira conditions has also been developed (eqn 9).[28]

$$R\text{-}C\equiv C\text{-}I \quad + \quad HC\equiv C\text{-}R^1 \xrightarrow{\text{Cu(I) / Pd(II)}} R\text{-}C\equiv C\text{-}C\equiv C\text{-}R^1 \qquad (9)$$

4. Selectivity

The difference in reactivity between iodide and bromide towards the coupling reaction allows selective substitutions to be performed. Thus, the iodide of the polyhalobenzene **7** reacts (eqn 10) under similar conditions to those given in Protocol 1, in preference to the bromide which would require conditions such as those given in Protocol 2 in order to react.[29] This selectivity allows the opportunity for different alkyne chains to be added to a ring system in a totally regiodefined manner.

F, Br, F, F, I (7) $\xrightarrow[20°C,\ 4\text{–}24\ h]{HC\equiv C\text{-}R,\ CuI,\ Pd(PPh_3)_2Cl_2,\ Et_3N}$ F, Br, F, F, C≡C-R (10)

R = Me_3Si, Me, 4-F-C_6H_4

Further selectivity in the reaction can be provided by consideration of the electronic nature of the substrate. This is nicely demonstrated by the reactions of the dibromides **8** (eqn 11) and **9** (eqn 12).[10] The electron-withdrawing nitro group of **8** activates the *para*-bromine atom to displacement by increasing the rate of oxidative addition to the palladium catalyst. Conversely, the amino function slows this addition in dibromide **9**, allowing the less-deactivated *meta*-bromine atom to be substituted.

Vinyl halides are generally more reactive than the corresponding aryl halides, but selectivity is equally possible as the same order of reactivity exists, i.e. I > Br > Cl. Vinyl triflates tend to react at a similar rate to iodides,

Br, Br, NO_2 (**8**) — $HC{\equiv}C\text{-}C_5H_{11}$, CuBr, $Pd(PPh_3)_4$, Et_3N; 5 h r.t/10 min reflux, 92% → $C{\equiv}C\text{-}C_5H_{11}$, Br, NO_2 (11)

Br, Br, NH_2 (**9**) — as above; 38 h reflux, 70% → Br, $C \equiv C\text{-}C_5H_{11}$, NH_2 (12)

but selectivity between the two would not normally be an issue as the required triflate could be generated from a carbonyl function at an appropriate point in a synthetic sequence. Acceptable monosubstitution of a vinyl dihalide can be achieved by using a large excess (more than two equivalents) of the dihalide over the alkyne (eqn 13).[30]

$C_6H_{13}-C{\equiv}CH$ + Cl, Cl — CuI, $Pd(PPh_3)_4$, PhH, $BuNH_2$; 5 h, 25°C, 87% → $C_6H_{13}-C{\equiv}C$—, Cl (13)

Protocol 4.
Coupling of a vinyl halide with an alkyne under Sonogashira conditions[30]

Caution! Carry out all procedures in a well-ventilated hood, and wear disposable vinyl or latex gloves and chemical-resistant safety goggles.

$C_6H_{13}-C{\equiv}CH$ + Cl, Cl — CuI, $Pd(PPh_3)_4$, PhH, $BuNH_2$ → $C_6H_{13}-C{\equiv}C$—, Cl

Equipment

- Two-necked, round-bottomed flask (250 mL) with magnetic stirrer bar and septa
- Nitrogen source and outlet

Materials

• 1-Octyne (FW 110.2) 1.18 mL, 8 mmol	**flammable, irritant**
• *cis*-1,2-Dichloroethene (FW 96.9), 3.4 mL, 45 mmol	**flammable, moisture sensitive**
• Copper(I) iodide[a] (FW 190.4) 77 mg, 0.4 mmol, 5 mol%	**irritant, light sensitive**
• Tetrakis(triphenylphosphine)palladium (0)[b] (FW 1155.6) 462 mg, 0.4 mmol, 5 mol%	**light sensitive**
• *n*-Butylamine[c] (FW 00.0) 1.19 mL, 12 mmol	**flammable, corrosive**
• Benzene[d] 75 mL	**flammable, cancer suspect agent**

1. Oven heat the apparatus (1 h, 140 °C), assemble hot, and flame dry the flask while purging with nitrogen.[e]
2. While maintaining a blanket of nitrogen, add the solvent followed by the alkyne, *n*-butylamine, palladium catalyst, dichloroethene, and finally the copper iodide.
3. Stir at room temperature for 5 h.
4. Wash the solution with brine (2 × 30 mL) and finally with water (30 mL).
5. Dry the solution (Na_2SO_4), filter through a short silica plug, and concentrate under reduced pressure to give a clear oil.
6. Distil the oil at water pump pressure (80 °C/*c*.20 mm Hg) to obtain the required product as a colourless liquid (1.19 g, 87%) displaying the appropriate spectroscopic and analytical data.

[a] Purified according to Protocol 2 in Chapter 2.
[b] From Aldrich.
[c] Distilled from CaH_2.
[d] Desulfurized and distilled from P_2O_5.
[e] Although rigorous drying has been used in this example, the Sonogashira reaction proceeds readily in the presence of moisture.

5. Functional group compatibility

5.1 Aromatic and vinylic substrates

The Sonogashira coupling reaction is one of the most functionally tolerant reactions available for carbon–carbon bond formation.[2] Although it may be possible to find unique combinations of functionality which inhibit the activity of the palladium catalyst, the reaction is generally compatible with all the commonly encountered functional groups. Some limitations in the nature of the alkyne exist (Section 5.2), but there are unlikely to be difficulties with the nature of the aromatic halide, except in the case of certain nucleophilic *ortho*-substituents (Section 6). Several examples of various functional groups reacting successfully are given in Table 10.1 to demonstrate the scope of the reaction.[16,31,32] Equally effective results can be found in the literature with many other types of functional group. The range of examples given also

Table 10.1. Functional group compatibility in aromatic substrates

$$p\text{-}X\text{-}C_6H_4\text{-}I + RC{\equiv}CH \xrightarrow[\text{Base, Solvent, r.t.}]{CuI,\ PdCl_2(PPh_3)_2} p\text{-}X\text{-}C_6H_4\text{-}C{\equiv}C\text{-}R$$

X	Base, solvent	Time (h)	Yield (%)[a]
OH	Et_3N, MeCN	6	37
NH_2	Et_2NH	24	64
$NHSO_2Me$	Et_2NH, THF	20	66
F	Et_2NH	18	34
CO_2Et	Et_2NH	16	77
$CONH_2$	Et_2NH, THF	18	96
CO_2H	Et_2NH	18	86
NO_2	Et_2NH	18	65

[a] Not optimized; isolated yields.

demonstrates that, when aromatic iodides are being reacted, the electronic nature of the substituent has little practical consequence for the rate of reaction. The same degree of functional group tolerance is also exhibited in the coupling of vinyl substrates.

The impressive functional group compatibility of the Sonogashira coupling gives the reaction a distinct advantage over many other methods for carbon–carbon bond formation, and allows the reaction to be performed very late in a synthetic sequence, without the requirement for multiple protecting groups. A typical example is shown in eqn (14), where compound **10**, designed as part of a series of potential β_2-stimulants for the relief of asthma, is prepared via a coupling reaction followed by hydrogenation.[31] The *N*-benzyl protection is necessary for previous reactions in the sequence and is not a requirement for the Sonogashira coupling to proceed.

5.2 Alkyne substrates

The range of alkynes which will participate in the Sonogashira reaction is almost as wide as that of the aromatic portion. Both alkyl- and arylalkynes with many different substituents will react, as will silylalkynes—precursors to terminal triple bonds. Acetylene itself will react at both ends to give symmetrical diarylalkynes.[1] There are, however, some specific alkynes which will not react or which may give difficulties. In particular, those alkynes which are conjugated to electron-withdrawing groups, e.g. $HC{\equiv}CCO_2Me$, will not take part in the coupling. Attempted reaction of this type of alkyne will

(i) CuI, $PdCl_2(PPh_3)_2$, $(C_6H_{11})_2NH$, MeCN
25°C, 3 h, 67%

(ii) H_2, Pd, HCl, EtOH, 90% (14)

10

generally lead only to products arising from Michael-type addition of the organic base to the unsaturated system. This restriction can, in some part, be overcome by coupling propargyl alcohol and subsequently oxidizing, or by preparing the terminal alkyne and performing a carboxylation. A recently developed modification involves the use of protected carbonyl functions, in the form of the acetal $HC{\equiv}CCH(OMe)_2$ or orthoester $HC{\equiv}CC(OMe)_3$, to circumvent this problem.[33] A slightly different problem exists for alkynes which contain electron-withdrawing groups β to the alkyne, e.g. $HC{\equiv}CCH_2CO_2H$. In this case, the highly acidic methylene group allows the rapid rearrangement of the alkyne under basic conditions to give the fully conjugated allene, i.e. $H_2C{=}C{=}CHCO_2H$. This process is normally more rapid than the coupling reaction and very little substituted allene arising from rearrangement of a coupled product is obtained.

The only other alkynes likely to give difficulties in the Sonogashira reaction are the short-chain alkynamines, i.e. $HC{\equiv}CCH_2NH_2$, $HC{\equiv}CCH_2CH_2NH_2$, and $HC{\equiv}CCH_2CH_2CH_2NH_2$. It has been shown that, under palladium catalysis, the amino group can undergo addition to triple bonds to give cyclic imines.[34] The consequence of this for the Sonogashira reaction is such that, for the short-chain alkynamines mentioned, serious problems are encountered on attempted coupling with aryl halides. Although the desired coupled products may be produced, they are invariably present in low yields and heavily contaminated with numerous by-products. If, however, the amino function is situated at a further distance from the triple bond, then the reactions proceed without difficulty. This problem is easily overcome by derivatizing the amines as benzamides or acetamides which then undergo reaction smoothly and can later be hydrolysed to the desired amines. Although the corresponding *N*-alkyl- and *N*,*N*-dialkylamines are less susceptible to this type of problem, in some cases similar results may be obtained. An unexpected example of this was obtained on the reaction of an aryl iodide with the dialkylpentynamine **11** (eqn 15).[17] Some of the desired product **12** was obtained but the main product isolated was the cyclic quaternary amine **13**.

$$\text{4-}IC_6H_4Br \xrightarrow[\text{CuI, } PdCl_2(PPh_3)_2\text{, } Et_3N\text{, } (C_6H_{11})_2NH\text{, MeCN; 25°C, 2 h}]{HC\equiv C\text{-}CH_2CH(OH)CH_2NR_1R_2\ \ \mathbf{11}} \mathbf{12}\ (20\%) + \mathbf{13}\ (72\%) \quad (15)$$

12: Br–C₆H₄–C≡C–CH₂CH(OH)CH₂NR¹R² (20%)

13: cyclic quaternary ammonium iodide, I^- (72%)

R^1R^2 = $-(CH_2)_5-$

These restrictions apart, the coupling reaction will tolerate a wide selection of functional groups on the alkyne portion and will therefore allow an almost limitless variety of chains to be introduced to an aromatic system. A list of examples is given in Table 10.2 to demonstrate the types of alkyne which will participate in the reaction. Of particular use is the ability to tolerate alkylating agents such as alkyl bromides which can later be transformed to other functions or used in other organometallic processes.

6. Ring formation

Although functional groups are extremely well tolerated in the Sonogashira coupling, it is possible to involve certain *ortho*-substituents on aromatic halides in a mild and efficient synthesis of benzofuran and indole ring systems.

Table 10.2. Functional group tolerance of alkyne substrates

$$\mathrm{ArX} \xrightarrow{\text{alkyne, Cu}^{\mathrm{I}}\text{, Pd}^{\mathrm{II}}\text{, amine}} \mathrm{ArC{\equiv}CR}$$

Alkyne	X	Yield (%)	References
$PhC{\equiv}CH$	I	90	1
$Me_3SiC{\equiv}CH$	Br	80	20
$HOCH_2C{\equiv}CH$	I	90	1
$F_3CCONHCH_2C{\equiv}CH$	I	90	35
$MeO_2C(CH_2)_3C{\equiv}CH$	Br	72	Protocol 2
$Br(CH_2)_6OCH_2C{\equiv}CH$	I	48	31
$PhCH_2NH(CH_2)_6O(CH_2)_2C{\equiv}CH$	I	77	Protocol 1

Thus, reaction of 2-iodophenol with a variety of alkynes under copper–palladium-catalysed conditions leads directly to the benzofuran ring substituted in the 2-position (eqn 16).[36] In this case piperidine–dimethylformamide is used as the solvent system with a palladium diacetate catalyst. This type of ring formation can also be performed by heating an alkyne with the iodophenol in the presence of copper(I) oxide–pyridine,[37] but the Sonogashira-type conditions are considerably milder and more tolerant of other functional groups. Isolation of the non-cyclized coupled product is not generally possible, although *meta-* and *para*-substituted iodophenols give arylalkynes via the normal mode of reaction.

I
OH
RC≡CH , CuI, piperidine ,
$Pd(OAc)_2(PPh_3)_2$, DMF
O
R
(16)
R = alkyl , aryl , silyl

In a similar manner, it has been shown that the indole ring system can be generated from *ortho*-iodo- or *ortho*-bromoanilines provided that the aniline is first derivatized as its methanesulfonamide (eqn 17).[38] Under standard coupling conditions the indole ring is produced directly, as is the case with the benzofuran system. Although the mechanism of cyclization is not clear, the authors have shown that copper iodide is crucially involved in the process. Other stepwise, but highly efficient, procedures to give the indole ring have also been developed using the Sonogashira coupling methodology.[17,39]

X
$NHSO_2Me$
X = Br, I
RC≡CH , CuI , $PdCl_2(PPh_3)_2$
Et_3N , DMF
R
N
SO_2Me
(17)
R = alkyl , aryl , silyl

7. Summary

The Sonogashira copper–palladium-catalysed coupling of alkynes to aromatic and vinylic halides has established itself as one of the most effective methods of carbon–carbon bond formation currently available. The technical simplicity and robust nature of the reaction allow the rapid and efficient preparation of series of small molecules of interest—a particularly desirable feature to chemists working in the pharmaceutical and agrochemical industries. Furthermore, the ability of the reaction to tolerate high levels of functional complexity has resulted in the coupling being used at crucial stages late in the construction of many complex natural products. Several recent publications illustrate the continuing use and growing potential of this elegant methodology.[40]

References

1. Sonogashira, K.; Tohda, Y.; Hagihara, N. *Tetrahedron Lett.* **1975**, *16*, 4467–4470.
2. Takahashi, S.; Kuroyama, Y.; Sonogashira, K.; Hagihara, N. *Synthesis* **1980**, 627–630.
3. Dieck, H. A.; Heck, F. R. *J. Organomet. Chem.* **1975**, *93*, 259–263.
4. Cassar, L. *J. Organomet. Chem.* **1975**, *93*, 253–257.
5. Heck, R. F. *Org. React.* **1982**, *27*, 345–390.
6. Castro, C. E.; Stephens, R. D. *J. Org. Chem.* **1963**, *28*, 3313–3315.
7. Sonogashira, K. In *Comprehensive Organic Synthesis*; Trost, B. M., Fleming, I., eds; Pergamon: Oxford, 1991, Vol. 3, pp. 521–548.
8. Viehe, H. G. *Chemistry of Acetylenes*; Marcel Dekker: New York, **1969**.
9. Fitton, P.; Rick, E. A. *J. Organomet. Chem.* **1971**, *28*, 287–291.
10. Casalnuovo, A. L.; Calabrese, J. C. *J. Am. Chem. Soc.* **1990**, *112*, 4324–4330.
11. Carpita, A.; Lezzi, A.; Rossi, R.; Marchetti, F.; Merlino, S. *Tetrahedron*, **1985**, *41*, 621–625.
12. Singh, R.; Just, G. *J. Org. Chem.* **1989**, *54*, 4453–4457.
13. Brandsma, L.; Van den Heuvel, H. G. M.; Verkruijsse, H. D. *Synth. Commun.*, **1990**, *20*, 1889–1892.
14. Hirota, K.; Kitade, Y.; Isobe, Y.; Maki, Y. *Heterocycles* **1987**, *26*, 355–358.
15. Bergbreiter, D. E.; Weatherford, D. A. *J. Org. Chem.*, **1989**, *54*, 2726–2730.
16. Campbell, I. B.; Finch, H.; Lunts, L. H. C.; Naylor, A.; Skidmore, I. F. UK Pat. GB 2 165 542 A, 1986 (*Chem. Abstr.* **1986**, *105*, 190 642t).
17. Campbell, I. B. Unpublished results.
18. Acheson, R. M.; Lee, G. C. M. *J. Chem. Res. (S)* **1986**, *10*, 380–383.
19. Tischler, A. N.; Lanza, T. J. *Tetrahedron Lett.* **1986**, *27*, 1653–1656.
20. Yamanaka, H.; Sakamoto, T.; Shiraiwa, M.; Kondo, Y. *Synthesis*, **1983**, 312–314.
21. Nicolaou, K. C.; Webber, S. E. *J. Chem. Soc., Chem. Commun.* **1986**, 1816–1817.
22. Nicolaou, K. C.; Ramphal, J. Y.; Petasis, N. A.; Serhan, C. N. *Angew. Chem., Int. Ed. Engl.* **1991**, *30*, 1100–1116.
23. Chemin, D.; Linstrumelle, G. *Tetrahedron* **1992**, *48*, 1943–1952.

24. Cacchi, S.; Morera, E.; Ortar, G. *Synthesis* **1986**, 320–322.
25. Brückner, R.; Scheuplein, S. W.; Suffert, J. *Tetrahedron Lett.* **1991**, *32*, 1449–1452.
26. Magnus, P.; Annoura, H.; Harling, J. *J. Org. Chem.* **1990**, *55*, 1709–1711.
27. Kende, A. S.; Smith, C. A. *Tetrahedron Lett.* **1988**, *29*, 4217–4220.
28. Wityak, J.; Chan, J. B. *Synth. Commun.* **1991**, *21*, 977–979.
29. Turner, W. R.; Suto, M. J. *Tetrahedron Lett.* **1993**, *34*, 281–284.
30. Kende, A. S.; Smith, C. A. *J. Org. Chem.* **1988**, *53*, 2655–2657. Ratovelomanana, V.; Linstrumelle, G. *Tetrahedron Lett.* **1981**, *22*, 315–318. Gamage, S.; McNaughton-Smith, G. A.; Taylor, R. J. K. University of East Anglia, unpublished observations.
31. Campbell, I. B.; Finch, H.; Lunts, L. H. C.; Naylor, A.; Skidmore, I. F. Netherl. Pat. NL 8 602 575, 1987 (*Chem. Abstr.* **1987**, *107*, 236 223d).
32. Campbell, I. B.; Finch, H.; Lunts, L. H. C.; Naylor, A.; Skidmore, I. F. Ger. Offen. DE 3 513 885, 1985 (*Chem. Abstr.* **1985**, *105*, 60 407j).
33. Sakamoto, T.; Shiga, F.; Yasuhara, A.; Uchiyama, D.; Kondo, Y.; Yamanaka, H. *Synthesis* **1992**, 746–748.
34. Fukuda, Y.; Matsubara, S.; Utimoto, K. *J. Org. Chem.* **1991**, *56*, 5812–5816.
35. Hobbs, Jr., F. W. *J. Org. Chem.* **1989**, *54*, 3420–3422.
36. Arcadi, A.; Marinelli, F.; Cacchi, S. *Synthesis* **1986**, 749–751.
37. Castro, C. E.; Gaughan, E. J.; Owsley, D. C. *J. Org. Chem.* **1966**, *31*, 4071–4078.
38. Sakamoto, T.; Kondo, Y.; Iwashita, S.; Nagano, T.; Yamanaka, H. *Chem. Pharm. Bull.* **1988**, *36*, 1305–1308.
39. Yamanaka, H.; Sakamoto, T.; Kondo, Y. *Heterocycles* **1986**, *24*, 31–32.
40. Arcadi, A.; Cacchi, S.; Marinelli, F. *Tetrahedron* **1993**, *49*, 4955–4964. Paley, R. S.; Lafontaine, J. A.; Ventura, M. P. *Tetrahedron Lett.* **1993**, *34*, 3663–3666. Suffert, J.; Eggers, A.; Scheuplein, S. W.; Bruckner, R. *Tetrahedron Lett.* **1993**, *34*, 4177–4180. Solooki, D.; Bradshaw, J. D.; Tessier, C. A.; Youngs, W. J. *Organometallics* **1994**, *13*, 451–455. Diederich, F.; Philp, D.; Seiler, P. *J. Chem. Soc., Chem. Commun.* **1994**, 205–208.

11

Alkyne carbocupration and polyene synthesis

J.-F. NORMANT

1. Introduction

1.1 Formation of vinylcopper reagents from alkynes

The addition of organocopper reagents to terminal alkynes and to acetylene represents an efficient way to synthesize vinylcopper reagents with a given geometry.[1–8] The addition proceeds in a Markownikov way in a purely *syn*-specific fashion. Two approaches may be considered. With terminal alkynes (eqn 1), RCu is preferably prepared from a Grignard reagent which can be primary or secondary, and the solvent can be ether, THF, or ether–dimethyl sulfide.

$$RMgX \xrightarrow{CuX} RCu{\cdot}MgX_2 \xrightarrow{R^1C{\equiv}CH} (R)(R^1)C{=}CH{-}Cu{\cdot}MgX_2\ (\mathbf{1}) \xrightarrow{E^+} (R)(R^1)C{=}CH{-}E \qquad (1)$$

The latter system provides a better stabilization of the vinylcopper reagent **1**, which should be kept at a temperature low enough to avoid decomposition (as seen by a black deposit or a mirror of copper on the walls of the flask) according to

$$(R)(R^1)C{=}CH{-}Cu \longrightarrow (R)(R^1)C{=}CH{-}CH{=}C(R^1)(R) + 2Cu^0\downarrow$$

With acetylene (eqn 2), a lithium cuprate in ether is more efficient and the reaction proceeds with double insertion to give reagent **2**, whereas a magnesium cuprate leads to single insertion only, whatever the ratio of the reagents.

A limitation is observed with carbocupration using vinylcopper reagents, but in this case it is possible to get dienylcopper species by double addition to

$$2RLi \xrightarrow{CuX} R_2CuLi \xrightarrow{2HC\equiv CH} (R\text{–CH=CH})_2CuLi \quad \mathbf{2} \qquad (2)$$

$$\xrightarrow{1\text{ eq } E^+} R\text{–CH=CH–}E + R\text{–CH=CH–}Cu \xrightarrow{1\text{ eq } E^+} 2\ R\text{–CH=CH–}E$$

acetylene (see Chapter 2, Protocol 11). The Markownikov regioselectivity pointed out above can be altered by the presence of heteroatoms on the acetylenic substrate, which can coordinate to the reagent or simply polarize[9] the carbon–carbon triple bond so that a total reversal of regioselectivity is observed. For example

$$HC\equiv C\text{–Z–R} \xrightarrow{'R^1Cu'} R^1CH{=}C('Cu')(ZR) \qquad Z = S, P$$

$$HC\equiv C\text{–Z–R} \xrightarrow{'R^1Cu'} 'Cu'CH{=}C(R^1)(ZR) \qquad Z = O, N$$

$$HC\equiv C\text{–CH(OEt)}_2 \xrightarrow{'R^1Cu'} R^1CH{=}C('Cu')(CH(OEt)_2)$$

Methyl cupration is not highly rewarding, although experiments in Me_2S as a solvent allow such reactions.[6] However, some substrates do add methylcopper species easily, for example[4] $HC\equiv CSR$ and $HC\equiv CCH(OR)_2$. Disubstituted acetylenes do not react, as a rule, except for the cases of $RC\equiv CSR^1$, $RC\equiv CPR^1{}_2$, and $RC\equiv CCH(OEt)_2$.[4]

The following protocols generally consist of three steps: (1) preparation of the organocopper or cuprate reagent; (2) addition to the alkyne; and (3) reaction with an electrophilic substrate, followed by hydrolysis and work-up. In the first three protocols, only the preparation of the vinylcopper species is described.

2. General considerations

All the organometallics used here are air sensitive (RLi, RMgX, RCu) and should be used in dry apparatus with dry solvents and reagents under an inert atmosphere (nitrogen or argon). A careful control of temperature is necessary since the thermal stability of the alkyl- or vinylcopper species is rather low (see above). All the following protocols start with the preparation of a vinylcopper–magnesium bromide or a lithium divinylcuprate in ether or ether–pentane. A further addition of THF may be necessary to improve the reactivity of these species. It is, however, possible to prepare the former reagents (but not the latter) in THF.[4] The copper salts are commercially available and they may be purified according to Chapter 2. Low temperatures are often necessary and for these a cryocool system is used.

Protocol 1.
Preparation of a divinylcuprate reagent from an organolithium reagent in Et_2O[4]

Caution! Carry out all procedures in a well-ventilated hood, and wear disposable vinyl or latex gloves and chemical-resistant safety goggles.

$$2EtLi\cdot LiBr \xrightarrow[\text{ether}]{CuI} Et_2CuLi\cdot LiBr \xrightarrow{HC\equiv CH} (Et\text{–}CH{=}CH)_2CuLi\cdot LiBr$$

Equipment

- One three-necked, round-bottomed flask (250 mL) equipped with a variable speed mechanical stirrer, a pressure-equalizing dropping funnel (100 mL) topped with a septum with a needle for gas inlet, a Claisen head bearing a low temperature thermometer (30 to −80°C), and a bubbler
- A cooling bath
- One cannula
- One column (30 cm long, 3 cm in diameter) packed with granulated $CaCl_2$
- Two traps and their cooling baths (Dewars filled with dry ice and acetone)
- One water gasometer filled with brine[a] (see Fig. 11.1)

Materials

- Ethyllithium[b] (FW 36.0) 1 M in ether, 50 mL, 50 mmol — **flammable, moisture sensitive**
- CuI[c] (FW 190.4) 5.35 g, 28 mmol — **irritant, light sensitive**
- Acetylene (FW 26.0) 2.0 L — **flammable gas**
- Dry ether 150 mL — **flammable, irritant**

1. Introduce a stream of nitrogen via the gas inlet. Dry the apparatus (flame or gun heater), place copper iodide and dry ether (100 mL) in the flask, and purge again with nitrogen.
2. Introduce the ethyllithium solution via a cannula into the dropping funnel.
3. Immerse the flask in a bath at −35°C (cryocool). Add the ethyllithium solution dropwise at −35°C, and stir for 20 min more at this temperature. A clear, blue-grey solution is formed.[b,c,d] Quickly remove the dropping funnel and replace with a septum.
4. Connect the acetylene cylinder to two traps immersed in a dry ice–acetone bath at −65°C to remove acetone. Connect the second trap to a water gasometer (see Fig. 11.1) filled with brine, where an exact volume of gas can be calibrated under normal pressure. Connect the outlet of the gasometer to a 50 cm column packed with anhydrous $CaCl_2$. Purge the whole manifold with acetylene, then connect to a needle fitted in the septum of the reaction flask and immersed below the level of the solution. Introduce 1.2 L (50 mmol)

Protocol 1. *Continued*

of acetylene while stirring at such a rate that the temperature never exceeds −25 °C (about 15 min). Remove the needle, and stir the pale-green solution at −25 °C (cryocool) for 30 min. The lithium divinylcuprate is now ready for further use.

[a] See Reichert, J. S.; Nieuwland, J. A. *Org. Synth. Coll. Vol.* **1941**, *1*, 230.
[b] If the lithium reagent is in solution in a hydrocarbon solvent, $CuBr{\cdot}Me_2S$ should be used instead of CuI.
[c] $CuBr{\cdot}Me_2S$ complex (5.75 g, 28 mmol) (see Chapter 2, Protocol 1) may be used instead.
[d] Regulate stirring to avoid splashing on the walls of the flask (which are warmer than the solution and promote the decomposition of the reagent with the formation of black Cu(0).

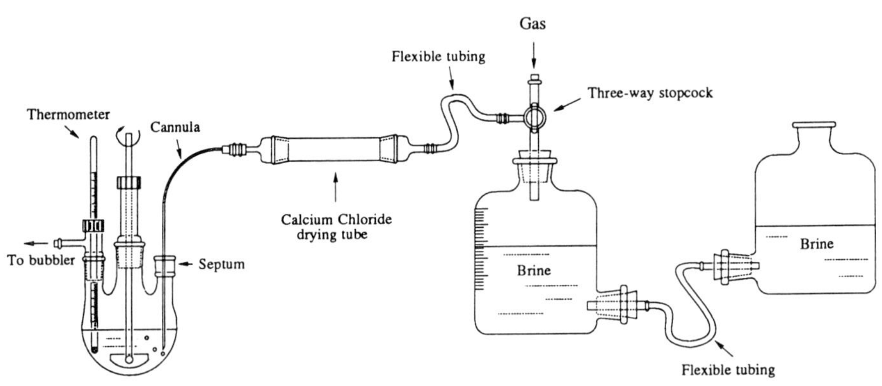

Figure 11.1 Water gasometer for acetylene carbocupration reactions.

Protocol 2. Preparation of a vinylcopper reagent from a Grignard reagent in Et_2O[4]

Caution! Carry out all procedures in a well-ventilated hood, and wear disposable vinyl or latex gloves and chemical-resistant safety goggles.

$$EtMgBr + CuBr \longrightarrow EtCu{\cdot}MgBr_2 \xrightarrow{MeC\equiv CH} (Me)(Et)C{=}CH{-}Cu{\cdot}MgBr_2$$

Equipment

- One three-necked, round-bottomed flask (250 ml) equipped with a variable speed mechanical stirrer, a pressure-equalizing dropping funnel (100 ml) topped with a septum with a needle for gas inlet, a Claisen head bearing a low temperature thermometer (30 to −80 °C), and a bubbler
- A cooling bath
- One cannula
- One column (30 cm long, 3 cm in diameter) packed with granulated $CaCl_2$
- One water gasometer filled with brine (see Fig. 11.1)

Materials

- Ethylmagnesium bromide[a] (FW 133.3) 1 M in ether, 50 mL, 50 mmol **flammable, moisture sensitive**
- CuBr[b] (FW 143.4) 7.89 g, 55 mmol **irritant, light sensitive**
- Propyne (FW 40.0) 2.0 L **flammable gas**
- Dry ether 150 mL **flammable, irritant**

1. Introduce a stream of nitrogen via the gas inlet. Dry the apparatus (flame or gun heater), place copper bromide (7.89 g, 55 mmol) and ether (50 mL) in the flask and purge again with nitrogen.
2. Introduce the ethereal Grignard solution (50 mmol, *c.* 1 M) via a cannula into the dropping funnel. Immerse the flask in a bath at −35°C (cryocool), and add the Grignard solution dropwise at −35°C. A yellow precipitate is formed.
3. Stir the suspension for another 30 min at this temperature. Replace the dropping funnel with a septum.
4. Connect the propyne cylinder to a water gasometer analogous to the one described in Protocol 1 (Fig. 11.1). Connect the outlet of the gasometer to a 50 cm column packed with anhydrous $CaCl_2$. Purge the whole manifold with propyne, then connect to a needle fitted in the septum of the reaction flask and immersed below the level of the solution. Introduce propyne (1.2 L, 50 mmol) while stirring at −40°C (cryocool). Stir for 1 h at −15°C (cryocool). A dark-green solution is obtained, ready for further use.

[a] Prepared from EtBr and Mg in ether and titrated (see Chapter 2, Protocol 6).[10]
[b] See Chapter 2, Protocol 1 for the preparation of pure CuBr. $CuBr{\cdot}Me_2S$ may be used instead.

Protocol 3.
Preparation of a vinylcopper reagent bearing a functionality[12]

Caution! Carry out all procedures in a well-ventilated hood, and wear disposable vinyl or latex gloves and chemical-resistant safety goggles.

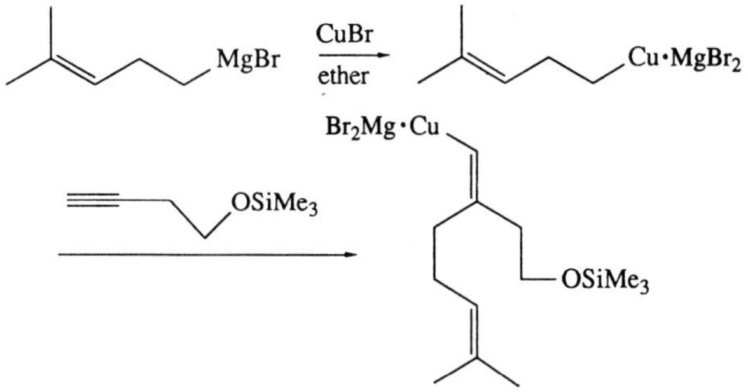

MgBr —CuBr / ether→ $Cu{\cdot}MgBr_2$

$Br_2Mg{\cdot}Cu$

$OSiMe_3$

$OSiMe_3$

Protocol 3. *Continued*

Equipment

- One three-necked, round-bottomed flask (250 mL) equipped with a variable speed mechanical stirrer, a pressure-equalizing dropping funnel (100 mL) topped with a septum with a needle for gas inlet, a Claisen head bearing a low temperature thermometer (30 to −80°C), and a bubbler
- A cooling bath

Materials

- 4-Methyl-3-penten-1-ylmagnesium bromide[a] (FW 187.2) 1 M in ether, 50 mL, 50 mmol — **flammable, moisture sensitive**
- CuBr[b] (FW 143.4) 7.89 g, 55 mmol — **irritant, moisture sensitive**
- 1-Trimethylsilyloxy-3-butyne (FW 142.2) 7.10 g, 50 mmol — **flammable**
- Dry pentane 250 mL — **flammable, irritant**
- Dry ether 150 mL — **flammable, irritant**

1. Introduce a stream of nitrogen via the gas inlet. Dry the apparatus (flame or gun heater), place copper bromide (7.89 g, 55 mmol) and ether (50 mL) in the flask, and purge again with nitrogen.
2. Introduce the ethereal Grignard solution (50 mmol, *c.* 1 M) via a cannula into the dropping funnel. Immerse the flask in a bath at −35°C (cryocool), and add the Grignard solution dropwise at −35°C. A yellow precipitate is formed.
3. Stir the suspension for a further 30 min at this temperature.
4. Add dry pentane (150 mL) at −40°C (cryocool).
5. Add dropwise a solution of 1-trimethylsilyloxy-3-butyne (7.10 g, 50 mmol) in ether (50 mL) at the same temperature.
6. Let the solution stir at −25°C (cryocool) for 20 h. The vinylcopper reagent is then ready for use.

[a] Prepared from the corresponding bromide[12] and Mg in ether.
[b] See Chapter 2, Protocol 1.

---

3. Trapping vinylcopper reagents

The vinylcopper reagents thus formed can be used for a host of synthetic transformations which are characterized by an excellent retention of configuration, leading to trisubstituted olefins (eqn 1) or pure (*Z*)-disubstituted olefins (eqn 2). The reactivity of cuprates such as **2** is much higher than that of a vinylcopper reagent in the presence of either MgX_2 or LiX, so that reaction of one equivalent of E^+ in eqn 1 or of the second equivalent of E^+ in eqn 2 requires additives that help to stabilize the reagent (HMPA, $P(OR)_3$, TMEDA, etc.) and which are necessary if the electrophilic partner is not very powerful. Heteroatoms can thus be introduced at the vinyl position (I, Br, PR_2, SR,

SO_2R, SiR_3, SnR_3),[4,13] at the allylic position (OH, OR, SR, NR_2, Cl),[4,13] and also at homoallylic and further positions. The well-known conjugate addition can also be performed with these reagents.

Protocol 4.
Hydrolysis of a vinylcopper reagent: preparation of 3-methylen-7-methyl oct-6-en-1-o1[11]

Caution! Carry out all procedures in a well-ventilated hood, and wear disposable vinyl or latex gloves and chemical-resistant safety goggles.

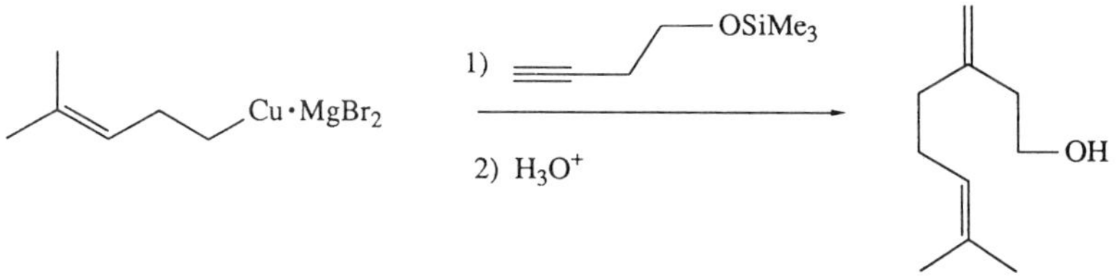

$Cu \cdot MgBr_2$

1) $\equiv\!\!-\!\!\diagup\!\!-OSiMe_3$

2) H_3O^+

OH

Equipment

See Protocol 3.

Materials

See Protocol 3.

1. Prepare the reagent according to Protocol 3.
2. Hydrolyse with a saturated ammonium chloride solution (50 mL) followed by a 1 M HCl solution (50 mL) to hydrolyse the silyl ether.
3. Extract with pentane (2 × 50 mL), wash the organic phases with brine (100 mL), and dry them over $MgSO_4$.
4. Evaporate the solvents *in vacuo*.
5. Distil the corresponding alcohol through a 10 cm Vigreux column (9.15 g, 81% yield; b.p. 74°C/0.05 mm Hg). Note that the high regioselectivity is due to the presence of pentane during the carbocupration step (Protocol 3).

Protocol 5.
Iodinolysis of a vinylcopper reagent: preparation of (*E*)-1-iodo-2-ethyl-1-hexene[14,15,a]

Caution! Carry out all procedures in a well-ventilated hood, and wear disposable vinyl or latex gloves and chemical-resistant safety goggles.

$$EtCu \cdot MgBr_2 \longrightarrow (Bu)(Et)C{=}CH{-}Cu \cdot MgBr_2 \xrightarrow{I_2} (Bu)(Et)C{=}CH{-}I$$

Equipment

- One three-necked, round-bottomed flask (250 mL) equipped with a variable speed mechanical stirrer, a pressure-equalizing dropping funnel (100 mL) topped with a septum with a needle for gas inlet, a Claisen head bearing a low temperature thermometer (30 to −80°C), and a bubbler
- A cooling bath

Materials

- Ethylmagnesium bromide (FW 133.3) 1 M in ether, 50 mL, 50 mmol **flammable, moisture sensitive**
- CuBr (FW 143.4) 7.89 g, 55 mmol **irritant, light sensitive**
- 1-Hexyne (FW 82.0) 4.10 g, 50 mmol **flammable, irritant**
- Iodine (FW 253.8) 13.2 g, 52 mmol **corrosive, lachrymator**
- Dry ether 150 mL **flammable, irritant**
- Triethyl phosphite (FW 166.2) 3.32 g, 20 mmol **moisture sensitive, irritant**

1. Prepare the vinylcopper reagent exactly as described for Protocol 2, except for the fact that 1-hexyne is a liquid. Instead of bubbling propyne add dropwise to the ethylcopper solution at −40°C (cryocool) a solution of 1-hexyne (4.10 g, 50 mmol) in ether (10 mL) and stir at −15°C (cryocool) for 1 h (deep-green solution).
2. Cool the vinylcopper solution to −50°C (cryocool) and add at once finely ground iodine (13.2 g, 52 mmol). Let the temperature rise to −30°C (cryocool) and maintain at this temperature until decolorization is observed (a white precipitate of copper iodide separates out). Add triethyl phosphite (3.32 g, 20 mmol) to destroy the excess iodine.
3. Hydrolyse the mixture at −30°C with 5 M HCl (75 mL). Separate the two layers and extract the aqueous layer once with ether (30 mL). Wash the organic phases successively with 5 M HCl (30 mL), water (30 mL), and a saturated solution of sodium hydrogencarbonate (30 mL), then dry over $MgSO_4$.
4. Remove the solvents in a rotary evaporator and distil the residue (8.92 g, 75% yield; b.p. 90–95°C).

[a] From BuMgBr and gaseous 1-butyne, (*Z*)-1-iodo-2-ethyl-1-hexene is obtained similarly (71% yield; b.p. 90–95°C). It is clearly distinguished from its (*E*)-isomer by GLC on a 20 m capillary column (OV 101).

Protocol 6. (Z)-1-Phenylthiobut-1-ene[16]

Caution! Carry out all procedures in a well-ventilated hood, and wear disposable vinyl or latex gloves and chemical-resistant safety goggles.

$$Et_2CuLi \xrightarrow{HC\equiv CH} (Et\text{–CH=CH})_2CuLi \xrightarrow{2PhSSPh} 2\ Et\text{–CH=CH–}SPh$$

Equipment

- One three-necked, round-bottomed flask (250 mL) equipped with a variable speed mechanical stirrer, a pressure-equalizing dropping funnel (100 mL) topped with a septum with a needle for gas inlet, a Claisen head bearing a low temperature thermometer (30 to −80°C), and a bubbler
- A cooling bath
- One cannula
- One column (30 cm long, 3 cm in diameter) packed with granulated $CaCl_2$
- Two traps and their cooling baths (Dewars filled with dry ice and acetone)
- One water gasometer filled with brine (see Fig. 11.1)

Materials

- $CuBr{\cdot}Me_2S$ (FW 205.5) 3.1 g, 15.1 mmol — **moisture sensitive**
- Ethyllithium (FW 35.9) 1 M in ether, 30 mL, 30 mmol — **flammable, moisture sensitive**
- Acetylene (FW 26.0) 2.0 L — **flammable gas**
- Hexamethylphosphoric triamide (HMPA)[a] (FW 179.0) 5.5 mL, 5.6 g, 31 mmol — **highly toxic, cancer suspect agent**
- Diphenyl disulfide (FW 218.3) 6.54 g, 30 mmol — **irritant**
- Ether 100 mL — **flammable, irritant**
- THF 70 mL — **flammable, irritant**
- Pentane 100 mL — **flammable, irritant**

1. Prepare lithium di-(Z)-butenylcuprate following Protocol 1, using the $CuBr{\cdot}Me_2S$ complex (3.08 g, 15 mmol) in ether (50 mL) and ethyllithium (1 M in ether, 30 mL, 30 mmol) at −25°C.
2. Add dropwise a solution of hexamethylphosphoric triamide (5.5 mL, 31 mmol) (**caution!** carcinogenic) in tetrahydrofuran (20 mL), then, immediately after, add a solution of diphenyl disulfide (6.54 g, 30 mmol) in tetrahydrofuran (40 mL). Let the mixture warm up to 20°C and stir for 3 h.
3. Cool to −50°C (cryocool) and hydrolyse with a saturated ammonium chloride solution (50 mL). Add pentane (100 mL) and filter off the salts on a sintered-glass funnel. Separate the organic phase, wash with a saturated ammonium chloride solution (3 × 30 mL), and dry over magnesium sulfate.
4. Evaporate the solvents *in vacuo* and distil the residue through a 10 cm Vigreux column (4.1 g, 83% yield; b.p. 63°C/0.1 mm Hg).

[a] If HMPA is omitted, only one vinyl group of the divinylcuprate reacts.

Protocol 7.
Alkynylation of a vinylcopper reagent: preparation of (*Z*)-5-methylhept-2-yn-4-en-1-ol[17,a]

Caution! Carry out all procedures in a well-ventilated hood, and wear disposable vinyl or latex gloves and chemical-resistant safety goggles.

$$\mathrm{EtCu\cdot MgBr_2} \xrightarrow{\mathrm{MeC{\equiv}CH}} \mathrm{(Me)(Et)C{=}CH{-}Cu\cdot MgBr_2} \xrightarrow[\text{2) } \mathrm{H_3O^+}]{\text{1) } \mathrm{BrC{\equiv}C{-}CH_2OSiMe_3}} \mathrm{(Me)(Et)C{=}CH{-}C{\equiv}C{-}CH_2OH}$$

Equipment

- One three-necked, round-bottomed flask (250 mL) equipped with a variable speed mechanical stirrer, a pressure-equalizing dropping funnel (100 mL) topped with a septum with a needle for gas inlet, a Claisen head bearing a low temperature thermometer (30 to −80°C), and a bubbler
- A cooling bath
- One cannula
- One column (30 cm long, 3 cm in diameter) packed with granulated $CaCl_2$
- One water gasometer filled with brine (see Fig. 11.1)

Materials

- Ethylmagnesium bromide (FW 133.3) 1 M in ether, 50 mL, 50 mmol — **flammable, moisture sensitive**
- CuBr (FW 143.4) 7.89 g, 55 mmol — **irritant, hygroscopic**
- Propyne (FW 40.0) 2.0 L — **flammable gas**
- TMEDA (FW 116.2) 5.22 g, 45 mmol — **flammable, corrosive**
- 1-Bromo-3-trimethylsilyloxyprop-1-yne[17] (FW 207.0) 6.21 g, 30 mmol — **flammable, harmful**
- Dry ether 150 mL — **flammable, irritant**
- THF 70 mL — **flammable, irritant**
- Pentane 100 mL — **flammable, irritant**

1. Prepare the vinylcopper reagent as described in Protocol 2.
2. To the stirred solution, add at −15°C (cryocool) TMEDA (5.22 g, 45 mmol) and THF (70 mL). Stir for 30 min at −15°C, then add 1-bromo-3-trimethylsilyloxyprop-1-yne (6.21 g, 30 mmol). Stir for 1 h at −15°C.
3. Hydrolyse with H_2SO_4 (2 M, 50 mL) at this temperature and extract with pentane (2 × 50 mL). Wash the organic phase with brine (100 mL) and dry over $MgSO_4$.
4. Evaporate the solvents *in vacuo* and distil the residue through a 10 cm Vigreux column (3.05 g, 82% yield; b.p. 57°C/0.4 mm Hg).

[a] This enynol can be reduced stereoselectively by lithium aluminum hydride to the pure (*E,Z*)-dienol (91%).[17]

Epoxides will react with only one vinyl moiety of a lithium divinylcuprate. It is also possible to use a vinylcopper-to-epoxide ratio of 1:1 if the cuprate reagent is a mixed lithium vinylalkynyl cuprate. More simply, as in the following protocol, it is possible to add an alkynyllithium equivalent after one equivalent of the divinylcuprate and two equivalents of epoxide have been mixed in the first step.

Protocol 8.
Preparation of (*Z*)-4-hepten-2-ol[18] using a mixed vinylalkynylcuprate

Caution! Carry out all procedures in a well-ventilated hood, and wear disposable vinyl or latex gloves and chemical-resistant safety goggles.

Et⁄=⁄)$_2$CuLi + 2 (propylene oxide) → Et⁄=⁄CH$_2$CH(CH$_3$)OLi + (propylene oxide) +

Et⁄=⁄Cu —1) BuC≡CLi; 2) H_2O→ 2 Et⁄=⁄CHOH-CH_3

Equipment

- One two-necked, round-bottomed flask (250 mL) equipped with a magnetic stirring bar, a low temperature thermometer, a Claisen head, a gas bubbler, and a dropping funnel (100 mL) topped with a septum in which a needle is inserted for a nitrogen inlet
- One three-necked, round-bottomed flask (250 mL) equipped with a variable speed mechanical stirrer, a pressure-equalizing dropping funnel (100 mL) topped with a septum with a needle for gas inlet, a Claisen head bearing a low temperature thermometer (30 to −80°C), and a bubbler
- A cooling bath
- One cannula
- One column (30 cm long, 3 cm in diameter) packed with granulated $CaCl_2$
- Two traps and their cooling baths (Dewars filled with dry ice and acetone)
- One water gasometer filled with brine (see Fig. 11.1)

Materials

- Ethyllithium (FW 35.9) 1 M in ether, 50 mL, 50 mmol — **flammable, moisture sensitive**
- *n*-Butyllithium (FW 64.1) 1 M in ether, 25 mL, 25 mmol — **flammable, moisture sensitive**
- CuI (FW 190.4) 5.35 g, 28 mmol — **irritant, light sensitive**
- 1-Hexyne (FW 82.0) 2.05 g, 25 mmol — **flammable, irritant**
- Propylene oxide (FW 58.0) 2.90 g, 50 mmol — **flammable, cancer suspect agent**
- Acetylene (FW 26.0) 2.0 L — **flammable gas**
- Dry ether 150 mL — **flammable, irritant**

1. Prepare separately, in the two-necked flask, a suspension of 1-lithio-1-hexynide (25 mmol) by addition of *n*-BuLi in ether (25 mmol) to 1-hexyne (2.05 g, 25 mmol) in ether (40 mL) at 0°C. Stir at 0°C for 1 h, then replace the Claisen head and the dropping funnel with septa.
2. Prepare the divinylcuprate according to Protocol 1 and cool to −30°C (cryocool).

Protocol 8. *Continued*

3. Add propylene oxide (2.90 g, 50 mmol) to the cuprate solution at −30 °C, then immediately transfer with a cannula the lithium hexynide suspension at −30 °C. Let the temperature rise to −15 °C (cryocool) and stir for 2 h at this temperature.
4. Hydrolyse with a mixture of saturated NH_4Cl solution (50 mL) and 10% HCl (20 mL). Add hexane (150 mL), filter the salts, and decant the organic layer. Wash the organic layer once with ammonia (50 mL), once with 10% HCl (10 mL), and dry over $MgSO_4$.
5. Remove the solvents *in vacuo* and distil the residue through a 10 cm Vigreux column (4.67 g, 82% yield; b.p. 61–62 °C/15 mm Hg).

Protocol 9. Carbonatation: preparation of neric acid[19]

Caution! Carry out all procedures in a well-ventilated hood, and wear disposable vinyl or latex gloves and chemical-resistant safety goggles.

MeC≡CH; Cu; 1) CO_2 2) H_3O^+; Cu; COOH

Equipment

- One three-necked, round-bottomed flask (250 mL) equipped with a variable speed mechanical stirrer, a pressure-equalizing dropping funnel (100 mL) topped with a septum with a needle for gas inlet, a Claisen head bearing a low temperature thermometer (30 to −80 °C), and a bubbler
- A cooling bath
- A 50 cm column packed with $CaCl_2$
- A water gasometer filled with brine (see Fig. 11.1)

Materials

- 4-Methyl-3-penten-1-ylmagnesium bromide[a] (FW 187.2) 1 M in ether, 50 mL, 50 mmol — **flammable, moisture sensitive**
- CuBr (FW 143.4) 7.89 g, 55 mmol — **irritant, hygroscopic**
- Propyne (FW 40.0) 1.2 L — **flammable gas**
- One cylinder of CO_2
- Triethyl phosphite (FW 166.2) 0.83 g, 5 mmol — **irritant, moisture sensitive**
- HMPA (FW 179.2) 43 mL — **highly toxic, cancer suspect agent**
- Dry pentane 250 mL — **flammable, irritant**
- Dry ether 150 mL — **flammable, irritant**

1. Prepare the organo copper reagent as described in Protocol 3 and keep at −25 °C (cryocool).
2. Introduce propyne (1.2 L, 50 mmol) as described in Protocol 2, step 4.

3. Add to the green solution, under stirring at −30°C, $P(OEt)_3$ (0.83 g, 5 mmol) and HMPA (43 mL). Replace the dropping funnel with a septum.
4. Pass a stream of CO_2 (from a cylinder) through a 50 cm column packed with $CaCl_2$ (which is thoroughly purged) and through the cannula into the flask, just above the surface of the stirred solution. After 1 h at −30°C, regulate the CO_2 flow so that a slow bubbling is maintained in the bubble. Warm the brown mixture slowly to room temperature over 15 h.
5. Hydrolyse at −30°C with 5 M HCl (70 mL) and extract with pentane (2 × 50 mL). Wash the organic layers with 5 M HCl (2 × 50 mL) and dry over $MgSO_4$.
6. Remove the solvents *in vacuo* and distil the residue through a 10 cm Vigreux column (7.81 g, 93% yield; b.p. 87–88°C/0.01 mm Hg).

[a] From the corresponding bromide[12] and Mg in ether.

Protocol 10.
Nucleophilic attack on propiolactone: preparation of (*Z*)-4-nonenoic acid[20,a–c]

Caution! Carry out all procedures in a well-ventilated hood, and wear disposable vinyl or latex gloves and chemical-resistant safety goggles.

$Bu_2CuLi \xrightarrow{HC\equiv CH} (Bu\text{–CH=CH})_2CuLi \xrightarrow[2)\ H_3O^+]{1)\ \text{propiolactone}} Bu\text{–CH=CH–CH}_2\text{CH}_2\text{–COOH}$

Equipment

- One three-necked, round-bottomed flask (250 mL) equipped with a variable speed mechanical stirrer, a pressure-equalizing dropping funnel (100 mL) topped with a septum with a needle for gas inlet, a Claisen head bearing a low temperature thermometer (30 to −80°C), and a bubbler
- A cooling bath
- One cannula
- One column (30 cm long, 3 cm in diameter) packed with granulated $CaCl_2$
- Two traps and their cooling baths (Dewars filled with dry ice and acetone)
- One water gasometer filled with brine (see Fig. 11.1)

Materials

- Butyllithium (FW 64.1) 1 M in ether, 50 mL, 50 mmol — **flammable, moisture sensitive**
- $CuBr{\cdot}Me_2S$ (FW 205.5) 5.75 g, 28 mmol — **moisture sensitive**
- Acetylene (FW 26.0) 2.0 L — **flammable gas**
- Me_2S (FW 62.1) 100 mL, 1.36 mol — **flammable, stench**
- Propiolactone (FW 72.0) 2.0 g, 27.5 mmol — **highly toxic, cancer suspect agent**
- Dry ether 150 mL — **flammable, irritant**
- Hexane 100 mL — **flammable, irritant**

Protocol 10. *Continued*

1. Prepare the vinylcuprate reagent as described in Protocol 1, but using a 1 M n-BuLi solution instead of ethyllithium and the $CuBr{\cdot}Me_2S$ complex in place of CuI. Once the vinylcuprate is formed, add Me_2S (100 mL) at −30°C (cryocool).
2. Add propiolactone (2.0 g, 27.5 mmol) in ether (10 mL) at −70°C with stirring, and let the temperature rise to −50°C. Stir for 30 min, then keep stirring at −25°C (cryocool) for 2 h.
3. Add HCl (5 M, 80 mL) and hexane (100 mL). Filter the precipitate and wash the organic layer with HCl (5 M, 50 mL). Neutralize the acid by washing the organic layer twice with an aqueous sodium carbonate solution (CO_2 will be evolved) (50 mL). Acidify the aqueous layer with 5 M HCl and extract it with ether (2 × 30 mL). Dry the ethereal phase over $MgSO_4$.
4. Remove the solvents *in vacuo* and distil the residue through a 10 cm Vigreux column (3.00 g, 69% yield; b.p. 114–115°C/1.6 mm Hg or 99°C/0.6 mm Hg).

[a] Only one vinyl moiety is transferred.
[b] A similar reaction may be performed from the magnesium divinylcuprate in a $THF–Et_2O–Me_2S$ mixture with a cuprate:lactone ratio of 2:1, leading to an 81% yield.[21]
[c] Organolithium and organomagnesium reagents react at the CO function of propiolactone and not in an S_N2 fashion as is the case here.

Protocol 11.
Conjugate addition to an α,β-acetylenic ester[18,a]

Caution! Carry out all procedures in a well-ventilated hood, and wear disposable vinyl or latex gloves and chemical-resistant safety goggles.

$$(C_5H_{11})_2CuLi \xrightarrow{2HC\equiv CH} (C_5H_{11}CH{=}CH)_2CuLi \xrightarrow{2HC\equiv CCOOEt} 2\,C_5H_{11}CH{=}CH{-}CH{=}CH{-}COOEt$$

Equipment

- One three-necked, round-bottomed flask (250 mL) equipped with a variable speed mechanical stirrer, a pressure-equalizing dropping funnel (100 mL) topped with a septum with a needle for gas inlet, a Claisen head bearing a low temperature thermometer (30 to −80°C), and a bubbler
- A cooling bath
- One cannula
- One column (30 cm long, 3 cm in diameter) packed with granulated $CaCl_2$
- Two traps and their cooling baths (Dewars filled with dry ice and acetone)
- One water gasometer filled with brine (see Fig. 11.1)

Materials

• *n*-Pentyllithium (FW 78.0) 1 M in ether, 50 mL, 50 mmol	**flammable, moisture sensitive**
• CuI (FW 190.4) 5.35 g, 28 mmol	**irritant, light sensitive**
• Acetylene (FW 26.0) 2.0 L	**flammable gas**
• Ethyl propiolate (FW 98.0) 4.9 g, 50 mmol	**flammable, lachrymator**
• Dry ether 150 mL	**flammable, irritant**
• Dry hexane 100 mL	**flammable, irritant**

1. Prepare the divinylcuprate as described in Protocol 1, but using *n*-pentyllithium instead of ethyllithium. Cool the solution to −70 °C (cryocool).
2. Add ethyl propiolate (4.9 g, 50 mmol) at −70 °C. Let the mixture warm to −20 °C (cryocool).
3. Hydrolyse at −10 °C with a mixture of saturated NH_4Cl solution (50 mL) and 20% HCl (30 mL). Add hexane (100 mL) and filter the solids. Wash the organic layer with saturated NH_4Cl solution (2 × 50 mL) and dry it over $MgSO_4$.
4. Remove the solvents *in vacuo* and distil the residue at 81–82 °C/0.01 mm Hg to obtain the 'pear ester' (7.65 g, 78% yield).

[a] Note that both vinyl moieties are transferred.

The reaction of vinylcopper derivatives with aryl or vinyl halides can be performed only in the presence of a Pd^0L_n catalyst and leads to styrenes or conjugated dienes. Starting from a lithium divinylcuprate, it is necessary to transmetallate it with one equivalent of zinc halide. Under these conditions, the palladium(0) catalysis will lead to the consumption of one vinyl moiety only; but if magnesium halides are also added, then both vinyl moieties of the cuprate can be transferred. Starting from a vinylcopper–magnesium halide reagent, vinylation (or arylation) under palladium(0) catalysis does not require the addition of zinc salts. This strategy has been used for the synthesis of a large array of pheromones of lepidopterae bearing an (*E*,*Z*)- or (*Z*,*Z*)-conjugated dienic unit.[22–25]

Protocol 12. Vinylation of a lithium divinylcuprate[22]

Caution! Carry out all procedures in a well-ventilated hood, and wear disposable vinyl or latex gloves and chemical-resistant safety goggles.

Et_2CuLi —(HC≡CH)→ $(Et\text{-}CH{=}CH)_2CuLi$ —($ZnBr_2$, 3%$Pd(PPh_3)_4$; I–CH=CH–(CH₂)₄–I)→ Et–CH=CH–CH=CH–(CH₂)₄–I

Protocol 12. *Continued*

Equipment

- One three-necked, round-bottomed flask (250 mL) equipped with a variable speed mechanical stirrer, a pressure-equalizing dropping funnel (100 mL) topped with a septum with a needle for gas inlet, a Claisen head bearing a low temperature thermometer (30 to −80°C), and a bubbler
- A cooling bath
- One cannula
- One column (30 cm long, 3 cm in diameter) packed with granulated $CaCl_2$
- Two traps and their cooling baths (Dewars filled with dry ice and acetone)
- One water gasometer filled with brine (see Fig. 11.1)

Materials

• Ethyllithium (FW 35.9) 1 M in ether, 50 mL, 50 mmol	**flammable, moisture sensitive**
• CuI (FW 190.4) 5.35 g, 28 mmol	**irritant, light sensitive**
• Acetylene (FW 26.0) 2.0 L	**flammable gas**
• $ZnBr_2$ (FW 225.2) 6.16 g, 27.4 mmol	**irritant, hygroscopic**
• $Pd(PPh_3)_4$ (FW 1155.6) 1.15 g, 1 mmol	**light sensitive**
• 1,5-Diiodopent-1-ene[a] (FW 321.9) 6.70 g, 20.8 mmol	**harmful**
• Dry ether 150 mL	**flammable, irritant**
• Dry THF 100 mL	**flammable, irritant**
• Pentane 200 mL	**flammable, irritant**

1. Prepare the lithium divinylcuprate following Protocol 1.
2. Cool the cuprate solution to −40°C (cryocool), then add THF (50 mL) and a solution of zinc bromide (6.16 g, 27.4 mmol) in THF (25 mL). The addition is exothermic, and the green solution turns brown-red.
3. Warm the solution to −20°C and add a solution of 1,5-diiodopent-1-ene (6.70 g, 20.8 mmol) and $Pd(PPh_3)_4$ (1.15 g, 1 mmol) in THF (25 mL). Remove the cooling bath and let the stirred solution gradually reach 15°C, then stir for 30 min at this temperature.
4. Hydrolyse with a saturated NH_4Cl solution (100 mL) at −10°C. Filter the salts on a sintered-glass funnel. Add pentane (200 mL) to the organic layer and wash it with a saturated solution of NH_4Cl (2 × 50 mL). Dry the organic layer over $MgSO_4$.
5. Remove the solvents *in vacuo*. Distil the residue through a 10 cm Vigreux column at 72–73°C/0.05 mm Hg to give the iododiene[b,c] (4.94 g, 95% yield).

[a] Prepared by hydroalumination of 5-chloro-1-pentyne[26] followed by iodination and halogen exchange with NaI in acetone.[27]
[b] One vinyl moiety of the cuprate is used here. Note the chemoselectivity of this substitution (inertness of the primary iodide).
[c] This strategy has been applied to the construction of (*Z,E*)-dienic pheromones such as bombykol.[23]

Acylation of vinylcopper reagents is also substantially improved by palladium catalysis and transmetallation with zinc salts, since the product enone will add the vinylcopper species (eqn 3) if the acetylation step is not speeded up by catalysis.

$$R\text{-CH=CH-'Cu'} + R^1COCl \longrightarrow R\text{-CH=CH-COR}^1 \xrightarrow[2)\ H_2O]{1)\ R\text{-CH=CH-'Cu'}} R\text{-CH=CH-CH(R)-CH}_2\text{-COR}^1 \quad (3)$$

Protocol 13. Palladium-catalysed acylation of a lithium divinylcuprate[28]

Caution! Carry out all procedures in a well-ventilated hood, and wear disposable vinyl or latex gloves and chemical-resistant safety goggles.

$$(C_7H_{15})_2CuLi \xrightarrow{HC\equiv CH} (C_7H_{15}\text{-CH=CH})_2CuLi \xrightarrow[3\%Pd(PPh_3)_4]{CH_3COCl} C_7H_{15}\text{-CH=CH-CO-CH}_3$$

Equipment

- One three-necked, round-bottomed flask (250 mL) equipped with a variable speed mechanical stirrer, a pressure-equalizing dropping funnel (100 mL) topped with a septum with a needle for gas inlet, a Claisen head bearing a low temperature thermometer (30 to −80°C), and a bubbler
- A cooling bath
- One cannula
- One column (30 cm long, 3 cm in diameter) packed with granulated $CaCl_2$
- Two traps and their cooling baths (Dewars filled with dry ice and acetone)
- One water gasometer filled with brine (see Fig. 11.1)

Materials

- *n*-Heptyllithium (FW 106.0) 1 M in ether, 50 mL, 50 mmol — **flammable, moisture sensitive**
- CuI (FW 190.4) 5.35 g, 28 mmol — **irritant, light sensitive**
- Acetylene (FW 26.0) 2.0 L — **flammable gas**
- $ZnBr_2$ (FW 225.2) 6.16 g, 27.4 mmol — **irritant, hygroscopic**
- $Pd(PPh_3)_4$ (FW 1155.6) 0.736 g, 0.64 mmol — **light sensitive**
- MeCOCl (FW 78.5) 1.63 g, 20.8 mmol — **flammable corrosive**
- Dry ether 150 mL — **flammable, irritant**
- Dry THF 100 mL — **flammable, irritant**
- Pentane 100 mL — **flammable, irritant**

1. Prepare the divinylcuprate following Protocol 1, but using a 1 M solution of *n*-heptyllithium instead of ethyllithium.
2. Cool the cuprate solution to −40°C (cryocool), then add THF (50 mL) and a solution of zinc bromide (6.16 g, 27.4 mmol) in THF (25 mL). The addition is exothermic, and the green solution turns brown-red.
3. Warm the solution to −20°C and add, with stirring, a solution of acetyl chloride (1.63 g, 20.8 mmol) and $Pd(PPh_3)_4$ (0.736 g, 0.64 mmol) in THF (15 mL). Let the mixture warm to room temperature over 1 h.
4. Hydrolyse with a saturated NH_4Cl solution (100 mL) at −10°C. Filter the salts on a sintered-glass funnel. Add pentane (200 mL) to the organic layer and

Protocol 13. *Continued*

wash it with a saturated solution of NH_4Cl (2 × 50 mL) Dry the organic layer over $MgSO_4$.

5. Remove the solvents *in vacuo*. Distil the residue through a 10 cm Vigreux column at 49°C/0.01 mm Hg to give the (*Z*)-enone (2.80 g, 80% yield).[a,b]

[a] One vinyl moiety is used. However, it is possible to use both of them if one equivalent of CuBr and two equivalents of $MgBr_2$ are added to one equivalent of the cuprate (after step 1). Then the addition of two equivalents of acid halide leads to a 76% yield of enone (calculated on both vinyl moieties).[28]
[b] These (*Z*)-enones are isomerized very rapidly and quantitatively, by treatment with 0.1 M HCl, into their (*E*)-isomers.

The formation of allenes from the reactions of organocuprates and propargylic acetates, mesylates, etc. results from a nucleophilic attack of the copper atom leading to an intermediate copper(III) species which collapses by reductive elimination in an overall *anti*-process. In the case of propargylic ethers, the reaction of organocopper reagents is a *syn*-carbocupration followed by *anti*-elimination of the copper alkoxide. It is even possible to perform this reaction with a Grignard reagent in the presence of a catalytic amount of a copper salt.

Protocol 14. Copper-catalysed carbometallation[29]

Caution! Carry out all procedures in a well-ventilated hood, and wear disposable vinyl or latex gloves and chemical-resistant safety goggles.

BuMgBr —(5% CuBr. 2PBu$_3$; (R)-3-methoxyhept-1-yne: ≡–C(Bu)(H)(OMe))→ Bu–CH=C=CH–Bu 95%

Optical yield: 90%

Equipment

- A three-necked, round-bottomed flask (100 mL) equipped with a low temperature thermometer (30 to −80°C), a stirring bar, a gas inlet for nitrogen, and a septum
- A round-bottomed flask (20 mL) equipped with a stirring bar, a gas inlet, and a septum
- A cooling bath

Materials

- CuBr (FW 143.5) 0.23 g, 1.6 mmol — **irritant, hygroscopic**
- PBu_3 (FW 202.1) 0.65 g, 3.2 mmol — **flammable, corrosive**
- A solution of BuMgBr (FW 161.3) 1 M in ether, 6.3 mL, 6.3 mmol — **flammable, moisture sensitive**
- (*R*)-3-Methoxyhept-1-yne[a] (FW 126.1) 400 mg, 3.17 mmol — **flammable**
- Dry ether 20 mL — **flammable, irritant**

1. Place 3-methoxyhept-1-yne (400 mg, 3.17 mmol) and ether (20 mL) in the three-necked flask.
2. Prepare separately in the small flask a 1 M solution of $CuBr.2PBu_3$ in ether (1.6 mL) from CuBr (230 mg, 1.6 mmol) and PBu_3 (646 mg, 3.2 mmol). Add 0.16 mL of this solution to the three-necked flask and cool the mixture to −78°C (cryocool).
3. Add the Grignard reagent (6.3 mmol) rapidly. Remove the cooling bath and follow the reaction by GLC (capillary column OV 101, 20 m). The reaction starts between −40 and −30°C.
4. Hydrolyse with a saturated solution of NH_4Cl (four parts) and aqueous NH_3 (one part) (50 mL). Extract the aqueous layer with ether (2 × 50 mL). Wash the organic phase with a mixture of NH_3 and NH_4Cl (3 × 50 mL) and dry it over $MgSO_4$.
5. Remove the solvents *in vacuo* with a cold-water bath. Chromatograph the residue on SiO_2 (60 mesh) using pentane as eluent to give undeca-5,6-diene (458 mg, 94%). Starting with the acetylenic ether of 48% enantiomeric purity, (*R*)-(+), the allene shows an $[\alpha]_D^{25}$ of +30° (c = 2.64, $CHCl_3$), corresponding to the (*S*)-isomer[30] with an optical yield of 90%.[b]

[a] The starting ether is made from the chiral alcohol hept-1-yn-3-ol by a non-racemizing procedure[31] of etherification. This alcohol is prepared by chiral reduction of the acetylenic ketone.[32]

[b] The overall process corresponds to a *syn*-addition–*anti*-elimination sequence. If the reaction is performed with RMgCl and 5% $CuBr{\cdot}2P(OEt)_3$ in ether, the process becomes a *syn*-addition–*syn*-elimination sequence, delivering the (*R*)-enantiomer of the undecadiene with a 60% optical yield. Modifications of the ether moiety improve the optical yield to 91%.[29] Thus, according to the procedure, it is possible to prepare at will both enantiomers of the allene from the same chiral propargylic ether.

References

1. Normant, J.-F.; Bourgain, M. *Tetrahedron Lett.* **1971**, 2583–2586.
2. Normant, J.-F.; Cahiez, G.; Bourgain, M.; Chuit, C.; Villieras, J. *Bull. Soc. Chim. Fr.* **1974**, 1656–1660.
3. Normant, J.-F. *J. Organomet. Chem. Library* **1976**, *1*, 219–256.
4. Normant, J.-F.; Alexakis, A. *Synthesis* **1981**, 841–870.
5. Normant, J.-F. In *Modern Synthetic Methods*; Scheffold, R., ed.; Wiley: New York, **1983**; pp. 139–171.
6. (a) Marfat, A.; McGuirk, P. R.; Kramer, R.; Helquist, P. *J. Am. Chem. Soc.* **1977**, *99*, 253–255. (b) Iyer, R. S.; Helquist, P. *Org. Synth.* **1985**, *64*, 1–9.
7. Knochel, P. In *Comprehensive Organic Synthesis*; Trost, B. M., Fleming, I., eds; Pergamon: Oxford, **1991**; Vol. 4, pp. 865–911.
8. Lipshutz, B. H.; Sengupta, S. *Org. React.* **1992**, *41*, 135–631.
9. Nakamura, E.; Miyashi, Y.; Koga, N.; Morokuma, K. *J. Am. Chem. Soc.* **1992**, *114*, 6686–6692.
10. Watson, S. C.; Eastham, J.-F. *J. Organomet. Chem.* **1967**, *9*, 165–168.

11. Alexakis, A.; Normant, J.-F.; Villieras, J. *J. Organomet. Chem.* **1975**, *96*, 471–485.
12. Biernacki, W.; Gdula, A. *Synthesis*, **1979**, 37–38.
13. (a) Westmijze, H.; Meijer, J.; Vermeer, P. *Recl. Trav. Chim. Pays-Bas* **1977**, *96*, 168–171. (b) Westmijze, H.; Meijer, J.; Vermeer, P. *Recl. Trav. Chim. Pays-Bas* **1977**, *96*, 194–196.
14. Normant, J.-F.; Cahiez, G.; Chuit, C.; Villieras, J. *J. Organomet. Chem.* **1974**, *77*, 269–279.
15. Alexakis, A.; Cahiez, G.; Normant, J.-F. *Org. Synth.* **1984**, *62*, 1–8.
16. Alexakis, A.; Normant, J.-F. *Synthesis* **1985**, 72–74.
17. Commerçon, A.; Normant, J.-F.; Villieras, J. *Tetrahedron* **1980**, *36*, 1215–1221.
18. Alexakis, A.; Cahiez, G.; Normant, J.-F. *Tetrahedron* **1980**, *36*, 1961–1969.
19. Normant, J.-F.; Cahiez, G.; Chuit, C.; Villieras, J. *J. Organomet. Chem.* **1974**, *77*, 281–287.
20. Normant, J.-F. Unpublished results.
21. Sato, T.; Kawara, T.; Sakata, K.; Fujisawa, T. *Bull. Chem. Soc. Jpn.* **1981,** *54*, 505–508.
22. Jabri, N.; Alexakis, A.; Normant, J.-F. *Bull. Soc. Chim. Fr. II* **1983**, 321–331, 332–338.
23. Gardette, M.; Jabri, N.; Alexakis, A.; Normant, J.-F. *Tetrahedron* **1984**, *40*, 2741–2750.
24. Gardette, M.; Alexakis, A.; Normant, J.-F. *J. Chem. Ecol.* **1983**, *9*, 219–223, 225–232.
25. Alexakis, A. *Actual. Chim.* **1987**, 203–210.
26. Zweifel, G.; Whitney, C. C. *J. Am. Chem. Soc.* **1967**, *89*, 2753–2754.
27. Harre, M.; Raddatz, P.; Walenta, R.; Winterfeldt, E. *Angew. Chem., Int. Ed. Engl.* **1982**, *21*, 480–492.
28. Jabri, N.; Alexakis, A.; Normant, J.-F. *Tetrahedron* **1986**, *42*, 1369–1380.
29. Alexakis, A.; Marek, I.; Mangeney, P.; Normant, J.-F. *J. Am. Chem. Soc.* **1990**, *112*, 8042–8047.
30. Pirkle, W. H.; Boeder, C. W. *J. Org. Chem.* **1978**, *43*, 1950–1952.
31. Brown, C. A.; Barton, D. *Synthesis* **1974**, 434–436.
32. (a) Vigneron, J. P.; Bloy, V. *Tetrahedron Lett.* **1979**, 2683–2686. (b) Brinckmeyer, R. S.; Kapoor, V. M. *J. Am. Chem. Soc.* **1977**, *99*, 8339–8341.

12

Synthetic applications of silylcuprates and stannylcuprates

IAN FLEMING

1. Introduction

Silylcopper[1] and stannylcopper[2] reagents and the corresponding cuprates react with many of the same substrates (Scheme 12.1) as their carbon-based counterparts. The products carry on their carbon framework a silyl or stannyl group, with which many transformations can be carried out, replacing these functional groups with carbon groups or with other functional groups. In addition, silylcuprates and stannylcuprates react with unactivated allenes to give vinyl or allyl derivatives depending upon the structure of the allene. In general, silylcopper and stannylcopper reagents and the corresponding cuprates appear to be thermodynamically more stable in THF solution than their carbon-based counterparts, allowing them to be treated with their substrates at temperatures of 0°C and above over several hours if necessary, although it rarely is, because they appear to be simultaneously kinetically more reactive than the carbon-based reagents, reacting with substrates such as allenes and α,β-unsaturated esters that carbon-based cuprates do not react with in the normal course of events.

2. Silylcuprates

Although the most simple reagents, trimethylsilylcopper and lithium bis(trimethylsilyl)cuprate, can be made, the most commonly used reagent is lithium bis(phenyldimethylsilyl)cuprate.[3] The disadvantage of using a phenyldimethylsilyl group is that the silicon-containing by-products, both of the silyl cupration steps illustrated in Scheme 12.1 and of any desilylative reactions carried out on the products, are necessarily relatively involatile. However, the phenyldimethylsilyl group in products like allyl- and vinylsilanes appears to impart very similar reactivity to that imparted by the trimethylsilyl group, and it has an advantage over the trimethylsilyl group in that the presence of the phenyl group allows the phenyldimethylsilyl group to be converted into a hydroxyl group **1** → **6** with retention of configuration at carbon (Scheme 12.2).

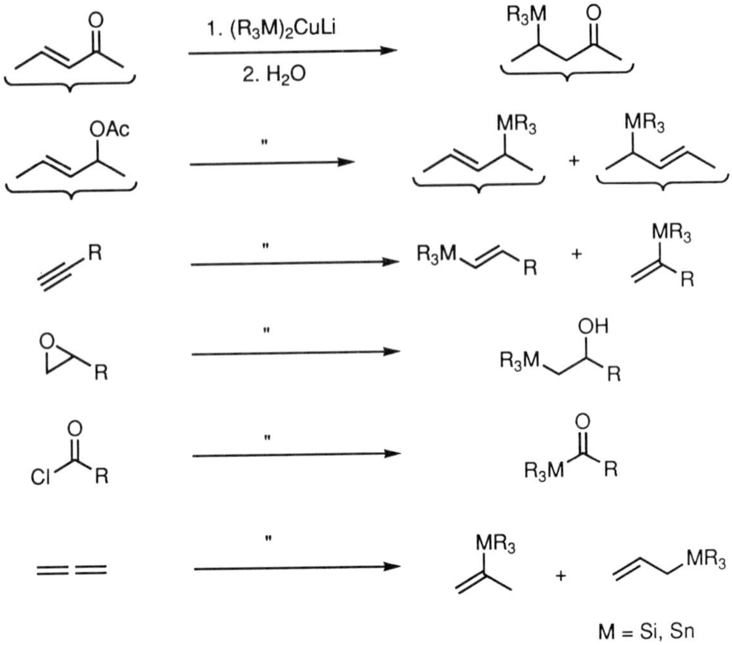

1. $(R_3M)_2CuLi$

2. H_2O

M = Si, Sn

Scheme 12.1

This powerful transformation requires first a reaction with an electrophile, such as a proton,[4] bromine, or the mercury(II) cation,[5] to remove the phenyl ring in an aromatic electrophilic substitution reaction **1** → **2** + **3**, thus placing a nucleofugal group X on the silicon atom. This step is followed by treatment either with peracid or with hydrogen peroxide and a base. The rearrangement step **4** → **5** resembles the well-known reaction of hydroperoxide on a borane, and the final hydrolysis **5** → **6** is unexceptional. If the electrophile is bromine or a mercury(II) cation, the two steps can be combined in one pot and bromine itself does not have to be used, since the peracetic acid oxidizes bromide ion to bromine *in situ*.

This capacity of the phenyldimethylsilyl group cannot be drawn upon, however, when there is a carbon–carbon double bond in the molecule because no matter which electrophile is used it attacks the double bond more rapidly than it removes the phenyl ring from the silyl group. This problem is overcome with the diethylaminodiphenylsilyl group, which can be introduced into organic structures using a cuprate reagent of formula $Et_2NPh_2SiCu(CN)Li$.[6] This silyl group already carries a nucleofugal group, which allows it to be converted directly into a hydroxyl group using alkaline hydrogen peroxide, without the need to remove a phenyl group by aromatic electrophilic substitution.

Two mixed cuprates, (phenyldimethylsilyl)methylcuprate[7] and (*t*-butyl-

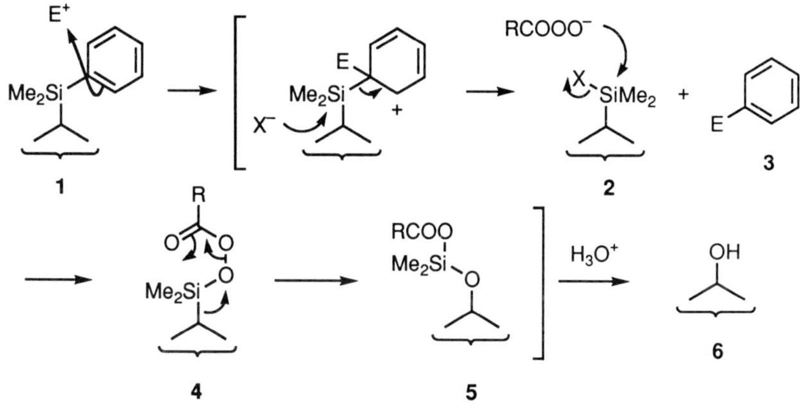

Scheme 12.2

dimethylsilyl)butylcuprate,[8] each containing one silyl and one alkyl group, have some advantages. Only the silyl group is transferred to the substrate from these reagents, and hence only one silyl group is needed, saving the expense of a second. Furthermore, the by-product of the silyl cupration step, methane or butane, is volatile. However, the former reagent is apt not to give quite such good yields in silyl cupration reactions carried out with only 1:1 stoichiometry. The latter reagent introduces a TBDMS group, which is not always suitable for the standard silicon-based chemistry that one might want to use the products for, since a TBDMS group is not as electrofugal as the trimethylsilyl and phenyldimethylsilyl groups.

Other silylcopper reagents and silylcuprates with more specific or limited applications are triphenylsilylcopper,[9] bis(*t*-butyldiphenylsilyl)cuprate,[10] bis[tris(trimethylsilyl)silyl]cuprate,[11] and bis(2-methylbut-2-enyldiphenylsilyl)-cuprate.[12]

2.1 Preparation of silylcopper reagents and silylcuprates

Silylcopper reagents and silylcuprates are usually prepared from the corresponding silyllithium reagent in THF solution simply by mixing it with the appropriate amount of a copper(I) salt. Because phenyldimethylsilyl-lithium is much easier to prepare than trimethylsilyllithium, the most commonly used silylcuprate reagent is derived from this silyl group. A reagent can be prepared using copper(I) iodide, the copper(I) bromide–dimethyl sulfide complex, or copper(I) cyanide. These three reagents appear to be very similar in their reactivity, except for the higher regioselectivity of the cyanide-derived reagent with terminal acetylenes (see Section 2.2.3). The cyanide-derived reagent also has the advantage that CuCN, although even more poisonous than the other copper salts, is easier to make and keep dry.

Protocol 1.
Preparation of the bis(phenyldimethylsilyl)cuprate reagent $(PhMe_2Si)_2Cu(CN)Li_2$[13]

Caution! Carry out all procedures in a well-ventilated hood, and wear disposable vinyl or latex gloves and chemical-resistant safety goggles.

$$2PhMe_2SiCl \xrightarrow{4Li} 2PhMe_2SiLi \xrightarrow{CuCN} (PhMe_2Si)_2Cu(CN)Li_2$$

Equipment

- One three-necked, round-bottomed flask (100 mL) and another three-necked, round-bottomed flask (either 100 mL or 250 mL) with a magnetic stirring bar in each and septa (it is also possible to use one-necked flasks in both cases)
- An argon (or nitrogen) gas supply and inlet
- Gas-tight syringes and a cannula previously oven dried
- An ice–water bath on a magnetic stirrer

Materials

- Chlorodimethylphenylsilane[a] (FW 170.7) 10 mL, 10.32 g, 60 mmoL — **corrosive moisture sensitive**
- Lithium shot (FW 6.9) 1 g, 143 mmol — **flammable, moisture sensitive**
- Anhydrous copper(I) cyanide (FW 89.6) 2 g, 22.3 mmol — **highly toxic, irritant**
- Dry, distilled THF 40 mL + 5 mL to form a CuCN slurry — **flammable, irritant**
- Dry hexane for washing lithium shot — **flammable, irritant**

1. Flame dry the 100 mL flask and purge it with the argon while allowing it to cool to room temperature. Add the stirrer bar, the lithium shot, and hexane (*c.* 10 mL), and stopper the outlets with septa. Stir for a few minutes, then remove the hexane by syringe. Repeat the washing procedure until the hexane is no longer discoloured. Flush the flask with argon to remove the last traces of hexane and add the THF and silyl chloride.
2. Stir steadily, cooling with an ice–water bath or in a cold room, for 12 h, when there will be a deep-red solution of the silyllithium reagent. Typically the colour appears at first transiently over the lithium, it becomes permanent after about 15 min, and increases in intensity after that, the solution becoming essentially opaque at 0.5 M. Longer stirring over two to five days does slowly increase the titre for silyllithium, but is rarely worth the effort. This reagent will keep for days at 0°C, but it is best to use it immediately. It is often convenient to make the silyllithium reagent over 5–6 h at 0°C, then store it in a fridge or freezer overnight and use it the next day.
3. Optionally, carry out a double titration of a sample of this solution as follows. Add aliquots (1 mL each) to water and to an excess of allyl bromide or 1,2-dibromoethane. Add water to the allyl bromide mixture and shake, and then titrate both mixtures against 0.1 M hydrochloric acid solution using phenolphthalein as an indicator. The difference in the two results measures the

molarity of the silyllithium solution, which is typically about 1.2 M (see Chapter 2, Protocol 5 for a detailed titration procedure).

4. Flame dry the second flask, and purge it with argon while allowing it to cool to room temperature. A 250 mL flask is sometimes needed to make room for the addition of the substrate in the next step of the synthesis (see, for example, Protocol 5). On other occasions, a 100 mL flask is big enough. Add the copper cyanide (typically 22.5 mmol) and gently flame dry under a stream of argon. Add the stirrer bar, purge with argon, and fit the septa, including one with a needle attached to a partially inflated balloon of argon. Add THF (*c.* 5 mL) to make a slurry of the copper cyanide. Immerse the flask in an ice–water bath and insert a cannula leading from the first flask into the second. Pump over the silyllithium solution (typically 45 mmol) onto the copper cyanide slurry using argon pressure, stirring vigorously. Optionally, add THF (2 mL) to the flask that contained the silyllithium solution and pump it through the cannula. Continue stirring for a further 20 min until the cyanide has all dissolved, before cooling it down to whatever temperature is needed for the next step. The mixture at this stage is a dark, reddish-brown. It is not normally necessary to standardize it. It is best used immediately, of course, but it is thermodynamically more stable than carbon-based cuprates, surviving at 0 °C for several hours, although with steady decomposition over this period. Many successful runs using this reagent have been carried out simply by injecting the silyllithium solution with a syringe instead of the cannula.

[a] From Aldrich, or prepared by the reaction of phenylmagnesium bromide on dichlorodimethylsilane.[14] Note that this compound, whether bought or made, is actually a mixture of the chloride and bromide in ratios ranging from 10:1 to 4:1. It is rarely worth allowing for this in calculating quantities needed. The presence of bromide ion neither assists nor hinders the reactions of the silylcuprate reagent.

The corresponding trimethylsilylcopper reagent and bis(trimethylsilyl) cuprate are prepared similarly, but the trimethylsilyllithium that is needed usually has to be prepared by treating hexamethyldisilane in neat HMPA with methyllithium at 0°C for a few minutes.[15] Fortunately, the HMPA, although an undesirable reagent, does not appear seriously to interfere with any of the reactions of these silylcopper reagents, but unfortunately DMPU does not appear to be a substitute for it in the preparation of trimethylsilyllithium. Trimethylsilyllithium can also be prepared from bis(trimethylsilyl)mercury,[16] but this is no more agreeable from the safety point of view, and is a great deal more cumbersome.

An alternative and cleaner preparation of the phenyldimethylsilyllithium, which may occasionally be useful, is to cleave diphenyltetramethyldisilane with small pieces of lithium over 12 h or, even faster (2 h), with lithium powder, in both cases assisted by ultrasound activation. This gives a deep-

Table 12.1. NMR characterization of silicon- and copper-containing reagents in THF[17] (chemical shifts in p.p.m.)

Reagent	^{29}**Si** ^{a}	^{13}**C** ^{a} **(SiMe)**	^{13}**C** ^{a} **(CuMe)**	^{13}**C** ^{a} **(*ipso*-SiPh)**	^{7}**Li** ^{b}	^{1}**H** ^{a} **(SiMe)**
$PhMe_2SiLi$	−28.5	7.5		166.0	−1.69	0.14
$PhMe_2SiCu(CN)Li$	−25.5	6.3		150.0		0.29
$(PhMe_2Si)_2Cu(CN)Li_2$	−24.4	5.1		157.4	−3.33	0.09
$(PhMe_2Si)_2CuLi$		6.0		156.0		
$(PhMe_2Si)_3CuLi_2$	−18.9	8.1				0.09
$PhMe_2Si(Me)Cu(CN)Li_2$	−20.6	6.4	−5.0			0.04

^{a} Reference signal Me_4Si.
^{b} Reference signal LiCl in CD_3OD in a capillary insert.

greenish silyllithium solution, which is free of halide ions. It was used to prepare some of the samples for NMR characterization of the various copper-containing species,[17] which showed that the 1:1 silicon-to-copper reagent, the 2:1 reagent, and the 3:1 reagent were distinct species, and that the cyanide-derived cuprate was distinct from the iodide-derived reagent (Table 12.1). Cuprates prepared from the disilane react in the usual way with the usual substrates. One disadvantage of this method for routine use is that finely divided lithium can interfere with the smooth use of a cannula, and any residual suspended lithium interferes with the titration.

The diethylaminodiphenylsilylcopper reagent has a special place because of its capacity to introduce a silyl group that can immediately be converted into a hydroxyl group. The reagent works best when prepared with a 1:1 silicon-to-copper stoichiometry.

Protocol 2.
Preparation of the diethylaminodiphenylsilylcopper reagent $Et_2NPh_2SiCu(CN)Li$[18]

Caution! Carry out all procedures in a well-ventilated hood, and wear disposable vinyl or latex gloves and chemical-resistant safety goggles.

$$Ph_2SiCl_2 \xrightarrow[Et_3N]{Et_2NH} Et_2NPh_2SiCl \xrightarrow{2Li} Et_2NPh_2SiLi \xrightarrow{CuCN} Et_2NPh_2SiCu(CN)Li$$

Equipment

- One two-necked, round-bottomed flask (300 mL) with a magnetic stirring bar, a stoppered, pressure-equalizing dropping funnel (50 mL), and a three-way stopcock attached to an N_2 supply and a pump
- One two-necked, round-bottomed flask (50 mL) fitted with a pressure-equalizing dropping funnel (10 mL), a magnetic stirring bar, and a three-way stopcock attached to an argon supply and a pump

- One two-necked, round-bottomed flask (100 mL) fitted with a three-way stopcock, a magnetic stirring bar, and a septum
- Nitrogen and argon gas supplies
- A rotary evaporator
- A gas-tight syringe (25 mL)
- An ice–water bath on a magnetic stirrer

Materials

- Triethylamine[a] (FW 101.2) 15.3 mL, 11.1 g, 110 mmol — **flammable, corrosive**
- Dichlorodiphenylsilane[b] (FW 253.2) 21.0 mL, 25.3 g, 100 mmol — **corrosive, moisture sensitive**
- Diethylamine[a] (FW 73.1) 11.4 mL, 8.04 g, 110 mmol — **flammable, corrosive**
- Dry, distilled THF approximately 200 mL in total — **flammable, irritant**
- Hexane 100 mL — **flammable, irritant**
- Dry toluene 6 mL — **flammable, irritant**
- Lithium pieces or wire (FW 6.9) 397 mg, 57.2 mmol — **flammable, moisture sensitive**
- Anhydrous copper(I) cyanide (FW 89.6) 1.32 g, 14.7 mmol — **highly toxic, irritant**

1. Flame dry the 300 mL flask fitted with accessories. Cool it to room temperature under nitrogen and add the stirrer bar. Evacuate the flask and fill it with nitrogen. Add successively the THF (150 mL), the triethylamine, and the dichlorodiphenylsilane.
2. Add the diethylamine dissolved in THF (10 mL) from the dropping funnel at room temperature over about 30 min, during which time a copious white precipitate of triethylammonium chloride will form. Continue to stir for a further 6 h. The larger the scale of this operation the harder it is to stir magnetically—use a mechanical stirrer with a motor if necessary.
3. Dilute the mixture with the hexane, and filter it through a Büchner funnel, washing the filter cake with a little more hexane. Concentrate the combined filtrates on the rotary evaporator, and distil the residue under reduced pressure to get diethylaminodiphenylsilyl chloride (24.9 g, 86%) as a pale-yellow oil, b.p. 129–133°C/0.55 mm Hg. This product is moisture sensitive, but otherwise quite stable under nitrogen over several months.
4. Flame dry and purge the second flask (50 mL) similarly, but using argon, and place the lithium and some THF (7 mL) in it. Add diethylaminodiphenylsilyl chloride (4.24 g, 14.6 mmol) from the dropping funnel with stirring at room temperature over about 5 min. An exothermic reaction starts, whereupon add more THF (7 mL). After about 20 min the solution turns blue. Cool it to 0°C and stir for 4 h, when it becomes dark green. The solution appears to be stable at 0°C over six days and can be used in the next step without titration, assuming that the reaction is essentially quantitative.
5. Flame dry the third flask (100 mL) as before and place the copper cyanide in it. Add half of the toluene and evaporate it under reduced pressure on a rotary evaporator using gentle heat to dry the copper cyanide. Repeat this operation, and then fill the flask with nitrogen and attach the funnel. Purge the flask with nitrogen and add THF (7 mL) from the funnel. Cool the flask and its contents to 0°C.

Protocol 2. *Continued*

6. Add the silyllithium solution prepared in step 4 above by syringe over a few minutes with stirring. Wash the residue of lithium metal left in the second flask with more THF (7 mL) and add this to the third flask, then stir the combined mixture for 1 h. Cool it down to whatever temperature is needed for the next stage (−20°C for Protocol 4) and use it immediately.

[a] Distilled from CaH_2.
[b] Freshly distilled.

The preparation of the mixed cuprate $PhMe_2SiCu(Me)(CN)Li_2$ is carried out in the same way as the synthesis of the bis(phenyldimethylsilyl)cuprate (Protocol 1), except that only one equivalent of the silyllithium is added to the copper cyanide, and this is followed by adding one equivalent of methyllithium. *n*-Butyllithium can also be used to make the corresponding mixed cuprate $PhMe_2SiCu(n\text{-}Bu)(CN)Li_2$, which is an equally effective reagent. The other mixed cuprate, $t\text{-}BuMe_2SiCu(n\text{-}Bu)(CN)Li_2$, is made differently by the action of dibutylcyanocuprate on the mixed silylstannane $t\text{-}BuMe_2SiSnMe_3$.[8] This method for making silylcuprates is restricted, however, to those with hindered silyl groups like *t*-butyldimethylsilyl and thexyldimethylsilyl. If the silyl group is smaller, the reaction gives the corresponding stannylcuprate (see Section 3.1).

Protocol 3.
Preparation of the *t*-butyldimethylsilyl(*n*-butyl)cuprate reagent $t\text{-}BuMe_2SiCu(n\text{-}Bu)(CN)Li_2$[8]

Caution! Carry out all procedures in a well-ventilated hood, and wear disposable vinyl or latex gloves and chemical-resistant safety goggles.

$$t\text{-}BuMe_2SiSnMe_3 + n\text{-}Bu_2Cu(CN)Li_2 \rightarrow t\text{-}BuMe_2SiCu(n\text{-}Bu)(CN)Li_2 + n\text{-}BuSnMe_3$$

Equipment

- One two-necked, round-bottomed flask (10 mL) with a magnetic stirring bar, a septum, and a three-way stopcock
- An argon gas supply and a pump fitted to the stopcock
- Gas-tight syringes
- A dry ice–acetone bath and an ice–water bath, each on a magnetic stirrer

Materials

- *t*-Butyldimethylsilyl(trimethyl)stannane[a] (FW 279.1) 0.11 mL, 154 mg, 0.55 mmol — **harmful, moisture sensitive**
- *n*-Butyllithium (FW 64.1) 1.6 M in hexane, 0.66 mL, 1.05 mmol — **flammable, moisture sensitive**
- Anhydrous copper(I) cyanide (FW 89.6) 0.045 g, 0.5 mmol — **highly toxic, irritant**
- Dry, distilled THF 1.5 mL — **flammable, irritant**

1. Flame dry the flask, cool it under nitrogen, and add the copper cyanide. Attach the septum and the stopcock, flame dry the flask again gently under vacuum, and refill with argon while allowing it to cool to room temperature. Repeat this operation three times. Add the stirrer bar, and purge again with argon. Add the THF by syringe, and cool the flask to −78°C in a dry ice–acetone bath.
2. Add the butyllithium solution by syringe, stirring steadily. Transfer the flask to an ice–water bath and stir for 3–5 min until a homogeneous, pale-yellow solution is obtained. It is important to make sure that all the copper cyanide dissolves.
3. Put the flask back in the dry ice–acetone bath, and add the neat silylstannane by syringe. Transfer the flask again to the ice–water bath and stir for 1 h to complete the transmetallation. Finally, put the flask containing the homogeneous, yellow solution of the cuprate back in the dry ice–acetone bath, ready for use.

[a] Prepared from trimethyltin hydride by treating it successively with LDA and *t*-butyldimethylsilyl chloride in THF at 0°C for 3 h.[19]

2.2 Reactions of silylcuprates

2.2.1 With enone systems

Unlike alkyllithiums, trimethylsilyllithium adds to α,β-unsaturated ketones like cyclohexenone at the β-position, even at −78°C under kinetic control,[13] so that it is not usually necessary to use a silylcuprate with these substrates. However, hindered α,β-unsaturated ketones like isophorone do not react with a silyllithium reagent but do with the corresponding cuprate (Scheme 12.3).[3] Similarly, α,β-unsaturated aldehydes, esters, amides, and nitriles do not undergo conjugate attack with silyllithium reagents, and the cuprates are needed for these substrates too, as illustrated for the cinnamate ester **7** in Scheme 12.4.[20] Although yields are usually high, the main cause of lower yields in these reactions is that the intermediate enolate created by the conjugate addition is sometimes as effective a nucleophile as the silylcuprate itself, leading to oligomerization of the unsaturated ester. This can largely be avoided by using, in place of the cuprate, the corresponding silyldimethylzincate, $PhMe_2SiZnMe_2Li$, for which the yields in the conjugate addition step are regularly >90%.[21] The enolate intermediates in the cuprate reactions, which are (*E*)-isomers **8**, can be used in alkylation **8** → **9**[20] and aldol **8** → **10** reactions,[22] which usually take place with high levels of stereocontrol, both in cyclic and, more remarkably, in the open-chain systems illustrated. Protonation of the enolate **8**, giving the simple product **11** of conjugate addition, and regeneration of an enolate with LDA gives the (*Z*)-isomer **12**, which also undergoes aldol reactions **12** → **13**, but to give a different stereoisomer with respect to the aldol geometry.

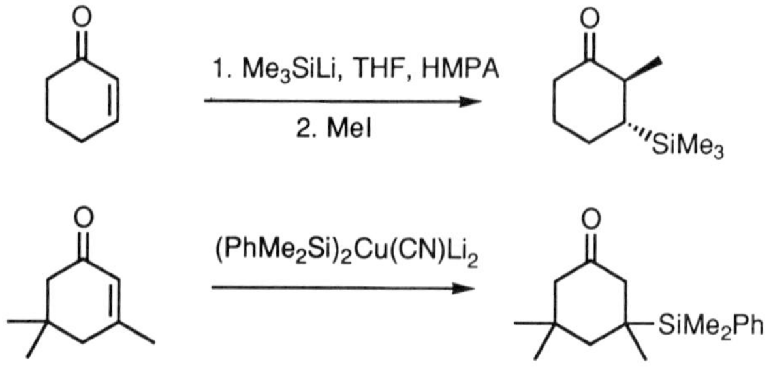

Scheme 12.3

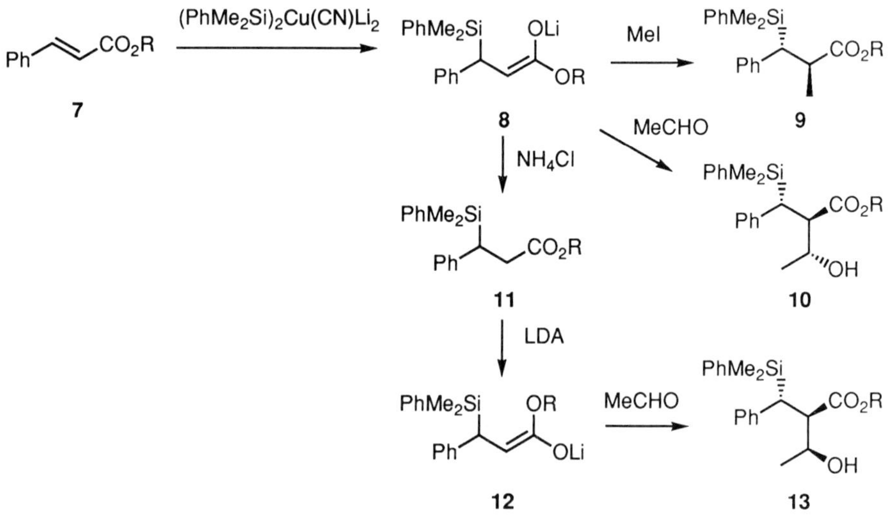

Scheme 12.4

When the enone system is attached to a suitable chiral auxiliary like Koga's triphenylmethyloxymethylpyrrolidone or Oppolzer's camphorsultam, the silylcuprate can add with high levels of diastereoselectivity[23,24] to create a stereogenic centre carrying a silyl group. Homochiral versions of β-silyl esters like **11** can be made in this way by removing the chiral auxiliary with a magnesium alkoxide.[25]

The β-silyl carbonyl compounds produced in all these reactions can be used in four ways for further synthetic transformations (Scheme 12.5): (1) the double bond can be restored by bromination followed by desilylation–debromination **14** → **15**;[3] (2) the silyl group can control the regiochemistry of a Baeyer–Villiger reaction **16** → **17** → **18**, leading to either of the unsaturated acids **20** from the β-hydroxysilane **19**;[26] (3) the silyl group can be converted, as described in Section 2, to the corresponding hydroxy compound **9** → **20**;[4,5] and (4) the aldol products such as **13** (R = allyl) can be converted into allylsilanes **22** of either double-bond geometry by decarboxylative elimination from the hydroxy acid **21**.[27]

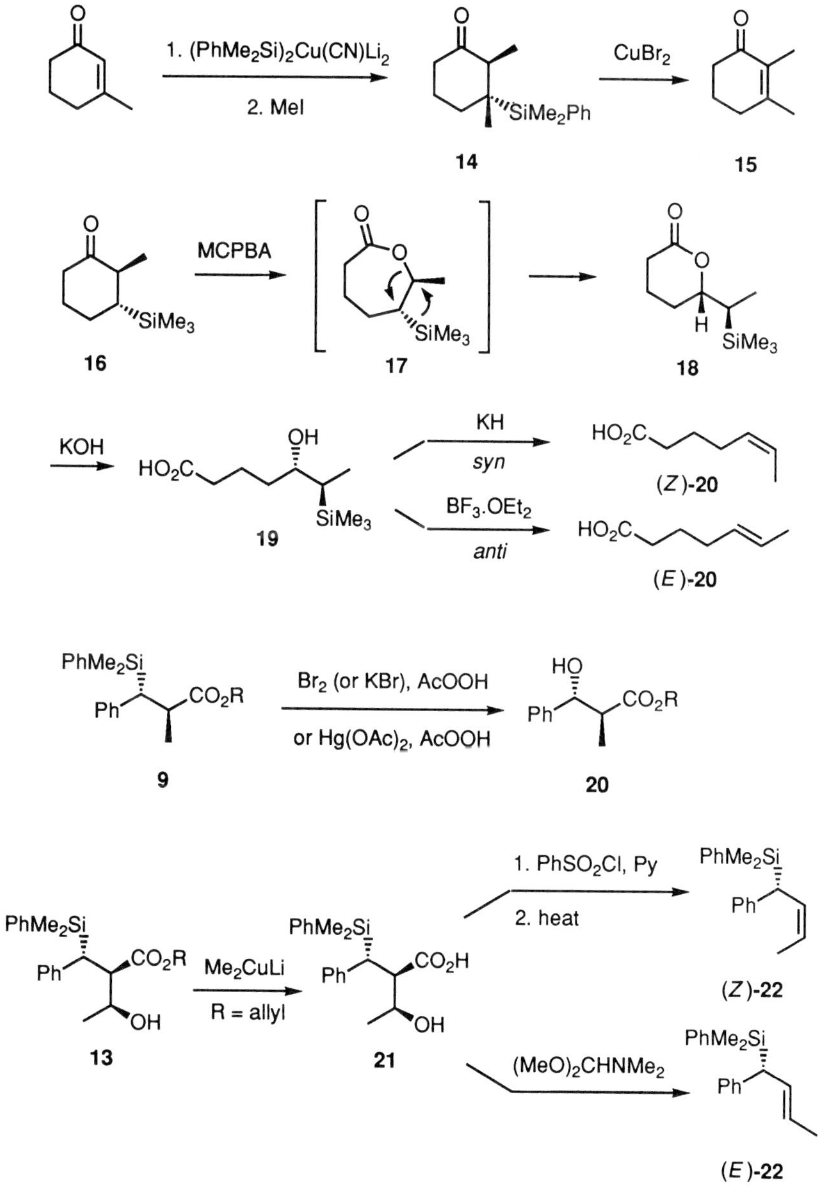

Scheme 12.5

2.2.2 With allylic acetates and related systems

Silylcuprates react with allylic acetates to give allylsilanes directly (Scheme 12.6). Allylic acetates that are secondary at both ends are apt to give both regioisomers, but some control of the regiochemistry is possible using a *cis*-double bond, which encourages reaction with allylic shift **23** → **24** + **25**, with **24** as the major product, especially when the silyl group is delivered to the less-hindered end of the allylic system as here. An alternative protocol, assembling a mixed silylcuprate on a carbamate group, is even better in controlling the regioselectivity, usually giving complete allylic shift **26** → **24**.[28] The acetate reaction **23** → **24** and the carbamate alternative **26** → **24** are complementary in their stereochemistry, the former taking place stereospecifically *anti* and the latter stereospecifically *syn*.[28,29] Tertiary acetates, on the other hand, are very well behaved regiochemically, giving only the product with the silyl group at the less-substituted end of the allylic system **27** → **28**.[6,30] Allylsilanes have many uses as carbon nucleophiles in organic synthesis.[31] In addition, the silyl group, provided that it is the diethylaminodiphenylsilyl group, can be converted into a hydroxyl group **28** → **29** even in the presence of a trisubstituted double bond, thus allowing overall the allylic inversion **27** → **29**.[6]

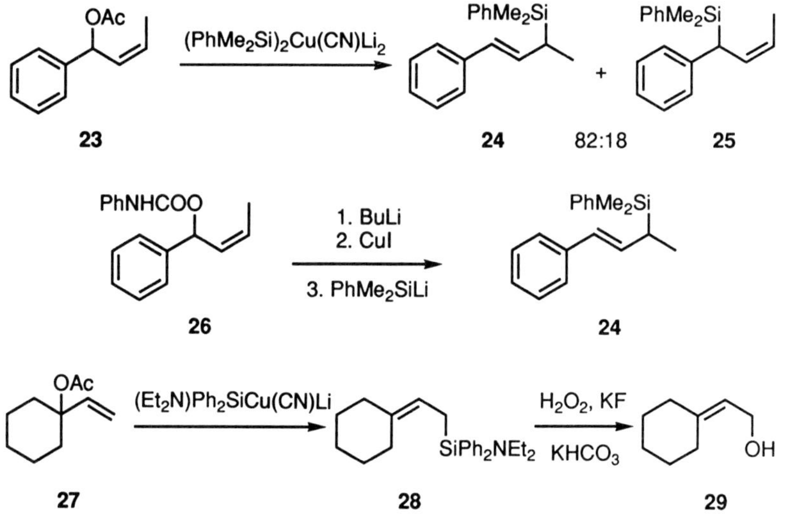

OAc; $(PhMe_2Si)_2Cu(CN)Li_2$; $PhMe_2Si$; $PhMe_2Si$; +

23 **24** 82:18 **25**

PhNHCOO; 1. BuLi 2. CuI 3. $PhMe_2SiLi$; $PhMe_2Si$

26 **24**

OAc; $(Et_2N)Ph_2SiCu(CN)Li$; $SiPh_2NEt_2$; H_2O_2, KF $KHCO_3$; OH

27 **28** **29**

Scheme 12.6

Protocol 4.
Addition of the diethylaminodiphenylsilylcopper reagent $Et_2NPh_2SiCu(CN)Li$ to the allylic acetate 27 and the oxidation of the allylsilane 28 to give the allylic alcohol 29[18]

Caution! Carry out all procedures in a well-ventilated hood, and wear disposable vinyl or latex gloves and chemical-resistant safety goggles.

OAc
$(Et_2N)Ph_2SiCu(CN)Li$
$SiPh_2NEt_2$
H_2O_2, KF
$KHCO_3$
OH

27 **28** **29**

Equipment

- One two-necked, round-bottomed flask (100 mL) fitted with a dropping funnel, a septum, and a thermometer, and containing a stirrer bar and the diethylaminodiphenylsilylcopper reagent prepared as described in Protocol 2 (14.7 mmol)
- An argon gas supply
- A round-bottomed flask (100 mL) containing a stirrer bar and fitted with a septum
- A separating funnel (250 mL)
- Three conical flasks (250 mL)
- A 1.5 × 50 cm chromatography column and a supply of conical flasks
- A dry ice–acetone bath on a magnetic stirrer

Materials

- 1-Vinylcyclohexyl acetate[32] (FW 168.2) 1.68 g, 10.1 mmol — **harmful**
- Ammonium chloride solution 200 mL, 5% — **corrosive**
- Ether for extraction 250 mL — **flammable, irritant**
- Brine 20 mL
- THF 15 mL — **flammable, irritant**
- Methanol 15 mL — **flammable, toxic**
- Hydrogen peroxide 7.4 mL, 30% — **oxidizer, corrosive**
- Potassium fluoride 6.1 g — **corrosive, toxic**
- Potassium hydrogencarbonate 5.89 g
- Sodium hydroxide solution 100 mL, 10% — **corrosive**
- Anhydrous magnesium sulfate 10 g
- Silica gel for chromatography (Merck Kieselgel 60, 70–230 mesh) 35 g — **irritant dust**
- A mixture of hexane and ethyl acetate (7:1) for chromatography — **flammable, irritant**

1. Cool the flask containing the silylcopper reagent to −20°C using the dry ice–acetone bath with enough dry ice to get this temperature, and stir the contents under argon. Add the allylic acetate dropwise over 3 min from the funnel, washing in the last of the ester with a little dry THF.
2. Stir the mixture at −20°C for 5 min. Remove the cooling bath and replace it with an ice–water bath, then stir for another hour, checking for the disappearance of the acetate by TLC.

Protocol 4. *Continued*

3. When the reaction is complete, add some of the ammonium chloride solution (30 mL) through the dropping funnel. Filter the mixture and extract the filtrate with ether (3 × 30 mL). Wash the organic layers with several 20 mL lots of ammonium chloride solution until the blue colour is completely removed and the aqueous layers are quite colourless. This is important in order to prevent the copper-catalysed decomposition of hydrogen peroxide in the next step. Wash the organic layer finally with brine, and dry the combined ether layers (Na_2SO_4). Filter and concentrate the solution in the 100 mL round-bottomed flask on a rotary evaporator to give a yellow oil.
4. Add a stirrer bar to the flask, then THF (15 mL), methanol (15 mL), potassium fluoride, and potassium hydrogencarbonate, and set it up for stirring with a thermometer in the liquid. Add the hydrogen peroxide to this mixture over 10 min from a pipette, watching the temperature. The mixture separates into two layers: a cloudy organic layer and a heavy, milky-white inorganic layer. There is a small amount of heat produced which raises the temperature to about 40 °C in about 30 min, after which it subsides. If the temperature threatens to rise above 40 °C, cool the flask in water to keep it just under 40 °C. Cool the mixture to room temperature and stir for 6 h, then raise the temperature to 35–40 °C for another 6 h. Cool the mixture to room temperature and pour it into water (50 mL). Decant the liquid off the inorganic solid and extract successively the solid and the aqueous layers with ether (5 × 30 mL). Combine the organic layers and wash successively with sodium hydroxide solution (3 × 30 mL) to remove phenol and water (50 mL). Dry the organic layer ($MgSO_4$), filter it, and concentrate the filtrate on a rotary evaporator.
5. Add the residue to the top of a chromatography column packed with silica gel (35 g) and elute with the mixture of hexane and ethyl acetate, evaporating the fractions containing the alcohol [R_f (hexane–EtOAc, 7:1) = 0.18]. The yield should be about 0.8 g (63%). The product has been characterized by ^{1}H and ^{13}C NMR spectroscopy.

2.2.3 With acetylenes

Silylcuprates react with acetylenes by *syn*-stereospecific metallometallation (Scheme 12.7).[33] Provided that the cuprate is derived from copper cyanide, the regioselectivity is highly in favour of the isomer **30** with the silyl group on the terminus of a terminal acetylene, in contrast both to carbon-based and stannylcuprates reacting with terminal acetylenes. The intermediate vinylcuprate **30** reacts with many substrates, familiar in carbon-based cuprate chemistry, to give overall *syn*-addition of a silyl group and an electrophile to the acetylene. A curious feature of this reaction is that the intermediate, although uncharacterized, has the stoichiometry of a mixed silicon–carbon cuprate, and yet it transfers to most substrates the carbon-based group, in contrast to the behaviour of mixed silylalkylcuprates.

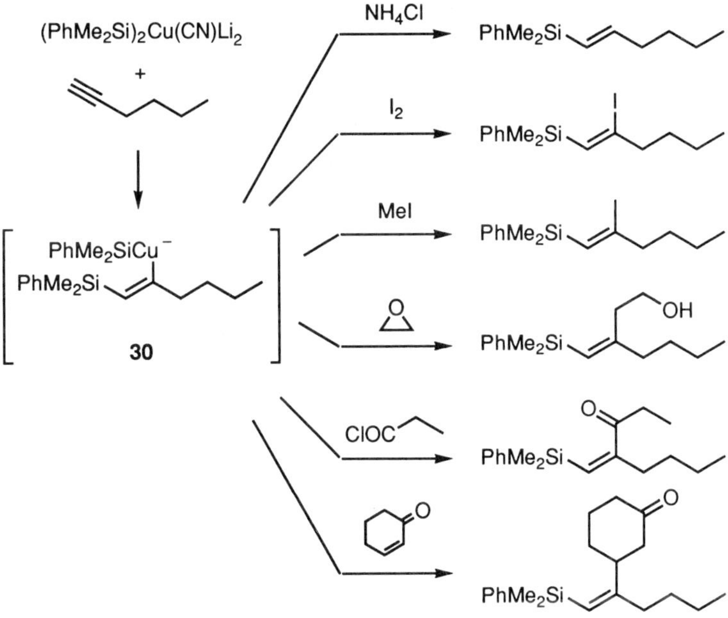

Scheme 12.7

Disubstituted acetylenes also react, and, if the two substituents are well differentiated sterically, as with a methyl group on one side and a branched chain on the other, the regioselectivity is highly in favour of the silyl group appearing at the less-hindered end.

The vinylsilanes prepared in this way have many uses as carbon nucleophiles in organic synthesis.[31] One particularly easy transformation is the desilylation with bromine **31** → **32**, which was used to provide a long-chain *cis*-vinylcuprate needed in a synthesis of tetrahydrolipstatin.[34] Desilylation with bromine is known to take place cleanly with inversion of configuration.[35]

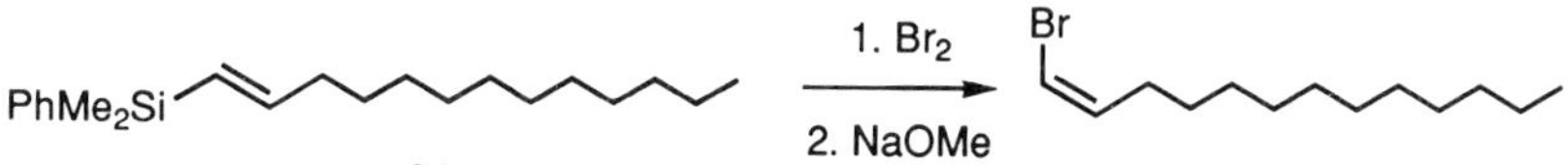

2.2.4 With allenes

Although silylcuprates react with propargylic acetates in the same way as they react with allylic acetates, the product allenylsilanes are themselves susceptible to attack by the silylcuprate reagent,[36] as are allenes in general.[37] Allene itself reacts with silylcuprates at low temperatures, in contrast to the reactions of carbon-based cuprates, where allenes must have either a coordinating or an electron-withdrawing group attached to them for any visible reaction to take place. With allene itself the regiochemistry overall places the silyl group on the central carbon atom and the added electrophile at the terminus (Scheme 12.8), suggesting that the intermediate is an allylcuprate **33**. One surprising exception to this rule is with iodine as the electrophile, when the product **34** has the opposite regiochemistry even from that of the reaction with chlorine, suggesting possibly that the regioisomeric intermediates are in equilibrium with each other. Since the iodide **34** can be converted into a lithium reagent **35** (and a cuprate that is not identical with **33**) and electrophiles added to it (Scheme 12.9), it is possible overall to achieve either regiochemistry in additions to allene. Monosubstituted allenes give mixtures of regioisomers,[37,38] and disubstituted and trisubstituted allenes give largely allylsilanes whatever electrophile is used. The metallometallation step is stereospecifically *syn*.[39] The *t*-butyldiphenylsilylcuprate shows a different regioselectivity in its kinetically controlled reactions with allene, giving allylsilane products with all electrophiles, with a selection shown in Scheme 12.10.[40] However, allowing the intermediate cuprate to warm to 0°C and then cooling it back down to −78°C before adding the electrophiles gives the regioisomers of the products shown in Scheme 12.10, indicating that the cuprate in Scheme 12.10 is the kinetic product, but that its isomer, the analogue of **33**, is the thermodynamic

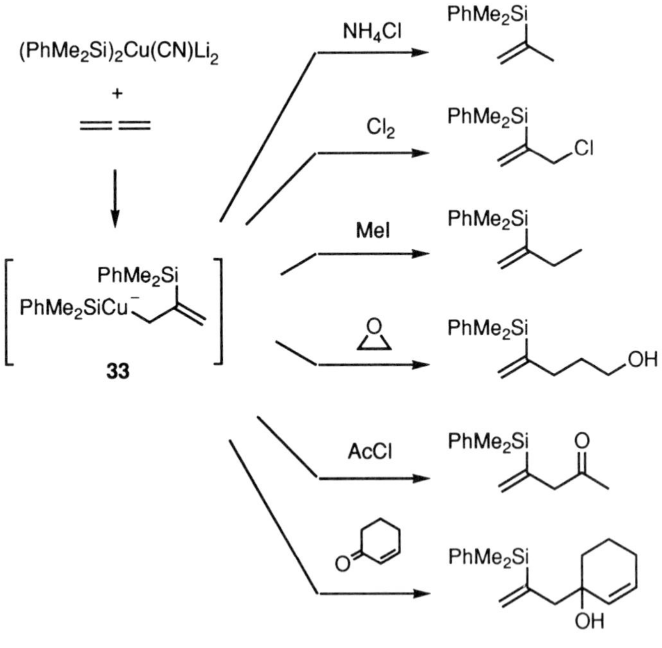

Scheme 12.8

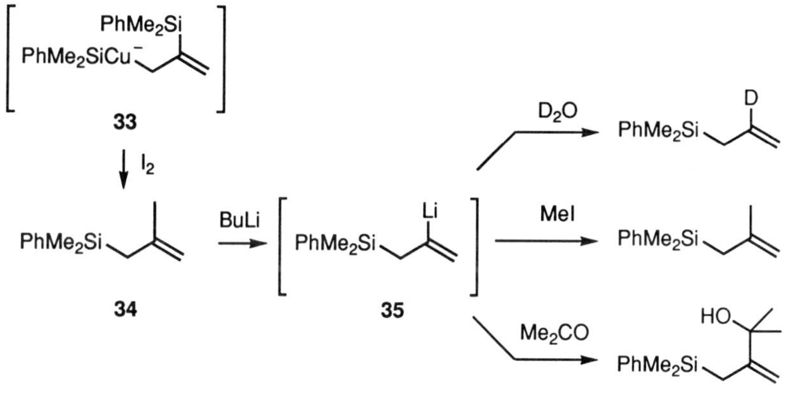

Scheme 12.9

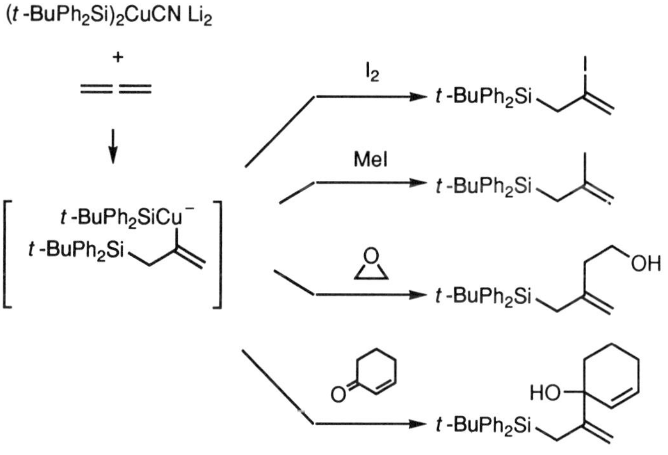

Scheme 12.10

product. These reactions are the only ones in which the substituents on the silyl group make a significant difference to the products obtained from a silylcuprate.

The products of all these reactions are functionalized (or not) allylsilanes and vinylsilanes, with many uses in organic synthesis.[31]

2.2.5 With other substrates

Other substrates that have been found to react unsurprisingly with silylcopper reagents and with silylcuprates (see Scheme 12.11) are allyl chlorides,[41] an alkyl bromide,[42] epoxides such as **38** (in which, using a mixed cuprate, there is unusually some transfer of the carbon ligand),[8,38] acid chlorides **36**,[9,43] a vinyl iodide **37**,[8] an aminomethyl acetate **39**,[44] ethyl tetrolate,[45] and a vinylsulfoxide (with some diastereocontrol).[46] More surprisingly, silylcuprates react with the allylic alcohol **40**,[47] which is used in a synthesis of the 2-stannylbutadiene **41**, and with a group of strained allylic ethers[48] such as **42** and **43**, among which the ketone group in the latter traps the silylcupration intermediate **44** before the nucleofugal group has time to depart.

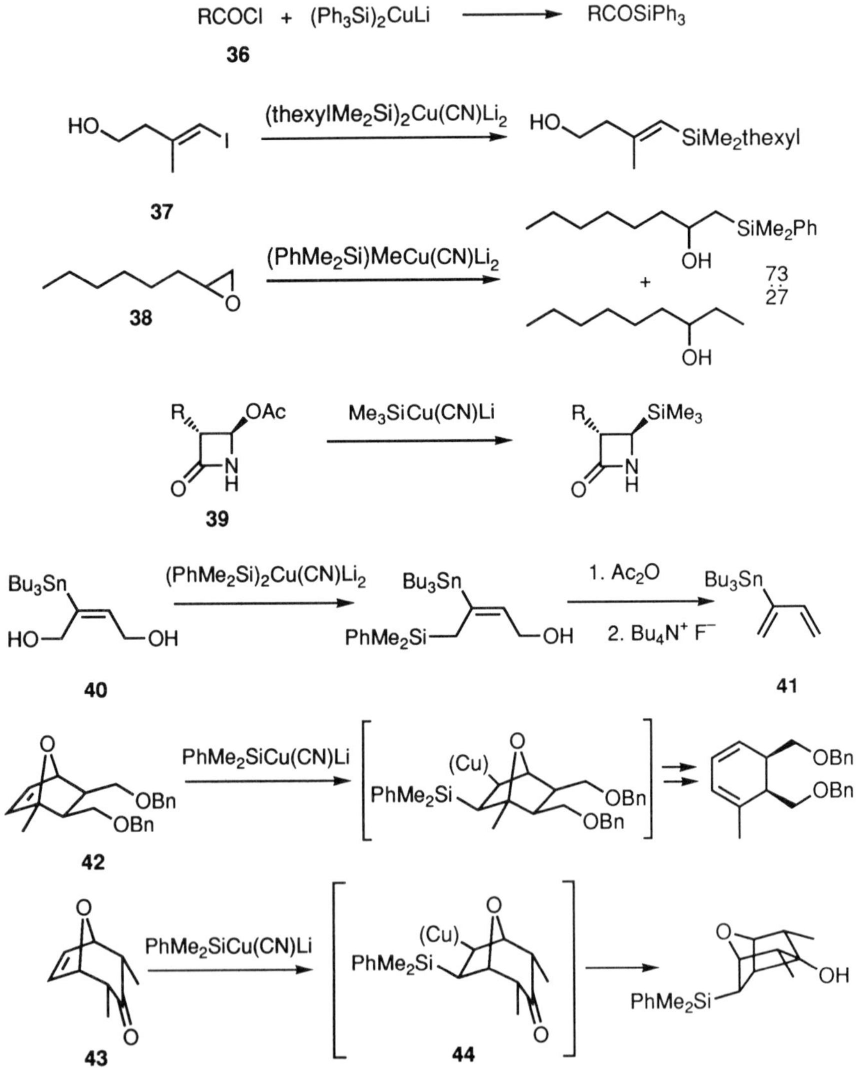

$RCOCl + (Ph_3Si)_2CuLi \longrightarrow RCOSiPh_3$

36

37 —$(thexylMe_2Si)_2Cu(CN)Li_2$→ HO…$SiMe_2$thexyl

38 —$(PhMe_2Si)MeCu(CN)Li_2$→ …$SiMe_2Ph$ (OH) + … (OH) 73 : 27

39 (R, OAc, N–H) —$Me_3SiCu(CN)Li$→ (R, $SiMe_3$, N–H)

40 (Bu_3Sn, HO, OH) —$(PhMe_2Si)_2Cu(CN)Li_2$→ Bu_3Sn, $PhMe_2Si$, OH —1. Ac_2O 2. $Bu_4N^+ F^-$→ Bu_3Sn **41**

42 (OBn, OBn) —$PhMe_2SiCu(CN)Li$→ [(Cu), $PhMe_2Si$, OBn, OBn] → OBn, OBn

43 —$PhMe_2SiCu(CN)Li$→ [(Cu), $PhMe_2Si$, O] **44** → $PhMe_2Si$, OH

Scheme 12.11

Protocol 5.
Addition of the bis(phenyldimethylsilyl)cuprate reagent $(PhMe_2Si)_2Cu(CN)Li_2$ to tridecyne to give the vinylsilane 31[34]

Caution! Carry out all procedures in a well-ventilated hood, and wear disposable vinyl or latex gloves and chemical-resistant safety goggles.

1. $(PhMe_2Si)_2Cu(CN)Li_2$
2. NH_4Cl, H_2O

$PhMe_2Si$ **31**

Equipment

- One two-necked, round-bottomed flask (100 or 250 mL) with a magnetic stirring bar, a septum, and a three-way stopcock, and containing a solution of bis(phenyldimethylsilyl)cuprate prepared as described in Protocol 1 (45 mmol, typically in 40–60 mL of THF). This preparation is also possible with a one-necked flask and without a three-way stopcock
- An argon gas supply and a pump fitted to the stopcock
- A gas-tight syringe
- A Büchner or sintered-glass funnel and associated filtration apparatus
- A separating funnel (250 mL)
- Three conical flasks (250 mL)
- A 2.5 × 30 cm chromatography column and a supply of conical flasks
- A dry ice–acetone bath on a magnetic stirrer

Materials

- Tridecyne[a] (FW 180.3) 5.7 g, 31.7 mmol — **flammable**
- Dry, distilled ether 15 mL — **flammable, irritant**
- A saturated aqueous solution of ammonium chloride (adjusted to pH 8 by addition of ammonia) 40 mL — **corrosive**
- Celite® 20 g
- Ether for extraction 100 mL — **flammable, irritant**
- Brine 20 mL
- Anhydrous sodium sulfate 10 g
- Silica gel for chromatography (230–400 mesh) 200 g — **irritant dust**
- Distilled hexane for chromatography — **flammable, irritant**

1. Cool the flask containing the silylcuprate solution to −78°C under argon using a dry ice–acetone bath and stir the contents. Add the acetylene in most of the dry ether by syringe (or cannula), washing in the last of the acetylene with the rest of the ether.
2. Stir the mixture in the dry ice–acetone bath for 2 h. Remove the cooling bath and replace it with an ice–water bath, then stir for another hour, by which time the solution will be black.
3. Add the ammonium chloride solution by syringe, or by pouring it directly into the flask, and stir for a few minutes. Filter the mixture through a pad of Celite® in a Büchner funnel (or sintered-glass funnel), washing through with more ether.

Protocol 5. *Continued*

4. Separate the layers, washing the organic layer with ammonium chloride solution (2 × 50 mL) and the combined aqueous layers with ether (2 × 50 mL). Wash the combined organic layers with saturated brine (20 mL). Dry the combined ether layers (Na_2SO_4), filter, and concentrate the filtrate on a rotary evaporator. Note that the aqueous layers will contain cyanide and should be disposed of accordingly. Leave the aqueous solution in the fume cupboard overnight to allow any remaining ether to evaporate, then quench the cyanide by the careful addition of bleach (**caution**! exothermic reaction).
5. Add the residue to the top of a chromatography column packed with silica gel and elute with hexane, evaporating the fractions containing the vinylsilane [R_f (hexane) = 0.51]. The yield should be about 10.5 g (95%). The product has been characterized by IR, ^{1}H NMR, and high resolution mass spectrometry.

[a] Prepared from bromoundecane and the lithium acetylide–ethylenediamine complex and freshly distilled before use.

3. Stannylcuprates

Stannyllithium reagents are even better than silyllithium reagents in undergoing conjugate addition to enone systems, reacting, for example, with α,β-unsaturated esters[49] as well as with α,β-unsaturated ketones.[50] In consequence, stannycuprate reagents are rarely needed for this type of reaction. Indeed, trimethylstannyllithium is actually cleaner in some conjugate addition reactions than the various trimethylstannylcuprates. Stannylcuprates are most often used with acetylenic substrates, where Piers has been particularly active in establishing how powerful this reaction is in creating bifunctional reagents. The stannylcopper reagents and stannylcuprates most often used are based on the tributyl- or trimethylstannyl group, with various combinations of other ligands on the copper. The otherwise similarly constituted tributyl- and trimethylstannyl reagents are sometimes noticeably different in their reactivities, with reagents based on the trimethylstannyl group, although more toxic and costly, usually giving products in higher yields and with relatively uncluttered NMR spectra. Triphenylstannylcopper reagents and cuprates have also been studied.[51] With three common stannyl groups, with a variety of ligands that can be used in addition to the stannyl group, and with several stoichiometric relationships and different methods of preparation, there is now a bewildering variety of known reagents containing both tin and copper. That some of these are structurally distinct species has been shown by their characterization by NMR spectroscopy (Table 12.2).[17] No general guidance can yet be given about how to choose a stannylcopper or stannylcuprate—it is probably best to follow the closest analogy in the literature, but there is much wisdom in being prepared to try a few of the other differently constituted reagents before giving up, since they do differ in their reactivity and selectivity.

Table 12.2. NMR characterization of tin- and copper-containing reagents in THF[17] (chemical shifts in p.p.m.)

Reagent	**^{13}C [a] (SnMe)**	**^{13}C [a] (CuMe)**
$Me_3SnCu(CN)Li$	−4.5	
$(Me_3Sn)_2Cu(CN)Li_2$	−0.04	
$Me_3Sn(Me)Cu(CN)Li_2$	−2.0	−9.3
$Me_3Sn(Me)_2CuLi_2$	−1.7	−9.3

[a] Reference signal Me_4Si.

3.1 Preparation of stannylcuprates

Like silylcuprates, stannylcuprates are usually made from the corresponding lithium reagents. Tributyl- and trimethylstannyllithium can easily be made by cleavage of the appropriate hexaalkylditin with methyl- or butyllithium[49] (with no need for HMPA) or, more cheaply, by cleaving the ditin with lithium metal (with no need for phenyl groups on the tin as there is with silicon).[52,19]

Protocol 6. Preparation of the trimethylstannylcopper reagent $Me_3SnCu{\cdot}SMe_2$[53]

Caution! Carry out all procedures in a well-ventilated hood, and wear disposable vinyl or latex gloves and chemical-resistant safety goggles.

$$Me_3SnSnMe_3 \xrightarrow{MeLi} Me_4Sn + Me_3SnLi \xrightarrow{CuBr{\cdot}SMe_2} Me_3SnCu{\cdot}SMe_2$$

Equipment

- One three-necked, round-bottomed flask (2 L) with a magnetic stirring bar, a gas inlet, a glass stopper, and a septum
- A dry argon gas supply
- A large, dry glass funnel
- Large PP/PE disposable syringes and stainless steel Luer hub syringe needles
- A dry ice–aqueous calcium chloride bath on a magnetic stirrer
- A dry ice–acetone bath

Materials

- Freshly distilled hexamethylditin (FW 327.6) 64 mL, 101.8 g, 0.311 mol — **highly toxic, irritant**
- Methyllithium (FW 22.0) 1.27 M in ether, 245 mL, 0.311 mol — **flammable, moisture sensitive**
- Anhydrous copper(I) bromide–dimethyl sulfide complex (FW 205.6) 63.9 g, 0.311 mol — **moisture sensitive**
- Dry, distilled THF 1.3 L — **flammable, irritant**

1. Flame dry the flask and purge it with the argon while allowing it to cool to room temperature. Add the stirrer bar, stopper the middle outlet, and put on

Protocol 6. *Continued*

the septum. Purge again through the septum. Add the hexamethylditin by syringe, add the THF by cannula, and cool the mixture to −20 °C. Add the methyllithium solution by syringe and stir at −20 °C for 30 min to get a pale-yellow solution of trimethylstannyllithium (the Teflon-coated stirring bar will turn blue).

2. Cool the flask to −78 °C. Remove the stopper, increase the flow of argon, and add the copper bromide using the large glass funnel. Return the stopper immediately, and stir the mixture at −78 °C for 30 min to get a reddish-brown solution of the cuprate reagent, which is ready for the next step (see Protocol 9).

Protocol 7. Preparation of the trimethylstannylcuprate reagent $Me_3SnCu(CN)Li$[54]

Caution! Carry out all procedures in a well-ventilated hood, and wear disposable vinyl or latex gloves and chemical-resistant safety goggles.

$$Me_3SnSnMe_3 \xrightarrow{MeLi} Me_4Sn + Me_3SnLi \xrightarrow{CuCN} Me_3SnCu(CN)Li$$

Equipment

- One three-necked, round-bottomed flask (2 L) with a magnetic stirring bar, a gas inlet, a glass stopper, and a septum
- A dry argon gas supply
- A large, dry glass funnel
- Large PP/PE disposable syringes and stainless steel Luer hub syringe needles
- A dry ice–aqueous calcium chloride bath on a magnetic stirrer

Materials

- Freshly distilled hexamethylditin (FW 327.6) 34 mL, 54.08 g, 165 mmol — **highly toxic, irritant**
- Methyllithium (FW 22.0) 1.41 M in ether, 117 mL, 165 mmol — **flammable, moisture sensitive**
- Anhydrous copper(I) cyanide (FW 89.6) 14.8 g, 165 mmol — **highly toxic, irritant**
- Dry, distilled THF 650 mL — **flammable, irritant**

1. Prepare the trimethylstannyllithium solution (165 mmol) as described in Protocol 6.
2. Cool the solution to −48 °C using the dry ice–aqueous calcium chloride bath. Remove the stopper, increase the flow of argon, and add the copper cyanide using the large glass funnel. Return the stopper immediately, and stir the mixture at −48 °C for 20 min to get an orange-red solution of the cuprate reagent, which is ready for the next step (see Protocol 10).

Tributylstannyllithium can also be made cheaply, but less conveniently, by deprotonation of tributyltin hydride with LDA.[55] Lipshutz has developed methods of preparing mixed alkylstannylcuprates, without using stannyllithium intermediates, by treating either tin hydrides[56] or, provided that the silyl group is not hindered, silylstannanes with alkylcuprates.[8,57] These mixed alkylstannylcuprates always transfer the stannyl group to the substrate. Similar reactions work, perhaps even more conveniently, on hexaalkylditins.[58]

Protocol 8.
Preparation of the tributylstannyl(butyl)cuprate reagent $Bu_3Sn(Bu)Cu(CN)Li_2$ from tributyltin hydride[56]

Caution! Carry out all procedures in a well-ventilated hood, and wear disposable vinyl or latex gloves and chemical-resistant safety goggles.

$$Bu_3SnH + Bu_2Cu(CN)Li_2 \longrightarrow Bu_3Sn(Bu)Cu(CN)Li_2$$

Equipment

- One round-bottomed flask (10 mL), with a magnetic stirring bar and a septum, containing the dibutylcyanocuprate [$Bu_2Cu(CN)Li_2$, 0.381 mmol] in THF (1 mL) under argon, as prepared in Protocol 3
- A gas-tight syringe
- A dry ice–acetone bath on a magnetic stirrer

Materials

- Redistilled tributyltin hydride (FW 291.0) 0.21 mL, 0.23 g, 0.78 mmol **irritant, moisture sensitive**

1. Make sure that all the copper cyanide has dissolved (see Protocol 3), and then cool the flask to −78°C using a dry ice–acetone bath.
2. Add the neat tin hydride and stir for 10 min. This is accompanied by hydrogen gas evolution[a] (**caution!** on a larger scale, this is dangerous and requires safe venting) and the formation of a bright-yellow, homogeneous solution of the mixed cuprate, ready for the injection of the electrophile.

[a] If dimethylcyanocuprate is used in place of dibutylcyanocuprate, the gas evolution does not take place unless the mixture is warmed to −50°C.

3.2 Reactions of stannylcuprates

3.2.1 With enone systems

Although stannyllithium reagents (and silylstannanes[59]) add easily to enones, the stannylcuprate may sometimes work better (Scheme 12.12). Thus the conjugate addition of Piers' reagent to the enone **45** appears to take place[60] with better stereochemical purity than with the lithium reagent.[61] It also works better in the conjugate addition to β-iodoenones **47** and to β-unsubstituted enoates like ethyl acrylate **48**. In the former case the lithium reagent gives some double addition of the trimethylstannyl group together with some unchanged starting material, and in the latter the intermediate enolate in the cuprate series is less apt to add to a second molecule of the enoate substrate than the lithium enolate. The related stannyl cuprations of α,β-acetylenic esters are discussed in Section 3.2.3.

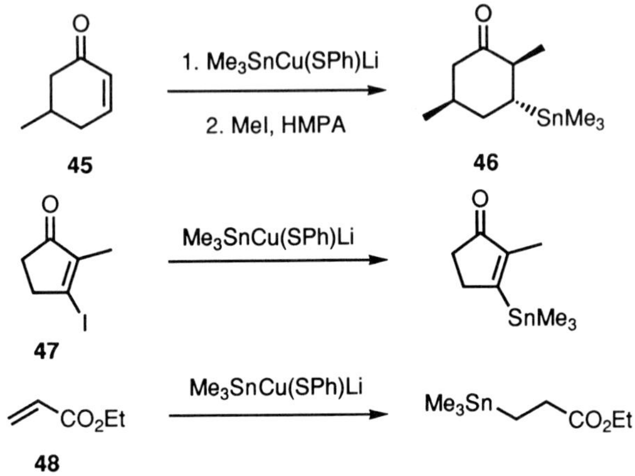

Scheme 12.12

β-Stannyl ketones and esters are useful in many ways (Scheme 12.13). The β-carbon atom is nucleophilic enough to react directly with the carbonyl group in the presence of a Lewis acid **49** → **50**.[62] The carbonyl group can be converted into a secondary or tertiary alcohol, which can then be used either to make cyclopropanes **51** → **52**[63] or in fragmentation reactions **53** → **54**.[64] The tin–carbon bond can be oxidized to an alcohol[65] (without the need for a nucleofugal group to be attached as it is with silicon) or directly to a ketone group **55** → **56**.[50] Alternatively, it can be reduced off **57** → **58**, the stannyl group having served its turn to control stereo- and regiochemistry.[60] In addition to the reactions illustrated, they can also be used as leaving groups in radical-based cyclizations[66] and fragmentations.[67]

3.2.2 With allylic and propargylic electrophiles

The reaction of stannylcuprates with allylic electrophiles has been little used. Whereas the allyl bromide **59** reacted with the stannyllithium reagent cleanly to give only the product **61** of direct displacement, the cuprate gave more of the product **60** of allylic rearrangement.[41,68] Propargylic mesylates can also be selective for conjugate displacement, with stannylcopper or stannyl cuprate reagents giving allenylstannanes, but the selectivity is affected by the substitution pattern (Scheme 12.4).[69] Allylstannanes and allenylstannanes are much used as carbon nucleophiles in organic synthesis.[70]

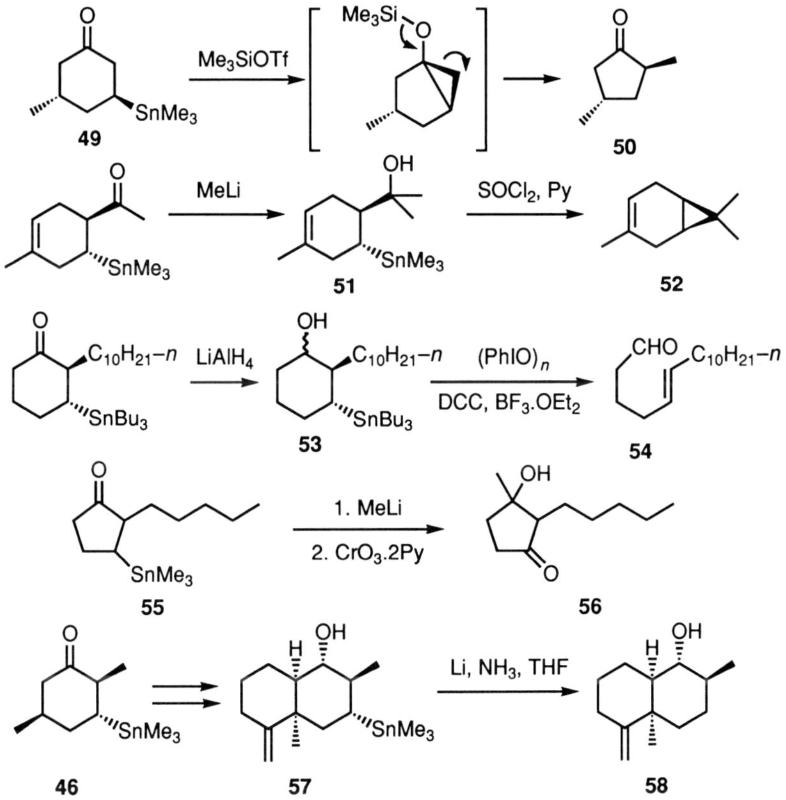

49 $\xrightarrow{Me_3SiOTf}$ [Me_3Si–O] → 50

→ 51 $\xrightarrow{SOCl_2,\ Py}$ 52 (MeLi; $SnMe_3$; OH)

$C_{10}H_{21}$–*n*, $SnBu_3$ $\xrightarrow{LiAlH_4}$ 53 $\xrightarrow[DCC,\ BF_3.OEt_2]{(PhIO)_n}$ 54 (CHO, $C_{10}H_{21}$–*n*)

55 ($SnMe_3$) $\xrightarrow[2.\ CrO_3.2Py]{1.\ MeLi}$ 56 (OH, O)

46 ($SnMe_3$) ⇉ 57 (H, OH, $SnMe_3$) $\xrightarrow{Li,\ NH_3,\ THF}$ 58 (H, OH)

Scheme 12.13

BnO–CH₂CH=CHCH₂–Br (59) $\xrightarrow{Me_3Sn^-}$ 60 (BnO, $SnMe_3$) + 61 (BnO, $SnMe_3$)

	60 : 61
Me_3Sn^- = Me_3SnLi	0:100
Me_3Sn^- = $(Me_3Sn)_2CuLi$	67:33

R^1–C≡C–CH(R^2)–OSO_2Me $\xrightarrow{(Ph_3Sn)_2CuLi}$ Ph_3Sn(R^1)C=C=CH–R^2 + R^1–C≡C–CH(R^2)–$SnPh_3$

R^1 = H	100:0	(R^2 = Me)
R^1 = Me	0:100	(R^2 = Pr)

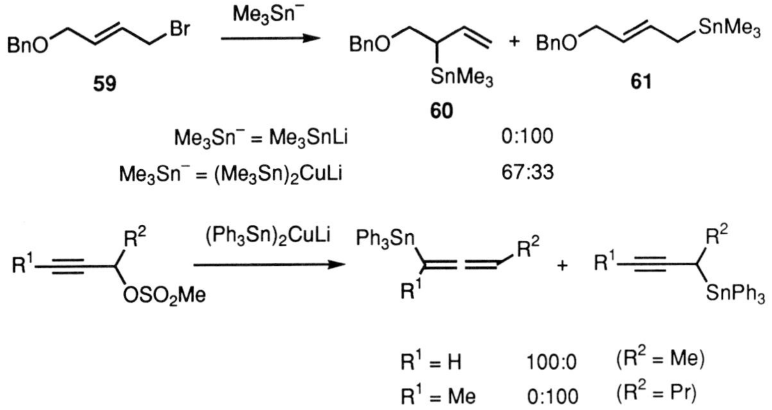

Scheme 12.14

3.2.3 With acetylenes

Stannylcopper reagents and stannylcuprates react with terminal acetylenes **62** with high but not complete regioselectivity to place the stannyl group on C-2 and the copper atom on C-1 **63**. The proportion of products derived from the regioisomer **64** is usually less than 10%. However, this reaction is easily reversible, and addition of an electrophile does not always give high yields of products **66**, because the stannylcopper or stannylcuprate is often so much more nucleophilic than the vinylcopper or cuprate intermediate that it reacts with the electrophile (Scheme 12.15) to give the alternative product **65**.[71] One of the few electrophiles that does not have this problem associated with it is the proton. Stannylcopper and stannylcuprate reagents are conspicuously poor bases, with the result that if a proton source like methanol is present during the stannyl cupration step, a good yield of the product **66** (E = H) of *syn*-addition of the stannyl group and a hydrogen atom to the triple bond can easily be achieved. With diynes, the addition leads to stannyl groups becoming attached to both atoms of the acetylene,[72a] and with ethynyl ethers the stannyl group can be attached to either end of the acetylene depending upon whether HMPA is present or not.[72b]

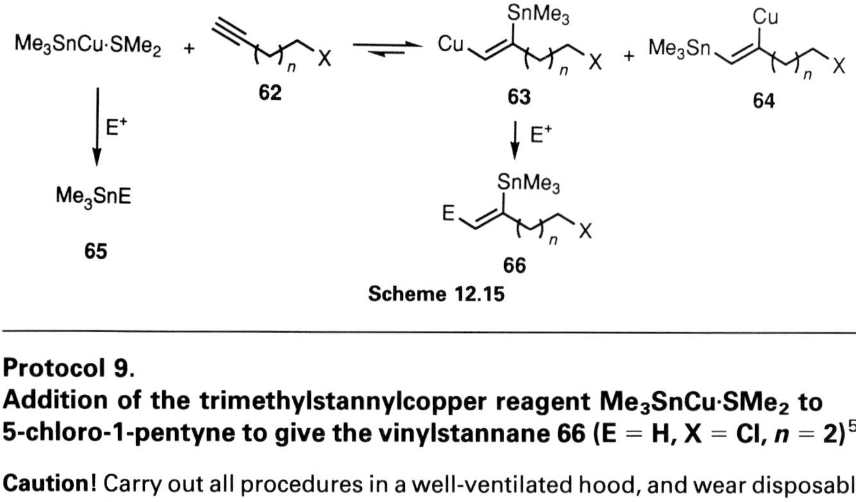

$Me_3SnCu{\cdot}SMe_2$ + **62** ⇌ **63** + **64**; $Me_3SnCu{\cdot}SMe_2$ —E^+→ Me_3SnE **65**; **63** —E^+→ **66**

Scheme 12.15

Protocol 9.
Addition of the trimethylstannylcopper reagent $Me_3SnCu{\cdot}SMe_2$ to 5-chloro-1-pentyne to give the vinylstannane 66 (E = H, X = Cl, n = 2)[53]

Caution! Carry out all procedures in a well-ventilated hood, and wear disposable vinyl or latex gloves and chemical-resistant safety goggles.

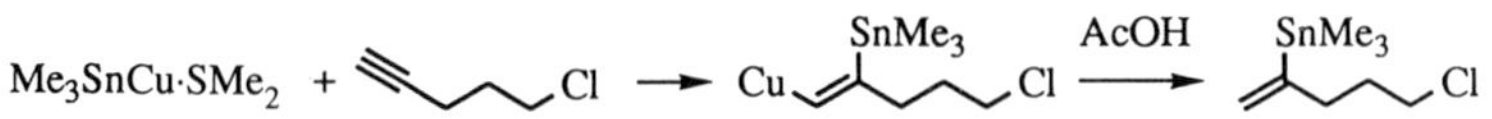

$Me_3SnCu{\cdot}SMe_2$ + 5-chloro-1-pentyne → (Cu, $SnMe_3$ vinyl intermediate) —AcOH→ ($SnMe_3$ vinylstannane)

Equipment

- One three-necked, round-bottomed flask (2 L) with a magnetic stirring bar, an argon gas inlet, a glass stopper in the middle port, and a septum, and containing the trimethylstannylcopper reagent (0.311 mol) under an Ar atmosphere from Protocol 6
- A dry argon gas supply and inlet
- Large PP/PE disposable syringes and stainless steel Luer hub syringe needles
- A syringe pump (or use a 50 mL dropping funnel in place of the syringes)

- Separating funnel (4 L)
- Two conical flasks (4 L)
- A dry ice–acetone bath on a magnetic stirrer
- An 8 × 65 cm chromatography column and a supply of conical flasks

Materials

• 5-Chloro-1-pentyne (FW 102.6) 32.9 mL, 31.8 g, 0.31 mol	**flammable**
• Glacial acetic acid (FW 60.1) 88.9 mL, 1.55 mol	**corrosive**
• Dry methanol 1 mL	**flammable, toxic**
• Ether 6L	**flammable, irritant**
• Saturated aqueous ammonium chloride solution (with ammonia added to pH 8) 5 L	**corrosive**
• Brine 2 L	
• Anhydrous magnesium sulfate 50 g	
• Silica gel for chromatography (230–400 mesh) 1.75 kg	**irritant dust**
• Light petroleum (b.p. 35–60 °C) for chromatography	**flammable, irritant**

1. Cool the flask containing the stannylcopper reagent under Ar to −78 °C and stir the solution while adding the freshly distilled acetylene over 40 min, using either the syringe and a pump or the dropping funnel. Rinse the dropping funnel with dry THF (5–10 mL).
2. Stir the mixture at −78 °C for 8 h.
3. Add the acetic acid to the solution at −78 °C and let the mixture stir for 20 min.
4. Add some of the ammonium chloride solution (200 mL), remove the cooling bath, and allow the mixture to warm to room temperature, stirring vigorously. Divide this brown solution between the two conical flasks, each containing ammonium chloride solution (1.5 L each) and ether (1 L each). Stir the contents vigorously until the aqueous phases become deep blue. This can conveniently be done overnight.
5. Pour the mixture of one of the flasks into the separating funnel and separate the layers. Wash the aqueous layer and the flask with ether (2 × 1 L). Wash the combined organic layers with ammonium chloride solution (1 × 1 L) and with brine (1 × 1 L). Dry the combined organic layers ($MgSO_4$). Filter and concentrate the filtrate. Repeat the whole procedure with the contents of the second flask, and combine the products to give a crude oil containing a little (<5%) hexamethylditin ($R_f \approx 1$ in light petroleum) as well as the two regioisomeric vinylstannanes in a ratio of about 10:1 in favour of the 2-stannylalkene.
6. Add the oil to the chromatography column and allow the light petroleum to drip through without pumping. The hexamethylditin elutes first, followed by the 2-stannylalkene, with the other isomer decomposing on the column. The whole operation takes 7–9 h. Concentrate the fractions containing the vinylstannane on the rotary evaporator, and distil the residue from bulb to bulb at 60 °C/12 mm Hg. The yield should be about 54.3 g (65%). The product has been characterized by IR and ^{1}H NMR spectroscopy and high resolution mass spectrometry.

The products of these reactions are especially useful because the vinylstannanes can be converted by transmetallation with methyllithium into vinyllithium reagents and hence into vinylcuprates, or directly into vinylcuprates.[73] Thus the vinylstannane **66** (E = H, X = Cl, n = 2) has been used in several natural product syntheses[74] as the C-2 nucleophilic and C-4 electrophilic bifunctional reagent **68** in the powerful cyclopentannelation **67** → **69** (Scheme 12.16). Its homologue **66** (E = H, X = Cl, n = 3) was similarly used in the annelation **46** → **57** (Scheme 12.13). An alternative annelation with these reagents uses the electrophilic group, as its iodide, first **70** → **71**, and then uses the vinylstannane group in an intramolecular coupling reaction with an enol triflate **71** → **72**.[75] Stannyl cupration has also been used to add a stannyl group to the triple bond of the butynediol **73**, for which there was no need to add an external alcohol to quench the intermediate.[76] The product **40** can be converted into 3-tributylstannylfuran **74**—its conversion into 2-stannylbutadiene **41** has already been mentioned (Scheme 12.11)—and it has also been used in a synthesis of octahydrobenzofurans.[77]

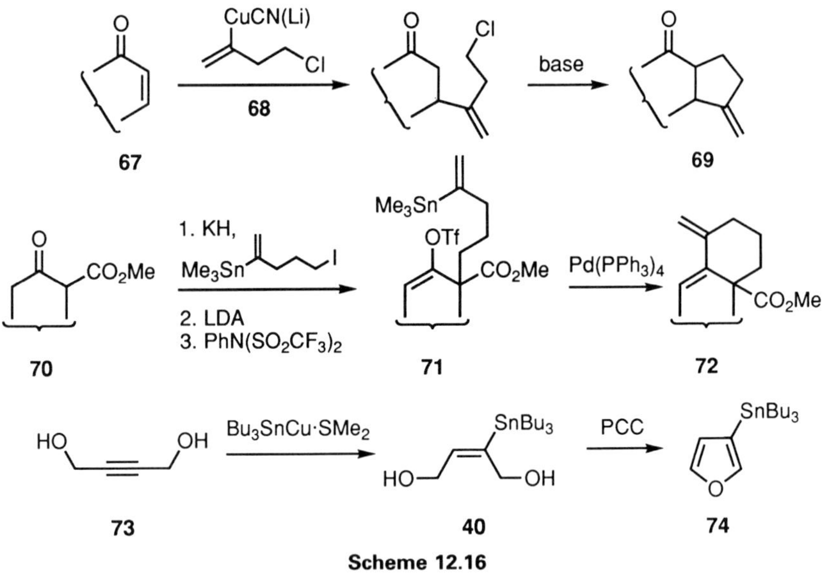

Scheme 12.16

Acetylenic esters have proved to be the most versatile of the acetylenic substrates so far for this type of reaction (Scheme 12.17).[78] The stannyl group adds to the β-position of α,β-acetylenic esters **75** (and amides[79]) to give the product **76** of *syn*-stannyl cupration. Protonation under kinetic control, by mixing an alcohol with the ester **75** *before* adding it to the cuprate solution at −100°C, takes place with retention of configuration, and gives (typically >99:1) the (*E*)-isomer **77**. Preparing the stannyl cupration mixture at −78°C and then warming it to −48 °C *before* adding the proton source gives (typically 98:2) the (*Z*)-isomer **79** from protonation *anti* to the stannyl group of the thermodynamically more stable *O*-tautomer **78**. It is wise to use the alcohol corresponding to the esterifying group in order to avoid complications from ester exchange. While the mixed cuprate $Me_3SnCu(SPh)Li$ prepared with phenylthiolate works to give both products, it is not needed for the formation of the (*E*)-isomer **77**, for which any of the common stannylcopper and stannylcuprate reagents is effective. Since it is not the most convenient of stannylcuprates to make, Piers has recently developed a different cuprate, $Me_3SnCu(CN)Li$, which works well in the synthesis of both products.[80]

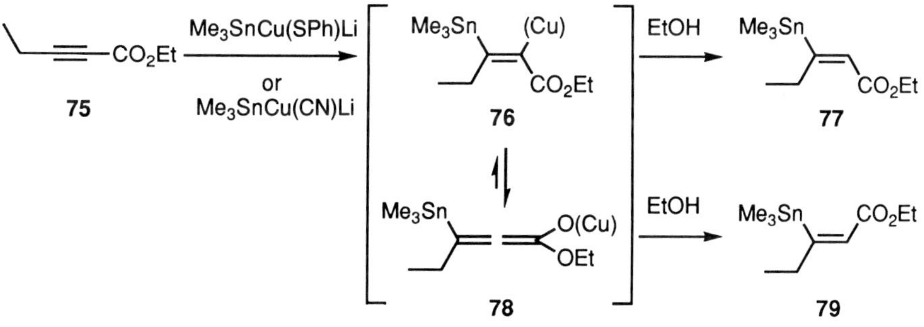

Me3SnCu(SPh)Li or Me3SnCu(CN)Li; 75; Me3Sn (Cu) CO2Et 76; Me3Sn O(Cu) OEt 78; EtOH; Me3Sn CO2Et 77; EtOH; Me3Sn CO2Et 79

Scheme 12.17

The products of these reactions have been used to prepare the chlorides **81** and **83** (Scheme 12.18) and others like them. They are valuable in natural product synthesis because they can be used, like the vinylstannanes **66** (E = H, X = Cl, *n* = 2 or 3), as bifunctional reagents for annelation, with the added advantage that an exocyclic double bond can now be prepared with complete stereocontrol because of the remarkable stereoselectivity of the deprotonation–reprotonation steps **77** → **80** and **79** → **82**.[81]

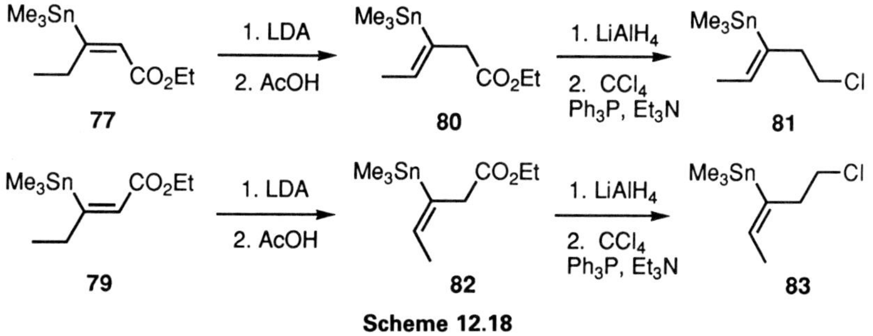

Me3Sn CO2Et 77; 1. LDA 2. AcOH; Me3Sn CO2Et 80; 1. LiAlH4 2. CCl4 Ph3P, Et3N; Me3Sn Cl 81

Me3Sn CO2Et 79; 1. LDA 2. AcOH; Me3Sn CO2Et 82; 1. LiAlH4 2. CCl4 Ph3P, Et3N; Me3Sn Cl 83

Scheme 12.18

Although there are sometimes difficulties in adding other electrophiles than a proton to intermediates like **63**, a number of stannyl cuprations of acetylenes have been reported where there was no problem in adding a wide variety of electrophiles. This appears to be especially true if cuprate reagents are used rather than a copper reagent and if the acetylene is unsubstituted (Scheme 12.19), although some of these reactions also worked well with 1-hexyne.[51,82]

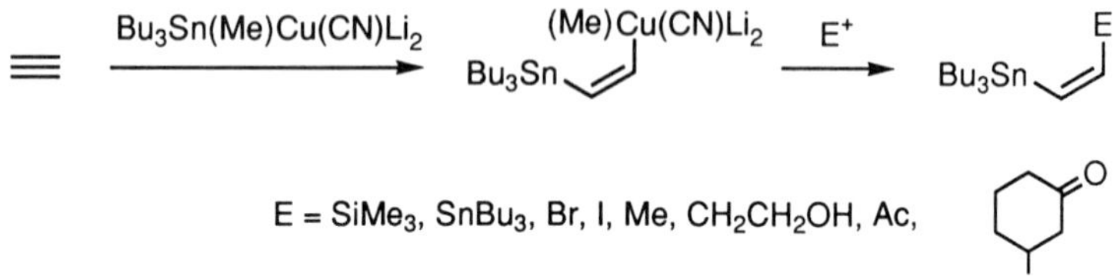

$Bu_3Sn(Me)Cu(CN)Li_2$; $(Me)Cu(CN)Li_2$; Bu_3Sn ; E^+ ; E ; Bu_3Sn

$E = SiMe_3$, $SnBu_3$, Br, I, Me, CH_2CH_2OH, Ac,

Scheme 12.19

Protocol 10.
Addition of the trimethylstannylcuprate reagent $Me_3SnCu(CN)Li$ to acetylenic esters[54]

Caution! Carry out all procedures in a well-ventilated hood, and wear disposable vinyl or latex gloves and chemical-resistant safety goggles.

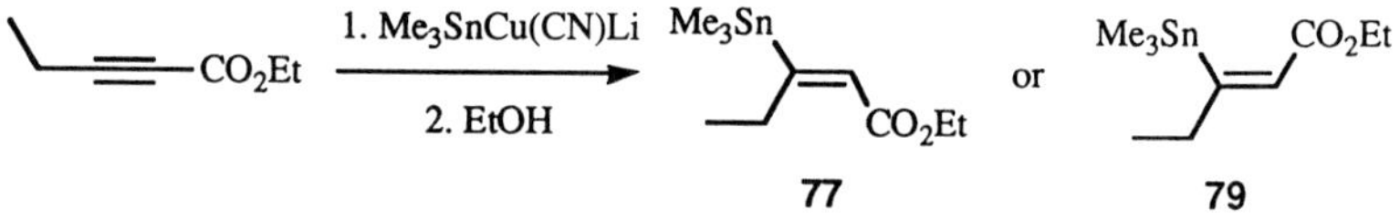

CO_2Et — 1. $Me_3SnCu(CN)Li$ 2. EtOH → Me_3Sn ... CO_2Et **77** or Me_3Sn ... CO_2Et **79**

Equipment

- One three-necked, round-bottomed flask (2 L) with a magnetic stirring bar, a gas inlet, a glass stopper, and a septum, and containing the stannylcuprate reagent (165 mmol) under Ar from Protocol 7
- A dry argon gas supply
- A cannula and syringes
- Separating funnel (2 L)
- Two 8 ×22 cm chromatography columns and a supply of conical flasks

Materials

- Ethyl pentynoate (FW 126.2) 16.4 g, 130 mmol — **irritant**
- Dry, distilled THF 10 mL — **flammable, irritant**
- Dry ethanol 10 mL, 170 mmol — **flammable, toxic**
- Ether 600 mL — **flammable, irritant**
- Saturated aqueous ammonium chloride solution (with ammonia added to pH 8) 325 mL — **corrosive**
- Brine 500 mL
- Anhydrous magnesium sulfate 50 g
- Silica gel for flash chromatography 230–400 mesh 2 × 350 g — **irritant dust**
- Light petroleum (b.p. 35–60°C) mixed with ether (200:3) for chromatography — **flammable, irritant**

For the (E)*-product* **77**

1. Cool the flask containing the trimethylstannylcuprate to −78°C using a dry ice–acetone bath, stir steadily, and add the ethanol by syringe. (Use ethanol for ethyl esters and methanol for methyl esters to minimize ester exchange.)

After 5 min, add the ester in the THF dropwise by cannula over about 15 min and stir at −78°C for 4 h.

2. Remove the stopper, add ammonium chloride solution (325 mL), remove the cooling bath, and allow the mixture to warm to room temperature with vigorous stirring, open to the atmosphere, until the aqueous phase becomes deep blue. Pour the mixture into the separating funnel and separate the layers. Wash the aqueous layer and the flask with ether (3 × 200 mL). Concentrate the combined organic layers to about 500 mL, wash with brine (2 × 250 mL), and dry ($MgSO_4$). Filter and concentrate the filtrate to give a crude oil containing hexamethylditin (R_f = 0.7 in light petroleum–ether, 200:3), the (*E*)-ester (R_f = 0.25), and a small amount (<5%) of the (*Z*)-ester (R_f = 0.27).
3. Divide the oil into two equal parts and add the portions one to each of the chromatography columns and elute with the light petroleum–ether mixture in the usual way for flash chromatography. Concentrate the fractions containing the (*E*)-ester (R_f = 0.25) on the rotary evaporator, and distil the residue from bulb to bulb at 50–70°C/0.6 mm Hg. The yield should be about 27.7 g (73%). This product has been characterized by IR, ^{1}H NMR, and ^{13}C NMR spectroscopy, and by combustion microanalysis and high resolution mass spectrometry.

For the (Z)-*product* **79**

1. Add the 2-pentynoate in the THF by cannula to the cold (−48°C using a dry ice–aqueous calcium chloride bath) solution of the trimethylstannylcuprate and stir at −48°C for 2 h. Replace the cooling bath with an ice–water cooling bath and stir for a further 2 h at 0°C.
2. Work up and flash chromatograph the product (R_f= 0.27) as described for the (*E*)-isomer above, and distil it from bulb to bulb at 54–84°C/0.6 mm Hg. The yield should be about 29 g (76%). This product has been characterized by IR, ^{1}H NMR and ^{13}C NMR spectroscopy, and by combustion microanalysis and high resolution mass spectrometry.

3.2.4 With allenes

Stannylcuprates react with allene at −100°C to give, under kinetic control, the stannyl cupration product **84** (Scheme 12.20). If this intermediate is warmed to 0°C, it rearranges into the thermodynamically more stable regioisomer **85**. These intermediates react with electrophiles to give allylstannanes **86** and vinylstannanes **87**, respectively. The range of products **86** is small, because only a few electrophiles react with the intermediate **84** fast enough to capture it before it rearranges to its more stable isomer **85**.[83] These products are powerful small synthons, possessing the usual capacity provided by the stannyl group for transmetallation and transition metal-catalysed coupling reactions. Substituted allenes also react, as in the conversion **88** → **89** →

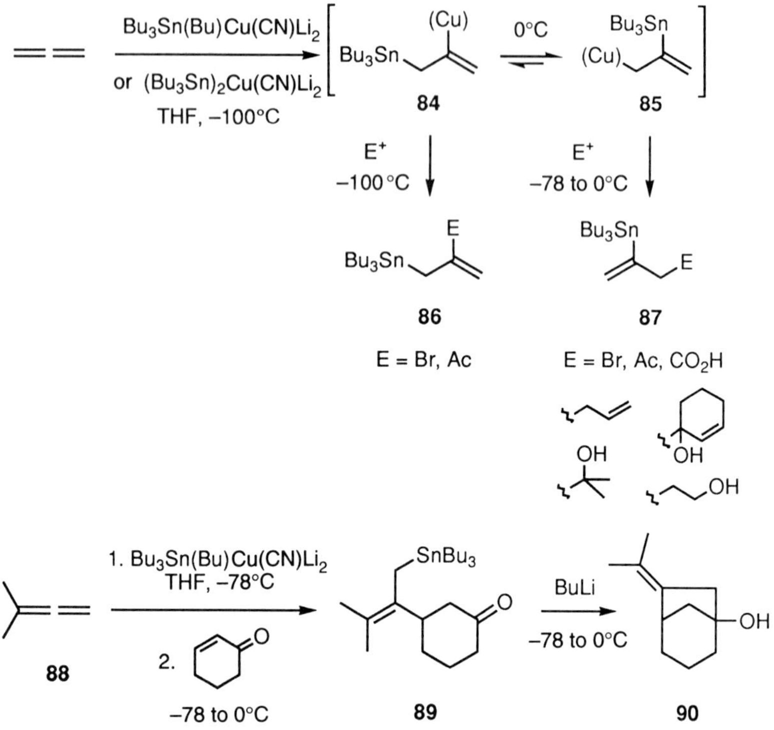

Scheme 12.20

90, in a sequence that illustrates that the remarkable rapidity of the transmetallation step allows the conversion **89** → **90** to take place without the butyllithium attacking the ketone group first.[84]

3.2.5 With vinyl triflates

The higher-order tributylstannyl cuprate reacts with vinyl triflates, which are themselves made from ketones, to give vinylstannanes.[57,85] The reaction typically takes 2 h at −20°C, and is conveniently worked up using silver acetate to remove the major by-product, hexabutyldistannane. This route to vinylstannanes can be used to give either regioisomer from an unsymmetrical ketone, but is especially useful for giving the regioisomer based on the thermodynamic enolate of the ketone (Scheme 12.21), since this is not easily prepared by the Shapiro reaction.

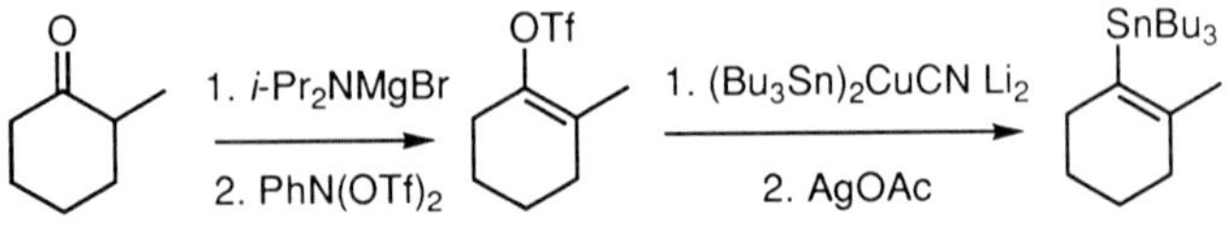

Scheme 12.21

References

1. First prepared by Gilman, H.; Dua, S. S. Unpublished results, quoted as a personal communication to: Brook, A. G.; Harris, J. W.; Lennon, J.; El Sheikh, M. *J. Am. Chem. Soc.* **1979**, *101*, 83–95.
2. First invoked by Hudec, J. *J. Chem. Soc., Perkin Trans. 1* **1975**, 1020–1023, but his copper-catalysed reaction might not actually have involved a cuprate reagent.
3. Ager, D. J.; Fleming, I. *J. Chem. Soc., Chem. Commun.* **1978**, 177–78. Ager, D. J.; Fleming, I.; Patel, S. K. *J. Chem. Soc., Perkin Trans. 1* **1981**, 2520–2526.
4. Fleming, I.; Henning, R.; Plaut, H. *J. Chem. Soc., Chem. Commun.* **1984**, 29–31.
5. Fleming, I.; Sanderson, P. E. J. *Tetrahedron Lett.* **1987**, *28*, 4229–4232.
6. Tamao, K.; Kawachi, A.; Ito, Y. *J. Am. Chem. Soc.* **1992**, *114*, 3989–3990.
7. Fleming, I.; Newton, T. W. *J. Chem. Soc., Perkin Trans. 1* **1984**, 1805–1808.
8. Lipshutz, B. H.; Reuter, D. C.; Ellsworth, E. L. *J. Org. Chem.* **1989**, *54*, 4975–4977.
9. Duffaut, N.; Dunoguès, J.; Biran, C.; Calas, R.; Gerval, J. *J. Organomet. Chem.* **1978**, *161*, C23–C24.
10. Cuadrado, P.; Gonzalez, A. M.; Gonzalez, B.; Pulido, F. J. *Synth. Commun.* **1989**, *19*, 275–283.
11. Chen, H.-M.; Oliver, J. P. *J. Organomet. Chem.* **1986**, *316*, 255–260.
12. Fleming, I.; Winter, S. B. D. *Tetrahedron Lett.* **1993**, *34*, 7287–7290.
13. Archibald, S. C.; Sabin, V.; Winter, S. B. D. Combined wisdom based on references 3 and 33 and other Fleming group experience.
14. Andrainov, K. A.; Delazari, N. V. *Dokl. Akad. Nauk SSSR* **1958**, *122*, 393 (*Chem. Abstr.* **1959**, *53*, 2133).
15. Still, W. C. *J. Org. Chem.* **1976**, *41*, 3063–3064.
16. Hengge, E.; Holtschmidt, N. *J. Organomet. Chem.* **1968**, *12*, P5–P7. Rösch, L.; Altnau, G.; Hahn, E.; Havemann, H. *Z. Naturforsch., Teil B* **1981**, *36*, 1234–1257. Rösch, L.; Erb, W. *Chem. Ber.* **1979**, *112*, 394–395.
17. Sharma, S.; Oehlschlager, A. C. *Tetrahedron* **1989**, *45*, 557–568. Sharma, S.; Oehlschlager, A. C. *J. Org. Chem.* **1991**, *56*, 770–776. Sharma, S. Ph.D. thesis, Simon Fraser University, 1989, p. 150.
18. Tamao, K. Personal communication based on reference 6.
19. Chenard, B. L.; Van Zyl, C. M. *J. Org. Chem.*. **1986**, *51*, 3561–3566.
20. Crump, R. A.; Fleming, I.; Hill, J. H. M.; Parker, D.; Reddy, N. L.; Waterson, D. *J. Chem. Soc., Perkin Trans. 1* **1992**, 3277–3294.
21. Crump, R. A. N. C.; Fleming, I.; Urch, C. J. *J. Chem. Soc., Perkin Trans. 1* **1994**, 701–706.
22. Fleming, I.; Kilburn, J. D. *J. Chem. Soc., Perkins Trans. 1* **1992**, 3295–3302.
23. Fleming, I.; Kindon, N. D. *J. Chem. Soc., Chem. Commun.* **1987**, 1177–1179.
24. Oppolzer, W.; Mills, R. J.; Pachinger, W.; Stevenson, T. *Helv. Chim. Acta* **1986**, *69*, 1542–1545.
25. Morgan, I. T. Ph.D. thesis, University of Cambridge, 1991. Archibald, S. C. Ph.D. thesis, University of Cambridge, 1993. Buckle, M. J. C. Ph.D. thesis, University of Cambridge, 1993.
26. Hudrlik, P. F.; Hudrlik, A. M.; Nagendrappa, G.; Yimenu, T.; Zellers, E. T.; Chin, E. *J. Am. Chem. Soc.* **1980**, *102*, 6894–6896. Hudrlik, P. F.; Hudrlik, A. M.; Nagendrappa, G.; Yimenu, T.; Zellers, E. T.; Chin, E. *J. Am. Chem.*

Soc. **1982**, *104*, 1157. Hudrlik, P. F.; Hudrlik, A. M.; Yimenu, T.; Waugh, M. A.; Nagendrappa, G. *Tetrahedron* **1988**, *44*, 3791–3803.

27. Fleming, I.; Gil, S.; Sarkar, A. K.; Schmidlin, T. *J. Chem. Soc., Perkin Trans. 1* **1992**, 3351–3361.
28. Fleming, I.; Higgins, D.; Lawrence, N. J.; Thomas, A. P. *J. Chem. Soc., Perkin Trans. 1* **1992**, 3331–3349.
29. Fleming, I.; Terrett, N. K. *J. Organomet. Chem.* **1984**, *264*, 99–118.
30. Fleming, I.; Marchi, D. *Synthesis* **1981**, 560–561.
31. Fleming, I.; Dunoguès, J.; Smithers, R. *Org. React.* **1989**, *37*, 57–575.
32. Johnson, C. R.; Cheer, C. J.; Goldsmith, D. J. *J. Org. Chem.* **1964**, *29*, 3320–3323. Höfle, G.; Steglich, W. *Synthesis* **1972**, 619–621.
33. Fleming, I.; Newton, T. W.; Roessler, F. *J. Chem. Soc., Perkin Trans. 1* **1981**, 2527–2532.
34. Fleming, I.; Lawrence, N. J. *Tetrahedron Lett.* **1990**, *31*, 3645–3648.
35. Miller, R. B.; McGarvey, G. *J. Org. Chem.* **1979**, *44*, 4623–4633.
36. Fleming, I.; Takaki, K.; Thomas, A. P. *J. Chem. Soc., Perkin Trans. 1* **1987**, 2269–2273.
37. Fleming, I.; Rowley, M.; Cuadrado, P.; González-Nogal, A. M.; Pulido, F. J. *Tetrahedron* **1989**, *45*, 413–424. Morizawa, Y.; Oda, H.; Oshima, K.; Nozaki, H. *Tetrahedron Lett.* **1984**, *25*, 1163–1166,
38. Singh, S. M.; Oehlschlager, A. C. *Can. J. Chem.* **1991**, *69*, 1872–1880.
39. Fleming, I.; Landais, Y.; Raithby, P. R. *J. Chem. Soc., Perkin Trans. 1* **1991**, 715–719.
40. Barbero, A.; Cuadrado, P.; González, A. M.; Pulido, F. J.; Fleming, I. *J. Chem. Soc., Perkin Trans. 1* **1991**, 2811–2816.
41. Smith, J. G.; Drozda, S. E.; Petraglia, S. P.; Quinn, N. R.; Rice, E. M.; Taylor, B. S.; Viswanathan, M. *J. Org. Chem.* **1984**, *49*, 4112–4120.
42. Singer, R. D.; Oehlschlager, A. C. *J. Org. Chem.* **1991**, *56*, 3510–3514. Fürstner, A.; Weidmann, H. *J. Organomet. Chem.* **1988**, *354*, 15–21.
43. Brook, A. G.; Harris, J. W.; Lennon, J.; Sheikh, M. E. *J. Am. Chem. Soc.* **1979**, *101*, 83–95.
44. Nativi, C.; Ricci, A.; Taddei, M. *Tetrahedron Lett.* **1990**, *31*, 2637–2640.
45. Audia, J. E.; Marshall, J. A. *Synth. Commun.* **1983**, *13*, 531–535.
46. Takaki, K.; Maeda, T.; Ishikawa, M. *J. Org. Chem.* **1989**, *54*, 58–62.
47. Taddei, M.; Mann, A. *Tetrahedron Lett.* **1986**, *27*, 2913. Fleming, I.; Taddei, M. *Synthesis* **1985**, 899–900.
48. Lautens, M.; Ma, S.; Belter, R. K.; Chiu, P.; Leschziner, A. *J. Org. Chem.* **1992**, *57*, 4065–4066. Lautens, M.; Belter, R. K.; Lough, A. J. *J. Org. Chem.* **1992**, *57*, 422–424.
49. Piers, E.; Morton, H. E.; Chong, J. M. *Can. J. Chem.* **1987**, *65*, 78–87.
50. Still, W. C. *J. Am. Chem. Soc.* **1977**, *99*, 4836–4838.
51. Westmijze, H.; Ruitenberg, K.; Meijer, J.; Vermeer, P. *Tetrahedron Lett.* **1982**, *23*, 2797–2798.
52. Tamborski, C.; Ford, F. E.; Soloski, E. J. *J. Org. Chem.* **1963**, *28*, 237–239.
53. Piers, E.; Oballa, R. M. Personal communication based on Piers and Chong, reference 71.
54. Piers, E.; Wong, T. Personal communication based on reference 80.
55. Still, W. C. *J. Am. Chem. Soc.* **1978**, *100*, 1481–1487.

56. Lipshutz, B. H.; Ellsworth, E. L.; Dimock, S. H.; Reuter, D. C. *Tetrahedron Lett.* **1989**, *30*, 2065–2068. Lipshutz, B. H.; Reuter, D. C. *Tetrahedron Lett.* **1989**, *30*, 4617–4620.
57. Lipshutz, B. H.; Sharma, S.; Reuter, D. C. *Tetrahedron Lett.* **1990**, *31*, 7253–7256.
58. Oehlschlager, A. C.; Hutzinger, M. W.; Aksela, R.; Sharma, S.; Singh, S. M. *Tetrahedron Lett.* **1990**, *31*, 165–168.
59. Chenard, B. L.; Laganis, E. D.; Davidson, F.; RajanBabu, T. V. *J. Org. Chem.* **1985**, *50*, 3666–3667.
60. Piers, E.; Roberge, J. Y. *Tetrahedron Lett.* **1991**, *32*, 5219–5222.
61. Wickham, G.; Olszowy, H. A.; Kitching, W. *J. Org. Chem.* **1982**, *47*, 3788–3793.
62. Sato, T.; Watanabe, T.; Hayata, T.; Tsukui, T. *J. Chem. Soc., Chem. Commun.* **1989**, 153–154.
63. Johnson, C. R.; Kadow, J. F. *J. Org. Chem.* **1987**, *52*, 1493–1500. Fleming, I.; Urch, C. J. *J. Organomet. Chem.* **1985**, *285*, 173–191.
64. Ochiai, M.; Ukita, T.; Iwaki, S.; Nagoa, Y.; Fujita, E. *J. Org. Chem.* **1989**, *54*, 4832–4840.
65. Herndon, J. W.; Wu, C. *Tetrahedron Lett.* **1989**, *30*, 6461–6464. Lautens, M.; Zhang, C. H.; Crudden, C. M. *Angew. Chem., Int. Ed. Engl.* **1992**, *31*, 232–234.
66. Baldwin, J. E.; Adlington, R. M.; Robertson, J. *Tetrahedron* **1991**, *47*, 6795–6812.
67. Baldwin, J. E.; Adlington, R. M.; Robertson, J. *J. Chem. Soc., Chem. Commun.* **1988**, 1404–1406.
68. Naruta, Y.; Maruyama, K. *J. Chem. Soc., Chem. Commun.* **1983**, 1264–1265. Takeda, T.; Ogawa, S.; Ohta, N.; Fujiwara, T. *Chem. Lett.* **1987**, 1967–1970.
69. Ruitenberg, K.; Westmijze, H.; Meijer, J.; Elsevier, C. J.; Vermeer, P. *J. Organomet. Chem.* **1982**, *241*, 417–422.
70. Fleming, I. In *Comprehensive Organic Synthesis*; Trost, B. M., Fleming, I., eds; Pergamon: Oxford, 1991; Vol. 2, Chapter 2.2. Eyley, S. C. In *Comprehensive Organic Synthesis*; Trost, B. M., Fleming, I., eds; Pergamon: Oxford, 1991; Vol. 2, Chapter 3.1. Kleinman, E. F.; Volkmann, R. A. In *Comprehensive Organic Synthesis*; Trost, B. M., Fleming, I., eds; Pergaman: Oxford, 1991; Vol. 2, Chapter 4.3.
71. Piers, E.; Chong, J. M. *Can. J. Chem.* **1988**, *66*, 1425–1429. Cox, S. D.; Wudl, F. *Organometallics* **1983**, *2*, 184–185. Hutzinger, M. W.; Singer, R. D.; Oehlschlager, A. C. *J. Am. Chem. Soc.* **1990**, *112*, 9397–9398.
72. (a) Zweifel, G.; Leong, W. *J. Am. Chem. Soc.* **1987**, *109*, 6409–6412. (b) Cabezas, J. A.; Oehlschlager, A. C. *Synthesis*, **1994**, 432–442.
73. Behling, J. R.; Babiak, K. A.; Ng, J. S.; Campbell, A. L.; Moretti, R.; Koerner, M.; Lipshutz, B. H. *J. Am. Chem. Soc.* **1988**, *110*, 2641–2643.
74. Piers, E.; Karunaratne, V. *Tetrahedron* **1989**, *45*, 1089–1104. Piers, E.; Karunaratne, V. *Can. J. Chem.* **1989**, *67*, 160–164. Piers, E.; Renaud, J. *Synthesis* **1992**, 74–82.
75. Piers, E.; Friesen, R. W.; Keay, B. A. *Tetrahedron* **1991**, *47*, 4555–4570.
76. Fleming, I.; Taddei, M. *Synthesis* **1985**, 898. See also reference 8. For a different synthesis see: Zhang, H. X.; Guibé, F.; Balavoine, G. *J. Org. Chem.* **1990**, *55*, 1857–1867.
77. Barrett, A. G. M.; Barta, T. E.; Flygare, J. A. *J. Org. Chem.* **1989**, *54*, 4246–4249.

78. Piers, E.; Chong, J. M.; Morton, H. E. *Tetrahedron* **1989**, *45*, 363–380.
79. Piers, E.; Chong, J. M.; Keay, B. A. *Tetrahedron Lett.* **1985**, *26*, 6265–6268.
80. Piers, E.; Wong, T.; Ellis, K. A. *Can. J. Chem.* **1992**, *70*, 2058–2064.
81. Piers, E.; Gavai, A. V. *J. Org. Chem.* **1990**, *55*, 2374–2390.
82. Barbero, A.; Cuadrado, P.; Fleming, I.; González, A. M.; Pulido, F. J. *J. Chem. Soc., Chem. Commun.* **1992**, 351–353. Full paper: Barbero, A.; Cuadrado, P.; Fleming, I.; González, A. M.; Pulido, F. J.; Rubio, R. *J. Chem. Soc., Perkin Trans. 1* **1993**, 1657–1662.
83. Barbero, A.; Cuadrado, P.; Fleming, I.; González, A. M.; Pulido, F. J. *J. Chem. Soc., Perkin Trans. 1* **1992**, 327–331.
84. Barbero, A.; Cuadrado, P.; González, A. M.; Pulido, F. J.; Rubio, R.; Fleming, I. *Tetrahedron Lett.* **1992**, *33*, 5841–5842.
85. Piers, E.; Tse, H. L. A. *Tetrahedron Lett.* **1984**, *25*, 3155–3158. Gilbertson, S. R.; Challener, C. A.; Bos, M. E.; Wulff, W. D. *Tetrahedron Lett.* **1988**, *29*, 4795.

13

Mechanistic studies of organocopper reactions

R. A. J. SMITH

1. Introduction

Many different types of organocopper reagents have been used in organic synthetic procedures and the variety of reactions observed is also extensive.[1] This synthetic efficiency has naturally promoted further extension and refinement and has also generated the inevitable questions regarding the mechanisms of these reactions. Mechanistic investigations in this area have therefore been undertaken for some considerable time and are aimed at rationalizing and directing further synthetic developments as well as increasing fundamental knowledge of a set of chemical reaction processes.

It should be noted that the chemical behaviour of one type of organocopper reagent is sometimes very different from that of another reagent, and also organocopper reagents often react differently in different solvents. Hence, for practical application to a desired transformation it is often necessary to examine several types of organocopper reagents under different reaction conditions in order to attain the optimum reaction efficiency. The wide variation in the types of organocopper reagents and reaction conditions successfully used in synthesis creates many opportunities for mechanistic investigation. However, in many cases reaction parameters uncovered for one reaction–reagent combination may not be applicable to another, apparently related, system. Some of the early mechanistic work[2] was associated with organocopper reagents which were relatively stable and therefore amenable to solid-state structural characterization. Experimental methodology in this area is very specific and is inappropriate for this type of publication. More recent mechanistic work has focused on one of the most generally used reaction systems, namely diorganocuprates (R_2CuLi) in ethereal solvents. Much of this mechanistic work has developed from a significant early study[3] which demonstrated a relationship between the success of the diorganocuprate reaction and the reduction potential of the substrate. Several reactions have been studied including substitution[4] and alkyne carbocupration;[5]

however, most work has concentrated on conjugate addition and this will be the topic of this chapter.

2. Product analysis studies

Product analysis from the reactions of organocopper reagents with various substrates which have been designed to test specific mechanistic points is a logical avenue to explore. These experiments require the standard techniques as described in Chapter 2 and will not be further elaborated. In addition, reactions with specific substrates designed to test the 'cuprate-like' electron-transfer ability of various RLi–copper(I) mixtures have been reported[6] (Scheme 13.1). The actual experimental techniques including the isolation of the product are quite standard.

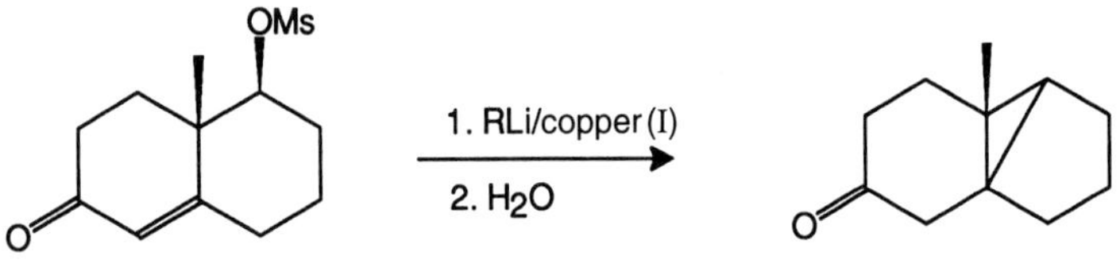

OMs

1. RLi/copper(I)

2. H_2O

Scheme 13.1

Experiments designed to monitor the *rate of production* of various products during the course of a reaction are also a useful source of mechanistic information. These types of experiments involving organocopper reagents are bedevilled by a lack of reproducibility. However, by careful attention to details such as low temperature sampling, as described in Protocol 1,[7] consistent results can be obtained. For example, it was discovered that the rates of production of the 1,4- and 1,2-addition products **2** and **3** from reaction of **1** with $Me_2CuLi{\cdot}LiI$ were not the same. This result indicates that **2** and **3** may not be derived from partitioning of a common intermediate species. By carrying out the reaction in sample vials and ensuring efficient stirring,[8] a set of related experiments can be carried out effectively in a short time. Product analysis using some form of automation is advantageous for a substantial study of this type. A gas chromatograph fitted with an autosampler and data manager is most useful in this regard. Gas chromatography of **3** leads to some sample loss owing to dehydration; however, this difficulty can be resolved by acid treatment prior to analysis which completely dehydrates **3** into quantifiable products and also converts **4** into **2**.

Protocol 1. Conjugate addition of a dialkylcuprate to an α,β-unsaturated ketone with the provision for monitoring the relative rates of formation of the various products[7]

Caution! Carry out all procedures in a well-ventilated hood, and wear disposable vinyl or latex gloves and chemical-resistant safety goggles.

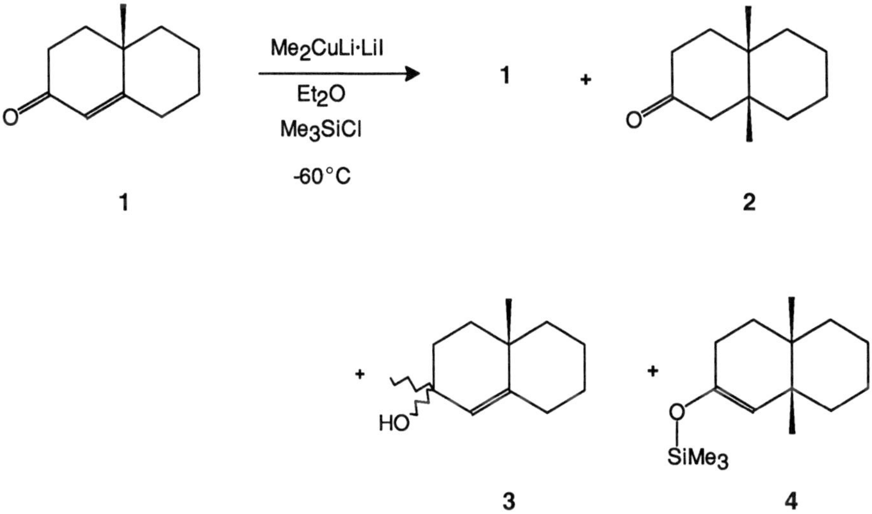

$Me_2CuLi{\cdot}LiI$, Et_2O, Me_3SiCl, -60°C

1 → 1 + 2 + 3 + 4

$SiMe_3$

Equipment

- Argon-filled glove box (e.g. Vacuum Atmospheres model TS 4000) containing an electronic balance (±0.1 mg)
- Glass sample vials (*c.* 12 mL) (e.g. Aldrich Z18,872–7) which have been thoroughly washed then held in an oven at 120°C for 12 h. The vials are allowed to cool under vacuum in the inlet port of the glove box, then introduced into the main body of the glove box under a slight positive argon pressure
- Teflon-covered, octagonal, 12 × 9 mm magnetic stirring bars (e.g. Aldrich Z12,695–0) washed with water followed by acetone and held in an oven at 120°C for 12 h
- White septa (e.g. Aldrich Z10,074–9)
- Nitrogen gas supply and manifold (see Section 6.2 in Chapter 2)
- New, freshly unpackaged, disposable plastic syringes (2 × 5 mL, 10 mL) fitted with 15 cm 20-gauge needles (e.g. Aldrich Z10,270–9). The needles are thoroughly rinsed with water followed by acetone and held in an oven at 120°C for 12 h before attachment to the syringe
- Low temperature refrigerator (e.g. Neslab, cryocool CC-60IIA) fitted with a flexible cooling head immersed in an acetone cooling bath controlled to −60°C
- Glass syringe (250 μL) fitted with a stainless steel plunger and 5 cm needle [e.g. Scientific Glass Engineering (SGE) 250R]. The syringe is washed with acetone and held, with the barrel separated from the plunger, in an oven at 90°C for 4 h. The two components are allowed to cool to room temperature before assembly
- Gas chromatograph fitted with a capillary column optimized for separation of the various reaction products and calibrated with the response factors of the product(s) against a dodecane internal standard. An autosampler and electronic integrator or data management system are desirable optional extras (e.g. Hewlett Packard model 5880 gas chromatograph, model 7673 automatic sampler, and model 3396 integrator)

Protocol 1. *Continued*

Materials

- Copper(I) iodide[a] (FW 190.4) 190.6 mg, 1.00 mmol — **irritant, light sensitive**
- Dry ether[b] 9.0 mL — **flammable, irritant**
- Methyllithium[c] (FW 22.0) 1.43 M in ether, 1.4 mL, 2.0 mmol — **flammable, moisture sensitive**
- Dry, distilled chlorotrimethylsilane (FW 108.6) 130 μL, 111 mg, 1.02 mmol — **flammable, corrosive**
- 10-Methyl-1(9)-octal-2-one[9] **1** (FW 164.3) 167.8 mg, 1.02 mmol — **harmful**
- Dodecane 37.5 mg — **hygroscopic**
- Nitrogen-purged, aqueous ammonium chloride 3 M, *c.* 50 mL — **corrosive**

1. Start up the gas chromatograph and establish retention times and response factors with known weighed mixtures. Using a 50 m × 0.2 mm SE-30 fused silica capillary column and column oven temperature programming from 40 to 250°C at 10°C min^{-1} the retention times are 20.15, 20.41, and 12.14 min for **1**, **2**, and **4**, respectively. Gas chromatography of **3** results in a chromatogram displaying three peaks from the derived dienes with retention times of 16.40, 17.10, and 17.20 min. These peaks are summed together for the quantification of **3**. The response factors for **1**, **2**, **3**, and **4** are 1.30, 1.27, 1.04, and 1.32, respectively, compared to the dodecane standard (retention time 15.81 min).
2. In the argon-filled glove box accurately weigh the copper(I) iodide into the dry vial containing a stirring bar and seal with a septum.
3. Remove the sealed vial from the glove box and connect to the static nitrogen gas supply with a syringe needle and rubber tubing.
4. Add dry ether (6.0 mL) to the vial using a 10 mL plastic syringe and cool to 0°C with an ice bath.
5. Using a 5 mL plastic syringe, add the ethereal methyllithium with vigorous stirring and stir for 6 min at 0°C to produce a clear, colourless solution.
6. Place the vial in the −60°C cold bath and stir for 15 min.
7. Add the chlorotrimethylsilane to the vial using the 250 μL syringe. After 1 min, add a solution of **1** and dodecane in dry ether (3.0 mL) with a 5 mL plastic syringe.
8. After stirring at −60°C for 30 min, flush the 10 mL plastic syringe and needle with nitrogen and insert the needle into the headspace of the reaction vial. Chill the syringe and needle by holding them in an insulated glove containing crushed dry ice for at least 2 min. Withdraw a sample (1.0 mL) of the reaction mixture into the cold syringe and add to nitrogen-purged aqueous ammonium chloride (3 M, 5 mL) contained in a septum-capped vial. Shake thoroughly at room temperature for 1 min.
9. Allow the contents of the sample vial to stand at room temperature for 5 min, then remove the upper ether layer with a Pasteur pipette and place in an autosampler vial containing anhydrous magnesium sulfate (0.3 g).

10. After 1 h at room temperature analyse the clear ethereal solution by gas chromatography to obtain the amounts of **1**, **2**, and **4** in the mixture.
11. Shake a sample of the ether solution obtained in step 9 with an equal volume of 1:9 10% aqueous sulfuric acid–tetrahydrofuran for 1 min in an autosampler vial. Allow the mixture to stand at room temperature for 1 h, then dry with magnesium sulfate and analyse as in step 10 to obtain the amount of **3** in the mixture.
12. Remove samples at various times and analyse as described in steps 8–11.

[a] Purified as described in Protocol 2, Chapter 2.
[b] Freshly opened, or purified as described in Table 2.3.
[c] Standardized as described in Chapter 2, Section 5.5.

3. Structure of organocopper reagents

The reactive organocopper reagents most commonly used for organic synthesis are not easily structurally defined, nor can the actual species in solution be readily represented. Solution structural studies are clearly important to determine the nature of the reacting species which are normally utilized synthetically in solution. NMR studies have been used most often in this regard and have revealed much useful information such as the complexity of the methyllithium–copper(I) iodide system in various solvents[10] and the nature of the metal–carbon bonding in allylcopper systems.[11]

The use of ^{13}C NMR in preference to ^{1}H, ^{6}Li or NMR offers many potential advantages for studies of these species in solution by virtue of the fact that carbon nuclei, being close to the metal site, should be more sensitive to electronic perturbations. The obvious cost of this useful information is the low sensitivity of the ^{13}C nucleus, particularly if used at natural abundance levels. The ^{13}C NMR spectra of ethereal solutions of $Me_2CuLi{\cdot}LiI$ prepared from the various forms of commercial methyllithium (low halide or lithium halide complexed) and copper(I) iodide generally show a ^{13}C NMR methyl resonance at *c.* −9 p.p.m. with half-height line widths of 15–25 Hz.[12] Preparation of 'halide-free' Me_2CuLi by separating and washing the insoluble MeCu, obtained from equimolar amounts of methyllithium and copper(I) iodide, before reaction with a second equivalent of methyllithium produces a solution which displays a resonance at −9.2 p.p.m. with a similar line width. In order to detect accurately the organometallic carbon resonances during reactions it is important to obtain material with as narrow a line width as possible. Halide-free Me_2CuLi displaying a sharp ^{13}C NMR resonance can be prepared from commercial methyllithium provided that care is taken to avoid the inclusion of impurities in the solid MeCu. The procedure (Protocol 2)[12] involves creating a homogeneous organocopper solution before precipitation of the MeCu, which avoids co-precipitation of impurities.

Protocol 2.
Preparation of halide-free lithium dimethylcuprate from commercial methyllithium which is suitable for spectroscopic studies[12]

Caution! Carry out all procedures in a well-ventilated hood, and wear disposable vinyl or latex gloves and chemical-resistant safety goggles.

$$2MeLi + CuI \xrightarrow[0\,°C]{Et_2O} Me_2CuLi{\cdot}LiI$$

$$\text{(cyclohexenone, } 0\,°C) \downarrow$$

$$Me_2CuLi \xleftarrow[Et_2O\text{-}d_{10}]{MeLi} [MeCu]_x(s) + \text{(lithium 3-methylcyclohexenolate, } O^- Li^+)$$

Equipment

- Centrifuge tube (*c.* 25 mL) containing a Teflon-covered magnetic stirring bar (12 mm) and capped with an appropriate size of white septum
- Vacuum/argon manifold (see Section 6.2 in Chapter 2)
- Argon gas supply
- New, freshly unpackaged, disposable plastic syringes (6 × 5 mL, 2 × 10 mL) fitted with 15 cm 20-gauge needles (e.g. Aldrich Z10,270–9). The needles are thoroughly rinsed with water followed by acetone and held in an oven at 120 °C for 12 h before attachment to the syringe
- Glass syringe (250 μL) fitted with a stainless steel plunger and 5 cm needle (e.g. SGE 250A-RN-5). The syringe is washed with acetone and held, with the barrel separated from the plunger, in an oven at 90 °C for 4 h. The two components are allowed to cool to room temperature before assembly
- Centrifuge
- Schlenk tube (*c.* 10 mL) containing a Teflon-covered magnetic stirring bar (9 mm) and capped with a white septum
- NMR tube (10 mm o.d.) (e.g. Wilmad 513–5PP) dried in an oven at 120 °C for 12 h and capped with a 7 mm white septum (e.g. Aldrich Z10,073–0)
- NMR spectrometer (e.g. Bruker AM 360 or Varian VXR 300) fitted with a variable temperature 10 mm $^1H/^{13}C$ probe

Materials

- Dry copper(I) iodide[a] (FW 190.4) 380 mg, 2 mmol — **irritant, light sensitive**
- Dry, distilled ether[b] *c.* 25 mL — **flammable, irritant**
- Methyllithium[c] (FW 22.0) 1.43 M in ether, *c.* 4 mL — **flammable, moisture sensitive**
- Cyclohexenone (FW 96.1) 194 μL, 192 mg, 2 mmol — **highly toxic**
- 2-Propanol *c.* 50 mL — **flammable, irritant**
- Ether-d_{10}[d] 4 mL — **flammable, irritant**

1. Flame dry the septum-capped centrifuge tube and magnetic stirring bar under vacuum by attaching to a vacuum/argon manifold via rubber tubing and a syringe needle. Cool the tube to room temperature under argon.
2. Add the copper(I) iodide to the centrifuge tube and remove any oxygen by alternate evacuation and filling with argon at least three times.
3. Cool the centrifuge tube in an ice bath for 15 min, then add dry ether (7 mL) using a 10 mL plastic syringe followed, with vigorous stirring, by ethereal methyllithium (2.80 mL, 4 mmol) via a 5 mL plastic syringe.
4. After stirring for 15 min at 0°C, add the cyclohexenone using the 250 μL syringe to the clear ethereal solution of $Me_2CuLi{\cdot}LiI$.
5. Remove the tube from the vacuum/argon manifold and centrifuge for 10 min at room temperature.
6. Reconnect the centrifuge tube to the vacuum/argon manifold and cool in an ice bath for 10 min.
7. Remove the clear supernatant using a 10 mL plastic syringe. Destroy any reactive species in the supernatant by adding it to 2-propanol (*c.* 50 mL) in a fume hood.
8. Add dry ether (5 mL) to the methylcopper residue and stir vigorously at 0°C for 5 min.
9. Centrifuge and remove the supernatant as in steps 5–7.
10. Repeat steps 8 and 9.
11. Remove the last traces of solvent from the solid methylcopper by passing a stream of argon gas over the yellow solid via a syringe needle at 0°C.
12. Syringe ethereal methyllithium (1.05 mL, 1.5 mmol) into the dry, septum-capped Schlenk tube previously prepared as described in step 1.
13. Evacuate the Schlenk tube at room temperature until all the solvents have been removed. A white, solid residue remains.
14. Add ether-d_{10} (0.5 mL) to the Schlenk tube, then remove all visible solvent by passing a stream of argon through the mixture as in step 11.
15. Repeat step 14.
16. Add ether-d_{10} (3.0 mL) to the residue in the Schlenk tube and stir vigorously at 0°C for 10 min.
17. Add the ethereal-d_{10} methyllithium solution to the methylcopper precipitate in the centrifuge tube at 0°C using a 5 mL plastic syringe. Stir vigorously at 0°C for 15 min.
18. Centrifuge the cloudy solution at room temperature for 10 min, then remove the supernatant with an argon-purged 5 mL plastic syringe and inject into the previously argon-purged 10 mm NMR tube.

Protocol 2. *Continued*

19. Obtain the NMR spectra by locking on to the higher field deuterium resonance of ether-d_{10}. Accumulate the spectral data using a pulse width corresponding to a 30–40° tip angle and a pulse repetition time of 2–3 s. Take up to 4096 scans as necessary to obtain an acceptable signal-to-noise ratio. Data massaging can be used to improve the signal-to-noise ratio provided that the associated line broadening is <0.2 Hz. Spectral windows of −2 to 10 p.p.m. (8k data points) for ^{1}H and −20 to 220 p.p.m. (64k data points) for ^{13}C are typical. The ^{1}H NMR spectrum shows a single line, 1.15 p.p.m. upfield from tetramethylsilane. The ^{1}H-decoupled ^{13}C NMR spectrum also shows a single, sharp resonance, with a half-height line width of 2.5 Hz, at 9.2 p.p.m. upfield from tetramethylsilane. The methyl resonance line width is comparable to that of the solvent.

[a] Purified as described in Protocol 2, Chapter 2.
[b] Purified as described in Table 2.3.
[c] Standardized as described in Chapter 2, Section 5.5.
[d] Dried over 4A molecular sieve beads (8–12 mesh).

Alternatively, utilization of purified methyllithium, prepared as described in Protocol 3,[13] allows access to methyl organocopper species by the usual routes which display sharp methyl resonances, comparable with those of the solvent. A bonus from these studies is the availability of ^{13}C-enriched methyllithium from the same reaction sequence using ^{13}C-enriched iodomethane.

Protocol 3.
Preparation of an ether solution of low halide methyllithium suitable for spectroscopic studies

Caution! Carry out all procedures in a well-ventilated hood, and wear disposable vinyl or latex gloves and chemical-resistant safety goggles.

$$\mathrm{MeI} + n\text{-BuLi} + \mathrm{Et_2O} \xrightarrow[-40\,^{\circ}\mathrm{C}]{\text{hexane}} [\mathrm{MeLi{\cdot}Et_2O}]_x(\mathrm{s}) + n\text{-BuI}$$

$$[\mathrm{MeLi{\cdot}Et_2O}]_x(\mathrm{s}) \xrightarrow[0\,^{\circ}\mathrm{C}]{\mathrm{Et_2O}} \text{(ethereal) MeLi}$$

Equipment

- Four Schlenk tubes (3×200 mL, 1×100 mL) fitted with 19/22 joints, containing Teflon-covered magnetic stirring bars (12 mm), and capped with white septa (Aldrich Z10,676–5)
- Graduated Schlenk tube (*c.* 75 mL) capped with a white septum (readily obtained by modification of a 100 mL measuring cylinder)
- Two 27 × 120 mm graduated centrifuge tubes (*c.* 70 mL), each containing a Teflon-covered magnetic stirring bar (12 mm) and capped with red septa (Aldrich Z10,145–1)
- Argon gas supply and inlet
- Vacuum/argon manifold (see *Section 6.2* in Chapter 2)
- Stainless steel double-tipped needles (60 cm × 1 mm and 60 cm × 2 mm) cleaned and dried after each use by successive passage of water (*c.* 10 mL), acetone (*c.* 10 mL), and finally dry air
- New, freshly unpackaged, disposable plastic syringes (5 mL and 10 mL) fitted with disposable 4 cm 20-gauge needles
- Low form Dewar flask (600 mL); i.d. 14 cm, depth 9 cm
- High form Dewar flask; i.d. 7 cm, depth 19 cm
- Low temperature refrigeration unit fitted with a flexible cooling head and temperature controller (e.g. Neslab, cryocool CC-60IIA) (optional)
- Low temperature (30 to −120°C) alcohol thermometer
- Centrifuge fitted with rotor buckets having at least 5 cm i.d.

Materials

- $CaCl_2$-dried iodomethane (FW 141.9) 4.87 mL, 11.1 g, 78 mmol — **highly toxic, corrosive**
- *n*-Butyllithium[a] (FW 64.1) 1.56 M in hexane, 50 mL, 78 mmol — **flammable, moisture sensitive**
- Sodium-dried hexane *c.* 150 mL — **flammable, irritant**
- Anhydrous ether[b] *c.* 65 mL — **flammable, irritant**
- Dry toluene *c.* 100 mL — **flammable, toxic**
- 2-Propanol *c.* 10 mL — **flammable, irritant**

1. Flame dry the septum-capped Schlenk tubes and stirring bars as required under vacuum and allow to cool under an atmosphere of argon.
2. Add hexane (150 mL) and ether (65 mL) to separate, dry 200 mL Schlenk tubes and deoxygenate the solvents by alternately evacuating and pressurizing with argon three times at room temperature.
3. Transfer the butyllithium to a dry 100 mL Schlenk tube and cool to −78°C by immersing in a dry ice–acetone bath contained in the high form Dewar flask. Add deoxygenated ether (8.65 mL, 78 mmol) to the butyllithium using the 10 mL syringe and stir the mixture at −78°C under argon for 15 min.
4. Using the 5 mL syringe, add the iodomethane to deoxygenated hexane (30 mL) in a dry 200 mL Schlenk tube under argon and place the tube in the low form Dewar flask containing acetone (*c.* 350 mL) held at −40°C either by intermittent addition of small pieces of dry ice or by the cooling head of the refrigeration unit.
5. Add the −78°C butyllithium–ether mixture to the iodomethane solution at −40°C dropwise over 15 min via the 1 mm double-tipped needle using an argon pressure differential. A white precipitate forms 5 min after the addition commences.
6. Stir the white suspension for 2 h at −40°C.

Protocol 3. *Continued*

7. Flame dry the two septum-capped centrifuge tubes and magnetic stirring bars under vacuum by attaching them to the vacuum/argon manifold via rubber tubing and a syringe needle. Cool the tubes to −78 °C under argon by placing them in the high form Dewar flask.
8. Using the 2 mm double-tipped needle, transfer approximately half of the white suspension under argon into each of the centrifuge tubes held at −78 °C. The transfer tube may be precooled by external massaging with a piece of dry ice *after both ends of the needle have pierced the appropriate septa.*
9. Disconnect the centrifuge tubes from the argon manifold and centrifuge for 10 min at −78 °C using rotors packed with dry ice.
10. Place the centrifuge tubes in the low form Dewar containing dry ice–acetone at −78 °C and connect each tube via rubber tubing and a syringe needle to the argon manifold as in step 7.
11. Remove the clear supernatant in the tubes using the 1 mm double-tipped needle and vent this solution into dry toluene (25 mL) containing 2-propanol (1 mL).
12. Transfer cold (−78 °C) deoxygenated hexane (*c.* 15 mL) to the white residue in each of the tubes using a double-tipped needle and agitate the contents for 15 s at −78 °C.
13. Centrifuge at −78 °C as in step 9.
14. Remove the supernatant at −78 °C as in steps 10 and 11.
15. Repeat steps 12–14 two more times.
16. Cool the Schlenk tube containing the remaining deoxygenated ether to −78 °C in the high form Dewar and, using a double-tipped needle, transfer *c.* 25 mL to the white slurry left in each of the centrifuge tubes after the decantation process.
17. Stir the ethereal mixtures in an ice bath for 1 h, then allow to stand at 0 °C for at least 16 h, maintaining an argon atmosphere.
18. Centrifuge the cloudy solution at room temperature for 15 min, then attach to the argon manifold as in step 10. Transfer the clear supernatant with a 1 mm double-tipped needle to the dry, graduated Schlenk tube by an argon pressure differential.
19. **Caution!** The small amount of residue remaining in the centrifuge tubes after decantation is *extremely pyrophoric. Carefully* dispose of the residues by dilution with dry toluene (20 mL) followed by dropwise addition of 2-propanol in a fume hood behind a robust safety shield.
20. Typical analysis of the product solution shows a methyllithium concentration of 1.03 M, free alkoxide 0.11 M (see Section 5.5 in Chapter 2) and halide

0.015 M. Add 1.0 mL of the methyllithium solution to water (10 mL), acidify with 10% aqueous nitric acid, and titrate[14] with standard aqueous silver nitrate to determine the halide content.

21. The long-term storage stability of this methyllithium solution is not as high as normal commercial material. Store the solution at 0°C under argon. Provided that due care is exercised in handling, the concentration of active reagent can be maintained at useful levels for up to four weeks.

[a] Standardized as described in Chapter 2, Section 5.5. Contains 6.8% free alkoxide.
[b] Freshly opened, or purified as described in Table 2.3.

4. Detection of reaction intermediates

Any comprehensive mechanistic examination is usually accompanied by a substantial kinetic investigation. The extremely sensitive and variable nature of many of the organocopper reagents in common use makes this type of investigation extremely technically demanding. Kinetic evidence[15] for equilibria between intermediate species on the reaction pathway has been further enhanced by IR[16] and NMR[17] spectroscopic observation of intermediates in reacting mixtures of some organocuprates and unsaturated esters. These measurements were made under reaction conditions arranged so that 1,4-addition proceeded at a relatively slow rate. The control of the reactivity was achieved by low temperature operation or by reaction solvent manipulation.

Extension of this approach to α,β-unsaturated ketones using Me_2CuLi in ether solvent has been achieved,[12] and Protocol 4[18] details a procedure which involves mixing two reactants at −80°C adjacent to the probe of an NMR spectrometer followed by continual monitoring of the reaction as it progresses. The simple, symmetrical substrate **5** is preferred as there are no stereochemical complications and it is amenable to selective ^{13}C isotope incorporation for unambiguous resonance assignment. Addition of one equivalent of **5** to $Me_2CuLi{\cdot}LiI$ in diethyl ether-d_{10} at −80°C gives a solution which shows resonances consistent with a lithium-coordinated α,β-unsaturated ketone **6** and a metal π-complex **7** in an approximate ratio of 1:2. The species are most readily identified from spectra taken with samples of **5** enriched with ^{13}C at either the carbonyl or 5-position. A solution of **5** and $Me_2CuLi{\cdot}LiI$ in ether appears to be stable at −80°C for several hours, but on gradual warming to −55°C the peaks assigned to **7** are much less apparent, the relative amount of the lithium complex **6** temporarily increases, and **8** can be detected. The relative intensities and line widths of all peaks in the ^{13}C NMR spectra taken at intervals during the reaction show no significant intensity variation or line broadening.

Protocol 4. Direct spectroscopic observation of lithium complexes and organocopper π-complexes in solutions of diorganocuprates and α,β-unsaturated ketones

Caution! Carry out all procedures in a well-ventilated hood, and wear disposable vinyl or latex gloves and chemical-resistant safety goggles.

$$2MeLi + CuI \xrightarrow[0\,°C]{Et_2O} Me_2CuLi{\cdot}LiI$$

5 → (−80 °C) → 6 + 7 → (−55 °C) → 8

Equipment

- Vacuum/argon manifold (see Section 6.2 in Chapter 2)
- Schlenk tube (*c.* 10 mL) containing a Teflon-covered magnetic stirring bar (9 mm) and capped with a white septum with an argon gas inlet
- Stainless steel double-tipped needles (60 cm × 1 mm and 60 cm × 0.8 mm), cleaned and dried after each use by successive passage of water (*c.* 10 mL), acetone (*c.* 10 mL), and finally dry air
- NMR tube (5 mm o.d.) (e.g. Wilmad 528-PP) with a calibration mark at 0.70 mL. The tube is dried in a 120°C oven for 12 h, then capped with a 4 mm white septum (e.g. Aldrich Z10,070–6)
- Glass syringe (100 μL) fitted with a stainless steel plunger and an 11.5 cm needle (e.g. SGE 100A-RN-11.5). The syringe is washed with acetone and held, with the barrel separated from the plunger, in an oven at 90°C for 4 h. The two components are allowed to cool to room temperature before assembly
- NMR spectrometer (e.g. Bruker AM 360 or Varian VXR 300) fitted with a variable temperature 5 mm ^{13}C probe capable of operating down to *c.* −100 °C

Materials

- Dry copper(I) iodide[a] (FW 190.4) 193 mg, 1.01 mmol — **irritant, light sensitive**
- Dry, distilled ether[b] *c.* 2.5 mL — **flammable, irritant**
- Methyllithium[c] (FW 22.0) 0.92 M in ether, 2.15 mL, 1.98 mmol — **flammable, moisture sensitive**
- Ether-d_{10}[d] 90 μL — **flammable, irritant**
- Distilled 2,2,5-trimethyl-4-hexen-3-one[19] **5** (FW 140.2) 26 mg, 0.19 mmol — **harmful, irritant**

1. Flame dry the septum-capped Schlenk tube and stirring bar under vacuum and allow to cool to room temperature under an atmosphere of argon via attachment to the argon manifold.
2. Add the copper(I) iodide to the Schlenk tube and remove any residual oxygen in the reaction vessel by alternate evacuation and filling with argon at least three times.
3. Cool the tube to 0°C in an ice bath and add dry ether (2 mL) followed by the methyllithium solution using the 1 mm double-tipped needle.
4. After stirring at 0°C under argon for 5 min, allow the solution to stand at 0°C for 15 min.
5. Attach the septum-capped 5 mm NMR tube to the vacuum/argon manifold via a 24-gauge syringe needle and rubber tubing. Evacuate and fill the tube with argon three times, then add the ether-d_{10} via the 100 μL syringe. Cool to 0°C in an ice bath.
6. Transfer 0.7 mL of the supernatant cuprate solution from the Schlenk tube to the cold NMR tube using the 0.8 mm double-tipped needle and an argon gas pressure differential.
7. Place the NMR tube in the NMR spinner turbine and thoroughly clean the external surfaces with a tissue moistened with methanol. Insert the tube and turbine into the probe of the spectrometer, cool to −80°C, and establish acceptable homogeneity. Accumulate the spectral data using a pulse width corresponding to a 90° tip angle and a pulse repetition rate of 3 s. Use a spectral window of −30 to −220 p.p.m. with 32k data points and optimize the data for maximum signal-to-noise ratio, maintaining the associated line broadening to <1 Hz. The ^{1}H-decoupled ^{13}C NMR resonance for the methyl carbon of $Me_2CuLi{\cdot}LiI$ appears 9.2 p.p.m. upfield from tetramethylsilane with a half-height line width of <4 Hz.
8. Eject the sample from the probe of the spectrometer and maintain it suspended in the cold gas plume within the insertion barrel.
9. Inject a solution of **5** in dry ether (50 μL) directly onto the top of the cuprate solution using the 100 μL syringe.
10. Rapidly insert the NMR tube back into the cold (−80°C) probe of the spectrometer and re-establish sample spinning and homogeneity. The ^{13}C NMR spectrum of this solution will show peaks assignable to $Me_2CuLi{\cdot}LiI$, **6**, and **7**. For example, the C-5 resonances for **6** and **7** appear at 158.9 and 67.7 p.p.m., respectively.
11. Increase the probe temperature to −55°C. After 30 min, peaks associated with **8** also become visible, e.g. 31.6 p.p.m. for C-5.

[a] Purified as described in Protocol 2, Chapter 2.
[b] Purified as described in Table 2.3.
[c] Prepared as described in Protocol 3.
[d] Dried over 4A molecular sieve beads (8–12 mesh).

Although the mechanism of conjugate addition reactions with diorganocuprates remains incomplete, reports of experiments such as described in Protocol 4 do help to clarify the picture somewhat. A partial scheme involving lithium coordination and the intermediacy of a cuprate–alkenyl complex appears well established and significant single-electron transfer is not supported. Despite many literature citations proposing a copper(III) intermediate,[20] the actual detection of this species has not yet been achieved and remains as the outstanding challenge in this area of research.

References

1. For a recent comprehensive review see: Lipshutz, B. H.; Sengupta, S. *Org. React.* **1992**, *41*, 135–631.
2. For a retrospective view of this area see: van Koten, G. *J. Organomet. Chem.* **1990**, *400*, 283–301.
3. House, H. O. *Acc. Chem. Res.* **1976**, *9*, 59–67.
4. For a recent example see: Bertz, S. H.; Dabbagh, G.; Mujsce, A. M. *J. Am. Chem. Soc.* **1991**, *113*, 631–636.
5. Review: Normant, J. F.; Alexakis, A. *Synthesis* **1981**, 841–870.
6. Hannah, D. J.; Smith, R. A. J.; Teoh, I.; Weavers, R. T. *Aust. J. Chem.* **1981**, *34*, 181–188.
7. Bertz, S. H.; Smith, R. A. J. *Tetrahedron* **1990**, *46*, 4091–4100.
8. Bertz, S. H.; Dabbagh, G. *Tetrahedron* **1989**, *45*, 425–434.
9. Heathcock, C. H.; Ellis, J. E.; McMurry, J. E.; Coppolino, A. *Tetrahedron Lett.* **1971**, 4995–4996.
10. House, H. O.; Respess, W. L.; Whitesides, G. M. *J. Org. Chem.* **1966**, *31*, 3128–3141. Pearson, R. G.; Gregory, C. D. *J. Am. Chem. Soc.* **1976**, *98*, 4098–4104. Ashby, E. C.; Watkins, J. J. *J. Am. Chem. Soc.* **1977**, *99*, 5312–5317. Ashby, E. C.; Watkins, J. J. *J. Chem. Soc., Chem. Commun.* **1976**, 784–785. Lipshutz, B. H.; Kozlowski, J. A.; Breneman, C. M. *J. Am. Chem. Soc.* **1985**, *107*, 3197–3204.
11. Lipshutz, B. H.; Ellsworth, E. L.; Dimock, S. H.; Smith, R. A. J. *J. Org. Chem.* **1989**, *54*, 4977–4979.
12. Bertz, S. H.; Smith, R. A. J. *J. Am. Chem. Soc.* **1989**, *111*, 8276–8277.
13. Vellekoop, A. S.; Smith, R. A. J.; Bertz, S. H. Manuscript in preparation.
14. Vogel, A. I. *Quantitative Inorganic Analysis*, 2nd edn; Longmans: London, **1960**; p. 251.
15. Krauss, S. R.; Smith, S. G. *J. Am. Chem. Soc.* **1981**, *103*, 141–148.
16. Berlan, J.; Battioni, J.-P.; Koosha, K. *Bull. Soc. Chim. Fr. II* **1979**, 183–190.
17. Hallnemo, G.; Olsson, T.; Ullenius, C. *J. Organomet. Chem.* **1984**, *265*, C22–C24. Hallnemo, G.; Olsson, T.; Ullenius, C. *J. Organomet. Chem.* **1985**, *282*, 133–144.
18. Vellekoop, A. S.; Smith, R. A. J. *J. Am. Chem. Soc.* **1994**, *116*, 2902–2913.
19. Dubois, J.-E.; Schutz, G.; Normant, J. -M. *Bull. Soc. Chim. Fr.* **1966**, 3578–3584.
20. For relevant comments regarding copper(III) intermediates see: Corey, E. J.; Boaz, N. W. *Tetrahedron Lett.* **1985**, *26*, 6015–6018.

A1

Compilation of organocopper preparations

RICHARD J. K. TAYLOR and JOHN M. HERBERT

The following compilation is not intended to be comprehensive, but rather to provide a convenient source of references to full procedures for the preparation of a representative sample of organocopper reagents. It must be emphasized that with the exception of *Organic synthesis* procedures, *the experimental procedures have not been checked independently*. Reagents are grouped by the number of carbons in the main transferable organic group bound directly to copper (excluding those in protecting groups) and, within these groups, they are organized by the molecular formula of the complete organic group. Abbreviated references, as used in *Comprehensive Heterocyclic Chemistry*,* are given alongside the reagents with full reference details listed at the end of the compilation. In generation the abbreviated references are self-evident (90T3667 = *Tetrahedron*, **1990**, 3667).* Where the reference is not in English, this is indicated by means of the letters Fr (French) or Ger (German) after it in the full listing.

In cases where reagents are available commercially, the suppliers' names are noted in italics (see Appendix 2 for a list of addresses). Where a procedure is available elsewhere in this book, the Chapter and Protocol numbers are given. Detailed lists of organocopper variants are provided for reagents based on methyl, butyl, and phenyl; smaller lists are provided for reagents based on *i*-propyl, *t*-butyl, and vinyl. These lists are intended as a source of representative procedures for a range of organocopper reagent types.

In addition to those used generally throughout this book, the following abbreviations are employed in this compilation:

bipy	2,2′-Bipyridyl
Cy	Cyclohexyl

* See *Comprehensive Heterocyclic Chemistry*; Katritzky, A. R.; Rees, C. W. ed.; Pergamon Press, Oxford, **1984**, *Vol. 8*, p. xiii for a full description of this system. Most abbreviations are straightforward; MI indicates a miscellaneous source such as a book or less common journal, TA is *Tetrahedron Asymmetry*.

DBU 1,8-Diazabicyclo[5.4.0]undec-7-ene
EE 1-Ethoxyethyl

n, used before SMe_2 or PBu_3, indicates that the ligand was added in excess

cat. Cu(I)/Pd(O) is used as an abbreviation for treatment of an acetylene with catalytic amounts (typically 5–10% and 0.1–1% respectively) of a copper(I) salt and a palladium(0) complex in the presence of a base (typically an aliphatic amine) which is present in excess.

C_0

CH_3O	$MeOCu.2PPh_3$	78BCJ2909
C_2H_5O	EtONa + 12% CuBr	92T3633
$C_3H_3N_2$	(imidazol-1-yl)NCu(CN)Li	92JOC5250; Chapter 5, Protocol 4
C_3H_9Si	$(Me_3Si)_2CuLi$	87JOC398
C_3H_9Sn	$Me_3SnCu(Me)(CN)Li_2$	91JOC4933
C_4H_9O	*n*- and *s*-BuOCu	74JA2829
C_4H_9O	*t*-BuOCu	73JA3076
C_4H_9S	BuSCu	73OSC107, *Alfa*
C_6H_5O	PhOCu	74JA2829
C_6H_5S	PhSCu	76OS(55)122, *Alfa, Fluka, K&K, Parish*
C_6H_5Se	PhSeCu	82S857, 87JOC4258
$C_8H_{11}Si$	$(PhMe_2Si)_2CuLi$	81JCS(P1)2520
	$(PhMe_2Si)_2Cu(CN)Li_2$	Chapter 9, Protocol 7
$C_8H_{18}N$	Bu_2NCu	81IC2728, *Alfa*
$C_8H_{19}Si$	$Me_2CHCMe_2Si(Me_2)Cu(Bu)(CN)Li$	89JOC4975
$C_9H_{27}Si_4$	$[(Me_3Si)_3Si]_2CuLi$	86JOM(316)255
$C_{12}H_{10}O_3P$	$(PhO)_2P(O)Cu$	83S68

$C_{12}H_{27}Sn$	$(Bu_3Sn)_2CuLi$	92JCS(P1)327
	$Bu_3SnCu(Imid)(CN)Li_2$	92JOC5250
$C_{13}H_{13}Si$	$(Ph_2MeSi)_2CuLi$	86JOM(316)255
$C_{16}H_{19}Si$	$(t\text{-}BuPh_2Si)_2CuLi$	92JCS(P1)327

C_1

CF_3	CF_3Cu	88JCS(P1)921
CH_2NO_2	$O_2NCH_2Cu.2PPh_3$	78BCJ2909
CH_3	Me_2Mg + 5% CuI	52JOC1630
	MeMgCl + 5% $CuBr.SMe_2$	84JOC3503
	MeMgBr+ 1% CuCl	76OS(55)122, 76OS(50)38
	MeMgI + 8% $Cu(OAc)_2$	66JOC1016
	Me_3Al + 10% or 20% CuBr	93AG1368
	MeLi + 25% (NH, NMe, OCu)	93JCS(P1)153
	MeCu	87OS(66)1
	MeCu, halide-free	52JOC1630
	$MeCu.SMe_2$	80MI118
	$MeCu.BF_3$	89JA1351
	$MeCu.AlCl_3$	83CPB128
	$MeCu.PCy_3$	85JOM(282)427
	$MeCu.3P(OMe)_3$	68JOC949
	$MeCu.SBu_2$	74JA2829
	$MeCu.Me_3SiI$	93JOC7238
	Me_2CuLi	78OS(58)158; Chapter 13 Protocol 1 Chapter 8, Protocol 4; Chapter 9, Protocol 2

Me_2CuLi, halide-free	69JA4871
	Chapter 13, Protocol 2
$Me_2CuLi.SMe_2$	75JOC1460
Me_2CuMgI	80JOM(188)293
$Me_2CuZnCl$	92JOC1024
$Me_2CuLi.2BF_3$	89JA1351
MeCu(C≡CBu)Li	74JA7138
MeCu(C≡CBu-*t*)Li	73JOC3893
$MeCu(C{\equiv}CCH_2NMe_2)Li$	79JOC1006
$MeCu(C{\equiv}CCMe_2OMe)Li$	89JA1351
MeCu(CN)Li	87JOC4258
$MeCu(CN)Li.2P(OEt)_3$	68JOC949
$MeCu(CN)Li.BF_3$	Chapter 7, Protocol 3
MeCu(CN)MgBr	86T2873
MeCu(2-Th)Li	89JA1351
MeCu(SPh)Li & MeCu(SePh)Li	87JOC4258
MeCu(2-Th)(CN)LiMgBr	86T2873
$MeCu(NCy_2)Li$	88JOC607
CH_2OLi NCu(Me)Li ; CH_2OMe NCu(Me)Li	87JA2040
$Me_2Cu(CN)Li_2$	85JA1034; Chapter 5, Protocol 6
$Me_2Cu(CN)(ZnCl)_2$	92JOC1024
$Me_2Cu(SCN)Li_2$	83JOC546
Me_3Cu_2Li	85JA3197
Me_3Cu_2MgCl	81RTC249
$Me_3Cu_2OLi_3.nEt_2O$	89T545
Me_3CuLi_2, $Me_5Cu_3Li_2$	77JOC1099

	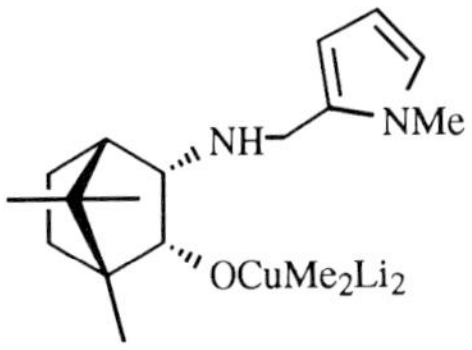 NH, NMe, OCuMe$_2$Li$_2$	92JCS(P1)1193
$C_4H_{11}Si$	$(Me_3SiCH_2)_2CuLi$	86T1389
$C_5H_{11}O_2$	*t*-BuCOOCH$_2$Cu(CN)ZnI	93JOC588
$C_5H_{12}PO_3$	$(EtO)_2P(O)CH_2Cu$	79SC287
C_8H_8NO	$[PhN(Me)CO]_2CuLi$	82CPB3395
$C_9H_6NO_2$	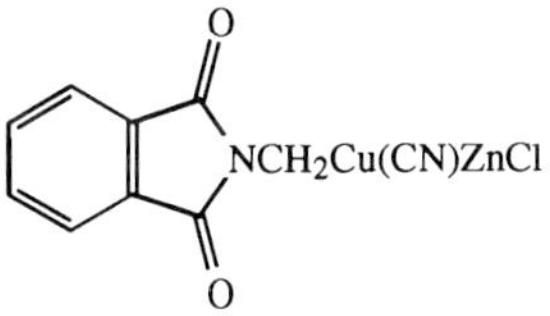 O, NCH$_2$Cu(CN)ZnCl, O	93JOC588
$C_{10}H_{15}SSi$	$[Me_3SiCH(SPh)]_2CuLi$	86JOC3983
$C_{11}H_{19}OS$	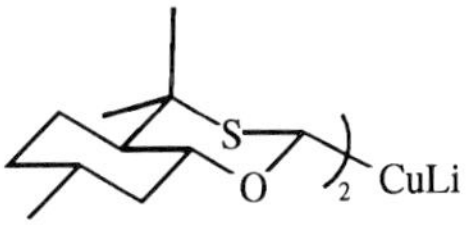 S, O, 2, CuLi	93JOC2920
$C_{14}H_{12}N$	$Ph_2C{=}NCH_2Cu.SMe_2$	82T363

C_2

C_2Cu	HC≡CH + cat. Cu(I)/Pd(0)	75TL4467
C_2F_5	C_2F_5Cu	88JCS(P1)921
C_2H	HC≡CMgBr + 2% CuBr	88MI226
C_2H_2N	$NCCH_2Cu$	72TL487
C_2H_3	$H_2C{=}CHMgBr$ + 3% CuI	82JA7609
	$(H_2C{=}CH)_2CuLi$	82JOC3333
	$(H_2C{=}CH)_2CuLi.SMe_2$	89JA2984
	$(H_2C{=}CH)_2CuMgBr$	80JA5253
	$(H_2C{=}CH)_2CuMgBr.SMe_2$	90JOC964

	$H_2C{=}CHCu(C{\equiv}CBu\text{-}t)Li$	73JOC3893
	$H_2C{=}CHCu(CN)Li$	82JOC5088
	$H_2C{=}CHCu(CN)Li.P(OEt)_3$	72JA7823
	$H_2C{=}CHCu(2\text{-}Th)(CN)Li_2$	90OS(69)80
	$(H_2C{=}CH)_2Cu(CN)Li_2$	87OS(66)52; Chapter 5, Protocol 1
	$(H_2C{=}CH)_2Cu(CN)(MgBr)_2.BF_3$	90JOC964
C_2H_5	EtCu	Chapter 8, Protocol 5; Chapter 11, Protocol 2
	Et_2CuLi	73JA7777; Chapter 11, Protocol 1
C_3H_5O	$[(E)\text{-}MeOCH{=}CH]_2CuLi$	87JOC4258
C_3H_5O	$[H_2C{=}C(OMe)]_2CuLi$	75JA3822
C_4H_5O	$EtOC{\equiv}CMgBr$ + 6% CuBr	81MI73
$C_4H_5O_2$	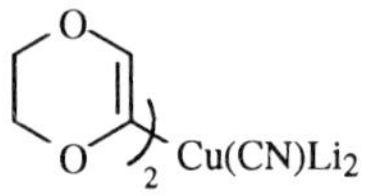 O O 2 $Cu(CN)Li_2$	92TL8087
C_4H_7O	$[H_2C{=}C(OEt)]_2CuLi$	79JOC4781
C_4H_7O	$[(Z)\text{-}EtOCH{=}CH]_2CuLi$	77JA7365
$C_4H_7O_2$	$EtOOCCH_2Cu$	78JOC555
C_5H_9Si	$Me_3SiC{\equiv}CH$ + cat. Cu(I)/Pd(0)	93OS104; Chapter 10, Protocol 3
$C_5H_{11}Si$	$[H_2C{=}C(SiMe_3)]_2CuLi.PBu_3$	91TL857
$C_6H_{11}NO_3P$	$(EtO)_2P(O)CH(CN)Cu$	85CL1779
$C_6H_{11}O_2$	$t\text{-}BuOOCCH_2Cu$	81MI170
$C_6H_{14}O_3P$	$(EtO)_2P(O)CH(Me)Cu$	79SC287
$C_8H_6NO_2S$	$PhSO_2CH(CN)Cu$	87CL887
$C_8H_{19}Si_2$	$(Me_3Si)_2C{=}CHCu(CN)(2\text{-}Th)Li_2$	88JA8129
$C_9H_{13}N_2O$	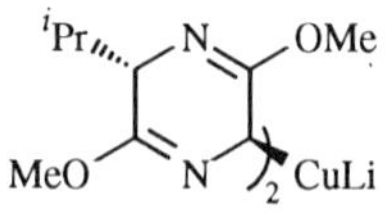 ^iPr N OMe MeO N 2 CuLi	88AG1194
$C_{14}H_{29}Sn$	$(E)\text{-}Bu_3SnCH{=}CHCu(C{\equiv}CPr)Li$	74JA5581
$C_{14}H_{29}Sn$	$(Z)\text{-}Bu_3SnCH{=}CHCu(CN)Li$	92TL49
$C_{22}H_{45}OSn$	$(Z)\text{-}Bu_3Sn(C_8H_{17}O)C{=}CHCu(CN)Li$	94S432
	$(Z)\text{-}Bu_3SnCH{=}C(OC_8H_{17})Cu(CN)Li$	94S432

C_3

C_3F_7	C_3F_7Cu	88JCS(P1)921
C_3H_3	$H_2C{=}C{=}CHMgBr$ + 2% CuCl	81OS(60)41
C_3H_3	$MeC{\equiv}CCu$	67JCS(C)578
C_3H_3O	$HOCH_2C{\equiv}CH$ + cat. Cu(I)/Pd(0)	84S728; Chapter 52 , Protocol 10
C_3H_4Cu	$H_2C{=}C(Cu)CH_2Cu$	92TL6575
C_3H_4N	$NCCH_2CH_2Cu(CN)ZnBr$	90MI3053
C_3H_4N	$NCCH_2CH_2Cu(CN)ZnI$	Chapter 4, Protocol 1
C_3H_5	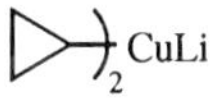 $(\triangleright)_2CuLi$	76BCJ1989
C_3H_5	$[H_2C{=}C(Me)]_2CuLi$	92JOC2960
C_3H_5	$[(E)\text{-}MeCH{=}CH]_2CuLi$	90TA237, 76OS(55)103
C_3H_5	$[(Z)\text{-}MeCH{=}CH]_2CuLi$	71HCA1939
C_3H_5	$[H_2C{=}CHCH_2]_2CuLi$	69JA4871
C_3H_6ClMgO	$ClMgO(CH_2)_3Cu.nSMe_2$	86JOC863
C_3H_7	PrMgBr + 5% $CuBr.SMe_2$	Chapter 6, Protocol 1
	Pr_2CuLi	78BSF(2)299
C_3H_7	*i*-Pr_2CuLi	70JOC1715
	i-PrMgCl + 5% 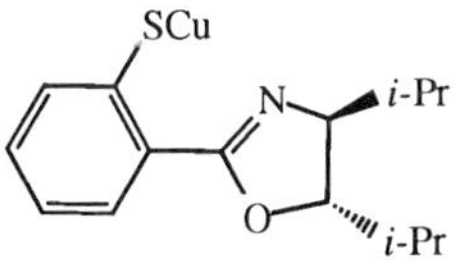 [structure: SCu, N, *i*-Pr, O, *i*-Pr]	93TL7725
C_4H_5O	$MeOCH_2C{\equiv}CMgBr$ + 5% CuCl	78LA658
C_4H_5O	$H_2C{=}C{=}C(OMe)Cu$	93JOC2694
$C_4H_7O_2$	$MeOOC(CH_2)_2Cu.SMe_2$	91JOC1083
C_4H_8NO	$[MeON{=}C(Me)CH_2]_2CuK.nSMe_2$	80TL3115
$C_5H_5O_2$	$EtOOCC{\equiv}CCu$	85S705
$C_5H_6BrO_4$	$(MeOOC)_2C(Br)K$ + *t*-BuOCu	94S242
$C_5H_6NO_2$	EtOOCCH(CN)Na + CuBr	86T2647
$C_5H_7O_2$	(Z)-$MeCH{=}C(COOMe)Cu(CN)Li$	81JOC3696

C_5H_8N	$Me_2NCH_2C{\equiv}CCu$	79JOC1006
$C_5H_9O_2$	(1,3-dioxolan-2-yl $CH_2CH_2)_2CuMgBr.nSMe_2$	84JA1443
$C_5H_9O_2$	$EtOOC(CH_2)_2ZnCl$ + 2% $CuBr.SMe_2$	84JA3368
	$[EtOOC(CH_2)_2]_2Zn$ + 8% $CuBr.SMe_2$	93JOC1038
$C_5H_9O_2$	$EtOOCCH(Me)Cu.P(OMe)_3$	82JOC3464
$C_5H_{11}N_2$	$[Me_2NN{=}C(Me)CH_2]_2CuLi.nSMe_2$	82JA1054
$C_5H_{11}O_2$	$(MeO)_2CHCH_2CH_2Cu.SMe_2$	87JOC1381
$C_6H_9S_2$	2-vinyl-1,3-dithian-2-yl $Cu.P(OMe)_3$	79TL4717
$C_6H_{10}MgBrOSi$	$[(E)\text{-}Me_3SiCH_2C(OMgBr){=}CH]_2CuLi$	86T1389
$C_6H_{11}O_2$	1,3-dioxan-2-yl CH_2CH_2MgBr + 5% CuI	84S278
$C_6H_{11}O_2$	*i*-$PrOOCCH_2CH_2ZnCl$ + 5% $CuBr.SMe_2$	87JA8056
$C_6H_{11}Si$	$(Me_3SiC{\equiv}CCH_2)_2CuMgBr$	81JA1831
$C_6H_{13}Si$	$[H_2C{=}C(SiMe_3)CH_2]_2CuMgBr.nSMe_2$	81JA1831
$C_7H_{11}O_2$	$EEOCH_2C{\equiv}CMgBr$ + 4% CuCl	88MI225
$C_7H_{11}O_4$	$(EtOOC)_2CH_2$ + NaH + CuBr	86T2647
$C_7H_{13}O_2$	$[H_2C{=}C[CH(OEt)_2]]_2CuLi$	92JOC2960
$C_7H_{13}S$	$H_2C{=}CHCH(SBu\text{-}t)Cu$	81JOC3790
$C_7H_{15}O_2$	$(EEOCH_2CH_2CH_2)_2CuLi$	72JOC1947
$C_8H_{11}O_2$	$THPOCH_2C{\equiv}CH$ + cat. Cu(I)/Pd(0)	72JOC1947
$C_8H_{17}OSi$	$(E)\text{-}Me_3SiCH_2C(OEt){=}CHCu$	86T1389,1399
C_9H_8NS	$NCCH_2CH(SPh)Cu(CN)ZnCl$	90TL7575
C_9H_9S	(1-(phenylthio)cyclopropyl$)_2CuLi$	84MI125
$C_9H_{11}S$	$PhSCH_2CH_2CH_2MgBr$ + 1% Li_2CuCl_4	78MI599
$C_9H_{16}N$	$[CyN{=}C(Me)CH_2]_2CuLi$	84TL2813
$C_9H_{19}OSi$	$(E)\text{-}t\text{-}BuMe_2SiOCH_2CH{=}CHCu(CN)(2\text{-}Th)Li_2$	Chapter 5, Protocol 3

$C_9H_{21}Si$	$(t\text{-}BuMe_2SiCH_2CH_2CH_2)_2CuLi$	85TL547
$C_{11}H_{10}NO_2$	(phthalimido) $NCH_2CH_2CH_2Cu(CN)ZnCl$	91OS(70)195
$C_{11}H_{15}Si$	$[(Z)\text{-}PhMe_2Si(Me)C{=}CH]_2Cu(C{\equiv}CBu)Li_2$	84JCS(P1)119
$C_{11}H_{15}Si$	$[H_2C{=}C(CH_2SiMe_2Ph)]_2Cu(CN)Li_2$	89T413
$C_{11}H_{21}O_2Si$	$(E)\text{-}THPOCH_2CH{=}C(SiMe_3)Cu.nP(OEt)_3$	86JCS(P1)1515
$C_{12}H_{23}Si$	$i\text{-}Pr_3SiC{\equiv}CCH_2Cu$	84JA6006
$C_{15}H_{31}Sn$	$[H_2C{=}C(SnBu_3)CH_2]_2CuLi$	92JCS(P1)327
$C_{15}H_{31}Sn$	$(Z)\text{-}MeCH{=}C(SnBu_3)Cu(2\text{-}Th)(CN)Li_2$	Chapter 5, Protocol 5
$C_{18}H_{20}NO$	(structure: Ph, OMe, N, Ph) $Cu(C{\equiv}CMe_2OMe)Li$	Chapter 8, Protocol 7

C_4

C_4H_3O	(2-furyl)$_2$ $CuLi.SMe_2$	79TL4577
C_4H_3S	(2-thienyl)$_2$ $CuLi$	70ACS2379; *Aldrich*
	(2-thienyl) $Cu(CN)Li$	Chapter 5, Protocol 3
C_4H_5	$H_2C{=}CHC(MgCl){=}CH_2$ + 8% $CuBr.SMe_2$	83TL1003
C_4H_5	$EtC{\equiv}CH$ + cat. Cu(I)/Pd(0)	84SC761
C_4H_5O	$H_2C{=}CHCH_2COCu(Me)(CN)Li_2$	90TL477
C_4H_6Cl	$[ClCH_2CH_2C({=}CH_2)]_2CuMgBr$	89T1089

	t-BuCu(OBu-*t*)Li	73JA3076
	t-BuCu(OPh)Li and *t*-BuCu(SBu-*t*)Li	73JA7788
	t-BuCu(SPh)Li	76OS(55)122
	t-$Bu_2Cu(CN)Li_2$	90JOC964
	t-BuCu(CN)(2-Th)Li_2	85JOM(285)437
C_5H_6N	MeN Cu	71ACS2596
$C_5H_9O_2$	O O Cu.TMEDA	91JOC5491
$C_5H_9O_2$	$MeOOCCH_2CH_2CH_2Cu.SMe_2$	91JOC1083
$C_6H_9O_2$	(Z)-MeCH=C(COOEt)Cu(C≡CBu)Li	83JOC4621
$C_6H_9O_3$	$EtOOCCH(COCH_3)Cu$	84S616
$C_6H_{11}O_2$	O O $CH_2CH_2CH_2MgBr$ + 25% $CuBr.SMe_2$	82JOC5045
$C_6H_{11}O_2$	$[AcO(CH_2)_4]_2Cu(CN)(MgCl)_2$	93JOC4781
$C_6H_{11}O_2$	*i*-PrCH(OAc)Cu(CN)ZnBr	90JOC4791
	$EtOOC(CH_2)_3Cu(CN)ZnI$	Chapter 4, Protocol 2
$C_7H_{13}O_2$	O O $)_2$ CuLi	82JOC4605
$C_7H_{13}Si$	$Me_3SiC≡CCH_2CH_2Cu.BF_3$	81T4431
$C_7H_{13}Sn$	(*E*,*E*)-$Me_3SnCH=CHCH=CHCu(Me)(CN)Li_2$	91TL7211
$C_7H_{15}Si$	$H_2C=C(CH_2SiMe_3)CH_2Cu$	81JA1831
$C_7H_{17}Si$	$PrCH(SiMe_3)Cu$	86T1389
$C_8H_{15}O_2$	*t*-$BuOOC(CH_2)_3Cu$	89T443; Chapter 3, Protocol 1
$C_8H_{17}O$	(S)-*t*-$BuOCH_2CH(Me)CH_2Cu(CN)Li$	90T4503
$C_8H_{17}O_2$	$(EtO)_2CH(CH_2)_3MgBr$ + 4% Li_2CuCl_4	82JOC4825
$C_9H_{13}O_2$	$THPOCH_2CH_2C≡CMgBr$ + 10% CuCl	86T3781

$C_9H_{15}O_2$	(Z)-EEOCH$_2$C(Me)C=CHCu(C≡CCMe$_2$OMe)Li	86T2831
$C_9H_{15}O_3$	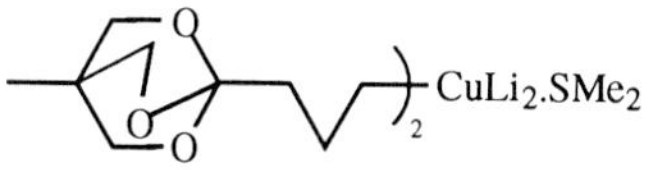 CuLi$_2$.SMe$_2$	93JOC2134
$C_9H_{17}OSi$	(*E*)-H$_2$C=C(SiMe$_3$)C(OEt)=CHCu	86T1389
$C_9H_{17}O_2$	THPO(CH$_2$)$_4$Cu.SMe$_2$	87JOC1381
$C_9H_{17}O_2$	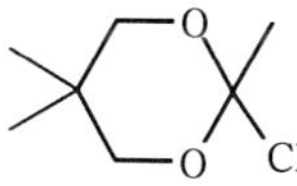 CH$_2$CH$_2$MgBr + 14% CuBr	90JOC4417
$C_9H_{17}O_2Si$	(Z)-Me$_3$SiCH$_2$CH=C(COOEt)Cu	86T1389
$C_{10}H_{21}OSi$	(*E*)-*t*-BuMe$_2$SiOCH$_2$CH$_2$CH=CHCu(CN)(2-Th)Li$_2$	91TL5647
$C_{10}H_{21}OSi$	[(*E*)-*t*-BuMe$_2$SiOCH$_2$CH=C(Me)]$_2$CuLi	76T2281
$C_{10}H_{21}OSi$	[(Z)-*t*-BuMe$_2$SiOCH$_2$C(Me)=CH]$_2$CuLi	92JA8008
$C_{11}H_{24}O_2Si$	[(Z)-Me$_3$SiCH$_2$CH=C[CH(OEt)$_2$]]$_2$CuLi	86T1389
$C_{16}H_{22}NO_4$	(S)-BocNHCH(COOBn)CH$_2$CH$_2$Cu	93SL219
$C_{17}H_{24}NO$	BnNH(CH$_2$)$_6$OCH$_2$CH$_2$C≡CH + cat. Cu(I)/Pd(O)	Chapter 10, Protocol 1
$C_{20}H_{27}Si$	*t*-BuPh$_2$Si(CH$_2$)$_4$ZrCp$_2$Cl + 10% CuBr.SMe$_2$	93S537

C_5

C_5H_4N	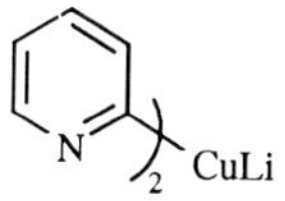 N, CuLi (2-pyridyl)$_2$CuLi	82T1509
C_5H_4N	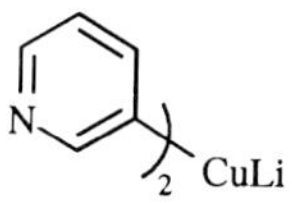 N, CuLi (3-pyridyl)$_2$CuLi	85H117
C_5H_5	H$_2$C=CMeC≡CCu	80JOC1158
C_5H_5O	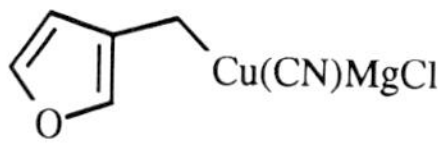 O, Cu(CN)MgCl	85JOC3988
C_5H_7	PrC≡CCu + Me$_3$SiI	93JOC7238
C_5H_7	(Me$_2$C=C=CH)$_2$CuLi	76TL275

C_5H_7	$HC{\equiv}CCH_2CH_2CH_2Cu(CN)ZnI$	91JOC5974; Chapter 4, Protocol 3
C_5H_7	$(MeC{\equiv}CCH_2CH_2)_2CuMgBr$	83JA2364
C_5H_7	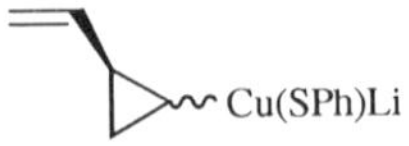 Cu(SPh)Li	83CJC1226
C_5H_7O	$HOCMe_2C{\equiv}CH + Et_3N$ + 6% CuI	90TL5161
C_5H_7O	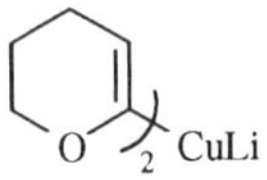 O 2 CuLi	83TL3905
$C_5H_7O_2$	$(CH_3CO)_2CHNa$ + 5% CuBr	78OS(58)52
C_5H_8Cl	(E)-$ClCH_2CH_2CH_2CH{=}CHCu(C{\equiv}CPr)Li$	80AG472
C_5H_8Cl	$H_2C{=}C(CH_2CH_2CH_2Cl)Cu.2PBu_3$	93CJC280
C_5H_9	$[H_2C{=}CH(CH_2)_3]_2CuLi.SMe_2$	86JOC1730
C_5H_9	$[(Z)$-$PrCH{=}CH]_2CuLi$	89T381
C_5H_9	(Z)-EtC(Me)=CHCu	Chapter11, Protocol 2
	(Z)-$EtC(Me){=}CHCu.SMe_2$	83JCS(P1)1387
C_5H_9	$[Me_2C{=}CHCH_2]_2Cu(CN)Li_2$	90JOC1695
C_5H_9O	 O $CH_2CH_2CH_2Cu$	89T443
C_5H_9O	$(t$-$BuCO)_2Cu(CN)Li_2$	85JA4551
C_5H_9O	$MeCO(CH_2)_3Cu$	91JOC4744
$C_5H_{10}Br$	$Br(CH_2)_5ZrCp_2Cl$ + 15% CuCN	92TL5857
$C_5H_{10}BrMgO$	$BrMgO(CH_2)_5MgBr$ + 34% Li_2CuCl_4	91JCS(P1)2639
$C_5H_{10}Cu$	$Li(PhS)Cu(CH_2)_5Cu(SPh)Li$	88JA2218
C_5H_{11}	$(C_5H_{11})_2CuLi$	87JOC1801
C_5H_{11}	$(i$-$PrCH_2CH_2)_2CuLi$	83JOC1404
C_5H_{11}	$(t$-$BuCH_2)_2CuLi$	84T1401
C_6H_9O	$MeOCMe_2C{\equiv}CCu$	78JOC3418
C_6H_9O	$H_2C{=}C(OMe)C(CH_2Cu){=}CH_2$	93JOC2694
$C_6H_{11}O_2$	$MeOOC(CH_2)_4Cu$	92MI1560

$C_7H_9O_2$	(spiro dioxolane cyclopentenyl) $Cu(C{\equiv}CCMe_2OMe)Li$	91TL605
$C_7H_{11}O_2$	(Z)-$EtCH{=}C(COOEt)Cu(Me)Li$	93SC143
$C_7H_{13}O_2$	(2-methyl-1,3-dioxolan-2-yl) $CH_2CH_2CH_2MgCl$ + 5% CuBr	85SC569
$C_7H_{13}O_2$	(Z)-$(MeO)_2CH(CH_2)_2CH{=}CHCu$	87JOC1381
$C_7H_{16}O_3P$	$(MeO)_2P(O)CH_2CH(Pr)Cu(CN)ZnBr$	90TL1833
$C_8H_{11}Si$	$Me_3SiC{\equiv}CCH_2C{\equiv}CH$ + cat. Cu(I)/Pd(O)	92JOC6853
$C_8H_{15}Si$	(Z)-$H_2C{=}CHCH_2CH{=}C(SiMe_3)MgBr$ + 10% CuI	82SC1027
$C_8H_{15}Si$	$Me_3SiC{\equiv}C(CH_2)_3MgI$+ 13% CuCN	81JCS(P1)1516
$C_8H_{17}Si$	(Z)-$PrCH{=}C(SiMe_3)Cu$	77TL1805
$C_9H_{13}O_2$	$EEOCH_2CH{=}CHC{\equiv}CMgBr$ + 4% CuBr	83JA3656
$C_9H_{19}O_2$	$EEO(CH_2)_2CH(Me)CH_2MgBr$ + 5% CuBr	85SC569
$C_{10}H_{15}O_2$	$THPOCMe_2C{\equiv}CCu$	78SC175
$C_{10}H_{17}O_2$	(*E*)-*t*-$BuCOO(CH_2)_3CH{=}CHCu(CN)ZnI$	92JA3983; Chapter 4, Protocol 6
$C_{10}H_{19}O_2$	$THPO(CH_2)_5MgCl$ + 2.5% Li_2CuCl_4	84SC761
$C_{10}H_{19}O_2$	$THPOCH_2CMe_2CH_2Cu.SMe_2$	85T3943
$C_{11}H_{19}BClO_2$	Cl(CH₂)₃CH=C(Bpin)Cu(CN)ZnI (structure: Cl, B, O, O, Cu(CN)ZnI)	92TL3717
$C_{12}H_{17}O$	$BnOCH_2CH_2CH(Me)CH_2MgBr$+ 2% Li_2CuCl_4	83S804
$C_{21}H_{25}OSi$	*t*-$BuPh_2SiO(CH_2)_3C{\equiv}CCu(CN)Li$	93JOC528

C_6

C_6Br_5	C_6Br_5Cu	72JOM(42)257
C_6Cl_5	C_6Cl_5Cu	73S170

C_6F_5	$(C_6F_5Cu)_2$.dioxane	79OS(59)122; *Alfa*
C_6H_4Cl	3-ClC_6H_4MgBr + 8% $Cu(OAc)_2$	67JOC784
C_6H_5	PhMgBr + 5% CuCl	90OS(69)1
	PhCu	52JOC1630; Chapter 8, Protocol 6
	PhCu (from active Cu)	88JOC4482
	$PhCu.SMe_2$	89T425
	$PhCu.BPh_3$	67JOM(8)339
	Ph_2CuLi	73JA7777
	$Ph_2CuMgBr$	80JOM(188)293
	$Ph_2CuMgBr.SMe_2$	90JOC964; Chapter 8, Protocol 3
	PhCu(CN)Li	93JOC36
	$PhCu(CN)MgBr.BF_3$	93JOC1207
	$Ph_2Cu(CN)Li_2$	89JOC1295
	$Ph_2Cu(SCN)Li_2$	83JOC546
	$Ph_2Cu(CN)(MgBr)_2$	90JOC964
	$PhCu(CN)(2\text{-}Th)Li_2$	85JOM(285)437
	PhCu(CN)(2-Th)MgBrLi	86T2873
$C_6H_6BrMgO_2$	$BrMgOOC(CH_2)_3C{\equiv}CMgBr$ + 14% CuCl	82JOC1221
C_6H_6N	N … Cu	84JOC3928
	N … $)_2$ $Cu(CN)Li_2$	Chapter 5, Protocol 2
C_6H_7O	O … $CH_2CH_2Cu(CN)MgBr$	85JOC3988
C_6H_7O	O … Cu(CN)ZnI	93T29; Chapter 4, Protocol 4
C_6H_9	$)_2$ $Cu(CN)Li_2$	90TL4539

$C_9H_{21}Si$	$(t\text{-}BuMe_2SiCH_2CH_2CH_2)_2CuLi$	85TL547
$C_{11}H_{10}NO_2$	(phthalimido)$NCH_2CH_2CH_2Cu(CN)ZnCl$	91OS(70)195
$C_{11}H_{15}Si$	$[(Z)\text{-}PhMe_2Si(Me)C{=}CH]_2Cu(C{\equiv}CBu)Li_2$	84JCS(P1)119
$C_{11}H_{15}Si$	$[H_2C{=}C(CH_2SiMe_2Ph)]_2Cu(CN)Li_2$	89T413
$C_{11}H_{21}O_2Si$	$(E)\text{-}THPOCH_2CH{=}C(SiMe_3)Cu.nP(OEt)_3$	86JCS(P1)1515
$C_{12}H_{23}Si$	$i\text{-}Pr_3SiC{\equiv}CCH_2Cu$	84JA6006
$C_{15}H_{31}Sn$	$[H_2C{=}C(SnBu_3)CH_2]_2CuLi$	92JCS(P1)327
$C_{15}H_{31}Sn$	$(Z)\text{-}MeCH{=}C(SnBu_3)Cu(2\text{-}Th)(CN)Li_2$	Chapter 5, Protocol 5
$C_{18}H_{20}NO$	(structure: Ph, OMe, N, Ph) $Cu(C{\equiv}CMe_2OMe)Li$	Chapter 8, Protocol 7

C_4

C_4H_3O	(2-furyl)$_2$CuLi.SMe$_2$	79TL4577
C_4H_3S	(2-thienyl)$_2$CuLi	70ACS2379; *Aldrich*
	(2-thienyl)Cu(CN)Li	Chapter 5, Protocol 3
C_4H_5	$H_2C{=}CHC(MgCl){=}CH_2$ + 8% $CuBr.SMe_2$	83TL1003
C_4H_5	$EtC{\equiv}CH$ + cat. Cu(I)/Pd(0)	84SC761
C_4H_5O	$H_2C{=}CHCH_2COCu(Me)(CN)Li_2$	90TL477
C_4H_6Cl	$[ClCH_2CH_2C({=}CH_2)]_2CuMgBr$	89T1089

C_4H_6N	$NCCH_2CH_2CH_2Cu(CN)ZnI$	89TL4799
C_4H_6OLi	(Z)-$MeCH{=}C(CH_2OLi)Cu(2\text{-}Th)(CN)Li_2$	93JA3966
C_4H_7	$(H_2C{=}CHCH_2CH_2)_2CuLi$	80JOC2229
C_4H_7	$H_2C{=}C(Me)CH_2Cu$	92JA5110; Chapter 3, Protocol 5
	$[H_2C{=}C(Me)CH_2]_2CuLi$	Chapter 9, Protocol 4
	$[H_2C{=}C(Me)CH_2]_2Cu(CN)Li_2$	Chapter 5, Protocol 6
C_4H_7	$(Me_2C{=}CH)_2CuLi$	90T4277
C_4H_7	[(Z)-$EtCH{=}CH]_2CuLi$	Chapter 11, Protocol 1
C_4H_8ClMgO	$ClMgO(CH_2)_4Cu$	79S885
C_4H_8Cu	$Li(PhS)Cu(CH_2)_4Cu(SPh)Li$	91OS(70)204
C_4H_8IZn	$IZn(CH_2)_4Cu(CN)ZnI$	91JOC4591
C_4H_9	BuMgBr + 5% CuI	87OS(66)116
	BuMgBr + 6% $Cu(OAc)_2.H_2O$	65JA82
	BuMgCl + 10% CuCN	89HCA1337
	BuMgBr + CuI	81JCS(P1)593
	BuMgCl + 30% $MnCl_2$ + 3% CuCl	93OS135
C_4H_9	$BuTi(OPr\text{-}i)_4Li$ + 5% CuI and $BuTi(OPr\text{-}i)_3$ + 5% CuI	91JOC5489
	BuCu	78IC275
	$BuCu.SMe_2$	91OS(70)215
	$BuCu.P(OEt)_3$	86T1389
	$BuCu.PBu_3$	82JCS(P1)1177
	$BuCu.2PBu_3$	87T813
	$BuCu.AlCl_3$	Chapter 7, Protocol 2
	$BuCu.BF_3$	77JA8068; Chapter 7, Protocol 1
	$BuCu.Me_3SiI$	91T9691; Chapter 8, Protocol 2
	Bu_2CuLi	73JA7777; Chapter 2, Protocol 11
	$Bu_2CuLi.SMe_2$	84OS(62)1
	$Bu_2CuLi.ClSiMe_2Bu\text{-}t$	89T349
	$Bu_2CuMgBr$	81JCS(P1)593
	$Bu_2CuMgBr.SMe_2$	89T403
	BuCu(OBu-*t*)Li	74JOC400

	BuCu(SPh)Li	81JCS(P1)593
	BuCu(PCy_2)Li	84JOC1119
	BuCu(C≡CPr)Li	72JA7210
	BuCu(C≡CPr)MgBr	82JCS(P1)1177
	BuCu(CH_2SOMe)(CN)Li_2	92OR135
	BuCu(CN)Li	82JOC5088
	N Ph N(Me)Cu(Bu)Li	93T965; Chapter 8, Protocol 8
	BuCu(CN)(2-Th)Li_2	85JOM(285)437
	BuCu(CN)(Imid)Li_2	Chapter 5, Protocol 4
	Bu_2Cu(CN)Li_2	84JOC3928
	Bu_2Cu(SCN)Li_2	83JOC546
	Bu_2Cu(CN)$(ZnCl)_2$	92JOC1024
C_4H_9	Ph Cu(Bu)Li N Li(Bu)Cu N Ph N N	94T4455
	Bu_3CuLi_2	Chapter 8, Protocol 1
	$Bu_5Cu_3Li_2$. 6 Me_2NP (Pr-*i* N, O, Ph)	Chapter 8, Protocol 9
C_4H_9	*i*-BuCu(SPh)Li	84JOC3183
C_4H_9	*s*-Bu_2CuLi	73JA7777
C_4H_9	*t*-BuMgCl + 5% CuBr	87JOC3901; Chapter 6, Protocol 2
	t-BuMgBr + 10% CuI	73JA7788
	t-BuCu	73JA7788
	t-Bu_2CuLi	73JA7777
	t-Bu_2CuLi.SMe_2	81MI361
	t-Bu_2CuMgCl	89T403
	t-BuCu(NEt_2)Li	73JA7788

	t-BuCu(OBu-*t*)Li	73JA3076
	t-BuCu(OPh)Li and *t*-BuCu(SBu-*t*)Li	73JA7788
	t-BuCu(SPh)Li	76OS(55)122
	t-$Bu_2Cu(CN)Li_2$	90JOC964
	t-BuCu(CN)(2-Th)Li_2	85JOM(285)437
C_5H_6N	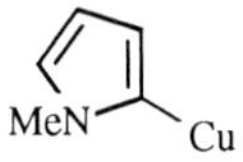 MeN Cu	71ACS2596
$C_5H_9O_2$	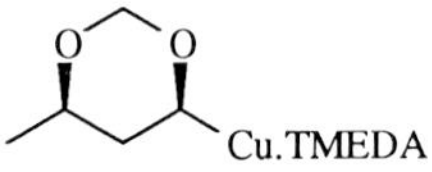 O O Cu.TMEDA	91JOC5491
$C_5H_9O_2$	$MeOOCCH_2CH_2CH_2Cu.SMe_2$	91JOC1083
$C_6H_9O_2$	(Z)-MeCH=C(COOEt)Cu(C≡CBu)Li	83JOC4621
$C_6H_9O_3$	$EtOOCCH(COCH_3)Cu$	84S616
$C_6H_{11}O_2$	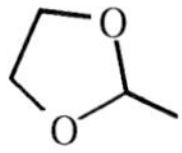 O O $CH_2CH_2CH_2MgBr$ + 25% $CuBr.SMe_2$	82JOC5045
$C_6H_{11}O_2$	$[AcO(CH_2)_4]_2Cu(CN)(MgCl)_2$	93JOC4781
$C_6H_{11}O_2$	*i*-PrCH(OAc)Cu(CN)ZnBr	90JOC4791
	$EtOOC(CH_2)_3Cu(CN)ZnI$	Chapter 4, Protocol 2
$C_7H_{13}O_2$	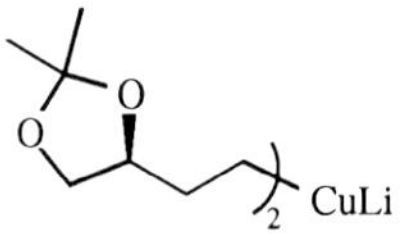 O O 2 CuLi	82JOC4605
$C_7H_{13}Si$	$Me_3SiC≡CCH_2CH_2Cu.BF_3$	81T4431
$C_7H_{13}Sn$	(*E*,*E*)-$Me_3SnCH=CHCH=CHCu(Me)(CN)Li_2$	91TL7211
$C_7H_{15}Si$	$H_2C=C(CH_2SiMe_3)CH_2Cu$	81JA1831
$C_7H_{17}Si$	$PrCH(SiMe_3)Cu$	86T1389
$C_8H_{15}O_2$	*t*-$BuOOC(CH_2)_3Cu$	89T443; Chapter 3, Protocol 1
$C_8H_{17}O$	(S)-*t*-$BuOCH_2CH(Me)CH_2Cu(CN)Li$	90T4503
$C_8H_{17}O_2$	$(EtO)_2CH(CH_2)_3MgBr$ + 4% Li_2CuCl_4	82JOC4825
$C_9H_{13}O_2$	$THPOCH_2CH_2C≡CMgBr$ + 10% CuCl	86T3781

$C_9H_{15}O_2$	(Z)-EEOCH$_2$C(Me)C=CHCu(C≡CCMe$_2$OMe)Li	86T2831
$C_9H_{15}O_3$	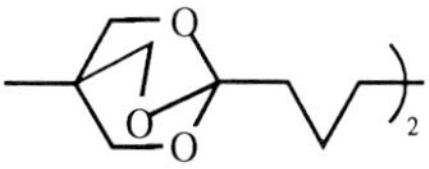 O O O ( )$_2$ CuLi$_2$.SMe$_2$	93JOC2134
$C_9H_{17}OSi$	(*E*)-H$_2$C=C(SiMe$_3$)C(OEt)=CHCu	86T1389
$C_9H_{17}O_2$	THPO(CH$_2$)$_4$Cu.SMe$_2$	87JOC1381
$C_9H_{17}O_2$	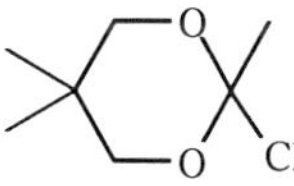 O O CH$_2$CH$_2$MgBr + 14% CuBr	90JOC4417
$C_9H_{17}O_2Si$	(Z)-Me$_3$SiCH$_2$CH=C(COOEt)Cu	86T1389
$C_{10}H_{21}OSi$	(*E*)-*t*-BuMe$_2$SiOCH$_2$CH$_2$CH=CHCu(CN)(2-Th)Li$_2$	91TL5647
$C_{10}H_{21}OSi$	[(*E*)-*t*-BuMe$_2$SiOCH$_2$CH=C(Me)]$_2$CuLi	76T2281
$C_{10}H_{21}OSi$	[(Z)-*t*-BuMe$_2$SiOCH$_2$C(Me)=CH]$_2$CuLi	92JA8008
$C_{11}H_{24}O_2Si$	[(Z)-Me$_3$SiCH$_2$CH=C[CH(OEt)$_2$]]$_2$CuLi	86T1389
$C_{16}H_{22}NO_4$	(S)-BocNHCH(COOBn)CH$_2$CH$_2$Cu	93SL219
$C_{17}H_{24}NO$	BnNH(CH$_2$)$_6$OCH$_2$CH$_2$C≡CH + cat. Cu(I)/Pd(O)	Chapter 10, Protocol 1
$C_{20}H_{27}Si$	*t*-BuPh$_2$Si(CH$_2$)$_4$ZrCp$_2$Cl + 10% CuBr.SMe$_2$	93S537

C_5

C_5H_4N	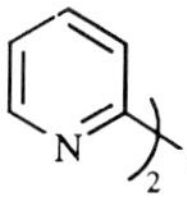 N ( )$_2$ CuLi	82T1509
C_5H_4N	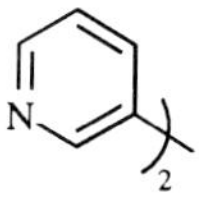 N ( )$_2$ CuLi	85H117
C_5H_5	H$_2$C=CMeC≡CCu	80JOC1158
C_5H_5O	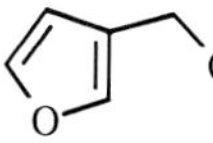 O Cu(CN)MgCl	85JOC3988
C_5H_7	PrC≡CCu + Me$_3$SiI	93JOC7238
C_5H_7	(Me$_2$C=C=CH)$_2$CuLi	76TL275

C_5H_7	$HC{\equiv}CCH_2CH_2CH_2Cu(CN)ZnI$	91JOC5974; Chapter 4, Protocol 3
C_5H_7	$(MeC{\equiv}CCH_2CH_2)_2CuMgBr$	83JA2364
C_5H_7	Cu(SPh)Li	83CJC1226
C_5H_7O	$HOCMe_2C{\equiv}CH + Et_3N$ + 6% CuI	90TL5161
C_5H_7O	O 2 CuLi	83TL3905
$C_5H_7O_2$	$(CH_3CO)_2CHNa$ + 5% CuBr	78OS(58)52
C_5H_8Cl	(*E*)-$ClCH_2CH_2CH_2CH{=}CHCu(C{\equiv}CPr)Li$	80AG472
C_5H_8Cl	$H_2C{=}C(CH_2CH_2CH_2Cl)Cu.2PBu_3$	93CJC280
C_5H_9	$[H_2C{=}CH(CH_2)_3]_2CuLi.SMe_2$	86JOC1730
C_5H_9	$[(Z)\text{-}PrCH{=}CH]_2CuLi$	89T381
C_5H_9	(Z)-EtC(Me)=CHCu	Chapter11, Protocol 2
	(Z)-EtC(Me)=$CHCu.SMe_2$	83JCS(P1)1387
C_5H_9	$[Me_2C{=}CHCH_2]_2Cu(CN)Li_2$	90JOC1695
C_5H_9O	O —$CH_2CH_2CH_2Cu$	89T443
C_5H_9O	$(t\text{-}BuCO)_2\,Cu(CN)Li_2$	85JA4551
C_5H_9O	$MeCO(CH_2)_3Cu$	91JOC4744
$C_5H_{10}Br$	$Br(CH_2)_5ZrCp_2Cl$ + 15% CuCN	92TL5857
$C_5H_{10}BrMgO$	$BrMgO(CH_2)_5MgBr$ + 34% Li_2CuCl_4	91JCS(P1)2639
$C_5H_{10}Cu$	$Li(PhS)Cu(CH_2)_5Cu(SPh)Li$	88JA2218
C_5H_{11}	$(C_5H_{11})_2CuLi$	87JOC1801
C_5H_{11}	$(i\text{-}PrCH_2CH_2)_2CuLi$	83JOC1404
C_5H_{11}	$(t\text{-}BuCH_2)_2CuLi$	84T1401
C_6H_9O	$MeOCMe_2C{\equiv}CCu$	78JOC3418
C_6H_9O	$H_2C{=}C(OMe)C(CH_2Cu){=}CH_2$	93JOC2694
$C_6H_{11}O_2$	$MeOOC(CH_2)_4Cu$	92MI1560

$C_7H_9O_2$	O, O (spiroketal cyclopentene) Cu(C≡CCMe$_2$OMe)Li	91TL605
$C_7H_{11}O_2$	(Z)-EtCH=C(COOEt)Cu(Me)Li	93SC143
$C_7H_{13}O_2$	O, O (dioxolane) CH$_2$CH$_2$CH$_2$MgCl + 5% CuBr	85SC569
$C_7H_{13}O_2$	(Z)-(MeO)$_2$CH(CH$_2$)$_2$CH=CHCu	87JOC1381
$C_7H_{16}O_3P$	(MeO)$_2$P(O)CH$_2$CH(Pr)Cu(CN)ZnBr	90TL1833
$C_8H_{11}Si$	Me$_3$SiC≡CCH$_2$C≡CH + cat. Cu(I)/Pd(O)	92JOC6853
$C_8H_{15}Si$	(Z)-H$_2$C=CHCH$_2$CH=C(SiMe$_3$)MgBr + 10% CuI	82SC1027
$C_8H_{15}Si$	Me$_3$SiC≡C(CH$_2$)$_3$MgI+ 13% CuCN	81JCS(P1)1516
$C_8H_{17}Si$	(Z)-PrCH=C(SiMe$_3$)Cu	77TL1805
$C_9H_{13}O_2$	EEOCH$_2$CH=CHC≡CMgBr + 4% CuBr	83JA3656
$C_9H_{19}O_2$	EEO(CH$_2$)$_2$CH(Me)CH$_2$MgBr + 5% CuBr	85SC569
$C_{10}H_{15}O_2$	THPOCMe$_2$C≡CCu	78SC175
$C_{10}H_{17}O_2$	(*E*)-*t*-BuCOO(CH$_2$)$_3$CH=CHCu(CN)ZnI	92JA3983; Chapter 4, Protocol 6
$C_{10}H_{19}O_2$	THPO(CH$_2$)$_5$MgCl + 2.5% Li$_2$CuCl$_4$	84SC761
$C_{10}H_{19}O_2$	THPOCH$_2$CMe$_2$CH$_2$Cu.SMe$_2$	85T3943
$C_{11}H_{19}BClO_2$	Cl (chain) B, O, O (pinacol boronate) Cu(CN)ZnI	92TL3717
$C_{12}H_{17}O$	BnOCH$_2$CH$_2$CH(Me)CH$_2$MgBr+ 2% Li$_2$CuCl$_4$	83S804
$C_{21}H_{25}OSi$	*t*-BuPh$_2$SiO(CH$_2$)$_3$C≡CCu(CN)Li	93JOC528

C_6

C_6Br_5	C_6Br_5Cu	72JOM(42)257
C_6Cl_5	C_6Cl_5Cu	73S170

C_6F_5	$(C_6F_5Cu)_2$.dioxane	79OS(59)122; *Alfa*
C_6H_4Cl	3-ClC_6H_4MgBr + 8% $Cu(OAc)_2$	67JOC784
C_6H_5	PhMgBr + 5% CuCl	90OS(69)1
	PhCu	52JOC1630; Chapter 8, Protocol 6
	PhCu (from active Cu)	88JOC4482
	PhCu.SMe_2	89T425
	PhCu.BPh_3	67JOM(8)339
	Ph_2CuLi	73JA7777
	$Ph_2CuMgBr$	80JOM(188)293
	$Ph_2CuMgBr.SMe_2$	90JOC964; Chapter 8, Protocol 3
	PhCu(CN)Li	93JOC36
	PhCu(CN)MgBr.BF_3	93JOC1207
	$Ph_2Cu(CN)Li_2$	89JOC1295
	$Ph_2Cu(SCN)Li_2$	83JOC546
	$Ph_2Cu(CN)(MgBr)_2$	90JOC964
	PhCu(CN)(2-Th)Li_2	85JOM(285)437
	PhCu(CN)(2-Th)MgBrLi	86T2873
$C_6H_6BrMgO_2$	BrMgOOC$(CH_2)_3$C≡CMgBr + 14% CuCl	82JOC1221
C_6H_6N	[2-pyridyl-CH_2Cu structure] Cu, N	84JOC3928
	[(2-pyridyl-CH_2)$_2$ structure] $Cu(CN)Li_2$, N	Chapter 5, Protocol 2
C_6H_7O	[3-furyl structure] O, $CH_2CH_2Cu(CN)MgBr$	85JOC3988
C_6H_7O	[3-oxocyclohex-1-enyl structure] O, Cu(CN)ZnI	93T29; Chapter 4, Protocol 4
C_6H_9	[(cyclohex-1-enyl)$_2$ structure] $Cu(CN)Li_2$	90TL4539

$C_9H_{21}Si$	$(t\text{-}BuMe_2SiCH_2CH_2CH_2)_2CuLi$	85TL547
$C_{11}H_{10}NO_2$	(phthalimido)$NCH_2CH_2CH_2Cu(CN)ZnCl$	91OS(70)195
$C_{11}H_{15}Si$	$[(Z)\text{-}PhMe_2Si(Me)C{=}CH]_2Cu(C{\equiv}CBu)Li_2$	84JCS(P1)119
$C_{11}H_{15}Si$	$[H_2C{=}C(CH_2SiMe_2Ph)]_2Cu(CN)Li_2$	89T413
$C_{11}H_{21}O_2Si$	$(E)\text{-}THPOCH_2CH{=}C(SiMe_3)Cu.nP(OEt)_3$	86JCS(P1)1515
$C_{12}H_{23}Si$	$i\text{-}Pr_3SiC{\equiv}CCH_2Cu$	84JA6006
$C_{15}H_{31}Sn$	$[H_2C{=}C(SnBu_3)CH_2]_2CuLi$	92JCS(P1)327
$C_{15}H_{31}Sn$	$(Z)\text{-}MeCH{=}C(SnBu_3)Cu(2\text{-}Th)(CN)Li_2$	Chapter 5, Protocol 5
$C_{18}H_{20}NO$	Ph, OMe, N, Ph; $Cu(C{\equiv}CMe_2OMe)Li$	Chapter 8, Protocol 7

C_4

C_4H_3O	(2-furyl)$_2$CuLi.SMe$_2$	79TL4577
C_4H_3S	(2-thienyl)$_2$CuLi	70ACS2379; *Aldrich*
	(2-thienyl)Cu(CN)Li	Chapter 5, Protocol 3
C_4H_5	$H_2C{=}CHC(MgCl){=}CH_2$ + 8% $CuBr.SMe_2$	83TL1003
C_4H_5	$EtC{\equiv}CH$ + cat. Cu(I)/Pd(0)	84SC761
C_4H_5O	$H_2C{=}CHCH_2COCu(Me)(CN)Li_2$	90TL477
C_4H_6Cl	$[ClCH_2CH_2C({=}CH_2)]_2CuMgBr$	89T1089

C_4H_6N	$NCCH_2CH_2CH_2Cu(CN)ZnI$	89TL4799
C_4H_6OLi	(Z)-MeCH=C(CH_2OLi)Cu(2-Th)(CN)Li_2	93JA3966
C_4H_7	(H_2C=$CHCH_2CH_2$)$_2$CuLi	80JOC2229
C_4H_7	H_2C=C(Me)CH_2Cu	92JA5110; Chapter 3, Protocol 5
	[H_2C=C(Me)CH_2]$_2$CuLi	Chapter 9, Protocol 4
	[H_2C=C(Me)CH_2]$_2$Cu(CN)Li_2	Chapter 5, Protocol 6
C_4H_7	(Me_2C=CH)$_2$CuLi	90T4277
C_4H_7	[(Z)-EtCH=CH]$_2$CuLi	Chapter 11, Protocol 1
C_4H_8ClMgO	ClMgO(CH_2)$_4$Cu	79S885
C_4H_8Cu	Li(PhS)Cu(CH_2)$_4$Cu(SPh)Li	91OS(70)204
C_4H_8IZn	IZn(CH_2)$_4$Cu(CN)ZnI	91JOC4591
C_4H_9	BuMgBr + 5% CuI	87OS(66)116
	BuMgBr + 6% Cu(OAc)$_2$.H_2O	65JA82
	BuMgCl + 10% CuCN	89HCA1337
	BuMgBr + CuI	81JCS(P1)593
	BuMgCl + 30% $MnCl_2$ + 3% CuCl	93OS135
C_4H_9	BuTi(OPr-*i*)$_4$Li + 5% CuI and BuTi(OPr-*i*)$_3$ + 5% CuI	91JOC5489
	BuCu	78IC275
	BuCu.SMe_2	91OS(70)215
	BuCu.P(OEt)$_3$	86T1389
	BuCu.PBu_3	82JCS(P1)1177
	BuCu.2PBu_3	87T813
	BuCu.$AlCl_3$	Chapter 7, Protocol 2
	BuCu.BF_3	77JA8068; Chapter 7, Protocol 1
	BuCu.Me_3SiI	91T9691; Chapter 8, Protocol 2
	Bu_2CuLi	73JA7777; Chapter 2, Protocol 11
	Bu_2CuLi.SMe_2	84OS(62)1
	Bu_2CuLi.$ClSiMe_2$Bu-*t*	89T349
	$Bu_2CuMgBr$	81JCS(P1)593
	$Bu_2CuMgBr$.SMe_2	89T403
	BuCu(OBu-*t*)Li	74JOC400

	BuCu(SPh)Li	81JCS(P1)593
	BuCu(PCy_2)Li	84JOC1119
	BuCu(C≡CPr)Li	72JA7210
	BuCu(C≡CPr)MgBr	82JCS(P1)1177
	BuCu(CH_2SOMe)(CN)Li_2	92OR135
	BuCu(CN)Li	82JOC5088
	Ph N N(Me)Cu(Bu)Li	93T965; Chapter 8, Protocol 8
	BuCu(CN)(2-Th)Li_2	85JOM(285)437
	BuCu(CN)(Imid)Li_2	Chapter 5, Protocol 4
	Bu_2Cu(CN)Li_2	84JOC3928
	Bu_2Cu(SCN)Li_2	83JOC546
	Bu_2Cu(CN)$(ZnCl)_2$	92JOC1024
C_4H_9	Ph Cu(Bu)Li N N Li(Bu)Cu N Ph N	94T4455
	Bu_3CuLi_2	Chapter 8, Protocol 1
	$Bu_5Cu_3Li_2$. 6 Me_2NP (Pr-*i* N, O, Ph)	Chapter 8, Protocol 9
C_4H_9	*i*-BuCu(SPh)Li	84JOC3183
C_4H_9	*s*-Bu_2CuLi	73JA7777
C_4H_9	*t*-BuMgCl + 5% CuBr	87JOC3901; Chapter 6, Protocol 2
	t-BuMgBr + 10% CuI	73JA7788
	t-BuCu	73JA7788
	t-Bu_2CuLi	73JA7777
	t-Bu_2CuLi.SMe_2	81MI361
	t-Bu_2CuMgCl	89T403
	t-BuCu(NEt_2)Li	73JA7788

	t-BuCu(OBu-*t*)Li	73JA3076
	t-BuCu(OPh)Li and *t*-BuCu(SBu-*t*)Li	73JA7788
	t-BuCu(SPh)Li	76OS(55)122
	t-$Bu_2Cu(CN)Li_2$	90JOC964
	t-BuCu(CN)(2-Th)Li_2	85JOM(285)437
C_5H_6N	MeN Cu	71ACS2596
$C_5H_9O_2$	O O Cu.TMEDA	91JOC5491
$C_5H_9O_2$	$MeOOCCH_2CH_2CH_2Cu.SMe_2$	91JOC1083
$C_6H_9O_2$	(Z)-MeCH=C(COOEt)Cu(C≡CBu)Li	83JOC4621
$C_6H_9O_3$	$EtOOCCH(COCH_3)Cu$	84S616
$C_6H_{11}O_2$	O O $CH_2CH_2CH_2MgBr$ + 25% $CuBr.SMe_2$	82JOC5045
$C_6H_{11}O_2$	$[AcO(CH_2)_4]_2Cu(CN)(MgCl)_2$	93JOC4781
$C_6H_{11}O_2$	*i*-PrCH(OAc)Cu(CN)ZnBr	90JOC4791
	$EtOOC(CH_2)_3Cu(CN)ZnI$	Chapter 4, Protocol 2
$C_7H_{13}O_2$	O O ( )$_2$ CuLi	82JOC4605
$C_7H_{13}Si$	$Me_3SiC≡CCH_2CH_2Cu.BF_3$	81T4431
$C_7H_{13}Sn$	(*E*,*E*)-$Me_3SnCH=CHCH=CHCu(Me)(CN)Li_2$	91TL7211
$C_7H_{15}Si$	$H_2C=C(CH_2SiMe_3)CH_2Cu$	81JA1831
$C_7H_{17}Si$	$PrCH(SiMe_3)Cu$	86T1389
$C_8H_{15}O_2$	*t*-$BuOOC(CH_2)_3Cu$	89T443; Chapter 3, Protocol 1
$C_8H_{17}O$	(S)-*t*-$BuOCH_2CH(Me)CH_2Cu(CN)Li$	90T4503
$C_8H_{17}O_2$	$(EtO)_2CH(CH_2)_3MgBr$ + 4% Li_2CuCl_4	82JOC4825
$C_9H_{13}O_2$	$THPOCH_2CH_2C≡CMgBr$ + 10% CuCl	86T3781

$C_9H_{15}O_2$	(Z)-$EEOCH_2C(Me)C{=}CHCu(C{\equiv}CCMe_2OMe)Li$	86T2831
$C_9H_{15}O_3$	(bicyclic ketal)–CH_2CH_2)$_2CuLi_2.SMe_2$	93JOC2134
$C_9H_{17}OSi$	(*E*)-$H_2C{=}C(SiMe_3)C(OEt){=}CHCu$	86T1389
$C_9H_{17}O_2$	$THPO(CH_2)_4Cu.SMe_2$	87JOC1381
$C_9H_{17}O_2$	(dioxane) CH_2CH_2MgBr + 14% CuBr	90JOC4417
$C_9H_{17}O_2Si$	(Z)-$Me_3SiCH_2CH{=}C(COOEt)Cu$	86T1389
$C_{10}H_{21}OSi$	(*E*)-*t*-$BuMe_2SiOCH_2CH_2CH{=}CHCu(CN)(2\text{-}Th)Li_2$	91TL5647
$C_{10}H_{21}OSi$	[(*E*)-*t*-$BuMe_2SiOCH_2CH{=}C(Me)]_2CuLi$	76T2281
$C_{10}H_{21}OSi$	[(*Z*)-*t*-$BuMe_2SiOCH_2C(Me){=}CH]_2CuLi$	92JA8008
$C_{11}H_{24}O_2Si$	[(Z)-$Me_3SiCH_2CH{=}C[CH(OEt)_2]]_2CuLi$	86T1389
$C_{16}H_{22}NO_4$	(S)-$BocNHCH(COOBn)CH_2CH_2Cu$	93SL219
$C_{17}H_{24}NO$	$BnNH(CH_2)_6OCH_2CH_2C{\equiv}CH$ + cat. Cu(I)/Pd(O)	Chapter 10, Protocol 1
$C_{20}H_{27}Si$	*t*-$BuPh_2Si(CH_2)_4ZrCp_2Cl$ + 10% $CuBr.SMe_2$	93S537

C_5

C_5H_4N	(2-pyridyl)$_2CuLi$	82T1509
C_5H_4N	(3-pyridyl)$_2CuLi$	85H117
C_5H_5	$H_2C{=}CMeC{\equiv}CCu$	80JOC1158
C_5H_5O	(3-furylmethyl)$Cu(CN)MgCl$	85JOC3988
C_5H_7	$PrC{\equiv}CCu + Me_3SiI$	93JOC7238
C_5H_7	$(Me_2C{=}C{=}CH)_2CuLi$	76TL275

C_5H_7	$HC{\equiv}CCH_2CH_2CH_2Cu(CN)ZnI$	91JOC5974; Chapter 4, Protocol 3
C_5H_7	$(MeC{\equiv}CCH_2CH_2)_2CuMgBr$	83JA2364
C_5H_7	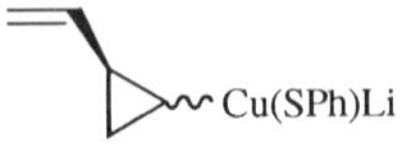 Cu(SPh)Li	83CJC1226
C_5H_7O	$HOCMe_2C{\equiv}CH + Et_3N$ + 6% CuI	90TL5161
C_5H_7O	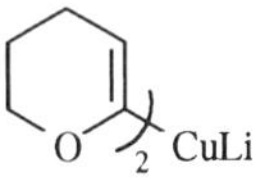 O 2 CuLi	83TL3905
$C_5H_7O_2$	$(CH_3CO)_2CHNa$ + 5% CuBr	78OS(58)52
C_5H_8Cl	(*E*)-$ClCH_2CH_2CH_2CH{=}CHCu(C{\equiv}CPr)Li$	80AG472
C_5H_8Cl	$H_2C{=}C(CH_2CH_2CH_2Cl)Cu.2PBu_3$	93CJC280
C_5H_9	$[H_2C{=}CH(CH_2)_3]_2CuLi.SMe_2$	86JOC1730
C_5H_9	[(Z)-$PrCH{=}CH]_2CuLi$	89T381
C_5H_9	(Z)-$EtC(Me){=}CHCu$	Chapter11, Protocol 2
	(Z)-$EtC(Me){=}CHCu.SMe_2$	83JCS(P1)1387
C_5H_9	$[Me_2C{=}CHCH_2]_2Cu(CN)Li_2$	90JOC1695
C_5H_9O	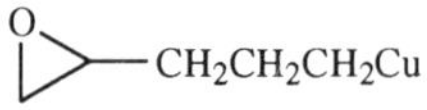 O $CH_2CH_2CH_2Cu$	89T443
C_5H_9O	(*t*-$BuCO)_2Cu(CN)Li_2$	85JA4551
C_5H_9O	$MeCO(CH_2)_3Cu$	91JOC4744
$C_5H_{10}Br$	$Br(CH_2)_5ZrCp_2Cl$ + 15% CuCN	92TL5857
$C_5H_{10}BrMgO$	$BrMgO(CH_2)_5MgBr$ + 34% Li_2CuCl_4	91JCS(P1)2639
$C_5H_{10}Cu$	$Li(PhS)Cu(CH_2)_5Cu(SPh)Li$	88JA2218
C_5H_{11}	$(C_5H_{11})_2CuLi$	87JOC1801
C_5H_{11}	(*i*-$PrCH_2CH_2)_2CuLi$	83JOC1404
C_5H_{11}	(*t*-$BuCH_2)_2CuLi$	84T1401
C_6H_9O	$MeOCMe_2C{\equiv}CCu$	78JOC3418
C_6H_9O	$H_2C{=}C(OMe)C(CH_2Cu){=}CH_2$	93JOC2694
$C_6H_{11}O_2$	$MeOOC(CH_2)_4Cu$	92MI1560

$C_7H_9O_2$	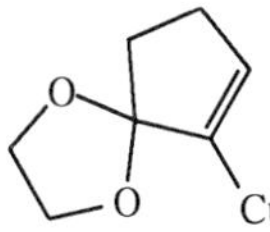 Cu(C≡CCMe$_2$OMe)Li	91TL605
$C_7H_{11}O_2$	(Z)-EtCH=C(COOEt)Cu(Me)Li	93SC143
$C_7H_{13}O_2$	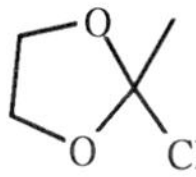 $CH_2CH_2CH_2MgCl$ + 5% CuBr	85SC569
$C_7H_{13}O_2$	(Z)-(MeO)$_2$CH(CH$_2$)$_2$CH=CHCu	87JOC1381
$C_7H_{16}O_3P$	(MeO)$_2$P(O)CH$_2$CH(Pr)Cu(CN)ZnBr	90TL1833
$C_8H_{11}Si$	Me$_3$SiC≡CCH$_2$C≡CH + cat. Cu(I)/Pd(O)	92JOC6853
$C_8H_{15}Si$	(Z)-H$_2$C=CHCH$_2$CH=C(SiMe$_3$)MgBr + 10% CuI	82SC1027
$C_8H_{15}Si$	Me$_3$SiC≡C(CH$_2$)$_3$MgI+ 13% CuCN	81JCS(P1)1516
$C_8H_{17}Si$	(Z)-PrCH=C(SiMe$_3$)Cu	77TL1805
$C_9H_{13}O_2$	EEOCH$_2$CH=CHC≡CMgBr + 4% CuBr	83JA3656
$C_9H_{19}O_2$	EEO(CH$_2$)$_2$CH(Me)CH$_2$MgBr + 5% CuBr	85SC569
$C_{10}H_{15}O_2$	THPOCMe$_2$C≡CCu	78SC175
$C_{10}H_{17}O_2$	(*E*)-*t*-BuCOO(CH$_2$)$_3$CH=CHCu(CN)ZnI	92JA3983; Chapter 4, Protocol 6
$C_{10}H_{19}O_2$	THPO(CH$_2$)$_5$MgCl + 2.5% Li$_2$CuCl$_4$	84SC761
$C_{10}H_{19}O_2$	THPOCH$_2$CMe$_2$CH$_2$Cu.SMe$_2$	85T3943
$C_{11}H_{19}BClO_2$	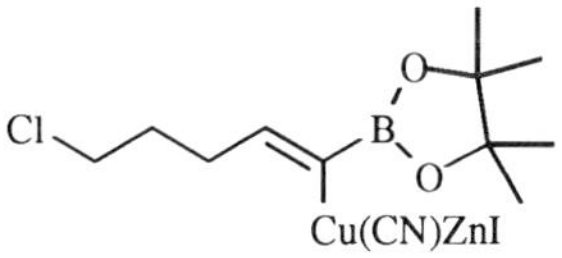 Cu(CN)ZnI	92TL3717
$C_{12}H_{17}O$	BnOCH$_2$CH$_2$CH(Me)CH$_2$MgBr+ 2% Li$_2$CuCl$_4$	83S804
$C_{21}H_{25}OSi$	*t*-BuPh$_2$SiO(CH$_2$)$_3$C≡CCu(CN)Li	93JOC528

C_6

C_6Br_5	C_6Br_5Cu	72JOM(42)257
C_6Cl_5	C_6Cl_5Cu	73S170

C_6F_5	$(C_6F_5Cu)_2$.dioxane	79OS(59)122; *Alfa*
C_6H_4Cl	3-ClC_6H_4MgBr + 8% $Cu(OAc)_2$	67JOC784
C_6H_5	PhMgBr + 5% CuCl	90OS(69)1
	PhCu	52JOC1630; Chapter 8, Protocol 6
	PhCu (from active Cu)	88JOC4482
	$PhCu.SMe_2$	89T425
	$PhCu.BPh_3$	67JOM(8)339
	Ph_2CuLi	73JA7777
	$Ph_2CuMgBr$	80JOM(188)293
	$Ph_2CuMgBr.SMe_2$	90JOC964; Chapter 8, Protocol 3
	PhCu(CN)Li	93JOC36
	$PhCu(CN)MgBr.BF_3$	93JOC1207
	$Ph_2Cu(CN)Li_2$	89JOC1295
	$Ph_2Cu(SCN)Li_2$	83JOC546
	$Ph_2Cu(CN)(MgBr)_2$	90JOC964
	$PhCu(CN)(2\text{-}Th)Li_2$	85JOM(285)437
	PhCu(CN)(2-Th)MgBrLi	86T2873
$C_6H_6BrMgO_2$	$BrMgOOC(CH_2)_3C{\equiv}CMgBr$ + 14% CuCl	82JOC1221
C_6H_6N	N, Cu	84JOC3928
	N, $Cu(CN)Li_2$, 2	Chapter 5, Protocol 2
C_6H_7O	O, $CH_2CH_2Cu(CN)MgBr$	85JOC3988
C_6H_7O	O, Cu(CN)ZnI	93T29; Chapter 4, Protocol 4
C_6H_9	2, $Cu(CN)Li_2$	90TL4539

C_6H_9	(ring structure)—Cu	92JA5110
C_6H_9	BuC≡CCu	74JA7138
C_6H_9	*t*-BuC≡CCu	73JOC3893
C_6H_9	(*E*)-H_2C=CHCH=CH$(CH_2)_2$MgBr + 25% CuI	83T3235
C_6H_9	[(Z,Z)-EtCH=CHCH=CH]$_2$CuLi	86JCS(P1)1809
$C_6H_9F_2$	BuC(Cu)=CF_2	93S537
C_6H_{11}	CyMgBr + 1.5% CuBr	87JOC3901
C_6H_{11}	[(Z)-BuCH=CH]$_2$CuLi	84OS(62)1
C_6H_{11}	(Z)-PrC(Me)=CHCu	79JOC1345
C_6H_{11}	(Z)-EtCH=CHCH_2CH_2MgBr + 10% CuBr	80G237
C_6H_{11}	Me_2C=CHCH_2CH_2Cu	Chapter 11, Protocol 3
	(Me_2C=CHCH_2CH_2)$_2$CuLi	81JOC1532
C_6H_{11}	H_2C=CHCH_2CH_2CH(Me)MgBr + 5% CuBr	87JCS(P1)1331
C_6H_{11}	(R) & (S)-H_2C=CHCH(Me)CH_2CH_2MgBr + 8% CuI	85JA2474
$C_6H_{11}LiNO_2$	C_5H_{11}C[=N(O)OLi]Cu	79HCA2239
$C_6H_{11}O$	MeCO$(CH_2)_4$Cu	91JOC4744
$C_6H_{12}Cl$	Cl$(CH_2)_6$Cu	93JOC2483
C_6H_{13}	$(C_6H_{13})_2$CuLi	90SC1989
C_6H_{13}	[Me_2CH$(CH_2)_3$]$_2$CuLi	83JOC1404
C_6H_{13}	*t*-BuCH(Me)Cu(Me)Li	80TL3151
C_7H_7O	3-MeOC_6H_4MgBr + 4% CuI	85T129
C_7H_7O	4-MeOC_6H_4MgBr + 10% CuI	79JA934
$C_7H_9O_2$	MeOOC$(CH_2)_3$C≡CH + cat. Cu(I)/Pd(0)	89JOC3635; Chapter 10, Protocol 2
$C_7H_{13}O_2$	MeOOC$(CH_2)_5$Cu(CN)ZnI	Chapter 6, Protocol 3
$C_8H_9O_2$	OMe / MeO (ring structure) MgCl + 3% Li_2CuCl_4	83S597

$C_8H_9O_2$	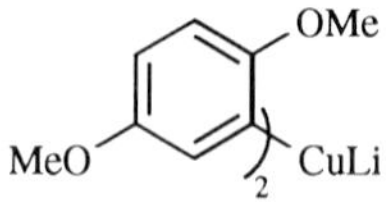OMe MeO CuLi 2	80JOC378
$C_8H_{11}O_2$	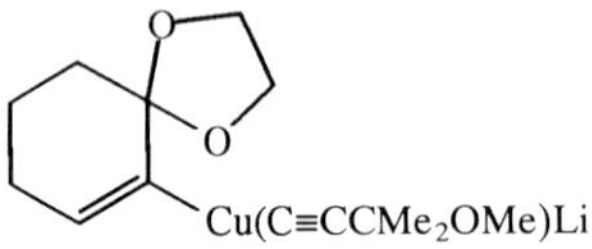O O Cu(C≡CCMe$_2$OMe)Li	91TL609
$C_8H_{13}S$	[(*Z*,*Z*)-EtCH=CHCH=C(SEt)]$_2$CuLi	80T1961
$C_8H_{15}N_2$	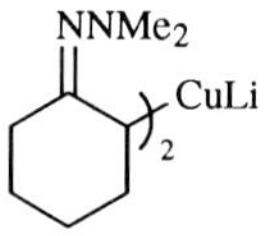NNMe$_2$ CuLi 2	92OR135
$C_8H_{15}O_2$	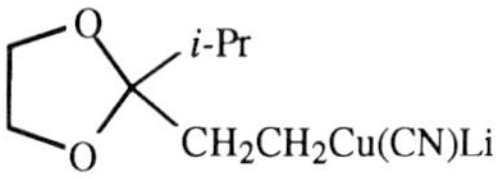O *i*-Pr O CH$_2$CH$_2$Cu(CN)Li	93T337
$C_9H_4CrFO_3$	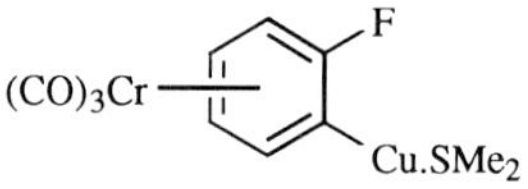F (CO)$_3$Cr Cu.SMe$_2$	85S492
$C_9H_{19}N_2O$	MeO(CH$_2$)$_4$C(=NNMe$_2$)CH$_2$Cu(SPh)Li	82JA1054
$C_9H_{19}Si$	(*E*)-BuC(SiMe$_3$)=CHCu.P(OEt)$_3$	86T1389
$C_9H_{21}Si$	C$_5$H$_{11}$CH(SiMe$_3$)Cu.SMe$_2$	91JOC638
$C_{10}H_{13}O_4$	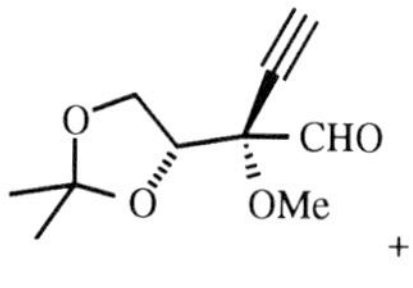O CHO O OMe + cat. Pd(O)/Cu(I)	93T1901
$C_{10}H_{17}O_2$	EEO(CH$_2$)$_4$C≡CMgBr + 4% CuI	78LA658
$C_{10}H_{17}O_2$	(EtO)$_2$CH(CH$_2$)$_3$C≡CMgBr + 12.5% CuI	78LA658
$C_{11}H_{19}O_2$	(Z)-THPO(CH$_2$)$_4$CH=CHCu	87JOC1381
$C_{11}H_{21}O_2$	THPO(CH$_2$)$_6$MgCl+ 2.6% Li$_2$CuCl$_4$	78S388
$C_{11}H_{23}OSi$	(Z)-Me$_3$SiCH(Pr)C(OEt)=CHCu	86T1389
$C_{12}H_{19}OSi$	[3-(*t*-BuMe$_2$SiO)C$_6$H$_4$]$_2$CuLi.PBu$_3$	78JOC2102
$C_{12}H_{25}N_2O_2$	EEO(CH$_2$)$_4$C(=NNMe2)CH2Cu(SPh)Li	82JA1054

$C_{13}H_{15}CrO_4Si$	$(OC)_3Cr$; $SiMe_3$; OMe; Cu	83JA2034

C_7

C_7H_4N	2-NCC_6H_4Cu	88JOC4482
C_7H_6Cl	2-$(ClCH_2)C_6H_4Cu$	90SC761
C_7H_7	Bn_2CuLi	75JOC1488
C_7H_7	(2-$MeC_6H_4)_2CuMgBr$	89T349
C_7H_7	4-$MeC_6H_4SnBu_3$ + cat. Cu(I)/Pd(0)	92TL919
C_7H_9O	(*E*)-EtCH(OMgBr)CH=CHC≡CMgBr + 7% CuI	87T4385
C_7H_9O	O; $CH_2CH_2CH_2Cu(CN)MgCl$	85JOC3988
$C_7H_{10}N$	CN; Cu(CN)ZnI	89TL5069
C_7H_{11}	$Cu.PBu_3$	72JOC3718
C_7H_{11}	$C_5H_{11}C≡CH$ + CuI.DBU	78LA658
C_7H_{11}	[(Z,Z)-EtCH=CHCH$_2$CH=CH]$_2$CuLi	93JOC5153
$C_7H_{11}O$	O; Cu(C≡CPr)Li	87JCS(P1)2183
$C_7H_{11}O$	O; Cu(CN)ZnI	93JOC2694
C_7H_{13}	[(Z)-C_5H_{11}CH=CH]$_2$CuLi	71HCA1939
C_7H_{13}	(*E*)-BuC(Me)=CHCu(CN)(C≡CBu)Li_2	90JOC1425

C_7H_{13}	$H_2C{=}C(C_5H_{11})MgBr$ + 5% CuI	84JCR(M)2327
C_7H_{13}	$[H_2C{=}CH(CH_2)_5]_2CuLi$	85T627
C_7H_{15}	$C_7H_{15}Cu$	90T3667
$C_8H_4LiO_4$	COOLi, $Cu(CN)Li_2$ & COOLi, $Cu(CN)Li_2$	93JCS(P1)1953
$C_8H_7O_2$	$CH_2Cu(CN)ZnOPO(OEt)_2$	Chapter 4, Protocol 5
$C_8H_7O_2$	4-(MeOOC)C_6H_4Cu	88JOC4482
C_8H_9O	2-(MeO)$C_6H_4CH_2MgBr$ + 3% CuBr	86JCS(P1)2151
$C_9H_9O_2$	4-(EtOOC)$C_6H_4Cu(CN)ZnI$	91JOC1445
$C_9H_{11}O_2$	MeO, Me, Cu, MeO	71JOC1143
$C_9H_{12}N$	CH_2NMe_2, Cu	75JOM(99)157
$C_9H_{15}O_2$	(Z)-BuCH=C(COOEt)Cu(C≡CBu)Li	83JOC4621
$C_{10}H_{21}Si$	(*E*)- & (Z)-$Me_3SiCH_2C(Bu){=}CHCu$	86T1389 & 1399
$C_{10}H_{23}OSi$	$Me_3SiO(CH_2)_7Cu(CN)Li$	81JOC4389
$C_{11}H_{17}O_5$	OH, OMe, OH + cat. Pd(O)/Cu(I)	93T1901
$C_{11}H_{21}O_2$	$[(Z)\text{-}BuCH{=}C[CH(OEt)_2]]_2CuLi$	84T715
$C_{12}H_{21}O_2$	$PrCH(OTHP)CH_2C(MgBr){=}CH_2$ + 5% CuI	84JCR(M)2327
$C_{13}H_{24}NSi_2$	2-[$(Me_3Si)_2N$]$C_6H_4CH_2Cu(CN)ZnBr$	89TL4795
$C_{13}H_{26}BO_2$	C_6H_{13}, B, $Cu(CN)ZnBr$	90JA7431

$C_{13}H_{27}OSi$	(*S*)-(*E*)-*t*-$BuMe_2SiOCH_2CH(Me)CH_2C(Me){=}CHCu(CN)(C{\equiv}CBu)LiAlMe_2$	93S537
$C_{13}H_{27}OSi$	(*E*)-$MeCH(OSiMe_2Bu$-*t*$)(CH_2)_3CH{=}CHCu(C{\equiv}CPr)Li$	82JA5473
$C_{13}H_{29}OSi$	*t*-$BuMe_2SiO(CH_2)_7MgBr$ + 50% CuI	81JCS(P1)599
$C_{13}H_{30}NSi_2$	(Z)-$(Me_3Si)_2NCH_2C(Bu){=}CHCu$	92T6231
$C_{14}H_{16}NO_2$	O, O, CH=NCy, $Cu.P(OEt)_3$	76JA8282
$C_{15}H_{20}NO_2$	MeO, MeO, CH=NCy, Cu	87OS(65)108
$C_{16}H_{33}OSi$	(*E*)-*i*-$Pr_3SiO(CH_2)_4C(Me){=}CHAlMe_2$ + 10% CuCN	91JOC5761

C_8

C_8H_5	$PhC{\equiv}CCu$	72OS(52)128,*Strem, Alfa, K&K, Parish*
C_8H_5S	S, Cu	93JOC2694
C_8H_7	$H_2C{=}C(Ph)Cu(SPh)Li$	80TL3849
C_8H_8Cu	2-$[Li(PhS)Cu]C_6H_4CH_2CH_2Cu(SPh)Li$	88JA2218
C_8H_9	$(PhCH_2CH_2)_2CuSmI_3$	93S537
C_8H_{13}	[(Z,Z)-$BuCH{=}CHCH{=}CH]_2CuLi$	Chapter 2, Protocol 11
C_8H_{13}	$C_6H_{13}C{\equiv}CH$ + cat. Cu(I)/Pd(O)	Chapter 10, Protocol 4
	$C_6H_{13}C{\equiv}CCu(CN)Li$	91TL1855
C_8H_{13}	Cu	82MI667
$C_8H_{13}O$	(*E*)-$PrC(Me){=}C(COMe)Cu(C{\equiv}CBu)Li$	83JOC4621

$C_8H_{13}O$	(structure: cyclohexane bearing $CH_2Cu(CN)ZnI$ and CHO)	93JOC2694
C_8H_{15}	$[(Z)\text{-}C_6H_{13}CH{=}CH]_2CuLi$	75HCA1016
C_8H_{15}	$[(E)\text{-}C_6H_{13}CH{=}CH]_2CuLi$	75JA857
C_8H_{15}	$Et_2C{=}CHCH_2CH_2Cu$	75JA1197
C_8H_{15}	(Z)-$PrC(Me){=}CHCH_2CH_2Cu.nSMe_2$	79JOC1345
C_8H_{15}	$[Me_2C{=}CHCH_2CH_2CH(Me)]_2CuMgCl$	86JA800
C_8H_{17}	$(C_8H_{17})_2CuLi$	85JA6028
$C_9H_{13}O$	(structure: OMe; Cu(CN)Li)	Chapter 4, Protocol 7
$C_{10}H_{11}O_2$	3-$EtOOCC_6H_4CH_2Cu(CN)ZnCl$	90MI3053
$C_{10}H_{14}N$	(structure: NMe_2; Cu)	89T569
$C_{10}H_{17}O$	$[(Z,E)\text{-}BuCH{=}CHC(OEt){=}CH]_2CuLi$	80T1961
$C_{10}H_{19}O_2$	(structure: O, O; MgBr + 13% CuI.bipy)	78BCJ547
$C_{10}H_{19}O_2$	$[(E)(3S)\text{-}C_5H_{11}CH(OMOM)CH{=}CH]_2CuLi$	72JA7827
$C_{11}H_{11}O_3S$	(structure: O, O; Cu; O, S)	80JA790
$C_{11}H_{23}Si$	(Z)-$C_5H_{11}CH(SiMe_3)CH{=}CHCu$	86T1389 & 1399
$C_{11}H_{23}Si$	(Z)-$Me_3SiCH_2C(C_5H_{11}){=}CHCu$	86T1399
$C_{12}H_{23}O_2$	(Z)(3R)-*i*-$PrCH_2CH(OCMe_2OMe)CH{=}C(Me)Cu.2PBu_3$	84TL1925
$C_{12}H_{23}O_2$	$[(E)(3S)\text{-}C_5H_{11}CH(OEE)CH{=}CH]_2CuLi.PBu_3$	75JA865
$C_{12}H_{23}O_2$	$[(E)(3S)\text{-}C_5H_{11}CH(OCMe_2OMe)CH{=}CH]_2CuLi$	72JA7827
$C_{12}H_{25}O$	$[t\text{-}BuO(CH_2)_8]_2CuLi$	90SC333
$C_{13}H_{18}NO_2S$	(structure: N$-SO_2$; MgBr + 19% CuI)	78LA658

Formula	Compound	Reference
$C_{13}H_{23}O_2$	(±)(*E*)-C_5H_{11}CH(OTHP)CH=CHCu	84MI118
	(3S)(*E*)-C_5H_{11}CH(OTHP)CH=CHCu	Chapter 9, Protocol 9
$C_{14}H_{23}OSi$	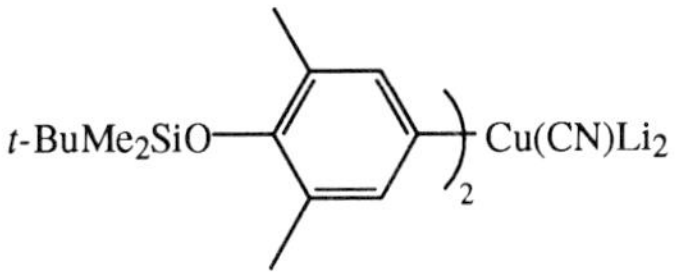 *t*-BuMe$_2$SiO– ; Cu(CN)Li$_2$; 2	92JOC6883
$C_{14}H_{27}OSi$	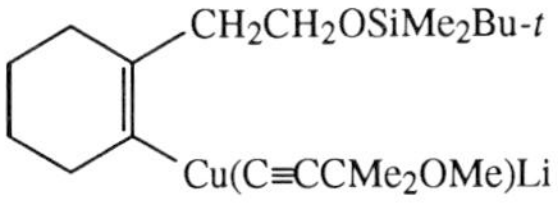 CH$_2$CH$_2$OSiMe$_2$Bu-*t* ; Cu(C≡CCMe$_2$OMe)Li	83JA3588
$C_{14}H_{29}OSi$	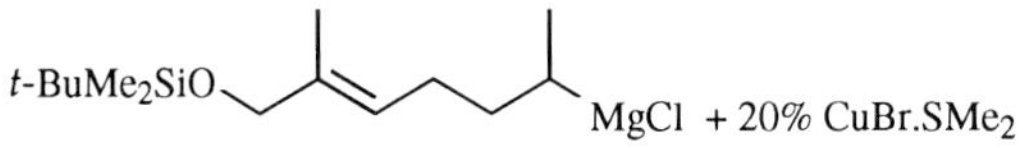 *t*-BuMe$_2$SiO ; MgCl + 20% CuBr.SMe$_2$	93JA1676
$C_{14}H_{29}OSi$	(*E*)(3S)-C_5H_{11}CH(OSiMe$_2$Bu-*t*)CH=CHCu(C≡CPr)Li.PBu$_3$	77T1105
	(*E*)(3S)-C_5H_{11}CH(OSiMe$_2$Bu-*t*)CH=CHCu.PBu$_3$	Chapter 9, Protocol 8
	(*E*)(±)-C_5H_{11}CH(OSiMe$_2$Bu-*t*)CH=CHCu.SMe$_2$	Chapter 2, Protocol 9
$C_{14}H_{31}OSi$	*t*-BuMe$_2$SiO(CH$_2$)$_8$MgBr + 2.5% CuBr.SMe$_2$	89T455; Chapter 2, Protocol 8
$C_{15}H_{20}NO$	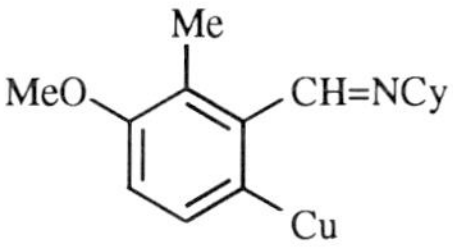 Me ; MeO ; CH=NCy ; Cu	79JA1857
$C_{16}H_{13}O_2$	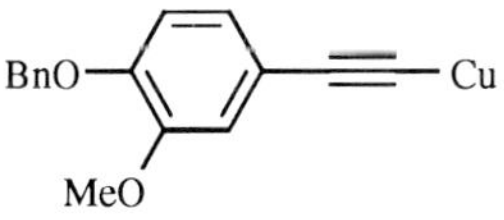 BnO ; Cu ; MeO	92JOC7248
$C_{16}H_{18}NO_3S$	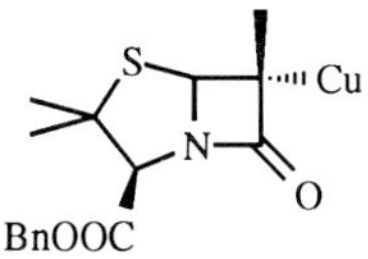 S ; Cu ; N ; O ; BnOOC	83T2469
$C_{18}H_{37}O_3Si$	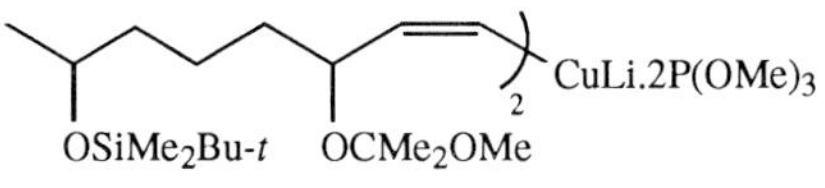 CuLi.2P(OMe)$_3$; 2 ; OSiMe$_2$Bu-*t* ; OCMe$_2$OMe	78JA6211

$C_{20}H_{26}CrNO_3Si$	[structure: 4-(Cu.SMe₂)-1-(Si-i-Pr₃)indole with (OC)₃Cr complexed to benzene ring]	88T7325
$C_{24}H_{40}NOSi$	[structure: bis[4-OCH(Me)C₅H₁₁-1-(Si-i-Pr₃)indol-7-yl]CuLi]	86JMC1457

C$_9$

C_9H_7S	[structure: benzothiophen-2-yl–$CH_2Cu.ZnI_2$]	93JOC2694
$C_9H_8BrMgO_2$	$BrMgOOC(CH_2)_3C{\equiv}CCH_2C{\equiv}CMgBr$ + 16% CuI	78LA658
$C_9H_9O_2$	[structure: trimethyl-1,4-benzoquinone–$SnBu_3$] + cat Pd(O)/Cu(I)	93JOC408
$C_9H_{10}Cu$	2-[Li(PhS)Cu]$C_6H_4CH_2CH_2CH_2Cu$(SPh)Li	88JA2218
C_9H_{11}	2,4,6-$Me_3C_6H_2Cu$	81JOC192, *Alfa*
C_9H_{13}	(*E*)-$C_5H_{11}CH{=}CHC{\equiv}CH$ + cat. Cu(I)/Pd(0)	87T4591
$C_9H_{16}LiO$	(*Z*)-$LiO(CH_2)_3CH{=}C$(*t*-Bu)Cu	91CC1595
C_9H_{17}	(*E*)-$C_6H_{13}C(Me){=}CHCu$	78TL1363
$C_9H_{18}Cu$	$BrMg(CH_2)_9MgBr$ + 2%CuCl	89JMC2072
C_9H_{19}	$C_9H_{19}MgBr$ + 21% CuBr	90JOC4417
$C_{11}H_{19}O_2$	[structure: 1,3-dioxolan-2-yl–$CH_2CH_2C(Et){=}CHCH_2CH_2Cu$]	72JA5374
$C_{13}H_{19}O_2$	$EEO(CH_2)_4C{\equiv}CCH_2C{\equiv}CMgBr$ + 10% CuI	78LA658
$C_{13}H_{19}O_2$	$(EtO)_2CH(CH_2)_3C{\equiv}CCH_2C{\equiv}CMgBr$ + 22% CuI	78LA658

$C_{15}H_{24}ClOSi$	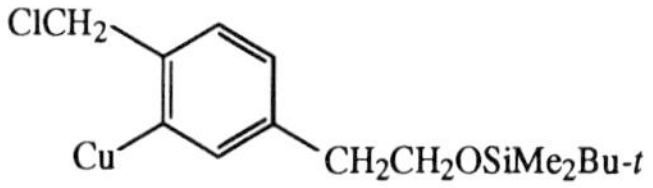 [structure: ClCH2 / Cu / $CH_2CH_2OSiMe_2Bu$-*t* aryl copper]	90SC761
$C_{16}H_{33}O_2Si$	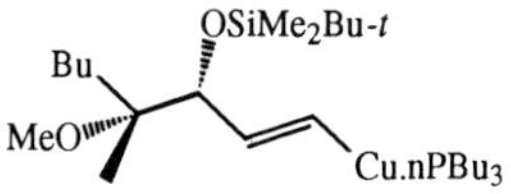 [structure: Bu, MeO, $OSiMe_2Bu$-*t*, $Cu.nPBu_3$]	92T6393
$C_{19}H_{38}LiO_3Si$	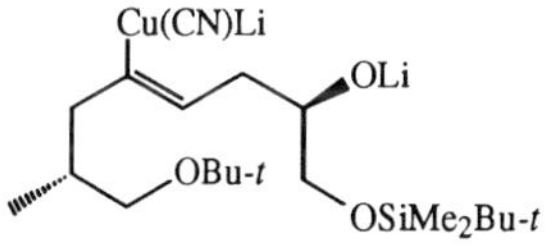 [structure: Cu(CN)Li, OLi, OBu-*t*, $OSiMe_2Bu$-*t*]	90T4503

C_{10}

$C_{10}H_7$	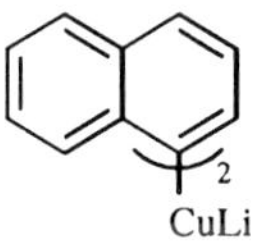 [structure: (1-naphthyl)$_2$CuLi]	90JOC3723
$C_{10}H_{12}LiO$	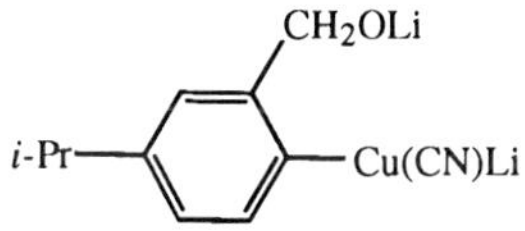 [structure: *i*-Pr, CH_2OLi, Cu(CN)Li aryl]	85JOC1987
$C_{10}H_{13}$	$PhCMe_2CH_2Cu$	72JA232
$C_{10}H_{13}O_2$	(Z)-$MeOOC(CH_2)_3CH=CHCH_2C\equiv CH$ + cat. Cu(I)/Pd(0)	92T4369
$C_{10}H_{17}$	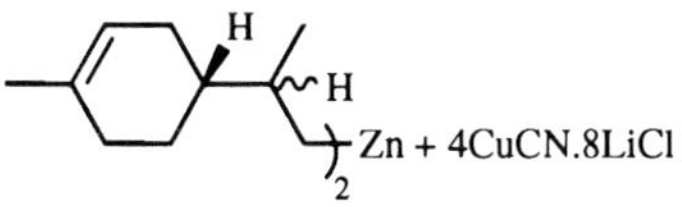 [structure: ( )$_2$Zn] + 4CuCN.8LiCl	93TL5261
$C_{10}H_{19}$	(Z)-$C_8H_{17}CH=CHMgBr$ + 5% CuI	86JOC4726
$C_{10}H_{19}$	(Z)-$C_6H_{13}C(Et)=CHCu.SMe_2$	79JOC3888
$C_{10}H_{19}$	(*E*)-$C_6H_{13}C(Et)=CHCu.SMe_2$	86OS(64)1
$C_{10}H_{19}$	$Me_2C=CH(CH_2)_2CH(Me)(CH_2)_2Cu$	75JA1197
$C_{10}H_{20}Cu$	$BrMg(CH_2)_{10}MgBr$ + 4% CuCl	85JBC8404
$C_{10}H_{21}$	$(C_{10}H_{21})_2CuLi$	86JOC3726
$C_{10}H_{21}$	(R)-$Me_2CH(CH_2)_3CH(Me)(CH_2)_2MgBr$ + 5% Li_2CuCl_4	87T4481

$C_{11}H_9O$	MeO–(naphthyl)–Cu(C≡CPr)MgBr	Chapter 9, Protocol 1
$C_{11}H_{17}O$	$MeOOC(CH_2)_7C{\equiv}CH$ + Et_2NH+ 13% CuI	92T4465
$C_{12}H_{12}N$	Me_2N, Cu (naphthalene)	87JOM(325)293
$C_{12}H_{19}O_2$	$AcO(CH_2)_8C{\equiv}CH$ + cat. Cu(I)/Pd(0)	81SC917
$C_{13}H_{24}NO_2$	$H_2C{=}C(Me)CH(Cu)CH_2CH_2CH(Me)CH_2CH_2OCONMe_2$	92JA5110
$C_{13}H_{25}OSi$	Cu, $OSiMe_3$	Chapter 11, Protocol 3
$C_{14}H_{19}OSi$	Me_3Si, Cu, MeO	84JOC1574
$C_{14}H_{27}O$	$[(Z)\text{-}t\text{-}BuO(CH_2)_8CH{=}CH]_2$ CuLi	86T4523

C_{11}

$C_{11}H_{13}$	Cu	86SC499
$C_{11}H_{17}O$	$(Z)\text{-}C_5H_{11}CH{=}CHCH_2CH(OH)C{\equiv}CH$ + cat. Cu(I)/Pd(0)	92T1943
$C_{11}H_{19}$	$(Z,Z)\text{-}EtC(Me){=}CH(CH_2)_2C(Et){=}CHCu$	76T1675
$C_{11}H_{21}$	$H_2C{=}CH(CH_2)_9Cu(C{\equiv}CBu\text{-}t)Li$	75JOC779
$C_{11}H_{21}$	$(Z)\text{-}C_9H_{19}CH{=}CHMgBr$ + 5% CuI	86JOC4726
$C_{11}H_{23}$	$(C_{11}H_{23})_2Cu(CN)Li_2$	91JOC4714
$C_{12}H_{23}O_2$	$MeOOC(CH_2)_{10}Cu(CN)ZnI$	93SC2547

Formula	Reagent	Reference
$C_{13}H_{19}O_2$	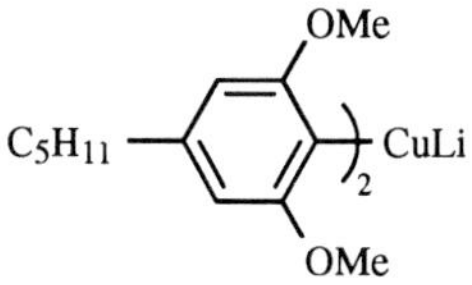 OMe; C_5H_{11}; CuLi; 2; OMe	92JOC3627
$C_{15}H_{31}O_4$	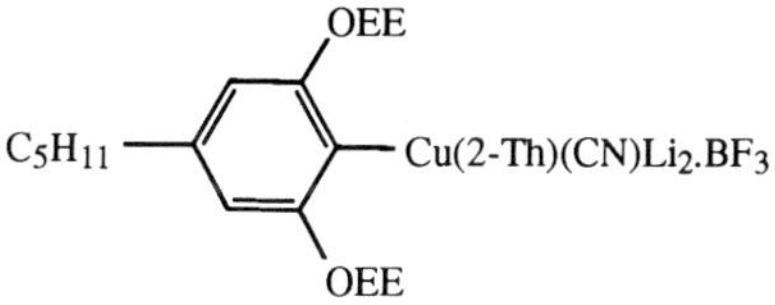 OEE; C_5H_{11}; $Cu(2\text{-}Th)(CN)Li_2.BF_3$; OEE	90JOC5711
$C_{18}H_{29}O$	$BnO(CH_2)_{11}MgBr$ + 1.5% Li_2CuCl_4	92S981
$C_{21}H_{31}O_4$	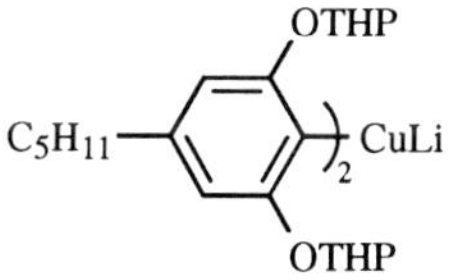 OTHP; C_5H_{11}; CuLi; 2; OTHP	79JCS(P1)201

C_{12}

Formula	Reagent	Reference
$C_{12}H_8Cu$	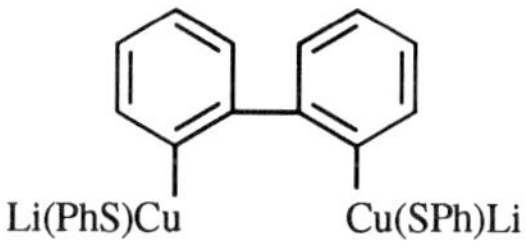 Li(PhS)Cu; Cu(SPh)Li	88JA2218
$C_{12}H_{15}$	(*E*)-PhC(Bu)=CHCu.nSMe_2	91OS(70)215
$C_{12}H_{21}$	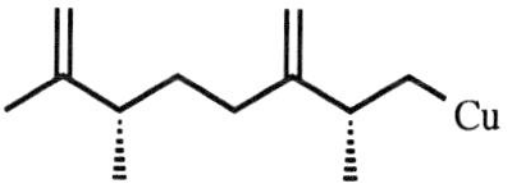 Cu	93JCS(P1)759
$C_{12}H_{23}$	$Me_2C=CH(CH_2)_2CH(Me)(CH_2)_4Cu$	75JA1197

C_{13}

Formula	Reagent	Reference
$C_{13}H_9O$	4-$PhCOC_6H_4Cu$	88JOC4482
$C_{13}H_{15}O$	2-$CyCOC_6H_4Cu(CN)ZnI$	90TL4413
$C_{13}H_{19}$	(Z,Z)-$C_5H_{11}CH=CHCH_2CH=CHCH_2C\equiv CH$ + cat. Cu(I)/Pd(0)	86T344
$C_{13}H_{23}$	(Z,Z)-$PrC(Me)=CH(CH_2)_2C(Pr)=CHCu.nSMe_2$	79JOC1345
$C_{13}H_{27}$	$[(R)\text{-}C_8H_{17}CH(Me)(CH_2)_3]_2CuLi$	79T1279

I + Zn(0)/Cu(I), ultrasound

OTf

$C_{14}H_{20}F_3O_3S$ 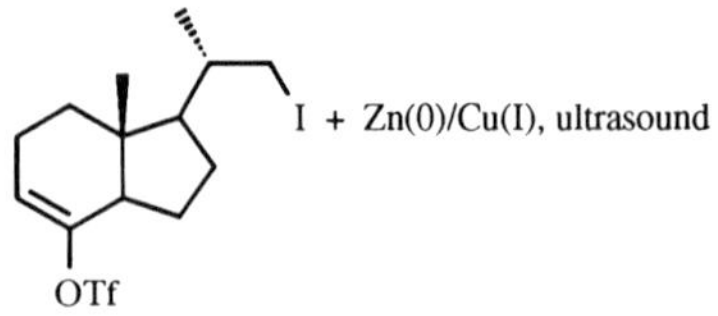 93JOC118

NMe_2

Cy

Cu(Me)Li

$C_{15}H_{22}N$ 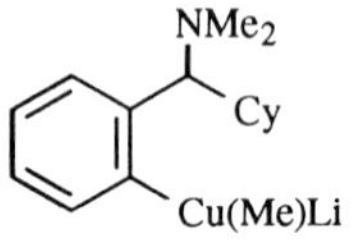 78T3023

$OSiMe_2Bu$-*t*

+ cat. Pd(O)/Cu(I)

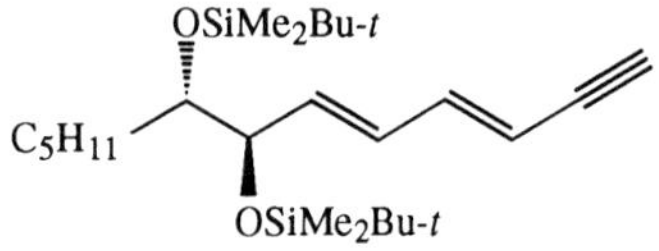

C_5H_{11}

$OSiMe_2Bu$-*t*

$C_{25}H_{47}O_2Si_2$ 86S453

C_{14}

t-Bu

Cu(*t*-Bu)Li

Bu-*t*

$C_{14}H_{21}$ 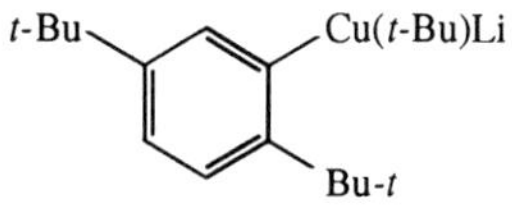 85JOC3222

C_{15}

$C_{15}H_{31}$ $C_{15}H_{31}MgBr$ + 0.2% Li_2CuCl_4 90T4473

Cu(C≡CCMe$_2$OMe)Li

O

O

t-BuMe$_2$SiO

$C_{21}H_{39}O_3Si$ 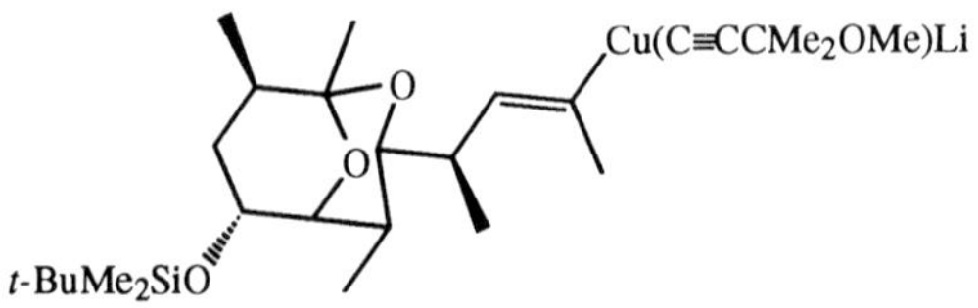 86JA5559

C_{16}

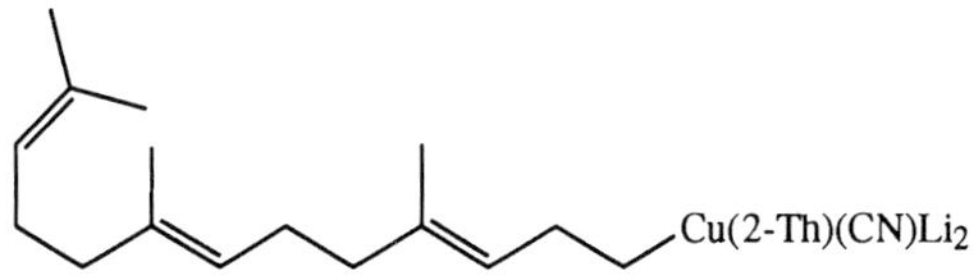

$C_{16}H_{27}$	Cu(2-Th)(CN)Li_2	92JOC2794
$C_{16}H_{33}$	$C_{16}H_{33}MgBr$ + 0.2% Li_2CuCl_4	90T4473

C_{20}

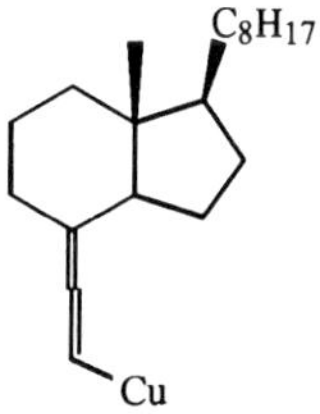

$C_{20}H_{33}$	C_8H_{17} Cu	83JOC4030

C_{21}

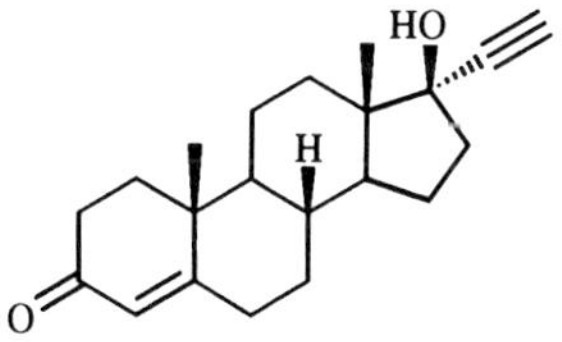

$C_{21}H_{27}O_2$	HO H O + cat. Cu(I)/Pd(O)	92S523
$C_{21}H_{35}$	Cu(2-Th)(CN)Li_2	92JOC2794

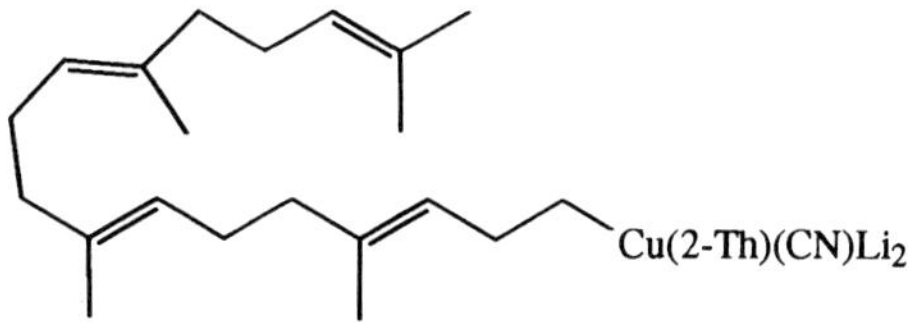

References

52JOC1630 Gilman, H.; Jones, R. G.; and Woods, L. A. *J. Org.; Chem.* **1952**, *17*, 1630–1634.

65JA82 Boatman, S.; Harris, T. M.; and Hauser, C. R. *J. Am. Chem. Soc.* **1965**, *87*, 82–86

66JOC1016 Marshall, J. A.; Fanta, W. I.; and Roebke, H. *J. Org. Chem.* **1966**, *31*, 1016–1020.

67JCS(C)578 Atkinson, R. E.; Curtis, R. F.; and Taylor, J. A. *J. Chem. Soc.(C)* **1967**, 578–582.

67JOC784 House, H. O.; and Bashe, R. W., II; *J. Org. Chem.* **1967**, *32*, 784–791

67JOM(8)339 Costa, G.; Camus, A.; Marsich, N.; and Gatti, L. *J. Organometal. Chem.* **1967**, *8*, 339–346.

68JOC949 House, H. O.; and Fischer, W. F., Jr. *J. Org. Chem.* **1968**, *33*, 949–956.

69JA4871 Whitesides, G. M.; Fischer, W. F., Jr.; San Filippo, J., Jr.; Bashe, R. W., II; and House, H. O. *J. Am. Chem. Soc.*; **1969**, *91*, 4871–4882.

70ACS2379 Nilsson, M.; and Ullenius, C. *Acta Chem. Scand.* **1970**, *24*, 2379–2388.

70JOC1715 Worm, A. T.; and Brewster, J. H. *J. Org. Chem.* **1970**, *35*, 1715–1716

70OS(50)38 Eliel, E. L.; Hutchins, R. O.; and Knoeber, Sr. M. *Org. Synth.* **1970**, *50*, 38–42 (**1988**, *Collective Vol. 6*, 442–445).

71ACS2596 Gjos, N.; and Gronowitz, S. *Acta Chem. Scand.* **1971**, *25*, 2596–2608.

71HCA1939 Näf, F.; and Degen, P. *Helv. Chim. Acta* **1971**, *54*, 1939–1949

71JOC1143 Büchi, G.; Klaubert, D. H.; Shank, R. C.; Weinreb, S. M. and Wogan, G. N. *J. Org. Chem.* **1971**, *36*, 1143–1147

72JA232 Whitesides, G. M.; Panek, E. J.; and Stedronsky, E. R. *J. Am. Chem. Soc.* **1972**, *94*, 232–239

72JA5374 Henrick, C. A.; Schaub, F.; and Siddall, J. B. *J. Am. Chem. Soc.* **1972**, *94*, 5374–5378.

72JA7210 Corey, E. J.; and Beames, D. J. *J. Am. Chem. Soc.* **1972**, *94*, 7210–7211

72JA7823 Alvarez, F. S.; Wren, D.; and Prince, A. *J. Am. Chem. Soc.* **1972**, *94*, 7823–7827.

72JA7827 Kluge, A. F.; Untch, K. G.; and Fried, J. H. *J. Am. Chem. Soc.* **1972**, *94*, 7827–7832.

72JOC1947 Eaton, P. E.; Cooper, G. F.; Johnson, R. C.; and Mueller, R. H. *J. Org. Chem.* **1972**, *37*, 1947–1950.

72JOC3718 Whitesides, G. M.; and Kendall, P. E. *J. Org. Chem.* **1972**, *37*, 3718–3725.

72JOM(42)257 Smith, C. F.; Moore, G. J.; and Tamborski, C. *J. Organometal. Chem.* **1972**, *42*, 257–265.

72OS(52)128 Owsley, D. C.; and Castro, C. E. *Org. Synth.* **1972**, 52, 128–131 (**1988**, *Collective Vol. 6*, 916–918).

72TL487 Corey, E. J.; and Kuwajima, I. *Tetrahedron Lett.* **1972**, 487–489.

73JA3076 Posner, G. H. and Sterling, J. J. *J. Am. Chem. Soc.* **1973**, *95*, 3076–3077.

73JA7777 Johnson, C. R.; and Dutra, G. A. *J. Am. Chem. Soc.* **1973**, *95*, 7777–7782.

73JA7788 Posner, G. H.; Whitten, C. E.; and Sterling, J. J. *J. Am. Chem. Soc.* **1973**, *95*, 7788–7800.

73JOC3893 House, H. O.; and Umen, M. J. *J. Org. Chem.* **1973**, *38*, 3893–3901.

73OSC107 Adams, R.; Reifschneider, W.; and Ferretti, A. *Org. Synth.* **1973**, *Collective Vol. 5*, 107–110.

73S170 Levy, L. A. *Synthesis*, **1973**, 170–171.

74JA2829 Whitesides, G. M.; Sadowski, J. S.; and Lilburn, J. *J. Am. Chem. Soc.* **1974**, *96*, 2829–2835.

74JA5581 Corey, E. J.; and Wollenberg, R. H. *J. Am. Chem. Soc.* **1974**, *96*, 5581–5583.

74JA7138 Marino, J. P.; and Floyd, D. M. *J. Am. Chem. Soc.* **1974**, *96*, 7138–7140

74JOC400 Mandeville, W. H.; and Whitesides, G. M. *J. Org. Chem.* **1974**, *39*, 400–405.

75HCA1016 Näf, F.; Decorzant, R.; Thommen, W.; Willhalm, B.; and Ohloff, G. *Helv. Chim. Acta* **1975**, *58*, 1016–1037.

75JA857 Sih, C. J.; Salomon, R. G.; Price, P.; Sood, R.; and Peruzzotti, G. *J. Am. Chem. Soc.* **1975**, *97*, 857–865.

75JA865 Sih, C. J.; Heather, J. B.; Sood, R.; Price, P.; Peruzzotti, P.; Hsu Lee, L. F.; and Lee, S. S. *J. Am. Chem. Soc.* **1975**, *97*, 865–874.

75JA1197 Anderson, R. J.; Corbin, V. L.; Cotterrell, G.; Cox, G. R.; Henrick, C. A.; Schaub, F.; and Siddall, J. B. *J. Am. Chem. Soc.* **1975**, *97*, 1197–1204.

75JA3822 Chavdarian, C. G.; and Heathcock, C. H. *J. Am. Chem. Soc.* **1975**, *97*, 3822–3823.

75JOC779 Bergbreiter, D. E.; and Whitesides, G. M. *J. Org. Chem.* **1975**, *40*, 779–782.

75JOC1460 House, H. O.; Chu, C-Y.; Wilkins, J. M.; and Umen, M. J. *J. Org. Chem.* **1975**, *40*, 1460–1469.

75JOC1488 Salomon, R. G.; and Salomon, M. F. *J. Org. Chem.* **1975**, *40*, 1488–1492.

75JOM(99)157 van Koten, G.; Schaap, C. A.; and Noltes, J. G. *J. Organometal. Chem.* **1975**, *99*, 157–170.

75TL4467 Sonogashira, K.; Tohda, Y.; and Hagihara, N. *Tetrahedron Lett.* **1975**, 4467–4470.

76BCJ1989	Willy, W. E.; McKean, D. R.; and Garcia, B. A. *Bull. Chem. Soc. Japan* **1976**, *49*, 1989–1995.
76JA8282	Ziegler, F. E.; Fowler, K. W.; and Kanfer, S. *J. Am. Chem. Soc.* **1976**, *98*, 8282–8283.
76OS(55)103	Linstrumelle, G.; Krieger, J. K.; and Whitesides, G. M. *Org. Synth.* **1976**, *55*, 103–113. Omitted from Collective Vol. 6, since a key reagent is no longer available commercially.
76OS(55)122	Posner, G. H.; and Whitten, C. E. *Org. Synth.* **1976**, *55*, 122–127 (**1988**, *Collective Vol. 6*, 248–252).
76T1675	Chuit, C.; Cahiez, G.; and Normant, J. F. *Tetrahedron*, **1976**, *32*, 1675–1680.
76T2281	Posner, G. H.; Ting, J.-S.; and Lentz, C. M. *Tetrahedron*, **1976**, *32*, 2281–2287.
76TL275	Michelot, D.; and Linstrumelle, G. *Tetrachedron Lett.* **1976**, 275–276 (Fr).
77JA7365	Wollenberg, R. H.; Albizati, K. F.; and Peries, R. *J. Am. Chem. Soc.* **1977**, *99*, 7365–7367.
77JA8068	Maruyama, K.; and Yamamoto, Y. *J. Am. Chem. Soc.* **1977**, *99*, 8068–8070.
77JOC1099	Ashby, E. C.; Lin, J. J.; and Watkins, J. J. *J. Org. Chem.* **1977**, *42*, 1099–1102.
77T1105	Tanaka, T.; Kurozumi, S.; Toru, T.; Kobayashi, M.; Miura, S.; and Ishimoto, S. *Tetrahedron*, **1977**, *33*, 1105–1112.
77TL1805	Obayashi, M.; Utimoto, K.; and Nozaki, H. *Tetrahedron Lett.* **1977**, 1805–1806.
78BCJ547	Tsuji, J.; Kaito, M.; and Takahashi, T. *Bull. Chem. Soc. Japan*, **1978**, *51*, 547–549.
78BCJ2909	Kubota, M.; and Yamamoto, A. *Bull. Chem. Soc. Japan*, **1978**, *51*, 2909–2915.
78BSF(2)299	Hammond, A.; and Descoins, C. *Bull. Soc. Chim. Fr.* **1978**, *II*-299–303 (Fr).
78IC275	San Filippo, J., Jr.; *Inorg. Chem.*; **1978**, *17*, 275–283.
78JA6211	Lüthy, C.; Konstantin, P.; and Untch, K. G. *J. Am. Chem. Soc.* **1978**, *100*, 6211–6217.
78JOC555	Amos, R. A.; and Katzenellenbogen, J. A. *J. Org. Chem.* **1978**, *43*, 555–560
78JOC2102	Morton, D. R.; and Thompson, J. L. *J. Org. Chem.* **1978**, **43**, 2102–2106.
78JOC3418	Corey, E. J.; Floyd, D.; and Lipshutz, B. H. *J. Org. Chem.* **1978**, *43*, 3418–3420.
78LA658	Eiter, K.; Lieb, F.; Disselnkötter, H.; and Oediger, H. *Liebigs Ann.* **1978**, 658–674 (Ger).
78MI599	Taber, D. F.; and Lee, C. H. *J. Labelled Comp. Radiopharm.* **1978**, *14*, 599–602.

78OS(58)52 Bruggink, A.; Ray, S. J.; and McKillop, A. *Org. Synth.* **1978**, *58*, 52–56 (**1988**, *Collective Vol. 6*, 36–39.

78OS(58)158 Boeckman, R. K.; Blum, D. M.; and Ganem, B. *Org. Synth.* **1978**, *58*, 158–161 (**1988**, *Collective Vol. 6*, 666–669).

78S388 Samain, D.; Descoins, C.; and Commerçon, A. *Synthesis*, **1978**, 388–389.

78SC175 Duffley, R. P.; and Stevenson, R. *Synth. Comm.* **1978**, *8*, 175–180.

78T3023 Gustafsson, B. *Tetrahedron* **1978**, *34*, 3023–3026.

78TL1363 Marfat, A.; McGuirk, P. R.; and Helquist, P. *Tetrahedron Lett.* **1978**, 1363–1366.

79HCA2239 Seebach, D.; and Lehr, F. *Helv. Chim. Acta* **1979**, *62*, 2239–2257 (Ger).

79JA934 Posner, G. H.; and Lentz, C. M. *J. Am. Chem. Soc.* **1979**, *101*, 934–946.

79JA1857 Kende, A. S.; and Curran, D. P. *J. Am. Chem. Soc.* **1979**, *101*, 1857–1864.

79JCS(P1)201 Luteijn, J. M.; and Spronck, H. J. W. *J. Chem. Soc.; Perkin 1* **1979**, 201–203.

79JOC1006 Ledlie, D. B. and Miller, G. *J. Org. Chem.* **1979**, *44*, 1006–1007.

79JOC1345 Marfat, A.; McGuirk, P. R.; and Helquist, P. *J. Org. Chem.* **1979**, *44*, 1345–1347.

79JOC3888 Marfat, A.; McGuirk, P. R.; and Helquist, P. *J. Org. Chem.* **1979**, *44*, 3888–3901.

79JOC4781 Boeckman, R. K.; Jr. and Bruza, K. J. *J. Org. Chem.* **1979**, *44*, 4781–4788.

79OS(59)122 Cairncross, A.; Sheppard, W. A.; and Wonchoba, E. *Org. Synth.* **1979**, *59*, 122–131 (**1988**, *Collective Vol. 6*, 875–882).

79S885 Calo, V.; Lopez, L.; Marchese, G.; Pesce, G. *Synthesis*, **1979**, 885–887.

79SC287 Savignac, P.; Breque, A.; Mathey, F.; Varlet, J.-M.; and Collignon, N. *Synth. Comm.* **1979**, *9*, 287–294.

79T1279 Mori, K. and Tamada, S. *Tetrahedron* **1979**, *35*, 1279–1284.

79TL4577 Kojima, Y.; Wakita, S.; and Kato, N. *Tetrahedron Lett.* **1979**, 4577–4580.

79TL4717 Ziegler, F. E. and Tam, C. C. *Tetrahedron Lett.* **1979**, 4717–4720.

80AG472 Köksal, Y.; Raddatz, P.; and Winterfeldt, E. *Angew. Chem. (Engl. Ed.)* **1980**, *19*, 472–473.

80G237 Rossi, R.; Carpita, A.; Gaudenzi, L.; and Quirici, M. G. *Gazz.* **1980**, *110*, 237–246.

80JA790 Ziegler, F. E.; Chliwner, L.; Fowler, K. W.; Kanfer, S. J.; Kuo, S. J.; and Sinha, N. D. *J. Am. Chem. Soc.* **1980**, *102*, 790–798.

80JA5253 Funk, R. L.; and Vollhardt, K. P. C. *J. Am. Chem. Soc.* **1980**, *102*, 5253–5261.

80JOC378 Chenard, B. L.; Manning, M. J.; Raynolds, P. W.; and Swenton, J. S. *J. Org. Chem.* **1980**, *53* 378–384.

80JOC1158 Oostveen, J. M.; Westmijze, H.; and Vermeer, P. *J. Org. Chem.* **1980**, *45* 1158–1160.

80JOC2229 Anderson, R. J.; Adams, K. G.; Chinn, H. R.; and Henrick, C. A. *J. Org. Chem.* **1980**, *45*, 2229–2236.

80JOM(188)293 Rahman, M. T.; Hoque, A. K. M. M.; Siddiqui, I.; Chowdhury, D. A. N.; Nahar, S. K.; and Saha, S. L. *J. Organometal. Chem.* **1980**, *188*, 293–300.

80MI118 Posner, G. H.; *An Introduction to Synthesis using Organocopper Reagents*. John Wiley & Sons, New York, **1980**, p. 118.

80T1961 Alexakis, A.; Cahiez, G.; and Normant, J. F. *Tetrahedron*, **1980**, *36*, 1961–1969.

80TL3115 Gawley, R. E.; Termine, E. J.; and Aube, J. *Tetrahedron Lett.* **1980**, *21*, 3115–3118.

80TL3151 Bertz, S. H. *Tetrahedron Lett.* **1980**, *21*, 3151–3154.

80TL3849 Kende, A. S.; and Jungheim, L. N. *Tetrahedron Lett.* **1980**, *21*, 3849–3852.

81IC2728 Tsuda, T.; Watanabe, K.; Miyata, K.; Yamamoto, H.; and Saegusa, T. *Inorg. Chem* **1981**, *20*, 2728–2730.

81JA1831 Paquette, L. A.; and Han, Y.-K. *J. Am. Chem. Soc.* **1981**, *103*, 1831–1835.

81JCS(P1)593 Christie, R. M.; Gill, M.; and Rickards, R. W. *J. Chem. Soc.; Perkin 1* **1981**, 593–598.

81JCS(P1)599 Gill, M.; and Rickards, R. W. *J. Chem. Soc.; Perkin 1* **1981**, 599–606.

81JCS(P1)1516 Jackson, W. P.; and Ley, S. V. *J. Chem. Soc.; Perkin 1* **1981**, 1516–1519.

81JCS(P1)2520 Ager, D. J.; Fleming, I.; and Patel, S. K. *J. Chem. Soc.; Perkin 1* **1981**, 2520–2526.

81JOC192 Tsuda, T.; Yazawa, T.; Watanabe, K.; Fujii, T.; and Saegusa, T. *J. Org. Chem.* **1981**, *46*, 192–194.

81JOC1532 Poulter, C. D.; Wiggins, P. L.; and Plummer, T. L. *J. Org. Chem.* **1981**, *46*, 1532–1538.

81JOC3696 Marino J. P.; and Linderman, R. J. *J. Org. Chem.* **1981**, *46*, 3696–3702.

81JOC3790 Binns, M. R.; and Haynes, R. K. *J. Org. Chem.* **1981**, *46*, 3790–3795.

81JOC4389 Marino, J. P.; and Kelly, M. C. *J. Org. Chem.* **1981**, *46*, 4389–4393.

81MI73 Brandsma, L.; and Verkruijsse, H. D. *Synthesis of Acetylenes, Allenes, and Cumulenes – A Laboratory Manual*, Elsevier, Amsterdam, **1981**, p. 73.

81MI170 Brandsma, L.; and Verkruijsse, H. D.; *Synthesis of Acetylenes,*

	Allenes, and Cumulenes – A Laboratory Manual, Elsevier, Amsterdam, **1981**, p. 170.
81MI361	Belanger, A.; Philibert, D.; and Teutsch, G. *Steroids* **1981**, *37*, 361–382.
81OS(60)41	Hopf, H.; Böhm, I.; and Kleinschroth, J. *Org. Synth.* **1981**, *60*, 41–48 (**1990**, *Collective Vol. 7*, 485–490).
81RTC249	Kleijn, H.; Westmijze, H.; Meijer, J.; and Vermeer, P. *Rec. Trav. Chim.* **1981**, *100*, 249–255.
81SC917	Ratovelomana, V.; and Linstrummelle, G. *Synth. Comm.* **1981**, *11*, 917–923.
81T4431	Schostarez, H.; and Paquette, L. A. *Tetrahedron* **1981**, *37*, 4431–4435.
82CPB3395	Wakita, Y.; Kobayashi, T.; Maeda, M.; and Kojima, M. *Chem. Pharm. Bull.* **1982**, *30*, 3395–3398.
82JA1054	Heathcock, C. H.; Kleinman, E. F.; and Binkley, E. S. *J. Am. Chem. Soc.* **1982**, *104*, 1054–1068.
82JA5473	Le Drian, C.; and Greene, A. E. *J. Am. Chem. Soc.* **1982**, *104*, 5473–5483.
82JA7609	Ito, Y.; Nakatsuka, M.; and Saegusa, T. *J. Am. Chem. Soc.* **1982**, *104*, 7609–7622.
82JCS(P1)1177	Batten, R. J.; Coyle, J. D.; Taylor, R. J. K.; and Vassiliou, S. *J. Chem. Soc.; Perkin 1* **1982**, 1177–1182.
82JOC1221	Jakubowski, A. A.; Guziec, F. S.; Jr. Sugiura, M.; Tam, C. C.; Tishler, M.; and Omura, S. *J. Org. Chem.* **1982**, *47*, 1221–1228.
82JOC3333	Wexler, B. A.; Toder, B. H.; Minaskanian, G.; and Smith, A. B., III *J. Org. Chem.* **1982**, *47*, 3333–3335.
82JOC3464	Fang, J. M. *J. Org. Chem.* **1982**, *47*, 3464–3478.
82JOC4605	Olsen, R. K.; Hennen, W. J.; and Wardle, R. B. *J. Org. Chem.* **1982**, *47*, 4605–4611.
82JOC4825	Roush, W. R.; and Gillis, H. R. *J. Org. Chem.* **1982**, *47*, 4825–4829.
82JOC5045	Bal, S. A.; Marfat, A.; and Helquist, P. *J. Org. Chem.* **1982**, *47*, 5045–5050.
82JOC5088	Kowalski, C. J.; Weber, A. E.; and Fields, K. W. *J. Org. Chem.* **1982**, *47*, 5088–5093.
82MI667	Walborsky, H. M.; Banks, R. B.; Banks, M. L. A.; and Duraisamy, M. *Organometallics* **1982**, *1*, 667–674.
82S857	Osuka, A.; Ohmasa, N.; and Suzuki, H. *Synthesis* **1982**, 857–858.
82SC1027	Yamaguchi, R.; Kawasaki, H.; and Kawanisi, M. *Synth. Comm.* **1982**, *12*, 1027–1037.
82T363	Claesson, A.; and Sahlberg, C. *Tetrahedron* **1982**, *38*, 363–368.
82T1509	Malmberg, H.; and Nilsson, M. *Tetrahedron* **1982**, *38*, 1509–1510.
83CJC1226	Piers, E.; Morton, H. E.; Nagakura, I.; and Thies, R. W. *Can. J. Chem.* **1983**, *61*, 1226–1238.

83CPB128 Ibuka, T.; Tabushi, E.; and Yasuda, M. *Chem. Pharm. Bull.* **1983**, *31*, 128–134.

83JA2034 Semmelhack, M. F.; and Zask, A. *J. Am. Chem. Soc.* **1983**, *105*, 2034–2043.

83JA2364 Snider, B. B.; and Kirk, T. C. *J. Am. Chem. Soc.* **1983**, *105*, 2364–2368.

83JA3588 Chandraratna, R. A. S.; Bayerque, A. L.; and Okamura, W. H. *J. Am. Chem. Soc.* **1983**, *105*, 3588–3594.

83JA3656 Rosenberger, M.; Newkom, C.; and Aig, E. R. *J. Am. Chem. Soc.* **1983**, *105*, 3656–3661.

83JCS(P1)1387 Baker, R.; Billington, D. C.; and Ekanayake, N. *J. Chem. Soc.; Perkin 1* **1983**, 1387–1393.

83JOC546 Lipshutz, B. H.; Kozlowski, J. A.; and Wilhelm, R. S. *J. Org. Chem,* **1983**, *48*, 546–550.

83JOC1404 Schmuff, N. R.; and Trost, B. M. *J. Org. Chem.* **1983**, *48*, 1404–1412.

83JOC4030 Gerdes, J. M.; and Okamura, W. H. *J. Org. Chem.* **1983**, *48*, 4030–4035.

83JOC4621 Marino, J. P.; and Linderman, R. J. *J. Org. Chem.* **1983**, *48*, 4621–4628.

83S68 Osuka, A.; Ohmasa, N.; Uno, Y.; and Suzuki, H. *Synthesis* **1983**, 68–69.

83S597 Novák, J.; and Salemink, C. A. *Synthesis* **1983**, 597–598.

83S804 Suzuki, S.; Shiono, M.; and Fujita, Y. *Synthesis* **1983**, 804–806.

83T2469 Arrowsmith, J. E.; Greengrass, C. W.; and Newman, M. J. *Tetrahedron* **1983**, *39*, 2469–2475.

83T3235 Dewanckele, J. M.; Zutterman, P.; and Vandewalle, M. *Tetrahedron* **1983**, *39*, 3235–3244.

83TL1003 Shea, K. J.; and Pham, P. Q. *Tetrahedron Lett.* **1983**, *24*, 1003–1006.

83TL3905 Kocieński, P.; and Yeates, C. *Tetrahedron Lett.* **1983**, *24*, 3905–3906.

84JA1443 Snider, B. B.; and Faith, W. C. *J. Am. Chem. Soc.* **1984**, *106*, 1443–1445.

84JA3368 Nakamura, E.; and Kuwajima, I. *J. Am. Chem. Soc.* **1984**, *106*, 3368–3370.

84JA6006 Marshall, J. A.; Peterson, J. C.; and Lebioda, L. *J. Am. Chem. Soc.* **1984**, *106*, 6006–6015.

84JCR(M)2327 Alexandre, C.; Chlyeh, M. A.; and Rouessac, F. *J. Chem. Res.* **1984**, *M*-2327–2342; *S*-247.

84JCS(P1)119 Fleming, I.; and Newton, T. W. *J. Chem. Soc.; Perkin 1* **1984**, 119–123.

84JOC1119 Bertz, S. H.; and Dabbagh, G. *J. Org. Chem.* **1984**, *49*, 1119–1122.

84JOC1574 Sternberg, E. D.; and Vollhardt, K. P. C. *J. Org. Chem.* **1984**, *49*, 1574–1583.

84JOC3183 Dieter, R. K.; Lin, Y. J.; and Dieter, J. W. *J. Org. Chem.* **1984**, *49*, 3183–3195.

84JOC3503 Edwards, M. P.; Ley, S. V.; Lister, S. G.; Palmer, B. D.; and Williams, D. J. *J. Org. Chem.* **1984**, *49*, 3503–3516.

84JOC3928 Lipshutz, B. H.; Wilhelm, R. S.; Kozlowski, J. A.; and Parker, D. J. *Org. Chem.* **1984**, *49*, 3928–3938.

84MI118 Suzuki, M.; Suzuki, T.; Kawagishi, T.; Morita, Y.; and Noyori, R. *Israel J. Chem.* **1984**, *24*, 118–124.

84MI125 Clarembeau, M.; Bertrand, J. L.; and Krief, A. *Israel J. Chem.* **1984**, *24*, 125–133.

84OS(62)1 Alexakis, A.; Cahiez, G.; and Normant, J. F. *Org. Synth.* **1984**, *62*, 1–8 (**1990**, *Collective Vol. 7*, 290–294).

84S278 Stowell, J. C.; and King, B. T. *Synthesis* **1984**, 278–280.

84S616 Suzuki, H.; Thiruvikraman, S. V.; and Osuka, A. *Synthesis* **1984**, 616–617.

84S728 Bumagin, N. A.; Ponamaryov, A. B.; and Beletskaya, I. P. *Synthesis* **1984** 728–729.

84SC761 Descoins, C.; Lettere, M.; Linstrumelle, G.; Michelot, D.; and Ratovelomanana, V. *Synth. Comm.* **1984**, *14*, 761–773.

84T715 Alexakis, A.; Commerçon, A.; Coulentianos, C.; and Normant, J. F. *Tetrahedron*, **1984**, *40*, 715–731.

84T1401 Posner, G. H.; Frye, L. L.; and Hulce, M. *Tetrahedron* **1984**, *40*, 1401–1407.

84TL1925 Takahashi, T.; Okumoto, H.; and Tsuji, J. *Tetrahedron Lett.* **1984**, *25*, 1925–1928.

84TL2813 Yamamoto, K.; Iijima, M.; Ogimura, Y.; and Tsuji, J. *Tetrahedron Lett* **1984**, *25*, 2813–2816.

85CL1779 Suzuki, H.; Watanabe, K.; and Yi, Q. *Chem. Lett.* **1985**, 1779–1780.

85H117 Ishikawa, M.; Kamada, M.; Oda, I.; and Terashima, M. *Heterocycles* **1985**, *23*, 117–120.

85JA1034 Okamura, W. H.; Peter, R.; and Reischl, W. *J. Am. Chem. Soc.* **1985**, *107*, 1034–1041.

85JA2474 Danishefsky, S.; Chackalamannil, S.; Harrison, P.; Silvestri, M.; and Cole, P. *J. Am. Chem Soc.* **1985**, *107*, 2474–2484.

85JA3197 Lipshutz, B. H.; Kozlowski, J. A.; and Breneman, C. M. *J. Am. Chem. Soc.* **1984**, *107*, 3197–3204.

85JA4551 Seyferth, D.; and Hui, R. C. *J. Am. Chem. Soc.* **1985**, *107*, 4551–4553.

85JA6028 Guo, C.; Brownawell, M. L.; and San Filippo, J., Jr.; *J. Am. Chem. Soc.* **1985**, *107*, 6028–6030.

85JBC8404 Bar-Tana, J.; Rose-Kahn, G.; and Srebnik, M. *J. Biol. Chem.* **1985**, *260*, 8404–8410.

85JOC1987 Taber, D. F.; Dunn, B. S.; Mack, J. F.; and Saleh, S. A. *J. Org. Chem.* **1985**, *50*, 1987–1988.

85JOC3222 Künzer, H.; and Berger, S. *J. Org. Chem.* **1985**, *50*, 3222–3223.

85JOC3988 Tanis, S. P.; and Herrinton, P. M. *J. Org. Chem.* **1985**, *50*, 3988–3996.

85JOM(282)427 Pasynkiewicz, S.; and Poplawska, J. *J. Organometal. Chem.* **1985**, *282*, 427–434.

85JOM(285)437 Lipshutz, B. H.; Kozlowski, J. A.; Parker, D. A.; Nguyen, S. L.; and McCarthy, K. E. *J. Organometal. Chem.* **1985**, *285*, 437–447.

85S492 Beswick, P. J.; and Widdowson, D. A.; *Synthesis* **1985**, 492–493.

85S705 Boland, W.; and Mertes, K. *Synthesis* **1985**, 705–708.

85SC569 Villiéras, J.; Rambaud, M.; and Graff, M. *Synth. Comm.* **1985**, *15*, 569–580.

85T129 Thorberg, S-O.; Gawell, L.; Csöregh, I.; and Nilsson, J. L. G. *Tetrahedron* **1985**, *41*, 129–139.

85T627 Rossi, R.; Carpita, A.; and Chini, M. *Tetrachedron* **1985**, *41*, 627–633.

85T3943 Curran, D. P.; and Rakiewicz, D. M. *Tetrahedron* **1985**, *41*, 3943–3958.

85TL547 Hua, D. H.; and Verma, A. *Tetrahedron Lett.* **1985**, *26*, 547–550.

86JA800 Crimmins, M. T.; and DeLoach, J. A. *J. Am. Chem. Soc.* **1986**, *108*, 800–806.

86JA5559 Neukom, C.; Richardson, D. P.; Myerson, J. H.; and Bartlett, P. A. *J. Am. Chem. Soc.* **1986**, *108*, 5559–5568.

86JCS(P1)1515 Bell, V. L.; Giddings, P. J.; Holmes, A. B.; Mock, G. A.; and Raphael, R. A. *J. Chem. Soc.; Perkin 1* **1986**, 1515–1522.

86JCS(P1)1809 Furber, M.; Taylor, R. J. K.; and Burford, S. C. *J. Chem. Soc.; Perkin 1* **1986**, 1809–1815.

86JCS(P1)2151 Ihara, M.; Toyota, M.; Fukumoto, K.; and Kametani, T. *J. Chem. Soc.; Perkin 1* **1986**, 2151–2161.

86JMC1457 Soll, R. M.; Humber, L. G.; Deininger, D.; Asselin, A. A.; Chau, T. T.; and Weichman, B. M. *J. Med. Chem.* **1986**, *29*, 1457–1460.

86JOC863 Marshall, J. A.; and DeHoff, B. S. *J. Org. Chem.* **1986**, *51*, 863–872.

86JOC1730 Marshall, J. A.; Audia, J. E.; and Shearer, B. G. *J. Org. Chem.* **1986**, *51*, 1730–1735.

86JOC3726 Gerth, D. B.; and Giese, B. *J. Org. Chem.* **1986**, *51*, 3726–3729.

86JOC3983 Ager, D. J.; and East, M. B. *J. Org. Chem.* **1986**, *51*, 3983–3992.

86JOC4726 Millar, J. G.; and Underhill, E. W. *J. Org. Chem.* **1986**, *51*, 4726–4728.

86JOM(316)255 Chen, H.-M.; and Oliver, J. P. *J. Organometal. Chem.* **1986**, *316*, 255–260.

86OS(64)1 Iyer, R. S.; and Helquist, P. *Org. Synth.* **1986**, *64*, 1–9 (**1990**, *Collective Vol. 7*, 236–240).

86S344	Nicolaou, K. C.; Ladduwahetty, T.; Taffer, I. M.; and Zipkin, R. E. *Synthesis* **1986**, 344–357.
86S453	Nicolaou, K. C.; and Webber, S. E. *Synthesis* **1986**, 453–461.
86SC499	Amano, T.; Yoshikawa, K.; Sano, T.; Ohuchi, Y.; Shiono, M.; Ishiguro, M.; and Fujita, Y. *Synth. Comm.* **1986**, *16*, 499–507.
86T1389	Foulon, J. P.; Bourgain-Commerçon, M.; and Normant, J. F. *Tetrahedron* **1986**, *42*, 1389–1397.
86T1399	Foulon, J. P.; Bourgain-Commerçon, M.; and Normant, J. F. *Tetrahedron* **1986**, *42*, 1399–1406.
86T2647	Setsune, J.; Ueda, T.; Shikata, K.; Matsukawa, K.; Iida, T.; and Kitao, T. *Tetrahedron* **1986**, *42*, 2647–2656.
86T2831	Funk, R. L.; Abelman, M. M.; and Munger, J. D., Jr. *Tetrahedron* **1986**, *42*, 2831–2846.
86T2873	Lipshutz, B. H.; Parker, D. A.; Nguyen, S. L.; McCarthy, K. E.; Barton, J. C.; Whitney, S. E.; and Kotsuki, H. *Tetrahedron Tetrahedron*, **1986**, *42*, 2873–2879.
86T3781	Kigoshi, H.; Shizuri, Y.; Niwa, H.; and Yamada, K. *Tetrahedron* **1986**, *42*, 3781–3787.
86T4523	Rama Rao, A. V.; Reddy, E. R.; Sharma, G. V. M.; Yadagiri, P.; and Yadav, J. S. *Tetrahedron* **1986**, *42*, 4523–4532.
87CL887	Suzuki, H.; Yi, Q.; Inoue, J.; Kusume, K.; and Ogawa, T. *Chem. Lett.* **1987**, 887–890.
87JA2040	Dieter, R. K.; and Tokles, M. *J. Am. Chem. Soc.* **1987**, *109*, 2040–2046.
87JA8056	Nakamura, E.; Aoki, S.; Sekiya, K.; Oshino, H.; and Kuwajima, I. *J. Am. Chem. Soc.* **1987**, *109*, 8056–8066.
87JCS(P1)1331	Ihara, M.; Kawaguchi, A.; Ueda, H.; Chihiro, M.; Fukumoto, K.; and Kametani, T. *J. Chem. Soc.; Perkin 1* **1987**, 1331–1337.
87JCS(P1)2183	Kocieński, P.; Yeates, C.; Street, S. D. A.; and Campbell, S. F. *J. Chem. Soc.; Perkin 1* **1987**, 2183–2187.
87JOC398	Ernst, L.; Hopf, H.; and Krause, N. *J. Org. Chem.* **1987**, *52*, 398–405.
87JOC1381	Cooke, M. P.; Jr. and Widener, R. K. *J. Org. Chem.* **1987**, *52*, 1381–1396.
87JOC1801	O'Connor, B.; and Just, G. *J. Org. Chem.* **1987**, *52*, 1801–1803.
87JOC3901	Weiberth, F. J.; and Hall, S. S. *J. Org. Chem.* **1987**, *52*, 3901–3904.
87JOC4258	Back, T. G.; Collins, S.; Krishna, M. V.; and Law, K-W. *J. Org. Chem.* **1987**, *52*, 4258–4264.
87JOM(325)293	Wehman, E.; van Koten, G.; Knotter, M.; Spelten, H.; Heijdenrijk, D.; Mak, A. N. S.; and Stam, C. H. *J. Organometal. Chem.* **1987**, *325*, 293–309.
87OS(65)108	Ziegler, F. E.; Fowler, K. W.; Rodgers, W. B.; and Wester, R. T. *Org. Synth.* **1987**, *65*, 108–118.

87OS(66)1 Danheiser, R. L.; Tsai, Y-M.; and Fink, D. M. *Org. Synth.* **1987**, *66*, 1–7.

87OS(66)52 Nugent, W. A.; and Hibbs, F.W., Jr.; *Org. Synth.* **1987**, *66*, 52–59.

87OS(66)116 Fujisawa, T.; and Sato, T. *Org. Synth.* **1987**, *66*, 116–120.

87T813 Tanaka, T.; Hazato, A.; Bannai, K.; Okamura, N.; Sugiura, S.; Manabe, K.; Toru, T.; Kurozumi, S.; Suzuki, M.; Kawagishi, T.; and Noyori, R. *Tetrahedron*, **1987**, *43*, 813–824.

87T4385 Rama Rao, A. V.; Reddy, E. R.; Purandare, A. V.; and Varaprasad, Ch. V. N. S. *Tetrahedron*, **1987**, *43*, 4385–4394.

87T4481 Gramatica, P.; Manitto, P.; Monti, D.; and Speranza, G. *Tetrahedron*, **1987**, *43*, 4481–4486.

87T4591 Andreini, B. P.; Benetti, M. Carpita, A.; and Rossi, R. *Tetrahedron*, **1987**, *43*, 4591–4600.

88AG1194 Schöllkopf, U.; Pettig, D.; Schulze, E.; Klinge, M.; Egert, E.; Benecke, B.; and Noltemeyer, M. *Angew, Chem. (Engl. Ed.)* **1988**, *27* 1194–1195.

88JA2218 Wender, P. A.; and White, A. W. *J. Am. Chem. Soc.* **1988**, *110* 2218–2223.

88JA8129 Danishefsky, S. J.; and Mantlo, N. *J. Am. Chem. Soc.* **1988**, *110*, 8129–8133.

88JCS(P1)921 Carr, G. E.; Chambers, R. D.; Holmes, T. F.; and Parker, D. G. *J. Chem. Soc.; Perkin I*, **1988**, 921–926.

88JOC607 Tsuda, T.; Yoshida, T.; and Saegusa, T. *J. Org. Chem.* **1988**, *53*, 607–610.

88JOC4482 Ebert, G. W.; and Rieke, R. D. *J. Org. Chem.* **1988**, 4482–4488.

88MI225 Brandsma, L. Preparative Acetylenic Chemistry. Second Edition. Elsevier, Amsterdam, **1988**, p. 225–226.

88MI226 Brandsma, L. *Preparative Acetylenic Chemistry*. Second Edition, Elsevier, Amsterdam, **1988**, p. 226–227.

88T7325 Beswick, P. J.; Greenwood, C. S.; Mowlem, T. J.; Nechvatal, G.; and Widdowson, D. A. *Tetrahedron*, **1988**, *44*, 7325–7334.

89HCA1337 Oppolzer, W.; and Kingma, A. J. *Helv. Chim. Acta* **1989**, *72*, 1337–1345.

89JA1351 Lipshutz, B. H.; Ellsworth, E. L.; and Siahaan, T. J. *J. Am. Chem. Soc.* **1989**, *111*, 1351–1358.

89JA2984 Roush, W. R.; Michaelides, M. R.; Tai, D. F.; Lesur, B. M.; Chong, W. K. M.; and Harris, D. J. *J. Am. Chem. Soc.* **1989**, *111*, 2984–2995.

89JMC2072 Bar-Tana, J.; Ben-Shoshan, S.; Blum, J.; Migron, Y.; Hertz, R.; Pill, J.; Rose-Khan, G.; and White, E.-C. *J. Med. Chem.* **1989**, *32*, 2072–2084.

89JOC1295 Klunder, J. M.; Onami, T.; and Sharpless, K. B. *J. Org. Chem.* **1989**, *54*, 1295–1304.

89JOC3635 Delorme, D.; Girard, Y.; and Rokach, J. *J. Org. Chem.* **1989**, *54*, 3635–3640.

89JOC4975 Lipshutz, B. H.; Reuter, D. C.; and Ellsworth, E. L. *J. Org. Chem.* **1989**, *54*, 4975–4977.

89T349 Matsuzawa, S.; Horiguchi, Y.; Nakamura, E.; and Kuwajima, I. *Tetrahedron* **1989**, *45*, 349–362.

89T381 Alexakis, A.; and Jachiet, D. *Tetrahedron* **1989**, *45*, 381–389.

89T403 Kawashima, M.; Sato, T.; and Fujisawa, T. *Tetrahedron* **1989**, *45*, 403–412.

89T413 Fleming, I.; Rowley, M.; Cuadrado, P.; González-Nogal, A. M.; and Pulido, F. J. *Tetrahedron*, **1989**, *45*, 413–424.

89T425 Bertz, S. H.; and Dabbagh, G. *Tetrahedron* **1989**, *45*, 425–434.

89T443 Rieke, R. D.; Wehmeyer, R. M.; Wu, T.-C.; and Ebert, G. W. *Tetrahedron* **1989**, *45*, 443–454.

89T455 Casy, G.; and Taylor, R. J. K. *Tetrahedron* **1989**, *45*, 455–466.

89T545 Corey, E. J.; Hannon, F. J.; and Boaz, N. W. *Tetrahedron* **1989**, *45*, 545–555.

89T569 van Koten, G.; and Jastrzebski, J. T. B. H. *Tetrahedron* **1989**, *45*, 569–578.

89T1089 Piers, E.; and Karunaratne, V. *Tetrahedron* **1989**, *45*, 1089–1104.

89TL4795 Chen, H. G.; Hoechstetter, C.; and Knochel, P. *Tetrahedron Lett.* **1989**, *30*, 4795–4798.

89TL4799 Yeh, M. C. P.; and Knochel, P. *Tetrahedron Lett.* **1989**, *30*, 4799–4802.

89TL5069 Majid, T. N.; Yeh, M. C. P.; and Knochel, P. *Tetrahedron Lett.* **1989**, *30*, 5069–5072.

90JA7431 Knochel, P. *J. Am. Chem. Soc.* **1990**, *112*, 7431–7433.

90JOC964 Cheng, M.; and Huce, M. *J. Org. Chem.* **1990**, *55*, 964–975.

90JOC1425 Ireland, R. E.; and Wipf, P. *J. Org. Chem.* **1990**, *55*, 1425–1426.

90JOC1695 Lipshutz, B. H.; and Elworthy, T. R. *J. Org. Chem.* **1990**, *55*, 1695–1696.

90JOC3723 Williams, R. M.; and Hendrix, J. A. *J. Org. Chem.* **1990**, *55*, 3723–3728.

90JOC4417 Kotsuki, H.; Kadota, I.; and Ochi, M. *J. Org. Chem.* **1990**, *55*, 4417–4422.

90JOC4791 Chou, T.-S.; and Knochel, P. *J. Org. Chem.* **1990**, *55* 4791–4793.

90JOC5711 Tius, M. A.; and Kannangara, G. S. K. *J. Org. Chem.* **1990**, *55*, 5711–5714.

90MI3053 Berk, S. C.; Yeh, M. C. P.; Jeong, N.; and Knochel, P. *Organometallics* **1990**, *9*, 3053–3064.

90OS(69)1 Schwartz, A.; Madan, P.; Whitesell, J. K.; and Lawrence, R. M. *Org. Synth.* **1990**, *69*, 1–9.

90OS(69)80 Lipshutz, B. H.; Moreitti, R.; and Crow, R. *Org. Synth.* **1990**, *69*, 80–88.

90SC333 Ramiandrasoa, F.; and Tellier, F. *Synth. Comm.* **1990**, *20*, 333–344.

90SC761 Nevill, C. R., Jr.; and Fuchs, P. L. *Synth. Comm.* **1990**, *20*, 761–772.

90SC1989 Ramiandrasoa, F.; and Descoins, C. *Synth. Comm.* **1990**, *20*, 1989–1999.

90TA237 Kolb, H. C.; and Hoffmann, H. M. R. *Tetrahedron: Asymmetry* **1990**, *1*, 237–250.

90T3667 Chattopadhyay, S.; Mamdapur, V. R.; and Chadha, M. S. *Tetrahedron* **1990**, *46*, 3667–3672.

90T4277 Larchevêque, M.; and Henrot, S. *Tetrahedron* **1990**, *46*, 4277–4282.

90T4473 Mori, K. and Takikawa, H. *Tetrahedron* **1990**, *46*, 4473–4486.

90T4503 Takle, A.; and Kocieński, P. *Tetrahedron* **1990**, *46*, 4503–4116.

90TL477 Lipshutz, B. H.; and Elworthy, T. R. *Tetrahedron Lett.* **1990**, *31*, 477–480.

90TL1833 Retherford, C.; Chou, T. S.; Schelkun, R. M.; and Knochel, P. *Tetrahedron Lett.* **1990**, *31*, 1833–1836.

90TL4413 Majid, T. N.; and Knochel, P. *Tetrahedron Lett.* **1990**, *31*, 4413–4416.

90TL4539 Lipshutz, B. H.; Ung, C.; Elworthy, T. R.; and Reuter, D. C. *Tetrahedron Lett.* **1990**, *31*, 4539–4542.

90TL5161 Mignani, G.; Chevalier, C.; Grass, F.; Allmang, G.; and Morel, D. *Tetrahedron Lett.* **1990**, *31*, 5161–5164.

90TL7575 AchyuthaRao, S.; Tucker, C. E.; and Knochel, P. *Tetrahedron Lett.* **1990**, *31*, 7575–7578.

91CC1595 Barber, P.; Bury, C.; Kocieński, P.; and O'Shea, M. *J. Chem. Soc.; Chem. Comm.* **1991**, 1595–1597.

91JCS(P1)2639 Moiseenkov, A. M.; Czeskis, B. A.; Ivanova, N. M.; and Nefedov, O. M. *J. Chem. Soc.; Perkin 1* **1991**, 2639–2649.

91JOC638 Barrett, A. G. M.; and Flygar, J. A. *J. Org. Chem.* **1991**, *56*, 638–642.

91JOC1083 Gooding, O. W.; Beard, C. C.; Jackson, D. Y.; Wren, D. L.; and Cooper, G. F. *J. Org. Chem.* **1991**, *56*, 1083–1088.

91JOC1445 Zhu, L.; Wehmeyer, R. M.; and Rieke, R. D. *J. Org. Chem.* **1991**, *56*, 1445–1453.

91JOC4591 AchyuthaRao, S.; and Knochel, P. *J. Org. Chem.* **1991**, *56*, 4591–4593.

91JOC4714 Chadha, N. K.; Batcho, A. D.; Courtney, L. F.; Cook, C. M.; Wovkulich, P. M.; and Uskokovic, M. R. *J. Org. Chem.* **1991**, *56*, 4714–4718.

91JOC4744 Ebert, G. W.; and Klein, W. R. *J. Org. Chem.* **1991**, *56*, 4744–4747.

91JOC4933 Singer, R. D.; Hutzinger, M. W.; and Oehlschlager, A. C. *J. Org. Chem.* **1991**, *56*, 4933–4938.

91JOC5489 Arai, M.; Nakamura, E.; and Lipshutz, B. H. *J. Org. Chem.* **1991**, *56*, 5489–5491.

91JOC5491	Linderman, R. J.; and Griedel, B. D. *J. Org. Chem.* **1991**, *56*, 5491–5493.
91JOC5761	Lipshutz, B. H.; and Dimock, S. H. *J. Org. Chem.* **1991**, *56*, 5761–5763.
91JOC5974	Knoess, H. P.; Furlong, M. T.; Rozema, M. J.; and Knochel, P. *J. Org. Chem.* **1991**, *56*, 5974–5978.
91OS(70)195	Yeh, M. C. P.; Chen, H. G.; and Knochel, P. *Org. Synth.* **1991**, *70*, 195–203.
91OS(70)204	Wender, P. A.; White, A. W.; and McDonald, F. E. *Org. Synth.* **1991**, *70*, 204–214.
91OS(70)215	Stang, P. J.; and Kitamura, T. *Org. Synth.* **1991**, *70*, 215–225.
91T9691	Bergdahl, M.; Nilsson, M. Olsson, T.; and Stern, K. *Tetrahedron* **1991**, *47*, 9691–9702.
91TL605	Majetich, G.; Leigh, A. J.; and Condon, S. *Tetrahedron Lett.* **1991**, *32*, 605–608.
91TL609	Majetich, G.; and Leigh, A. J. *Tetrahedron Lett.* **1991**, *32*, 609–610.
91TL857	Burke, S. D.; Piscopio, A. D.; Marron, B. E.; Matulenko, M. A.; and Pan, G. *Tetrahedron Lett.* **1991**, *32*, 857–858.
91TL1855	Rozema, M. J.; and Knochel, P. *Tetrahedron Lett.* **1991**, *32*, 1855–1858.
91TL5647	Lipshutz, B. H.; and Kato, K. *Tetrahedron Lett.* **1991**, **32**, 5647–5650.
91TL7211	Lipshutz, B. H.; and Lee, J. I. *Tetrahedron Lett.* **1991**, *32*, 7211–7214.
92JA3983	Tucker, C. E.; Majid, T. N.; and Knochel, P. *J. Am. Chem. Soc.* **1992**, *114*, 3983–3985.
92JA5110	Stack, D. E.; Dawson, B. T.; and Rieke, R. D. *J. Am. Chem. Soc.* **1992**, *114*, 5110–5116.
92JA8008	Smith, A. B., III; Rano, T. A.; Chida, N.; Sulikowski, G. A., and Wood, J. L. *J. Am. Chem. Soc.* **1992**, *114*, 8008–8022.
92JCS(P1)327	Barbero, A.; Cuadrado, P.; Fleming, I.; González, A. M.; and Pulido, F. J. *J. Chem. Soc.; Perkin 1* **1992**, 327–331.
92JCS(P1)1193	Tanaka, K.; Matsui, J.; Suzuki, H.; and Watanabe, A. *J. Chem. Soc.; Perkin 1* **1992**, 1193–1194.
92JOC1024	Yamamoto, Y.; Chounan, Y.; Tanaka, M.; and Ibuka, T. *J. Org. Chem.* **1992**, *57*, 1024–1026.
92JOC2794	Dodd, D. S.; and Oehlschlager, A. C. *J. Org. Chem.* **1992**, *57*, 2794–2803.
92JOC2960	Caine, D. Venkataramu, S. D.; and Kois, A. *J. Org. Chem.* **1992**, *57*, 2960–2963.
92JOC3627	Vaillancourt, V.; and Albizati, K. F. *J. Org. Chem.* **1992**, *57*, 3627–3631.
92JOC5250	Dodd, D. S.; Pierce, H. D., Jr.; and Oehlschlager, A. C. *J. Org. Chem.* **1992**, *57*, 5250–5253.

92JOC6853 Rowley, E. G.; and Schore, N. E. *J. Org. Chem.* **1992**, *57*, 6853–6861.

92JOC6883 Angle, S. R.; and Rainier, J. D. *J. Org. Chem.* **1992**, *57* 6883–6890.

92JOC7248 Yang, Z.; Liu, H. B.; Lee, C. M.; Chang, H. M.; and Wong, H. N. C. *J. Org. Chem.* **1992**, *57*, 7248–7257.

92MI1560 Ebert, G. W.; Cheasty, J. W.; Tehrani, S. S.; and Aouad, E. *Organometallics* **1992**, *11*, 1560–1564.

92OR135 Lipshutz, B. H.; and Sengupta, S. *Org. React.* **1992**, *41*, 135–631.

92S523 Kumar, V.; and Todaro, A. B. *Synthesis* **1992**, 523–525.

92S981 Dutoit, J.-C. *Synthesis* **1992**, 981–984.

92T1943 Chemin, D.; and Linstrumelle, G. *Tetrahedron* **1992**, *48*, 1943–1952.

92T3633 Keegstra, M. A.; Peters, T. H. A.; and Brandsma, L. *Tetrahedron* **1992**, *48*, 3633–3652.

92T4369 Chemin, D.; Gueugnot, S.; and Linstrumelle, G. *Tetrachedron* **1992**, *48*, 4369–4378.

92T4465 Yadav, J. S.; Deshpande, P. K.; and Sharma, G. V. M. *Tetrahedron*, **1992**, *48*, 4465–4474.

92T6231 Corriu, R. J. P.; Huynh, V.; Iqbal, J.; Moreau, J. J. E.; and Vernhet, C. *Tetrahedron* **1992**, *48*, 6231–6244.

92T6393 Van Hijfte, L.; and Kolb, M. *Tetrahedron* **1992**, *48*, 6393–6402.

92TL49 Marino, J. P.; Emonds, M. V. M.; Stengel, P. J.; Olivira, A. R. M.; Simonelli, F.; and Ferreira, J. T. B. *Tetrahedron Lett.* **1992**, *33*, 49–52.

92TL919 Johnson, C. R.; Adams, J. P.; Braun, M. P.; and Senanayake, C. B. W. *Tetrahedron Lett.* **1992**, *33*, 919–922.

92TL3717 Waas, J. R.; Sidduri, A.; and Knochel, P. *Tetrahedron Lett.* **1992**, *33*, 3717–3720.

92TL5857 Venanzi, L. M.; Lehmann, R.; Keil, R. and Lipshutz, B. H. *Tetrahedron Lett.* **1992**, *33*, 5857–5860.

92TL6575 Stack, D. E.; and Rieke, R. D. *Tetrahedron Lett.* **1992**, *33* 6575–6578.

92TL8087 Blanchot-Courtois, V.; and Hanna, I. *Tetrahedron Lett.* **1992**, *33*, 8087–8090.

93AG1368 Westermann, J.; and Nickisch, K. *Angew. Chem.* (Engl. Ed.) **1993**, *32*, 1368–1370.

93CJC280 Piers, E.; Yeung, B. W. A.; and Fleming, F. F. *Can. J. Chem.* **1993**, *71*, 280–286.

93JA1676 Paquette, L. A.; Wang, T.-Z.; and Vo, N. H. *J. Am. Chem. Soc.* **1993**, *115*, 1676–1683.

93JA3966 Angle, S. R.; Fevig, J. M.; Knight, S. D.; Marquis, R. W.; Jr. and Overman, L. E. *J. Am. Chem. Soc.* **1993**, *115*, 3966–3976.

93JCS(P1)153 Tanaka, K.; Matssui, J.; and Suzuki, H. *J. Chem. Soc.; Perkin 1* **1993**, 153–157.

93JCS(P1)759 White, J. D.; Reddy, G. N.; and Spessard, G. O. *J. Chem. Soc.; Perkin 1* **1993**, 759–767.

93JCS(P1)1953 Astley, S. T.; and Stephenson, G. R. *J. Chem. Soc.; Perkin 1* **1993**, 1953–1955.

93JOC36 Meyers, A. I.; and Snyder, L. *J. Org. Chem.* **1993**, *58*, 36–42.

93JOC118 Sestelo, J. P.; Mascarenas, J. L.; Castedo, L.; and Mourino, A. *J. Org. Chem.* **1993**, *58*, 118–123.

93JOC408 Liebeskind, L. S.; and Reisinger, S. W. *J. Org. Chem.* **1993**, *58*, 408–413.

93JOC528 Burns, M. R.; and Coward, J. K. *J. Org. Chem.* **1993**, *58*, 528–532

93JOC588 Knochel, P.; Chou, T.-S.; Jubert, C.; and Rajagopal, D. *J. Org. Chem.* **1993**, *58*, 588–599.

93JOC1038 Crimmins, M. T.; Nantermet, P. G.; Trotter, B. W.; Vallin, I. M.; Watson, P. S.; McKerlie, L. A.; Reinhold, T. L.; Cheung, A. W.-H.; Stetson, K. A.; Dedopoulou, D.; and Gray, J. L. *J. Org. Chem.* **1993**, *58*, 1038–1047.

93JOC1207 Ibuka, T.; Taga, T.; Habashita, H.; Nakai, K.; Tamamura, H.; Fujii, N.; Chounan, Y.; Nemoto, H.; and Yamamoto, Y. *J. Org. Chem.* **1993**, *58*, 1207–1214.

93JOC2134 Tao, C.; and Donaldson, W. *J. Org. Chem.* **1993**, *58*, 2134–2143.

93JOC2483 Rieke, R. D.; Stack, D. E.; Dawson, B. T.; and Wu, T.-C. *J. Org. Chem.* **1993**, *58*, 2483–2491.

93JOC2694 Sidduri, A.; Rozema, M. J.; and Knochel, P. *J. Org. Chem.* **1993**, *58*, 2694–2713.

93JOC2920 Wei, J.; Hutchins, R. O.; and Prol, J., Jr.; *J. Org. Chem.* **1993**, *58*, 2920–2922.

93JOC4781 Tucker, C. E.; and Knochel, P. *J. Org. Chem.* **1993**, *58*, 4781–4782.

93JOC5153 Bell, T. W.; and Ciaccio, J. A. *J. Org. Chem.* **1993**, *58*, 5153–5162.

93JOC7238 Bergdahl, M.; Eriksson, M.; Nilsson, M.; and Olsson, T. *J. Org. Chem.* **1993**, *58*, 7238–7244.

93OS104 Myers, A. G.; and Dragovich, P. S. *Org. Synth.* **1993**, *72*, 104–111.

93OS135 Alami, M.; Marquais, S.; and Cahiez, G. *Org. Synth.* **1993**, *73*, 135–146.

93S537 Wipf, P. *Synthesis* **1993**, 537–557.

93SC143 Rossi, R.; Carpita, A.; and Cossi, P. *Synth. Commun.* **1993**, *23*, 143–152.

93SC2547 Bricard, L.; and Kunesch, G. *Synth. Commun.* **1993**, *23*, 2547–2558.

93SL219 Jackson, R. F. W.; Wishart, N.; and Wythes, M. J. *Synlett* **1993**, 219–220.

93T29 Knochel, P.; and Rao, C. J. *Tetrahedron* **1993**, *49*, 29–48.

93T337	Back, T. G.; and Hu, N-X. *Tetrahedron*, **1993**, *49*, 337–348.
93T965	Rossiter, B. E.; Eguchi, M.; Miao, G.; Swingle, N. M.; Hernández, A. M.; Vickers, D.; Fluckiger, E.; Patterson, R. G.; and Reddy, K. V. *Tetrahedron* **1993**, *49*, 965–986.
93T1901	Nakatani, K.; Arai, K.; and Terashima, S. *Tetrahedron* **1993**, *49*, 1901–1912.
93TL5261	Langer, F.; Waas, J.; and Knochel, P. *Tetrahedron Letters* **1993**, *34*, 5261–5264.
93TL7725	Zhou, Q.-L.; and Pfaltz, A. *Tetrahedron Letters* **1993**, *34*, 7725–7728.
94S242	Haglund, O.; and Nilsson, M. *Synthesis* **1994**, 242–244.
94S432	Cabezas, J. A.; and Oehlschlager, A. C. *Synthesis* **1994**, 432–442.
94T4455	Swingle, N. M.; Reddy, K. V.; and Rossiter, B. E. *Tetrahedron* **1994**, *50*, 4455–4466.

A2

List of suppliers

Acros Chimica
see **Janssen Chimica**

Aldrich Chemical Co. Ltd.
France: 80 Rue de Lozais, BP 701, 38070, St Quentin, Fallavier Cedex, LYON. Tel. 74822800
Germany: SAF, Messerschmitt Strasse 17, D-7910 Neu-Ulm. Tel. 0731-9733640
Japan: Aldrich Chemical Co., Inc. (Japan), Kyodo-building Shinkanda, 10 Kandamikura-chou, Chiyoda-ku, Tokyo 101. Tel. 03-54344712
UK: The Old Brickyard, New Road, Gillingham, Dorset SP8 4JL. Tel. 0800-717181
USA: PO Box 355, Milwaukee, WI 53201. Tel. 0414-2733850.

Alfa
France: Johnson Matthey SA, BP 50240, Rue de la Perdix, Z1 Paris Nord LL, 95956 Roissy, Charles De Gaulle Cedex. Tel. 1-48632299
Germany: Johnson-Metthey GmbH, Zeppelinstrasse 7, D-7500 Karlsruhe-1. Tel. 0721-840070
UK: Catalogue Sales, Materials Technology Division, Orchard Road, Royston, Herts. SG8 5HE. Tel. 0763-253715
USA: Alfa/Johnson Matthey, PO Box 8247, Ward Hill, MA 01835-0747. Tel. 0508-5216300

BDH
UK: (Head Office and International Sales): Merck Ltd, Merck House, Poole, Dorset BH15 1TD. Tel. 0202-665599

Boulder Scientific Co
USA: 598 3rd Street, PO Box 548, Mead, CO 80542. Tel. 03035354494

Fluka Chemika-BioChemika
France: Fluka S.a.r.l., F-38297 St Quentin, Fallavier Cedex, Lyon. Tel. 74822800
Germany: Fluka Feinchemikalien GmbH, D-7910 Neu-Ulm. Tel. 0731–729670
Japan: Fluka Fine Chemical, Chiyoda-Ku, Tokyo. Tel. 03-32554787
UK: Fluka Chemicals Ltd, Gillingham, Dorset SP8 4JL. Tel. 0747-823097

USA: Fluka Chemical Corp., Ronkonkoma, NY 11779-7238. Tel. 0516-4670980

FMC Corporation, Lithium Division
Japan: Asia Lithium Corporation (ALCO), Shin-Osaka Daiichi-Seimei Building 11F, 5-24, Miyahara 3-Chome, Yodogawa-Ku, Osaka. Tel. 06-3992331
UK and Mainland Europe: Commercial Road, Bromborough, Merseyside L62 3NL. Tel. 051-3348085
USA: 449 North Cox Road, Gastonia, NC 28054. Tel. 0704-8685300

Heraeus
Germany: Alter Weinberg, D-7500, Karlsruhe 41-Ho. Tel. 0721-4716769

ICN Biomedicals/K and K Rare and Fine Chemicals
France: ICN Biomedicals France, Parc Club Orsay, 4 rue Jean Rostand, 91893 Orsay Cedex. Tel. 1-60193460
Germany: ICN Biomedicals, GmbH, Mühlgrabenstrasse 10, Postfach 1249, D-5309, Meckenheim. Tel. 02225-88050
Japan: ICN Biomedicals Japan Co., Ltd, 8th Floor, Iidabashi Central Building, 4-7-10 Iidabashi, Chiyoda-ku, Tokyo 102. *Tel. 03-32370938*
UK: ICN Biomedicals, Ltd, Eagles House, Peregrine Business Park, Gomm Road, High Wycombe, Bucks. HP13 7DL. Tel. 04940443826
USA: ICN Biomedicals, Inc., 3300 Hyland Avenue, Costa Mesa, CA 92626. Tel. 0800-8540530

Janssen Chimica (Acros Chimica)
Mainland Europe, Central Offices Belgium: Janssen Pharmaceuticalaan 3, 2440 Geel. Tel. 014-604200
UK: Hyde ParkHouse, Cartwright Street, Newton, Hyde, Cheshire SK14 4EH. Tel. 0613-244161
USA: Spectrum Chemical Mgf. Corp., 14422 South San Pedro Street, Gardena, CA 90248. Tel. 0800-7728786

Johnson Matthey Chemical Products
France: Johnson Matthey SA, BP 50240, Rue de la Perdix, Z1 Paris Nord LL, 95956 Roissy, Charles De Gaulle Cedex. Tel. 48632299
Germany: Johnson-Matthey GmbH, Zeppelinstrasse 7, D-7500 Karlsruhe-1. Tel. 0721-840070
UK: Catalogue Sales, Materials Technology Division, Orchard Road, Royston, Herts. SG8 5HE. Tel. 0763-253715
USA: Alfa/Johnson Matthey, PO Box 8247, Ward Hill MA 01835-0747. Tel. 0508-5216300

Kanto Chemical Co., Inc
Japan: 2-8, Nihonbashi-honcho-3-chome, Chuo-ku, Tokyo 103. Tel. 03-2791751

Lancaster Synthesis
France: Lancaster Synthesis Ltd, 15 Rue de l'Atome, Zone Industrielle, 67800 Bischheim, Strasbourg. Tel. 05035147.
Germany: Lancaster Synthesis GmbH, Postfach 15 18, D-63155 Mülheim am Main. Tel. 0130-6562
Japan: Hydrus Chemical Inc., Tomitaka Building, 8-1, Uchikanda 2-chome, Chiyoda-ku, Tokyo 101. Tel. 03-32585031
UK: Lancaster Synthesis, Eastgate, White Lund, Morecambe, Lancashire LA3 3DY. Tel. 0800-262336
USA: Lancaster Synthesis Inc, PO Box 1000, Windham, NH 03087-9977. Tel. 0800-2382324

Merck
Germany: Promochem, PO Box 101340, Mercatorstrasse 51, D-46469 Wesel. Tel. 0281-530081
Japan: Merck Japan Ltd, ARCO Tower, SF, 8-1, Shimomeguro-1-chome, Merguro-ku, Tokyo 153. Tel. 03-54344712
UK: Merck Ltd, Merck House, Poole, Dorset BH15 1TTD. Tel. 0202-669700
USA: Gallard Schlesinger Companies, 584 Mineola Avenue, Carle Place, NY. Tel. 0516-3335600

Mitsuwa Scientific Corp.
Japan: 11-1, Tenma-1-Chome, Kita-ku, Osaka 530. Tel. 06-3519631

Nacalai Tesque, Inc.
Japan: Karasuma-nishi-iru, Nijo-dori, Nakagyo-Ku, Kyoto 604. Tel. 075-2315301

Organometallics, Inc.
USA: PO Box 287, East Hampstead, NH. Tel. 0603-3296021

Parish Chemical Company
USA: 145 North Geneva Road, Orem. UT 84057. Tel. 0801-2262018

Prolabo
UK etc.: See **Rhône-Poulenc**
France: 12, Rue Pelee, BP 369, 75526, Paris Cedex 11. Tel. 1-49231500

Rhône-Poulenc
France: Rhône-Poulenc SA, 25 Quai Paul Donmer, F-92408 Courbezoie Cedex. Tel. 1-47681234

Germany: Rhône-Poulenc GmbH, Staedelstrasse 10, Postfach 700862, Frankfurt Am Main 70. Tel. 069-60930
Japan: Rhône-Poulenc Japan Ltd, 15 Kowa Building Annexe, Central PO Box 1649, Tokyo 107. Tel. 03-35854691
UK: Rhône-Poulenc Chemicals Ltd, Laboratory Products, Liverpool Road, Barton Moss, Eccles, Manchester M30 7RT. Tel. 061-7895878
USA: Rhône-Poulenc Basic Chemicals Co., 1 Corporate Dr., Shelton, CT 06484. Tel. 0203-9253300 *or* Rhône-Poulenc Inc., Fine Organics, CN 7500, Cranbury, NJ 08512-7500. Tel. 0609-860400

Riedel de Haen
Germany: Wunstorfer Strasse 40, Postfach, D-3016 Seelze 1. Tel. 05137-7070

Strem
France: Strem Chemicals, Inc., 15 Rue de l'Atome, Zone Industrielle, 67800 Bischheim. Tel. 88625260
Germany: Strem Chemicals GmbH, Querstrasse 2, D-7640 Karlsruhe. Tel. 0721-75879
Japan: Hydrus Chemical Co., Tomikata Building, 8-1, Uchkanda, 2-Chome, Chiyoda-ku, Tokyo 101. Tel. 03-2585031
UK: Fluorochem Limited, Wesley Street, Old Glossop, Derbyshire SK13 9RY. Tel. 0457-868921
USA: Strem Chemicals, Inc., Dexter Industrial Park, 7 Mulliken Way, Newburyport, MA 01950-4098. Tel. 0508-4623191

Tokyo Kasei Kogyo Co., Ltd
Japan: 3-1-13, Nihonbashi-Honcho, Chuo-ku, Tokyo 103. Tel. 03-38082821
UK: Fluorochem Limited, Wesley Street, Old Glossop, Derbyshire SK13 9RY. Tel. 0457-868921
USA: PCI America, 9211 North Harbourgate Street, Portland, OR 97203. Tel. 503-2831681

Ventron
See **Alfa**

Wako
Japan: 3-10 Dosho-Machi, Higashi-Ku, Osaka 541. Tel. 06-2033741

Index

Page entries in *italics* refer to protocols on those pages.

Index

Index